D1686624

Edited by
Gerhard Dehm, James M. Howe,
and Josef Zweck

In-situ Electron Microscopy

Related Titles

Van Tendeloo, G., Van Dyck, D., Pennycook, S. J. (Eds.)

Handbook of Nanoscopy

2012

Hardcover

ISBN: 978-3-527-31706-6

Tsukruk, V. V., Singamaneni, S.

Scanning Probe Microscopy of Soft Matter

Fundamentals and Practices

2012

Hardcover

ISBN: 978-3-527-32743-0

Baró, A. M., Reifenberger, R. G. (Eds.)

Atomic Force Microscopy in Liquid

Biological Applications

2012

Hardcover

ISBN: 978-3-527-32758-4

Bowker, M., Davies, P. R. (Eds.)

Scanning Tunneling Microscopy in Surface Science

2010

Hardcover

ISBN: 978-3-527-31982-4

García, R.

Amplitude Modulation Atomic Force Microscopy

2010

Hardcover

ISBN: 978-3-527-40834-4

Codd, S., Seymour, J. D. (Eds.)

Magnetic Resonance Microscopy

Spatially Resolved NMR Techniques and Applications

2009

Hardcover

ISBN: 978-3-527-32008-0

Stokes, D.

Principles and Practice of Variable Pressure

Environmental Scanning Electron Microscopy (VP-ESEM)

2009

Hardcover

ISBN: 978-0-470-06540-2

*Edited by
Gerhard Dehm, James M. Howe,
and Josef Zweck*

In-situ Electron Microscopy

Applications in Physics, Chemistry and Materials Science

WILEY-VCH

WILEY-VCH Verlag GmbH & Co. KGaA

The Editors

Prof. Dr. Gerhard Dehm
Montanuniversität Leoben
Dept. Materialphysik
Jahnstr. 12
8700 Leoben
Austria

Prof. Dr. James M. Howe
University of Virginia
Dept. of Mat. Science & Engin.
116 Engineer's Way
Charlottesville, VA 22904-4745
USA

Prof. Dr. Josef Zweck
Universität Regensburg
Fak. für Physik
93040 Regensburg
Germany

All books published by **Wiley-VCH** are carefully produced. Nevertheless, authors, editors, and publisher do not warrant the information contained in these books, including this book, to be free of errors. Readers are advised to keep in mind that statements, data, illustrations, procedural details or other items may inadvertently be inaccurate.

Library of Congress Card No.: applied for

British Library Cataloguing-in-Publication Data
A catalogue record for this book is available from the British Library.

Bibliographic information published by the Deutsche Nationalbibliothek
The Deutsche Nationalbibliothek lists this publication in the Deutsche Nationalbibliografie; detailed bibliographic data are available on the Internet at http://dnb.d-nb.de.

© 2012 Wiley-VCH Verlag & Co. KGaA, Boschstr. 12, 69469 Weinheim, Germany

All rights reserved (including those of translation into other languages). No part of this book may be reproduced in any form – by photoprinting, microfilm, or any other means – nor transmitted or translated into a machine language without written permission from the publishers. Registered names, trademarks, etc. used in this book, even when not specifically marked as such, are not to be considered unprotected by law.

Cover Design   Adam-Design, Weinheim
Typesetting   Thomson Digital, Noida, India
Printing and Binding   Strauss GmbH, Mörlenbach

Printed in the Federal Republic of Germany
Printed on acid-free paper

Print ISBN: 978-3-527-31973-2
ePDF ISBN: 978-3-527-65219-8
ePub ISBN: 978-3-527-65218-1
mobi ISBN: 978-3-527-65217-4
oBook ISBN: 978-3-527-65216-7

Contents

List of Contributors XIII
Preface XVII

Part I **Basics and Methods** 1

1 **Introduction to Scanning Electron Microscopy** 3
Christina Scheu and Wayne D. Kaplan
1.1 Components of the Scanning Electron Microscope 4
1.1.1 Electron Guns 6
1.1.2 Electromagnetic Lenses 9
1.1.3 Deflection System 13
1.1.4 Electron Detectors 13
1.1.4.1 Everhart–Thornley Detector 13
1.1.4.2 Scintillator Detector 15
1.1.4.3 Solid-State Detector 16
1.1.4.4 In-Lens or Through-the-Lens Detectors 16
1.2 Electron–Matter Interaction 16
1.2.1 Backscattered Electrons (BSEs) 20
1.2.2 Secondary Electrons (SEs) 22
1.2.3 Auger Electrons (AEs) 25
1.2.4 Emission of Photons 25
1.2.4.1 Emission of X-Rays 25
1.2.4.2 Emission of Visible Light 26
1.2.5 Interaction Volume and Resolution 26
1.2.5.1 Secondary Electrons 27
1.2.5.2 Backscattered Electrons 27
1.2.5.3 X-Rays 27
1.3 Contrast Mechanisms 28
1.3.1 Topographic Contrast 28
1.3.2 Composition Contrast 31
1.3.3 Channeling Contrast 31

1.4	Electron Backscattered Diffraction (EBSD)	31
1.5	Dispersive X-Ray Spectroscopy	34
1.6	Other Signals	36
1.7	Summary	36
	References	37

2 Conventional and Advanced Electron Transmission Microscopy 39
Christoph Koch

2.1	Introduction	39
2.1.1	Introductory Remarks	39
2.1.2	Instrumentation and Basic Electron Optics	40
2.1.3	Theory of Electron–Specimen Interaction	42
2.2	High-Resolution Transmission Electron Microscopy	48
2.3	Conventional TEM of Defects in Crystals	54
2.4	Lorentz Microscopy	55
2.5	Off-Axis and Inline Electron Holography	57
2.6	Electron Diffraction Techniques	59
2.6.1	Fundamentals of Electron Diffraction	59
2.7	Convergent Beam Electron Diffraction	61
2.7.1	Large-Angle Convergent Beam Electron Diffraction	63
2.7.2	Characterization of Amorphous Structures by Diffraction	63
2.8	Scanning Transmission Electron Microscopy and Z-Contrast	63
2.9	Analytical TEM	66
	References	67

3 Dynamic Transmission Electron Microscopy 71
*Thomas LaGrange, Bryan W. Reed, Wayne E. King,
Judy S. Kim, and Geoffrey H. Campbell*

3.1	Introduction	71
3.2	How Does Single-Shot DTEM Work?	72
3.2.1	Current Performance	74
3.2.2	Electron Sources and Optics	75
3.2.3	Arbitrary Waveform Generation Laser System	80
3.2.4	Acquiring High Time Resolution Movies	81
3.3	Experimental Applications of DTEM	82
3.3.1	Diffusionless First-Order Phase Transformations	82
3.3.2	Observing Transient Phenomena in Reactive Multilayer Foils	85
3.4	Crystallization Under Far-from-Equilibrium Conditions	88
3.5	Space Charge Effects in Single-Shot DTEM	90
3.5.1	Global Space Charge	90
3.5.2	Stochastic Blurring	91
3.6	Next-Generation DTEM	91
3.6.1	Novel Electron Sources	91
3.6.2	Relativistic Beams	92
3.6.3	Pulse Compression	93

3.6.4	Aberration Correction 93
3.7	Conclusions 94
	References 95

4 Formation of Surface Patterns Observed with Reflection Electron Microscopy 99

Alexander V. Latyshev

4.1	Introduction 99
4.2	Reflection Electron Microscopy 102
4.3	Silicon Substrate Preparation 107
4.4	Monatomic Steps 109
4.5	Step Bunching 111
4.6	Surface Reconstructions 114
4.7	Epitaxial Growth 115
4.8	Thermal Oxygen Etching 116
4.9	Conclusions 119
	References 119

Part II Growth and Interactions 123

5 Electron and Ion Irradiation 125

Florian Banhart

5.1	Introduction 125
5.2	The Physics of Irradiation 126
5.2.1	Scattering of Energetic Particles in Solids 126
5.2.2	Scattering of Electrons 128
5.2.3	Scattering of Ions 129
5.3	Radiation Defects in Solids 129
5.3.1	The Formation of Defects 129
5.3.2	The Migration of Defects 130
5.4	The Setup in the Electron Microscope 131
5.4.1	Electron Irradiation 131
5.4.2	Ion Irradiation 132
5.5	Experiments 132
5.5.1	Electron Irradiation 133
5.5.2	Ion Irradiation 140
5.6	Outlook 141
	References 142

6 Observing Chemical Reactions Using Transmission Electron Microscopy 145

Renu Sharma

6.1	Introduction 145
6.2	Instrumentation 146
6.3	Types of Chemical Reaction Suitable for TEM Observation 150

6.3.1	Oxidation and Reduction (Redox) Reactions	150
6.3.2	Phase Transformations	151
6.3.3	Polymerization	151
6.3.4	Nitridation	152
6.3.5	Hydroxylation and Dehydroxylation	152
6.3.6	Nucleation and Growth of Nanostructures	153
6.4	Experimental Setup	154
6.4.1	Reaction of Ambient Environment with Various TEM Components	154
6.4.2	Reaction of Grid/Support Materials with the Sample or with Each Other	154
6.4.3	Temperature and Pressure Considerations	155
6.4.4	Selecting Appropriate Characterization Technique(s)	156
6.4.5	Recording Media	156
6.4.6	Independent Verification of the Results, and the Effects of the Electron Beam	157
6.5	Available Information Under Reaction Conditions	157
6.5.1	Structural Modification	158
6.5.1.1	Electron Diffraction	158
6.5.1.2	High-Resolution Imaging	158
6.5.2	Chemical Changes	161
6.5.3	Reaction Rates (Kinetics)	164
6.6	Limitations and Future Developments	164
	References	165
7	**In-Situ TEM Studies of Vapor- and Liquid-Phase Crystal Growth**	**171**
	Frances M. Ross	
7.1	Introduction	171
7.2	Experimental Considerations	172
7.2.1	What Crystal Growth Experiments are Possible?	172
7.2.2	How Can These Experiments be Made Quantitative?	173
7.2.3	How Relevant Can These Experiments Be?	175
7.3	Vapor-Phase Growth Processes	175
7.3.1	Quantum Dot Growth Kinetics	176
7.3.2	Vapor–Liquid–Solid Growth of Nanowires	177
7.3.3	Nucleation Kinetics in Nanostructures	180
7.4	Liquid-Phase Growth Processes	183
7.4.1	Observing Liquid Samples Using TEM	183
7.4.2	Electrochemical Nucleation and Growth in the TEM System	184
7.5	Summary	187
	References	188
8	**In-Situ TEM Studies of Oxidation**	**191**
	Guangwen Zhou and Judith C. Yang	
8.1	Introduction	191
8.2	Experimental Approach	192

8.2.1	Environmental Cells	*192*
8.2.2	Surface and Environmental Conditions	*193*
8.2.3	Gas-Handling System	*194*
8.2.4	Limitations	*195*
8.3	Oxidation Phenomena	*196*
8.3.1	Surface Reconstruction	*196*
8.3.2	Nucleation and Initial Oxide Growth	*197*
8.3.3	Role of Surface Defects on Surface Oxidation	*198*
8.3.4	Shape Transition During Oxide Growth in Alloy Oxidation	*199*
8.3.5	Effect of Oxygen Pressure on the Orientations of Oxide Nuclei	*202*
8.3.6	Oxidation Pathways Revealed by High-Resolution TEM Studies of Oxidation	*203*
8.4	Future Developments	*205*
8.5	Summary	*206*
	References	*206*
Part III	**Mechanical Properties** *209*	
9	**Mechanical Testing with the Scanning Electron Microscope** *211*	
	Christian Motz	
9.1	Introduction	*211*
9.2	Technical Requirements and Specimen Preparation	*212*
9.3	*In-Situ* Loading of Macroscopic Samples	*214*
9.3.1	Static Loading in Tension, Compression, and Bending	*214*
9.3.2	Dynamic Loading in Tension, Compression, and Bending	*216*
9.3.3	Applications of *In-Situ* Testing	*216*
9.4	*In-Situ* Loading of Micron-Sized Samples	*217*
9.4.1	Static Loading of Micron-Sized Samples in Tension, Compression, and Bending	*218*
9.4.2	Applications of *In-Situ* Testing of Small-Scale Samples	*220*
9.4.3	*In-Situ* Microindentation and Nanoindentation	*222*
9.5	Summary and Outlook	*223*
	References	*223*
10	***In-Situ* TEM Straining Experiments: Recent Progress in Stages and Small-Scale Mechanics** *227*	
	Gerhard Dehm, Marc Legros, and Daniel Kiener	
10.1	Introduction	*227*
10.2	Available Straining Techniques	*228*
10.2.1	Thermal Straining	*228*
10.2.2	Mechanical Straining	*229*
10.2.3	Instrumented Stages and MEMS/NEMS Devices	*230*
10.3	Dislocation Mechanisms in Thermally Strained Metallic Films	*233*
10.3.1	Basic Concepts	*233*
10.3.2	Dislocation Motion in Single Crystalline Films and Near Interfaces	*235*

10.3.3	Dislocation Nucleation and Multiplication in Thin Films	236
10.3.4	Diffusion-Induced Dislocation Plasticity in Polycrystalline Cu Films	239
10.4	Size-Dependent Dislocation Plasticity in Metals	239
10.4.1	Plasticity in Geometrically Confined Single Crystal fcc Metals	241
10.4.2	Size-Dependent Transitions in Dislocation Plasticity	243
10.4.3	Plasticity by Motion of Grain Boundaries	244
10.4.4	Influence of Grain Size Heterogeneities	245
10.5	Conclusions and Future Directions	247
	References	248

11 In-Situ Nanoindentation in the Transmission Electron Microscope 255
Andrew M. Minor

11.1	Introduction	255
11.1.1	The Evolution of *In-Situ* Mechanical Probing in a TEM	255
11.1.2	Introduction to Nanoindentation	256
11.2	Experimental Methodology	260
11.3	Example Studies	263
11.3.1	*In-Situ* TEM Nanoindentation of Silicon	263
11.3.2	*In-Situ* TEM Nanoindentation of Al Thin Films	269
11.4	Conclusions	272
	References	274

Part IV Physical Properties 279

12 Current-Induced Transport: Electromigration 281
Ralph Spolenak

12.1	Principles	281
12.2	Transmission Electron Microscopy	283
12.2.1	Imaging	283
12.2.2	Diffraction	288
12.2.3	Convergent Beam Electron Diffraction (CBED): Measurements of Elastic Strain	288
12.3	Secondary Electron Microscopy	289
12.3.1	Imaging	289
12.3.2	Elemental Analysis	291
12.3.3	Electron Backscatter Diffraction (EBSD)	292
12.4	X-Radiography Studies	292
12.4.1	Microscopy and Tomography	292
12.4.2	Spectroscopy	293
12.4.3	Topography	294
12.4.4	Microdiffraction	294
12.5	Specialized Techniques	295
12.5.1	Focused Ion Beams	295

12.5.2	Reflective High-Energy Electron Diffraction (RHEED)	296
12.5.3	Scanning Probe Methods	296
12.6	Comparison of *In-Situ* Methods	297
	References	299

13 Cathodoluminescence in Scanning and Transmission Electron Microscopies 303
Yutaka Ohno and Seiji Takeda

13.1 Introduction 303
13.2 Principles of Cathodoluminsecence 304
13.2.1 The Generation and Recombination of Electron-Hole Pairs 304
13.2.2 Characteristic of CL Spectroscopy 305
13.2.3 CL Imaging and Contrast Analysis 306
13.2.4 Spatial Resolution of CL Imaging and Spectroscopy 306
13.2.5 CL Detection Systems 307
13.3 Applications of CL in Scanning and Transmission Electron Microscopies 307
13.3.1 Assessments of Group III–V Compounds 308
13.3.1.1 Nitrides 308
13.3.1.2 III–V Compounds Except Nitrides 309
13.3.2 Group II–VI Compounds and Related Materials 310
13.3.2.1 Oxides 310
13.3.2.2 Group II–VI Compounds, Except Oxides 312
13.3.3 Group IV and Related Materials 313
13.4 Concluding Remarks 313
References 313

14 *In-Situ* TEM with Electrical Bias on Ferroelectric Oxides 321
Xiaoli Tan

14.1 Introduction 321
14.2 Experimental Details 323
14.3 Domain Polarization Switching 324
14.4 Grain Boundary Cavitation 326
14.5 Domain Wall Fracture 331
14.6 Antiferroelectric-to-Ferroelectric Phase Transition 335
14.7 Relaxor-to-Ferroelectric Phase Transition 341
References 345

15 Lorentz Microscopy 347
Josef Zweck

15.1 Introduction 347
15.2 The *In-Situ* Creation of Magnetic Fields 350
15.2.1 Combining the Objective Lens Field with Specimen Tilt 351
15.2.2 Magnetizing Stages Using Coils and Pole-Pieces 352
15.2.3 Magnetizing Stages Without Coils 356

15.2.3.1	Oersted Fields 356
15.2.3.2	Spin Torque Applications 358
15.2.3.3	Self-Driven Devices 361
15.3	Examples 362
15.3.1	Demagnetization and Magnetization of Ring Structures 362
15.3.2	Determination of Wall Velocities 364
15.3.3	Determination of Stray Fields 365
15.4	Problems 366
15.5	Conclusions 367
	References 367

Index 371

List of Contributors

Florian Banhart
Université de Strasbourg
Institut de Physique et Chimie des
Matériaux, UMR 7504
23 rue du Loess
67034 Strasbourg
France

Nigel D. Browning
Lawrence Livermore National
Laboratory
Physical and Life Sciences Directorate
7000 East Avenue
Livermore
California 94550
USA

Geoffrey H. Campbell
Lawrence Livermore National
Laboratory
Physical and Life Sciences Directorate
7000 East Avenue
Livermore
California 94550
USA

Gerhard Dehm
Austrian Academy of Sciences
Erich Schmid Institute of Materials
Science
Jahnstr. 12
8700 Leoben
Austria

and
Montanuniversität Leoben
Department Materials Physics
Franz-Josef-Str. 18
8700 Leoben
Austria

Wayne D. Kaplan
Technion - Israel Institute of Technology
Department of Materials Engineering
Haifa 32000
Israel

Daniel Kiener
Montanuniversität Leoben
Department Materials Physics
Franz-Josef-Str. 18
8700 Leoben
Austria

Judy S. Kim
Lawrence Livermore National
Laboratory
Physical and Life Sciences Directorate
7000 East Avenue
Livermore
California 94550
USA

and
University of California
Department of Chemical Engineering
and Materials Science
One Shields Avenue
Davis
California 95616
USA

Wayne E. King
Lawrence Livermore National
Laboratory
Physical and Life Sciences Directorate
7000 East Avenue
Livermore
California 94550
USA

Christoph Koch
Max-Planck-Institut für
Metallforschung
Heisenbergstr. 3
70569 Stuttgart
Germany

Thomas LaGrange
Lawrence Livermore National
Laboratory
Physical and Life Sciences Directorate
7000 East Avenue
Livermore
California 94550
USA

Alexander V. Latyshev
Siberian Branch of Russian Academy of
Sciences
Institute of Semiconductor Physics
Prospect Lavrent'eva 13
630090 Novosibirsk
Russia

Marc Legros
CEMES-CNRS
29 Rue Jeanne Marvig
31055 Toulouse
France

Andrew M. Minor
University of California, Berkeley and
National Center for Electron Microscopy
Department of Materials Science and
Engineering, Lawrence Berkeley
National Laboratory
One Cyclotron Road, MS 72
Berkeley
CA 94720
USA

Christian Motz
Österreichische Akademie der
Wissenschaften
Erich Schmid Institut für
Materialwissenschaft
Jahnstr. 12
8700 Leoben
Austria

Yutaka Ohno
Tohoku University
Institute for Materials Research
Katahira 2-1-1
Aoba-ku
Sendai 980-8577
Japan

Bryan W. Reed
Lawrence Livermore National
Laboratory
Physical and Life Sciences Directorate
7000 East Avenue
Livermore
California 94550
USA

List of Contributors

Frances M. Ross
IBM T. J. Watson Research Center
1101 Kitchawan Road
Yorktown Heights
NY 10598
USA

Christina Scheu
1Ludwig-Maximilians-Universität
München
Department Chemie & Center for
NanoScience (CeNS)
Butenandstr. 5-13, Gerhard-Ertl-
Gebäude (Haus E)
81377 München
Germany

Renu Sharma
National Institute of Science and
Technology
Center for Nanoscale Science and
Technology
100 Bureau Drive
Gaithersburg
MD 20899-6201
USA

Ralph Spolenak
ETH Zurich
Laboratory of Nanometallurgy,
Department of Material
Wolfgang-Pauli-Str. 10
8093 Zurich
Switzerland

Seiji Takeda
Osaka University
The Institute of Scientific and Industrial
Research
Mihogaoka 8-1
Ibaraki
Osaka 567-0047
Japan

Xiaoli Tan
Iowa State University
Department of Materials Science and
Engineering
2220 Hoover Hall
Ames
IA 50011
USA

Judith C. Yang
University of Pittsburgh
Department of Chemical and Petroleum
Engineering
1249 Benedum Hall
Pittsburgh
PA 15261
USA

Guangwen Zhou
P. O. Box 6000
85 Murray Hill Road
Binghampton
NY 13902
USA

Josef Zweck
University of Regensburg
Physics Faculty
Physics Building Office Phy 7.3.05
93040 Regensburg
Germany

Preface

Today, transmission electron microscopy (TEM) represents one of the most important tools used to characterize materials. Electron diffraction provides information on the crystallographic structure of materials, conventional TEM with bright-field and dark-field imaging on their microstructure, high-resolution TEM on their atomic structure, scanning TEM on their elemental distributions, and analytical TEM on their chemical composition and bonding mechanisms. Each of these techniques is explained in detail in various textbooks on TEM techniques, including *Transmission Electron Microscopy: A Textbook for Materials Science* (D.B. Williams and C.B. Carter, Plenum Press, New York, 1996), and *Transmission Electron Microscopy and Diffractometry of Materials* (3rd edition, B. Fultz and J. M. Howe, Springer-Verlag, Berlin, Heidelberg, 2008).

Most interestingly, however, TEM also enables dynamical processes in materials to be studied through dedicated *in-situ* experiments. To watch changes occurring in a material of interest allows not only the development but also the refinement of models, so as to explain the underlying physics and chemistry of materials processes. The possibilities for *in-situ* experiments span from thermodynamics and kinetics (including chemical reactions, oxidation, and phase transformations) to mechanical, electrical, ferroelectric, and magnetic material properties, as well as materials synthesis.

The present book is focused on the state-of-the-art possibilities for performing dynamic experiments inside the electron microscope, with attention centered on TEM but including scanning electron microscopy (SEM). Whilst *seeing is believing* is one aspect of *in-situ* experiments in electron microscopy, the possibility to obtain *quantitative data* is of almost equal importance when accessing critical data in relation to physics, chemistry, and the materials sciences. The equipment needed to obtain quantitative data on various stimuli – such as temperature and gas flow for materials synthesis, load and displacement for mechanical properties, and electrical current and voltage for electrical properties, to name but a few examples – are described in the individual sections that relate to *Growth and Interactions* (Part Two), *Mechanical Properties* (Part Three), and *Physical Properties* (Part Four).

During the past decade, interest in *in-situ* electron microscopy experiments has grown considerably, due mainly to new developments in quantitative stages and micro-/nano-electromechanical systems (MEMS/NEMS) that provide a *"lab on chip"* platform which can fit inside the narrow space of the pole-pieces in the transmission electron microscope. In addition, the advent of imaging correctors that compensate for the spherical and, more recently, the chromatic aberration of electromagnetic lenses has not only increased the resolution of TEM but has also permitted the use of larger pole-piece gaps (and thus more space for stages inside the microscope), even when designed for imaging at atomic resolution. Another driving force of *in-situ* experimentation using electron probes has been the small length-scales that are accessible with focused ion beam/SEM platforms and TEM instruments. These are of direct relevance for nanocrystalline materials and thin-film structures with micrometer and nanometer dimensions, as well as for structural defects such as interfaces in materials.

This book provides an overview of dynamic experiments in electron microscopy, and is especially targeted at students, scientists, and engineers working in the fields of chemistry, physics, and the materials sciences. Although experience in electron microscopy techniques is not a prerequisite for readers, as the basic information on these techniques is summarized in the first two chapters of Part One, *Basics and Methods*, some basic knowledge would help to use the book to its full extent. Details of specialized *in-situ* methods, such as *Dynamic TEM* and *Reflection Electron Microscopy* are also included in Part One, to highlight the science which emanates from these fields.

Gerhard Dehm, Leoben, Austria
James M. Howe, Charlottesville, USA
Josef Zweck, Regensburg, Germany
January 2012

Part I
Basics and Methods

1
Introduction to Scanning Electron Microscopy

Christina Scheu and Wayne D. Kaplan

The scanning electron microscope is without doubt one of the most widely used characterization tools available to materials scientists and materials engineers. Today, modern instruments achieve amazing levels of resolution, and can be equipped with various accessories that provide information on local chemistry and crystallography. These data, together with the morphological information derived from the sample, are important when characterizing the microstructure of materials used in a wide number of applications. A schematic overview of the signals that are generated when an electron beam interacts with a solid sample, and which are used in the scanning electron microscope for microstructural characterization, is shown in Figure 1.1. The most frequently detected signals are high-energy backscattered electrons, low-energy secondary electrons and X-rays, while less common signals include Auger electrons, cathodoluminescence, and measurements of beam-induced current. The origin of these signals will be discussed in detail later in the chapter.

Due to the mechanisms by which the image is formed in the scanning electron microscope, the micrographs acquired often *appear* to be directly interpretable; that is, the contrast in the image is often directly associated with the microstructural features of the sample. Unfortunately, however, this may often lead to gross errors in the measurement of microstructural features, and in the interpretation of the microstructure of a material. At the same time, the fundamental mechanisms by which the images are formed in the scanning electron microscope are reasonably straightforward, and a little effort from the materials scientist or engineer in correlating the microstructural features detected by the imaging mechanisms makes the technique of scanning electron microscopy (SEM) being extremely powerful.

Unlike conventional optical microscopy or conventional transmission electron microscopy (TEM), in SEM a focused beam of electrons is rastered across the specimen, and the signals emitted from the specimen are collected as a function of position of the incident focused electron beam. As such, the final image is collected in a sequential manner across the surface of the sample. As the image in SEM is formed from signals emitted due to the interaction of a focused incident electron probe with the sample, two critical issues are involved in understanding SEM images, as well as in the correlated analytical techniques: (i) the nature of the incident electron probe; and (ii) the manner by which incident electrons interact with matter.

In-situ Electron Microscopy: Applications in Physics, Chemistry and Materials Science, First Edition.
Edited by Gerhard Dehm, James M. Howe, and Josef Zweck.
© 2012 Wiley-VCH Verlag GmbH & Co. KGaA. Published 2012 by Wiley-VCH Verlag GmbH & Co. KGaA.

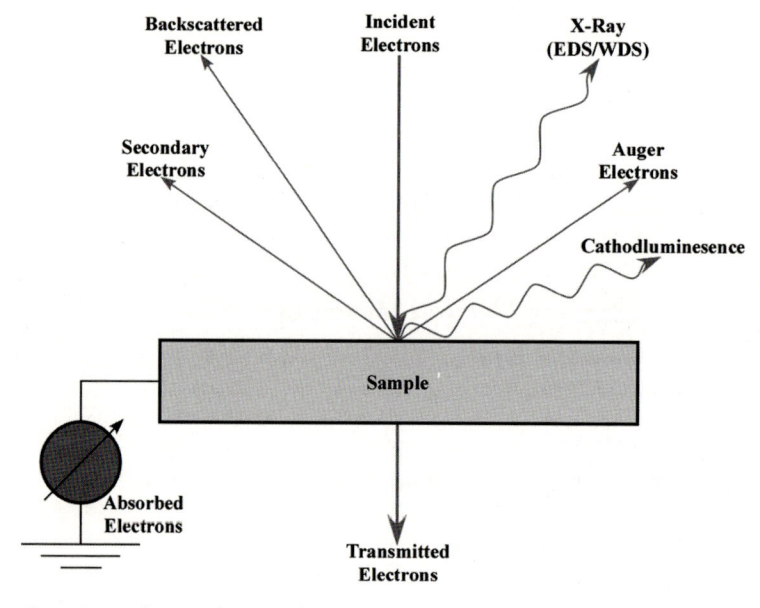

Figure 1.1 Schematic drawing of possible signals created when an incident electron beam interacts with a solid sample. Reproduced with permission from Ref. [4]; © 2008, John Wiley & Sons.

The electron–optical system in a scanning electron microscope is actually designed to demagnify rather than to magnify, in order to form the small incident electron probe which is then rastered across the specimen. As such, the size of the incident probe depends on the electron source (or gun), and the electromagnetic lens system which focuses the emitted electrons into a fine beam that then interacts with the sample. The probe size is the first parameter involved in defining the spatial resolution of the image, or of the analytical measurements. However, the signals (e.g., secondary electrons, backscattered electrons, X-rays) that are used to form the image emanate from regions in the sample that may be significantly larger than the diameter of the incident electron beam. Thus, electron–matter interaction must be understood, together with the diameter of the incident electron probe, to understand both the resolution and the contrast in the acquired image.

The aim of this chapter is to provide a fundamental introduction to SEM and its associated analytical techniques (further details are available in Refs [1–5]).

1.1
Components of the Scanning Electron Microscope

It is convenient to consider the major components of a scanning electron microscope as divided into four major sections (see Figure 1.2):

- The electron source (or electron gun).

- The electromagnetic lenses, which are used to focus the electron beam and demagnify it into a small electron probe.
- The deflection system.
- The detectors, which are used to collect signals emitted from the sample.

Before discussing these major components, a few words should be mentioned regarding the vacuum system. Within the microscope, different levels of vacuum are required for three main reasons. First, the electron source must be protected against

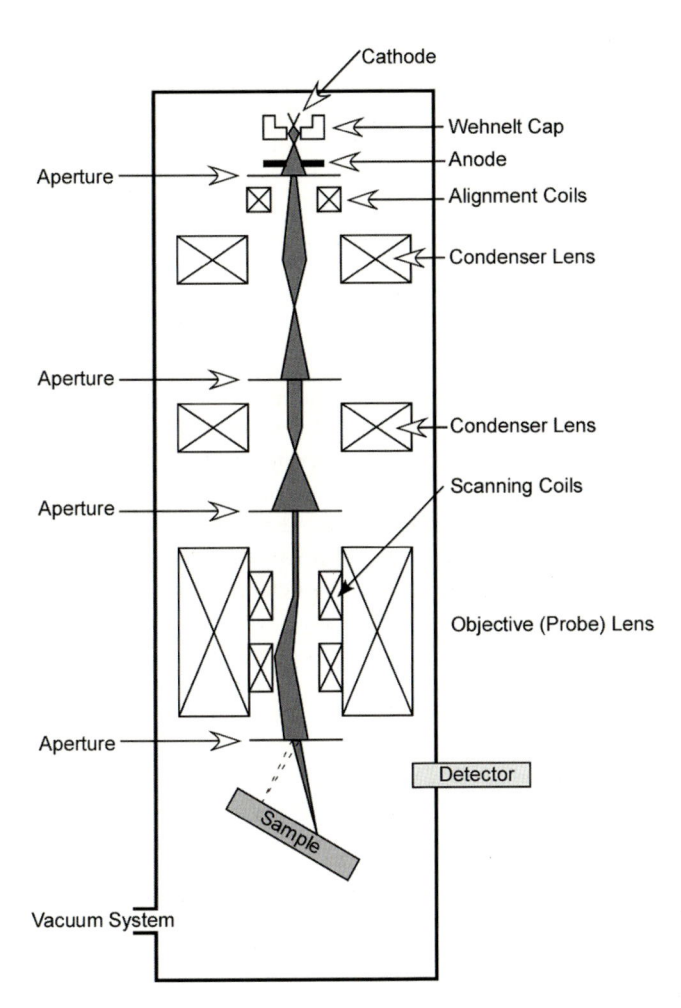

Figure 1.2 Schematic drawing of the major components of a scanning electron microscope. The electron lenses and apertures are used to demagnify the electron beam that is emitted from the electron source into a small probe, and to control the beam current density. The demagnified beam is than scanned across the sample. Various detectors are used to register the signals arising from various electron–matter interactions.

oxidation, which would limit the lifetime of the gun and may cause instabilities in the intensity of the emitted electrons. Second, a high level of vacuum is required to prevent the scattering of electrons as they traverse the column from the gun to the specimen. Third, it is important to reduce the partial pressure of water and carbon in the vicinity of the sample, as any interaction of the incident electron beam with such molecules on the surface of the sample may lead to the formation of what is commonly termed a "carbonaceous" (or contamination) layer, which can obscure the sample itself. The prevention of carbonaceous layer formation depends both on the partial pressure of water and carbon in the vacuum near the sample, and the amount of carbon and water molecules that are adsorbed onto the surface of the sample prior to its introduction into the microscope. Thus, while a minimum level of vacuum is always required to prevent the scattering of electrons by molecules (the concentration of which in the vacuum is determined from a measure of partial pressure), it is the partial pressure of oxygen in the region of the electron gun, and the partial pressure of carbon and water in the region of the specimen, that are in fact critical to operation of the microscope. Unfortunately, most scanning electron microscopes do not provide such measures of partial pressure, but rather maintain different levels of vacuum in the different regions of the instrument. Normally, the highest vacuum (i.e., the lowest pressure) is in the vicinity of the electron gun and, depending on the type of electron source, an ultra-high-vacuum (UHV) level (pressure $<10^{-8}$ Pa) may be attained. The nominal pressure in the vicinity of the specimen is normally in the range of 10^{-3} Pa. Some scanning electron microscopes that have been designed for the characterization of low-vapor pressure liquids, "moist" biological specimens or nonconducting materials, have differential apertures between the regions of the microscope. This allows a base vacuum as high as approximately 0.3 Pa close to the sample. These instruments, which are often referred to as "environmental" scanning electron microscopes, offer unique possibilities, but their detailed description is beyond the scope of the present chapter.

1.1.1
Electron Guns

The role of the electron gun is to produce a high-intensity source of electrons which can be focused into a fine electron beam. In principle, free electrons can be generated by thermal emission or field emission from a metal surface (Figure 1.3). In thermal emission, the energy necessary to overcome the work function is supplied by heating the tip. In order to reduce the work function an electric field is applied ("Schottky effect"). If the electric field is of the order of $10\,\mathrm{V\,nm^{-1}}$, the height and width of the potential barrier is strongly reduced, such that the electrons may leave the metal via field emission.

Although several different electron sources have been developed, their basic design is rather similar (see Figure 1.4). In a *thermionic source*, the electrons are extracted from a heated filament at a low bias voltage that is applied between the source and a cylindrical cap (the Wehnelt cylinder). This beam of thermionic

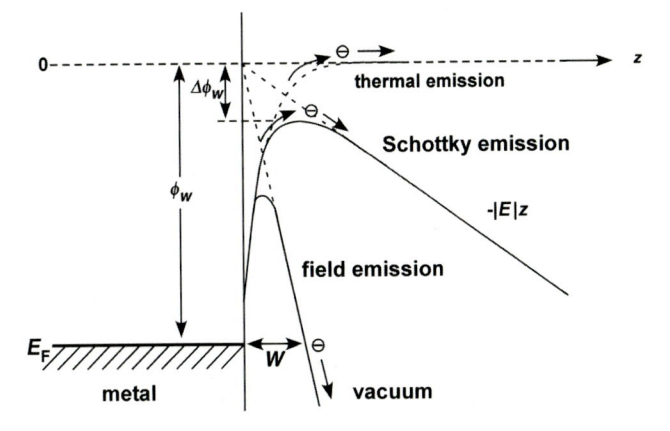

Figure 1.3 Schematic drawing of the electrostatic potential barrier at a metal surface. In order to remove an electron from the metal surface, the work function must be overcome. The work function can be lowered by applying an electric field (Schottky effect). If the field is very high, the electrons can tunnel through the potential barrier. Redrawn from Ref. [1].

electrons is brought to a focus by the electrostatic field and then accelerated by an anode beneath the Wehnelt cylinder.

The beam that enters the microscope column is characterized by the effective source size d_{gun}, the divergence angle of the beam α_0, the energy of the electrons E_0, and the energy spread of the electron beam ΔE.

An important quantity here is the axial gun brightness (β), which is defined as the current ΔI passing through an area ΔS into a solid angle $\Delta\Omega = \pi\alpha^2$, where α is the angular spread of the electrons. With $j = \Delta I/\Delta S$ being the current density in A cm^{-2}, the following is obtained:

$$\beta = \frac{\Delta I}{\Delta S \Delta \Omega} = \frac{j}{\pi \alpha^2} = \text{const.} \tag{1.1}$$

The brightness is a conserved quantity, which means that its value is the same for all points along the optical axis, independent of which apertures are inserted, or how many lenses are present.

Currently, three different types of electron sources are in common use (Figure 1.4); the characteristics of these are summarized in Table 1.1. A heated tungsten filament is capable of generating a brightness of the order of 10^4 A cm^{-2} sr^{-1}, from an effective source size, defined by the first cross-over of the electron beam, approximately 15 μm across. The thermionic emission temperatures are high, which explains the selection of tungsten as the filament material. A lanthanum hexaboride LaB$_6$ crystal can generate a brightness of about 10^5 A cm^{-2} sr^{-1}, but this requires a significantly higher vacuum level in the vicinity of the source, and is now infrequently used in SEM instruments. The limited effective source size of thermionic electron guns, which must be demagnified by the electromagnetic lens system before impinging on the sample, leads to microscopes equipped with thermionic sources being defined as *conventional* scanning electron microscopes.

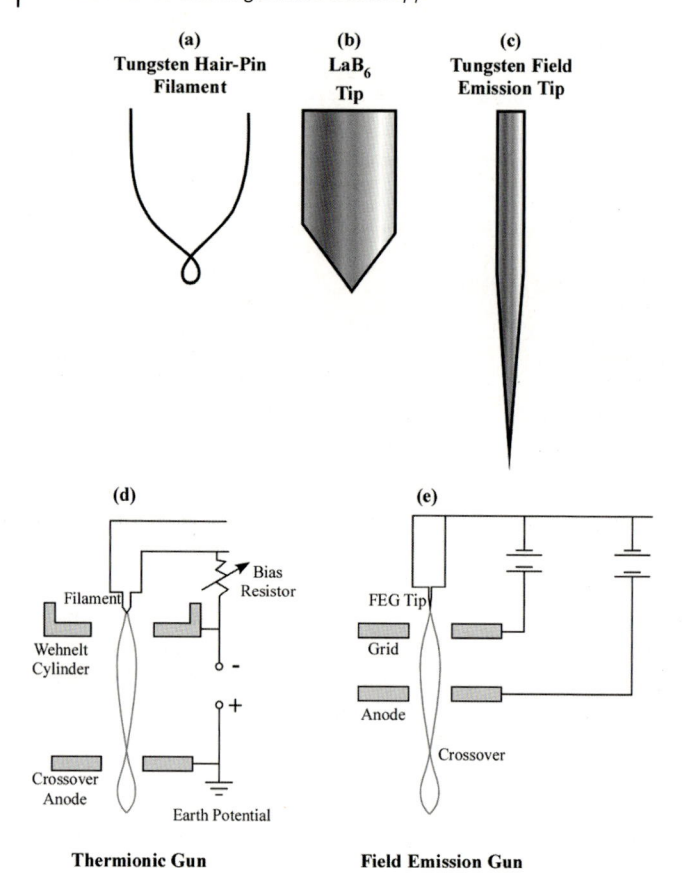

Figure 1.4 Schematic drawings of (a) a tungsten filament and (b) a LaB$_6$ tip for thermionic electron sources. (c) For a field-emission gun (FEG) source, a sharp tungsten tip is used. (d) In thermionic sources the filament or tip is heated to eject electrons, which are then focused with an electrostatic lens (the Wehnelt cylinder). (e) In FEGs, the electrons are extracted by a high electric field applied to the sharp tip by a counterelectrode aperture, and then focused by an anode to image the source. Reproduced with permission from Ref. [4]; © 2008, John Wiley & Sons.

The effective source size can be significantly reduced (leading to the term *high-resolution* SEM) by using a "cold" field emission gun (FEG), in which the electrons "tunnel" out of a sharp tip under the influence of a high electric field (Figures 1.3 and 1.4). Cold FEG sources can generate a brightness of the order of 10^7 A cm^{-2} sr^{-1}, and the sharp tip of the tungsten needle that emits the electrons is of the order of 0.2 μm in diameter; hence, the effective source size is less than 5 nm. More often, a "hot" source replaces the "cold" source, in which case a sharp tungsten needle is heated to enhance the emission (this is termed a "thermal" field emitter, or TFE). The heating of the tip leads to a self-cleaning process; this has proved to be another benefit of TFEs in that they can be operated at a lower vacuum level (higher pressures). In the

Table 1.1 A comparison of the properties of different electron sources.

Source type	Thermionic	Thermionic	Schottky	Cold FEG
Cathode material	W	LaB$_6$	W(100) + ZrO	W(310)
Work function [eV]	4.5	2.7	2.7	4.5
Tip radius [μm]	50–100	10–20	0.5–1	<0.1
Operating temperature [K]	2800	1900	1800	300
Emission current density [A cm^{-2}]	1–3	20–50	500–5000	10^4–10^6
Total emission current [μA]	200	80	200	5
Maximum probe current [nA]	1000	1000	>20	0.2
Normalized brightness [A cm^{-2} sr^{-1}]	10^4	10^5	10^7	2×10^7
Energy spread at gun exit [eV]	1.5–2.5	1.3–2.5	0.4–0.7	0.3–0.7

so-called "Schottky emitters," the electrostatic field is mainly used to reduce the work function, such that electrons leave the tip via thermal emission (see Figure 1.3). A zirconium-coated tip is often used to reduce the work function even further. Although Schottky emitters have a slightly larger effective source size than cold field emission sources, they are more stable and require less stringent vacuum requirements than cold FEG sources. Equally important, the probe current at the specimen is significantly larger than for cold FEG sources; this is important for other analytical techniques used with SEM, such as energy dispersive X-ray spectroscopy (EDS).

1.1.2 Electromagnetic Lenses

Within the scanning electron microscope, the role of the general lens system is to demagnify an image of the initial crossover of the electron probe to the final size of the electron probe on the sample surface (1–50 nm), and to raster the probe across the surface of the specimen. As a rule, this system provides demagnifications in the range of 1000- to 10 000-fold. Since one is dealing with electrons rather than photons the lenses may be either electrostatic or electromagnetic. The simplest example of these is the electrostatic lens that is used in the electron gun.

Electromagnetic lenses are more commonly encountered, and consist of a large number of turns of a copper wire wound around an iron core (the pole-piece). A small gap located at the center of the core separates the upper and lower pole-pieces. The magnetic flux of the lens is concentrated within a small volume by the pole-pieces, and the stray field at the gap forms the magnetic field. The magnetic field distribution is inhomogeneous in order to focus electrons traveling parallel to the optical axis onto a point on the optical axis; otherwise, they would be unaffected. Thereby, the radial component of the field will force these electrons to change their direction in such a way that they possess a velocity component normal to the optical axis; the longitudinal component of the field would then force them towards the optical axis. Accordingly, the electrons move within the lens along screw trajectories about the optical axis due to the Lorentz force associated with the longitudinal and radial magnetic field components.

Generally, in order to determine the image position and magnification (demagnification) for the given position of the object, it is possible to use the lens formula:

$$\frac{1}{F} = \frac{1}{U} + \frac{1}{V} \quad (1.2)$$

where F is the focal length of the lens, U is the distance between the object and the lens, and V is the distance between the image and the lens. The magnification (demagnification) of the image – that is, the ratio of the linear image size h to the corresponding linear size of the object H – is equal to (see Figure 1.5):

$$M = \frac{h}{H} = \frac{V}{U}. \quad (1.3)$$

If $U \gg F$, then for the total demagnification of a three-lens system a spot is obtained with a geometric diameter of

$$d_0 = \frac{F_1 F_2 F_3}{U_1 U_2 U_3} d_{gun} = M d_{gun} \quad (1.4)$$

where d_{gun} is the initial crossover diameter. To obtain $d_0 \leq 10$ nm for a thermionic cathode, which possesses an initial crossover d_{gun} of ≈ 20–$50\,\mu$m, the total demagnification must be $\leq 1/5000$. A Schottky or field-emission gun can result in $d_{gun} \leq 10$ nm, such that only one probe-forming (objective) lens is necessary to demagnify the electron probe to $d_0 \approx 1$ nm. The distance between the objective lens and the sample surface is termed the "working distance" of the microscope. From the above discussion, it follows that a short working distance will lead to a stronger demagnification and thus to a smaller electron probe size.

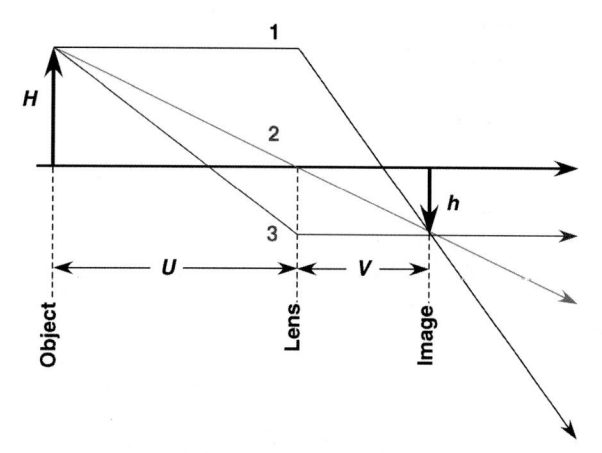

Figure 1.5 Schematic drawing of the relationship between focal length and magnification for a ideal "thin" lens. Reproduced with permission from Ref. [4]; © 2008, John Wiley & Sons.

As with any lens system, the final size (and shape) of the electron probe will also depend on aberrations intrinsic to the electromagnetic lenses used in the scanning electron microscope. In a simplistic approach, the three main lens aberrations are spherical and chromatic aberrations (Figure 1.6) and astigmatism (Figure 1.7):

- *Spherical aberration* results in electrons traversing different radial distances in the lens (r_1 and r_2 in Figure 1.6a), to be focused at different focal lengths; this will result in a blurring of the image (and a finite resolution).
- Due to *chromatic aberrations*, electrons having a difference in energy (wavelength) are focused to different focal lengths along the optical column (Figure 1.6b). In contrast to optical microscopy, electrons with shorter wavelengths (i.e., higher energy) will reach a focal point at larger focal lengths.
- Finally, *astigmatism* results in different focal lengths for electrons entering the lens at different tangential angles about the optical axis (Figure 1.7).

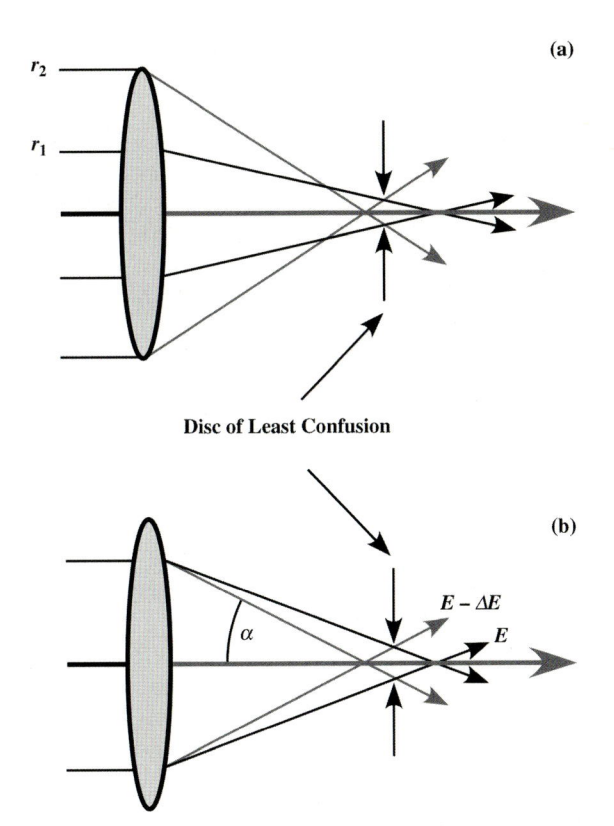

Figure 1.6 (a, b) Schematic drawings of the influence of (a) spherical and (b) chromatic aberrations on the focused electron probe. In this schematic drawing the angles of deflection are exaggerated. Reproduced with permission from Ref. [4]; © 2008, John Wiley & Sons.

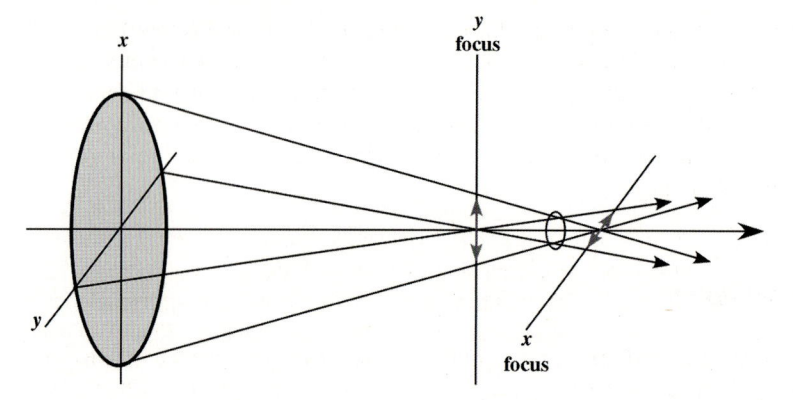

Figure 1.7 Schematic drawing of the influence of astigmatism on size of a focused electron probe. Reproduced with permission from Ref. [4]; © 2008, John Wiley & Sons.

The electron current density j_p and the probe aperture-dependent semi-convergence angle α_p are linked via the gun brightness, β:

$$j_p = \pi \beta \alpha_p^2. \tag{1.5}$$

If it is assumed, for simplicity, that the current density is uniform over a circle of diameter d_0, then the total probe current will be given by:

$$i_p = \frac{\pi}{4} d_0^2 j_p. \tag{1.6}$$

Then:

$$d_0 = \sqrt{\frac{4i_p}{\beta}} \cdot \frac{1}{\pi \alpha_p} = C_0 \frac{1}{\alpha_p}. \tag{1.7}$$

It is important to note that the parameters i_P, d_0 and α_P cannot be varied independently since, as mentioned above, the brightness remains constant. For a fixed vale of d_0 and α_P, a large value of β will be required to obtain a large probe current, i_P.

This geometric probe diameter d_0 is broadened by the action of the lens aberrations. Assuming a Gaussian distribution for both the geometric electron probe profile and all the aberrations, one obtains for the probe size:

$$d_p^2 = d_0^2 + d_d^2 + d_s^2 + d_c^2 \tag{1.8}$$

$$d_p^2 = \left[C_0^2 + (0.6\lambda)^2 \right] \alpha_p^{-2} + 0.25 C_s^2 \alpha_p^6 + \left(C_c \frac{\Delta E}{E} \right)^2 \alpha_p^2 \tag{1.9}$$

where C_0 contains the probe current and the gun brightness, d_d is the diffraction limit due to the apertures, and C_S and C_C are the spherical and chromatic aberration coefficients, respectively. When using a scanning electron microscope with a thermionic cathode, the constant C_0 is much larger than λ, which means that the diffraction error can be neglected. The dominant terms are those containing C_0 and C_S because, for energies in the 10 to 20 kV range, the term that contains C_C becomes small due to the presence of $\Delta E/E$. When operating with $E < 5$ keV, the chromatic error term dominates and C_0 is increased owing to the decrease in β (which is proportional to E).

1.1.3 Deflection System

As mentioned above, the image is formed by scanning a focused electron beam along a raster where, at each point, a signal produced by the interaction between the incident electron beam and the sample is detected, amplified, and displayed. Scanning over a raster is accomplished by two pairs of scanning coils which deflect the electron beam along a line; the coils then move the beam to the beginning of the next line where it is again deflected. By repeating this process the entire rastered area can be scanned. Simultaneously, a spot is scanned over the viewing screen, and displays the detected signal at each point. The viewing screen is either a cathode ray tube (these are rarely used in modern systems) or a liquid crystal display (LCD) computer monitor-based system.

Due to the image formation process, the magnification M of a scanning electron microscope is given by the ratio of the length of the raster on the viewing screen L_{Screen} and the length of the raster on the sample surface L_{Sample}:

$$M = L_{Screen}/L_{Sample}. \qquad (1.10)$$

1.1.4 Electron Detectors

As will be described in detail later, electron–matter interaction can lead to the emission of secondary electrons (SEs) and backscattered electrons (BSEs). These are distinguished by their energy, with electrons having energies <50 eV being considered as SEs, while those with energies close to that of the incident electron beam are labeled BSEs. Both can be used in the imaging process in the scanning electron microscope, while several different types of detectors are required to differentiate between them.

1.1.4.1 Everhart–Thornley Detector

One of the most frequently used detectors, the Everhart–Thornley (ET), can be used to detect both SEs and BSEs. The basic components of an ET detector (see Figure 1.8) include a scintillator which is surrounded by a metal collector grid, a light guide, and a photomultiplier system. Any electrons that enter the detector are collected if their energy is sufficient to create photons in the scintillator; the photons are then guided via a light guide to a photomultiplier system where the photon causes electrons to be

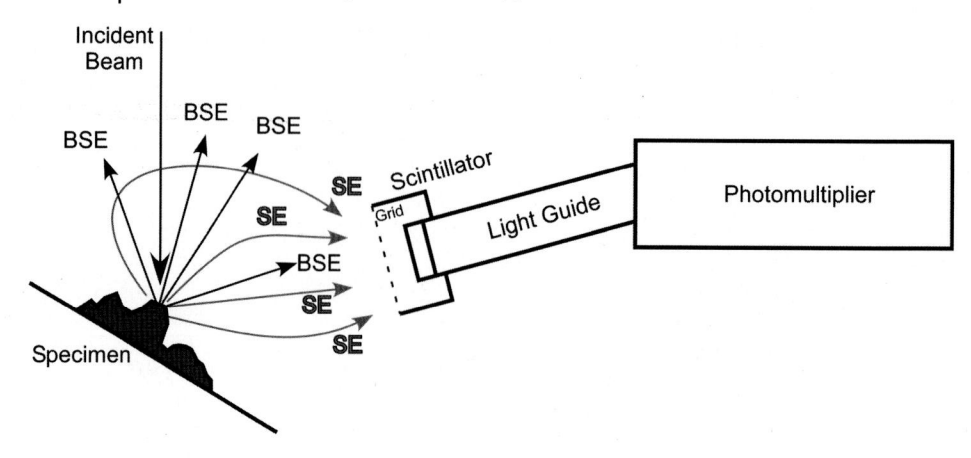

Figure 1.8 Schematic drawing of an Everhart–Thornley (ET) detector. The scintillator is biased to attract the electrons, and a separate bias on the grid can be used to screen against low-energy SE electrons. Modified from Ref. [2].

ejected from a photocathode and accelerated to the first of a series of positively biased dynodes of the photomultiplier systems, where they cause further electrons to be ejected. These latter electrons, and also those originally impinging on the first dynode, are accelerated to subsequent dynodes where the process of electron ejection is repeated. In this way, a large amplification of the incoming signal is obtained, depending on the number of dynodes present and the voltage applied. At the last dynode, which serves as the anode, the incoming electron current pulse is converted to a voltage pulse, with the help of a resistor. The voltage pulse is then further amplified by an electronic system and used to generate the signal which, after conversion to a digital signal in the case of a modern scanning electron microscope, is displayed on the viewing screen. Consequently, the brightness of each image point (pixel) in the image will be directly related to the number of SEs or BSEs detected.

Although, the energy of a SE is not sufficient to create photons in the scintillator, this problem can be overcome by coating the scintillator with a thin, positively biased ($\sim$10 keV) aluminum film. This causes the incoming SE electrons to be accelerated to a value necessary to create photons. A biased (-200 to $+200$ V) metal collector grid is located immediately in front of the scintillator. If a positive bias is applied to the grid, then the SEs emitted from the sample in directions not towards the detector will be strongly attracted towards the metal grid, thus increasing the efficiency of collection. This also means that any BSEs traveling in the direction of the ET detector will always contribute to the signal (albeit to only a small degree). In order to obtain only the signal from the BSEs, either the metal collector grid must be negatively biased, or the voltage applied to the scintillator turned off. Generally, the ET detector can be used for rapid acquisition (i.e., fast scan rates), and is usually located at an angle inclined to the sample surface at one side, so that it has only a limited solid angle detection range. Nevertheless, this geometric arrangement leads we want this effect, it is not unfortunately to shadowing effects, and to a three-dimensional effect (3-D) in the final image (see below).

1.1.4.2 Scintillator Detector

It is worth mentioning that BSEs are only weakly deflected by the electrical field associated with the collector bias voltage of the ET detector, and that they basically move without being disturbed in the direction that they are emitted. Accordingly, as the detection efficiency for BSEs will be low for an ET detector, dedicated scintillator detectors for BSE detection have been developed which have a large solid angle of detection and an annular detection area which is located above the sample surface (Figure 1.9). In a manner similar to the ET detector, the BSEs first create photons in the scintillator; the photons are then guided to a photomultiplier system, where they produce photoelectrons which are further amplified. The annular detector has an opening in the center to allow the incident electron beam to reach the sample surface.

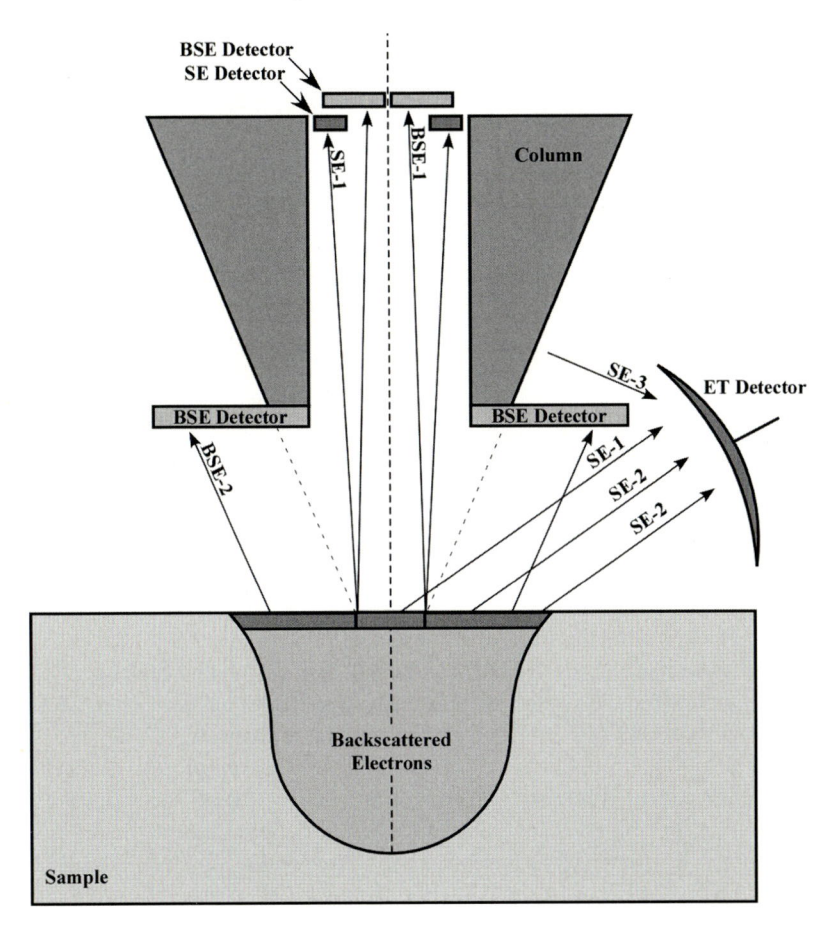

Figure 1.9 Schematic drawing of an in-lens detector system, combined with a standard BSE and ET detector. The in-lens detectors are efficient only over small working distances, where the ET and conventional detectors have low detector efficiencies. Reproduced with permission from Ref. [4]; © 2008, John Wiley & Sons.

However, as the BSE scintillator detectors are usually rather thick in terms of their dimensions, they may limit the available working distance of the SEM.

1.1.4.3 Solid-State Detector

Another possible approach to detecting BSEs is the solid-state detector (Figure 1.9). In this case, the annular semiconductor detector is placed above the sample and has, again, a hole through which the incident electron beam can pass. The active area is often separated into different segments, from which the signal can be read out separately. A BSE that strikes the active area generates electron-hole pairs at the pn-junction of the detector, which is formed at the interface between p- and n-type doped Si. The electron and holes are separated in the electrical field of the space charge region; this leads to a current pulse which is transformed to a voltage pulse with the help of an external resistor. While solid-state detectors usually have slow acquisition rates compared to scintillators and ET detectors, their main advantage is that signals from the individual segments can be combined to obtain various contrast mechanisms. A second benefit is that they are rather thin in terms of their dimensions, which allows smaller working distances.

1.1.4.4 In-Lens or Through-the-Lens Detectors

As their name implies, "in-lens" or "through-the-lens" detectors are placed directly inside the SEM lens system (Figure 1.9). Those SEs that are emitted to a certain solid angular regime from the sample surface are subjected to the magnetic field of the lens pole piece and reach the detector which, again, is based on a scintillator-light guide-photomultiplier system. Any BSEs and SEs that are emitted at larger angles will not be detected. The spatial resolution obtained by using such an in-lens detector system is significantly improved compared to that of conventional detectors (down to <1 nm for SEs), as only those electrons emitted from a region that is directly defined by the incident beam size are collected.

1.2
Electron–Matter Interaction

An incident electron beam that impinges on a sample surface can undergo both elastic and inelastic scattering events as it penetrates the solid sample. *Inelastic scattering* refers to any process in which the primary electron loses a detectable amount of energy ΔE, whereas in *elastic scattering* the electron energy remains unchanged.

All backscattered electrons lose some energy during the scattering process. Although the analytical treatment of inelastic scattering is beyond the scope of this chapter, an outline will be provided of BSEs, following an elastic model. *Elastic scattering* occurs due to the coulombic interaction of the incident charged electron with the electrical field of the atomic nucleus, which is screened by the inner-shell electrons. This scattering process can be described as "Rutherford scattering" at a screened nucleus:

$$\frac{d\sigma}{d\Omega} \approx \frac{Z^2}{E_0^2[\mathrm{Sin}^2(\theta/2) + \mathrm{Sin}^2(\theta_0/2)]^2} \tag{1.11}$$

where $\frac{d\sigma}{d\Omega}$ is the differential cross-section which gives the probability that an electron is scattered into a solid angle element $d\Omega$, θ is the scattering angle, θ_0 the characteristic angle or screening parameter which depends on the wavelength and the screening radius of the atom, Z the atomic number, and E_0 is the incident beam energy. For large scattering angles, and in particular backscattering (>90°), Eq. 1.11 is not valid and the electron spin must be considered. This so-called "Mott scattering" leads to a more complicated mathematical description of the differential cross-section that is beyond the scope of this chapter. Nevertheless, for the most probable elastic scattering angles, which are between 3° and 5°, the concept of Rutherford scattering from a screened nucleus can be used as an approximation. Multiple elastic scattering events can result in large scattering angles, and cause the electron trajectories to spread relative to the incident beam position. A cumulative change in direction may lead to electrons which finally can escape from the sample surface. These electrons, together with those that are directly scattered to angles >90°, are termed "backscattered" electrons and are used for imaging (as discussed below).

Inelastic scattering may occur as a result of several processes, including the excitation of phonons, plasmons, single-valence electrons, or inner-shell electrons (Figure 1.10). The average amount of energy that is transferred from the incident electron to the sample is different for these various events. For example, the excitation of phonons which are atomic vibrations in the solid are associated with an energy loss $\Delta E < 1$ eV, and lead to a slight heating of the sample, whereas the excitation of plasmons, which occurs via a collective excitation of the electron gas, is related to an energy loss of $\Delta E \approx$ 5–30 eV. Single-valence electron excitation requires an energy transfer of a few eV up to a few tens of eV. In order to excite the inner-shell electrons of the atoms, the incident electron must lose a larger amount of energy, typically in the range of hundreds to several thousands of eV. However, the probability that an electron loses a large quantity of its energy is low, and decreases rapidly with increasing energy loss. Consequently, processes which involve an energy transfer in the range of 5 to 50 eV dominate and, accordingly, the electrons lose their energy continuously in small quantities. The average inelastic mean free path (MFP), which describes the average distance that the

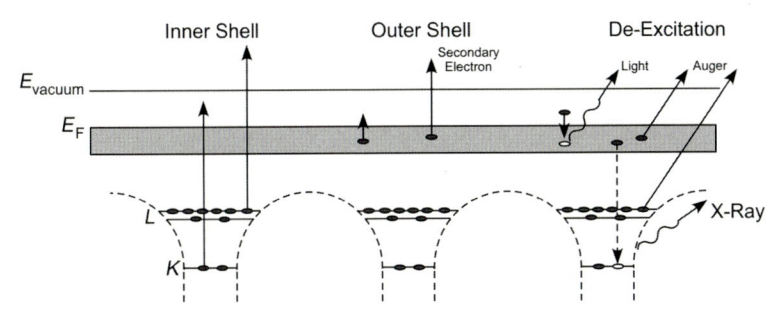

Figure 1.10 Schematic drawing of an energy diagram of a solid. Possible primary and secondary effects are indicated. Redrawn from Ref. [6].

electron can travel before being inelastically scattered, depends on the specific sample material and the incident electron energy, but typically is of the order of 100 nm. The scattering angle for inelastic scattering is small (usually <1°).

Although all inelastic scattering events have a different differential cross-section which considers the underlying physics, the Bethe formula can be used as a first approximation to estimate the average energy loss dE that occurs when the electron has traveled a distance ds in the material:

$$\frac{dE}{ds} \approx \frac{Z\varrho}{AE_i} \ln\left(\frac{E_i}{I}\right) \quad (1.12)$$

where Z is the atomic number, ϱ the density, A the atomic weight, E_i the electron energy at point i in the specimen, and I the average energy loss per scattering event. The quantity I is often estimated by an average ionization energy of the atom.

For thick samples, as are usually investigated using SEM, inelastic scattering processes dominate and the electron energy is reduced gradually. Due to multiple inelastic scattering processes, the electrons reach an average thermal kinetic energy of kT (where k is the Boltzmann constant and T the temperature) and are finally absorbed in the sample. Therefore, most of the energy of the incident electron will result in heating of the sample, though a small quantity will be used to generate secondary electrons, X-rays, or light that, eventually, can escape from the solid (Figure 1.10). These effects are termed "secondary" as they can be detected outside the sample (these will be discussed further below).

Multiple elastic and inelastic scattering events result in a lateral and vertical spreading of the electron beam relative to the incident direction, and to a maximum distance which the electron can travel before it is absorbed ("penetration depth"). The associated volume, which is termed the "interaction volume," is typically pear-shaped in thick samples. The volume size is defined by an envelope which fulfills a specific condition; for example, the electron energy has been reduced to a specific value, or that the volume contains 95% of all incident electrons. The interaction volume may be calculated and visualized using Monte Carlo electron trajectory simulations, and various programs are available (e.g., Casino [7]). For these simulations, independent scattering centers are assumed within the solid where the electron can undergo either elastic or inelastic scattering in a random, statistical fashion. This is necessary to account for the large range of scattering angles and energy loss rates which are possible. Although Monte Carlo simulations usually ignore crystallo-graphic effects, they can provide very useful information on the general aspects of the interaction volume, such as its dependence on the incident beam energy and on the atomic number.

The influence of the incident beam energy on the size of the interaction volume is shown in Figure 1.11, where the interaction volume is larger for electrons with a higher energy; that is, with a higher acceleration voltage, both in the lateral and in the vertical directions. This can be explained with assistance from the above-described Rutherford and Bethe descriptions. Elastic scattering is inversely proportional to the square of the incident beam energy (Eq. 1.11); hence, with increasing beam energy

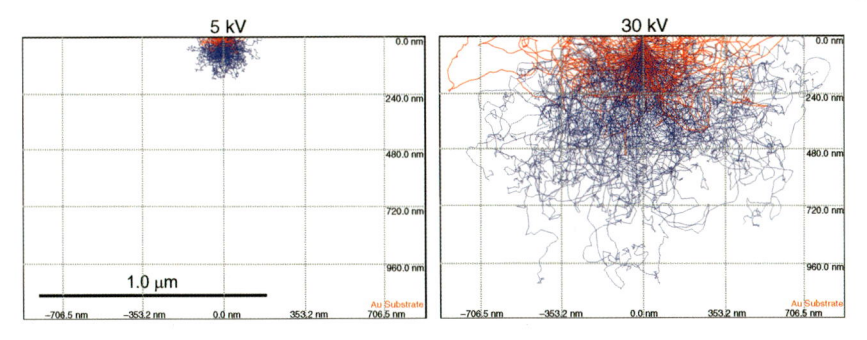

Figure 1.11 Monte Carlo simulation of electron trajectories in Au, at an incident electron energy of 5 kV and 30 kV and using a 10 nm probe diameter. The red trajectories are for electrons which eventually escaped the sample. The scale bar is identical for both calculations. Calculated using CASINO [7].

the electrons are scattered less and can penetrate more deeply into the material. In addition, the average energy loss per distances traveled (Eq. 1.12) depends inversely on the beam energy. Accordingly, electrons with a higher energy can travel for a greater distance before being inelastically scattered and before (after multiple scattering) being absorbed by the sample.

The interaction volume also depends strongly on the atomic number Z, and its size is drastically reduced with increasing Z (Figure 1.12). Besides this change in size, a change in shape is also observed, from pear-like to a more spherical shape with increasing atomic number. Again, this general trend may be better understood by

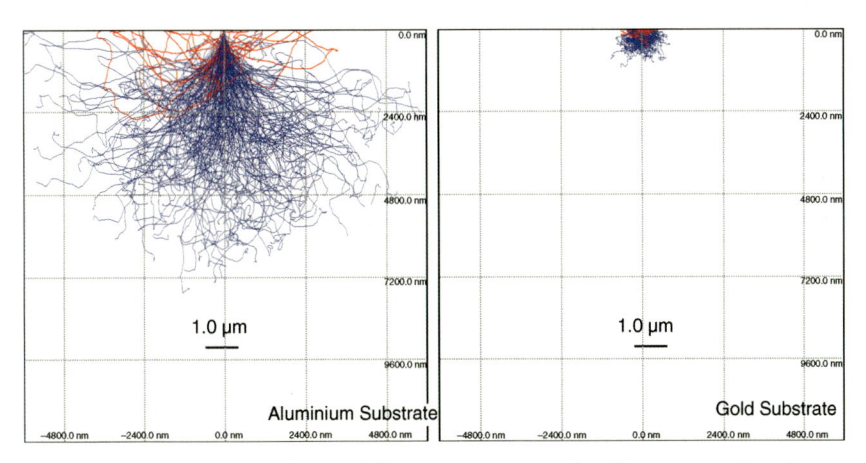

Figure 1.12 Monte Carlo simulation of electron trajectories in Al and Au, at an incident electron energy of 30 kV and using a 10 nm probe diameter. The red trajectories are for electrons which eventually escaped the sample. Calculated using CASINO [7].

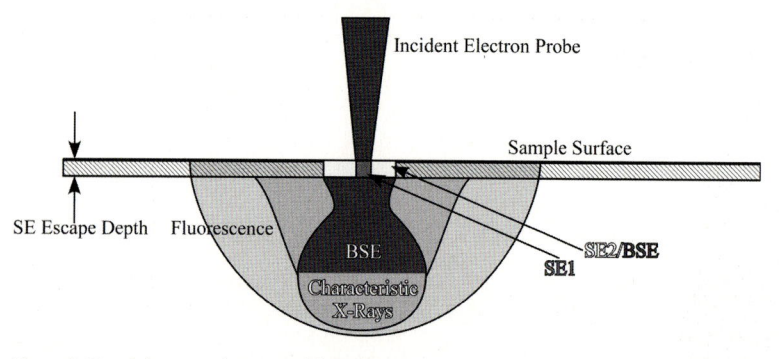

Figure 1.13 Schematic drawing of the interaction volume for a given material and given energy of the incident beam electrons which, together with the escape depth, define the spatial resolution of each signal.

examining the probability for elastic scattering, which scales with the square of the atomic number (Eq. 1.11).

The critical finding from this discussion is that *"the interaction volume is much larger in dimensions compared to the incident beam diameter."* Typical values for a 10 nm beam diameter can be lateral and vertical spread into a volume which has a length of 1 μm or more. This has a very strong impact on the spatial resolution of SEM, and for associated analytical techniques, as the various signals (SEs, BSEs, and X-rays) are generated within this interaction volume as long as the energy of the electrons is sufficient for the process. The interaction volume is a critical parameter for defining the spatial resolution which, for different imaging and spectroscopic techniques, is also governed by the *escape depth*. This is the maximum distance that the generated electrons or photons can travel (and leave the solid to reach the detector). As will be seen, the escape depth varies significantly for the different signals, and as such so does the corresponding spatial resolution (Figure 1.13).

1.2.1
Backscattered Electrons (BSEs)

After a brief introduction to elastic and inelastic scattering effects, and the correlated interaction volume, it is now possible to discuss the secondary effects. First, the characteristics of BSEs will be described, followed by an explanation of the generation of SEs and Auger electrons. Finally, the basics regarding photon emission will be discussed.

As noted above, all electrons that leave the surface with energies >50 eV are termed BSEs. The contribution of the BSEs to the energy distribution of all electrons emitted from the sample surface forms a continuum with a large peak centered near E_0 (the energy of the primary beam) and a tail towards lower energies (Figure 1.14). The peak near E_0 possesses a higher intensity and smaller width for heavier elements compared to elements with a low atomic number. The BSEs are primary electrons (PEs) that originate in the incident electron beam, and are scattered in the reverse

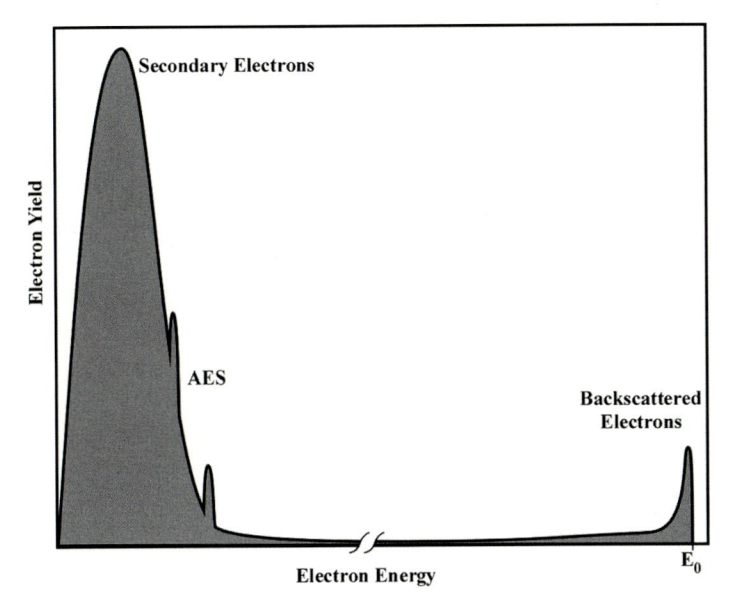

Figure 1.14 Energy distribution of electrons emitted from the sample surface. Reproduced with permission from Ref. [4]; © 2008, John Wiley & Sons.

direction (deflected by >90°), either by a single scattering event or, more likely, by multiple scattering events. The efficiency η of BSE generation is defined as:

$$\eta \approx \frac{n_{BSE}}{n_{PE}} \tag{1.13}$$

where n_{BSE} and n_{PE} are the numbers of BSEs and PEs, respectively. The efficiency depends greatly on the atomic number Z and the tilt angle θ of the incident beam direction relative to the sample normal. The detector position relative to the incident beam direction is also important.

The atomic number dependency shows a monotonic increase of η with increasing Z (Figure 1.15), giving rise to the so-called "atomic number contrast" or "compositional contrast." The efficiency also depends heavily on the tilt angle θ; with an increasing tilt angle the efficiency (η) increases monotonically as more electrons can escape from the sample surface. It is this dependency on tilt angle that gives rise to a topological contrast in the BSE images. The effect of the incident beam energy on η is much less pronounced. For incident beam energies >5 keV, η is basically independent of the beam energy, whereas for lower beam energies a weak dependence occurs. Such dependence is often obscured, however, by contamination layers which prevent successful BSE image acquisition in the low-energy regime.

At this point, it should be mentioned again that BSEs continue to travel along nearly straight trajectories in the direction that they are emitted, even when a collector bias voltage is applied, as the bias is often too weak to have any significant effect on the

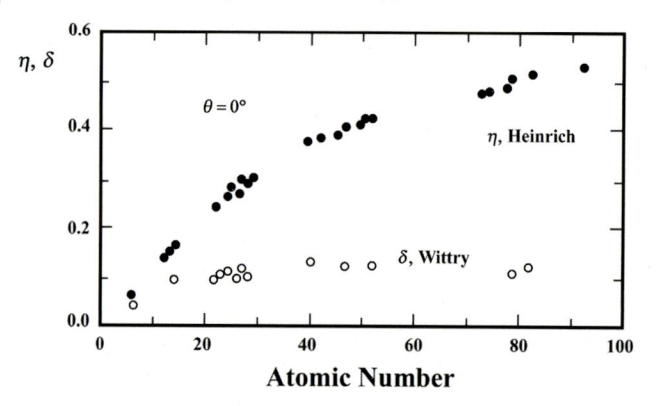

Figure 1.15 Schematic drawing of the efficiency of BSE (η, after Heinrich, 1966) and SE (δ, after Wittry, 1966) generation as a function of atomic number (Z) at an incident beam energy of 30 keV. Reproduced with permission from Ref. [3].

BSEs' trajectory. For image interpretation this directionality of motion is important, and the detector position relative to the BSEs' trajectories when they leave the sample must be considered. In fact, two cases should be distinguished: (i) when the incident beam direction is parallel to the surface normal; and (ii) when the incident beam direction possesses an angle relative to the surface normal. In the first case ($\theta = 0°$), η follows a cosine function, being largest when the detector is placed parallel to the incident beam direction:

$$\eta(\phi) = \eta_n \cos \phi. \tag{1.14}$$

Here, η_n is the value measured when the detector is placed parallel to the incident beam direction, and ϕ is the angle between the sample normal vector and the direction of the detector axis. Equation 1.14 implies that the signal decreases when ϕ is increased, reaching approximately 70% of η_n for $\phi = 45°$. In the second case, and if θ is larger than ~45°, the angular distribution changes from a symmetrical cosine function to an asymmetric ellipsoidal distribution, such that backscattering is favored in directions away from the incident beam direction, and is highest in the forward direction. Thereby, the long axis of the ellipsoid is approximately at ($90° - \theta$) above the sample surface as the incident beam. This dependency on the detector position also contributes to topological contrast in BSE images.

It is also important to note that the escape depth of BSEs can be in the range of microns, depending of course on their energy (see Figure 1.13). In addition, the interaction volume becomes asymmetric when θ is larger than 45°.

1.2.2
Secondary Electrons (SEs)

As noted above, all electrons with energies between 1 eV and ~50 eV are termed SEs. The secondary electron yield δ is defined as the number of SE (n_{SE}) released per number of incident high-energy electrons (n_{PE}):

$$\delta = \frac{n_{SE}}{n_{PE}}. \qquad (1.14)$$

It is important to remember that, in this equation, n_{PE} *includes* the contribution from backscattered electrons which are traversing the solid and continue to generate SEs when they interact inelastically with the sample. As such, the SE yield is much higher than unity (typically >100%), and consequently most of the electrons detected are SEs. The secondary electron energy distribution shows a very large peak which is centered around 2–5 eV; this means that 90% of all detected SEs have an energy less than 10 eV (Figure 1.14). It is useful here to distinguish between two different types of SE, as they stem from different regions. The SE1 electrons are the outer-shell electrons of the sample that are directly excited by the incident beam electrons as they enter the surface and are able to escape (Figure 1.13). This is only possible if the SE1 electrons are excited above the vacuum level and are close to the sample surface. Their signal originates from an area which is approximately the diameter of the incident electron beam, and may be just under 1 nm. The SE2 electrons are generated by backscattered electrons after several inelastic scattering events. Consequently, the resolution of images formed by SE2 electrons is much worse than that of SE1 electrons, as the range of the backscattered electrons laterally across the sample may be on the order of microns (Figure 1.13). The advantage of "in-lens" detectors is now obvious, as most of the SE2 electrons are removed from the signal, such that the image is formed primarily from SE1 electrons.

The secondary electron yield δ depends on the incident beam energy, the work function of the surface, the incident beam angle relative to the sample normal (specimen tilt), and the local curvature of the sample. The dependency of the secondary electron yield δ on the incident beam energy E_0 is shown in Figure 1.16. For low incident beam energies, δ first increases to reach a maximum (at about 1 to 5 keV, depending on the material); above this value, δ slowly decreases with increasing E_0. The reason for this behavior can be explained as follows. For higher-keV incident electrons, the SEs are generated at a greater depth and thus cannot escape from the surface, whereas for lower-keV primary electrons the SEs are created closer to the sample surface, which makes it easier for them to escape. The incident beam energy has also another effect on δ; with increasing beam energy, the brightness – and thus the incident beam current i_P – is increased, which in turn leads to a higher SE current.

In general, the SE yield depends on the work function of the material, which is the energy barrier that an electron at the Fermi level must overcome to reach the vacuum level. The work function is of the order of a few eV, and depends on the material composition and the atomic packing at the surface – that is, the crystal structure and orientation. However, this dependency is usually obscured by contamination layers or any conductive layers (Au or C) deposited to prevent charging.

The atomic number Z also influences the SE yield since, with increasing Z, a larger number of electrons will be backscattered and this will lead to a greater number of SEs. In addition, for samples with a higher mean Z, a larger percentage of SEs will be created near the surface, which will result in a greater probability of their escape

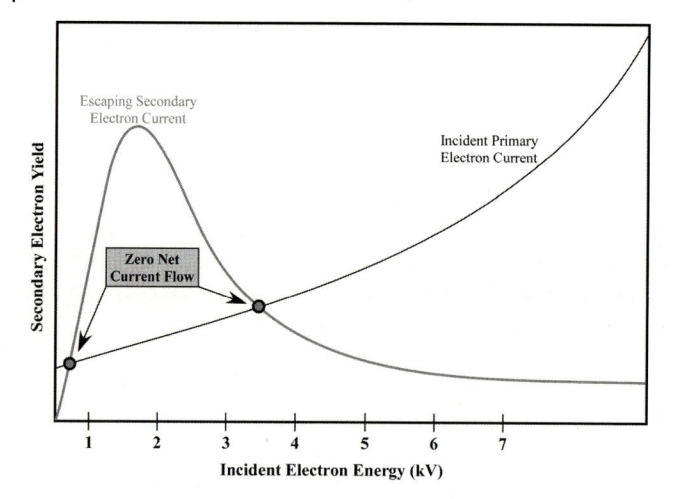

Figure 1.16 Secondary electron yield (black curve) and incident primary electron current (gray curve) as a function of incident beam energy. Electrostatic charging of the sample is prevented at two points, which are marked by circles. At these points the total net current is zero. Reproduced with permission from Ref. [4]; © 2008, John Wiley & Sons.

(to reach the detector). This dependency on Z is higher for low incident beam energies where the escape depth of the SEs is of the order of the penetration depth of the incident beam electrons. For high accelerating voltages and non-UHV conditions, δ is often found to be relatively independent of the atomic number, which is again attributed to the fact that contamination or deposited conductive layers may obscure this effect. Nevertheless, the secondary electron yield δ is less strongly affected by Z compared to the efficiency η of BSE generation, making the Z dependency in the SE images less pronounced (Figure 1.15).

The secondary electron yield δ is dependent on the angle θ of the incident beam relative to the sample normal. With increasing tilt angle, δ is increased and follows a sectant function:

$$\delta(\theta) = \delta_0 \sec \theta \tag{1.16}$$

where δ_0 is the value measured when the surface normal is parallel to the incident beam direction. This can be explained by an increased path length of the incident beam electrons within the surface region if the beam/sample is tilted. In addition, δ is also dependent on the local curvature of the surface which determines the probability that a SE can escape. A region that possesses a positive radius of curvature, such as a region protruding from the surface, will enhance δ, whereas regions with a negative radius of curvature will lead to a lower δ as the electrons will be trapped. Both dependencies – beam/sample tilt and local curvature – give rise to topological contrast in SE images.

The detector position also affects the number of SEs detected. When the incident beam direction is parallel to the surface normal, the same cosine dependency as for BSEs (Eq. 1.14) is observed; that is, the largest signal is detected when the detector is placed parallel to the incident beam. This cosine dependency remains for large values of θ, which contrasts with the strong anisotropy effect observed for BSEs.

1.2.3
Auger Electrons (AEs)

When a high-energy electron excites an inner-shell electron to an unoccupied state above the Fermi level, the atom remains in an excited state. An empty inner-shell state exists which is subsequently filled by an electron from a higher state. During this relaxation process, a specific amount of energy is released that is used to generate characteristic X-rays; alternately, the energy can be transferred to an electron of the sample possessing a lower binding energy (see Figure 1.10). If the amount of energy transferred is large enough to excite this electron above the vacuum level, it can eventually leave the solid as a characteristic Auger electron, in what is termed a "nonradiative" process. These two possibilities are alternative processes by which the energy can be released. The probability whether X-ray generation or the emission of AEs dominate depends on the material. It is more likely that AEs are created for light elements, whereas for heavier elements the emission of X-rays is the dominant process. The AEs have a characteristic energy which is given by the difference of the binding energies of the electron states involved, and the work function of the material. Therefore, they can be seen as characteristic peaks in the electron yield versus electron energy curve (Figure 1.14); moreover, if an appropriate detector is used the AEs can be detected and quantified, so as to provide important information on the surface chemistry of the sample. Since AEs have energies comparable to SEs (they are, in fact, a type of secondary electron), their mean free path in the sample is extremely limited (1–2 nm). Although AEs contribute to the total SE signal acquired in SEM, specialized detectors for energy analysis – and thus the chemical analysis of AEs in SEM – are not often employed, as UHV and clean surfaces are required. As a result, modern AE systems are usually dedicated UHV scanning electron microscopes.

1.2.4
Emission of Photons

1.2.4.1 Emission of X-Rays

As noted above, if an inner-shell electron (energy level E_1) has been excited to an unoccupied state above the Fermi level, the empty state can be filled with an electron from a higher state (energy level E_2), and the energy difference ΔE can be released by the (radiative) emission of X-rays with a characteristic energy (or wavelength) (see Figure 1.10). The wavelength can be calculated from the de Broglie relationship:

$$\lambda = \frac{hc}{\Delta E} \tag{1.17}$$

where h is Planck's constant and c is the speed of light. The transition from a higher level to the unoccupied inner-shell state is only allowed if the dipole selection rule is fulfilled; that is, when the dipole moment number l is changing by ± 1 (e.g., transitions form p-states to s-states are allowed while s-states to s-states are forbidden). If an unoccupied state in the K-shell is filled by an electron from the L-shell, the characteristic photon will contribute to the K_α-line, if it is filled by an electron from the M-shell to the K_β line, and so on. If the hole is in the L- or M-shell, then L and M characteristic radiation series will occur.

In addition to characteristic X-rays, the incident electron beam also generates X-ray radiation with a continuous energy distribution that ranges up to an energy equivalent to the total incident beam energy. When an electron passes through the Coulomb field of an atom it experiences a deceleration, which reduces the magnitude of the electron velocity. The energy released can be of any amount, and is emitted as electromagnetic radiation. Due to its origin such radiation is referred to as "Bremsstrahlung" or "braking radiation."

The escape depth of characteristic X-ray radiation depends on the interaction volume, and can be up to several micrometers in dimension; the escape depth for Bremsstrahlung is only slightly larger. The characteristic X-ray radiation can be used for quantitative chemical composition analysis where the nonlinear background in the X-ray spectrum, which stems from the Bremsstrahlung, must be removed prior to the analysis.

1.2.4.2 Emission of Visible Light

Besides electromagnetic radiation in the X-ray range, the emission of light can also be detected. The emission of photons in the visible range may occur due to different effects. For example, if a high-energy electron does not excite an inner-shell electron but rather a weakly bound valence electron, then the amount of energy released when the atom relaxes will be small. This process may lead to the emission of photons in the visible range (see Figure 1.10). In addition, X-rays generated in the solid by a high-energy electron can excite other atoms by themselves. The subsequent relaxation of these atoms may also lead to the emission of electromagnetic radiation in the visible or X-ray range; this process is termed "fluorescence."

1.2.5
Interaction Volume and Resolution

Based on the above discussion on electron–matter interaction and the generation of the various signals, together with the volume of interaction, it is now possible to summarize some important points regarding the resolution capabilities of SEM. First, while its clear that the demagnification of an incident beam to as small a diameter as possible is important, the interaction volume, the escape depth of the signal of interest, and the type of detector each play critical roles in defining the resolution of the final image. Thus, unlike optical microscopy or TEM, it is not possible to define the resolution of the microscope alone, but rather the resolution of SEM for a particular sample investigated under specific operating conditions. As such, the main points defining the resolution for each particular signal available in SEM are worthy of summary.

1.2.5.1 Secondary Electrons

As noted above, the mean free path of SEs in solids is extremely limited, and consequently SEs generated far below the surface of the sample, either by the incident electrons or by BSEs, will not escape to the surface and reach the detector. If a conventional ET detector is used, then SEs generated by BSEs (SE2) will reach the detector, and the resolution will be defined by the interaction volume of BSEs reaching the surface of the sample (see Figure 1.13). The signal-to-noise ratio (SNR) of the image will depend on the yield of SEs and the detector position, but since the yield of SEs is always greater than 1.0, the SNR of SE images will usually be very good. This can be improved by increasing the current, and as a larger spot size will not significantly influence the backscattered-limited resolution, this is usually a good option if the SNR is important. For conducting specimens, a slower scan rate will also increase the SNR. This is especially important if the incident accelerating voltage of the electron beam is reduced in an attempt to limit the contribution of SE2, which will in turn decrease the SNR.

For FEG sources, the incident accelerating voltage can be reduced while maintaining a reasonable beam current, thus further reducing the contribution of the SE2 electrons. The use of an "in-lens" detector significantly reduces the contribution of SE2 electrons when operating at small working distances, thus improving the resolution of the image by using mainly SE1 electrons. A combination of a FEG source and "in-lens" detectors provides the definition of "high-resolution SEM."

1.2.5.2 Backscattered Electrons

As the mean free path of BSEs is significantly larger than that of SEs, it is the interaction volume with the sample which defines the resolution. "In-lens" detection systems operating at small working distances will improve the resolution by preventing BSEs from extended regions to reach the detector, but BSEs from the sample depth cannot be removed from the signal (Figure 1.13), and the SNR will always be limited due to the poor BSE yield compared to that of SEs. Reducing the accelerating voltage, and thus reducing the interaction volume (see Figure 1.11) is a good way of improving BSE resolution, bearing in mind the limited SNR. As the energy of the BSEs falls with the distance traversed in the sample, an energy-filtered BSE detector might improve the BSE resolution, and this is a possible direction for future SEM systems.

1.2.5.3 X-Rays

Since the mean free path of X-rays is significantly larger than that of electrons (due to a lower cross-section), the interaction volume within a specific material defines the resolution of X-ray images and the region for which the local concentration is measured (see Figure 1.13). One option would be to decrease the incident accelerating voltage, but only if it can be maintained at a value *greater* than the energy required for electron excitation and generation of a photon (in this case it will be necessary to know beforehand which elements are in the sample). In addition, while the interaction volume roughly defines the source of the X-rays (depending on the energy of the X-ray of interest), the fluorescence of X-rays can lead to a significantly larger volume of material from which X-rays are emitted.

The maximum depth in the sample from which the characteristic X-ray signal originates, ignoring fluorescence, can be approximated by:

$$Z_r \approx 0.033\left(E_0^{1.7} - E_C^{1.7}\right)\left(\frac{\bar{A}}{\varrho \bar{Z}}\right) \tag{1.15}$$

where $\bar{A}$ and $\bar{Z}$ are the average atomic weight and average atomic number, respectively, in the volume of material being excited by the incident beam, ϱ is the density, E_0 is the incident electron energy, and E_C is the critical incident energy required for excitation. The maximum diameter of the excited volume generating the X-ray signal is usually approximated by:

$$D \approx \frac{0.231}{\varrho}\left(E_0^{1.5} - E_C^{1.5}\right). \tag{1.19}$$

1.3
Contrast Mechanisms

1.3.1
Topographic Contrast

From the discussion above, it follows that SE and BSE images may each show topographic contrast. The topographic contrast is mainly due to shadowing effects and to enhanced SE and BSE production at the edges and surface perturbations. Both effects are strongly affected by the angle of incidence of the primary electron and the position of the detector relative to the surface normal. Sample regions that are tilted towards the detector possess a much higher signal than those tilted away, giving rise to shadowing effects (Figure 1.17). In general, topographic contrast is usually associated with SE images, and this effect is strongly visible in images acquired

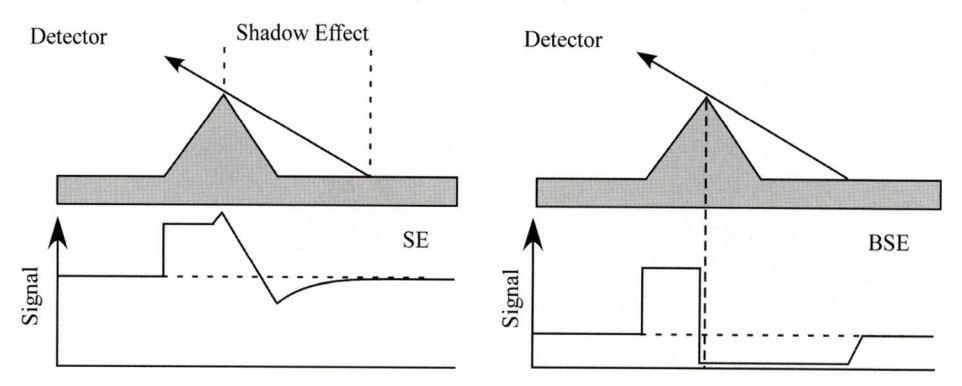

Figure 1.17 Effect of local curvature and detector position on the SE and BSE signals. Only those electrons which are drawn with a solid line reach the detector. Modified after Ref. [8]

using an ET detector located at one side of the sample. For large surface perturbations (i.e., rough surfaces), the BSE signal in the region behind the topographic feature (relative to the detector position) is very low, as most generated BSEs do not reach the detector because their trajectories are basically unaffected by the applied bias (see Figure 1.17). This situation is different for the SEs, however, which are attracted and move towards the detector. Accordingly, the SE signal behind large surface perturbations will be higher than that of the BSE signal. As a result, low- to medium-magnification BSE images may have a stronger topographic contrast than SE images, if an off-axis detector is used. In general, images acquired using an ET detector appear as if the object is illuminated from one side, which results in a three-dimensional (3-D) effect (Figure 1.18). Some of this 3-D shadowing effect associated with the

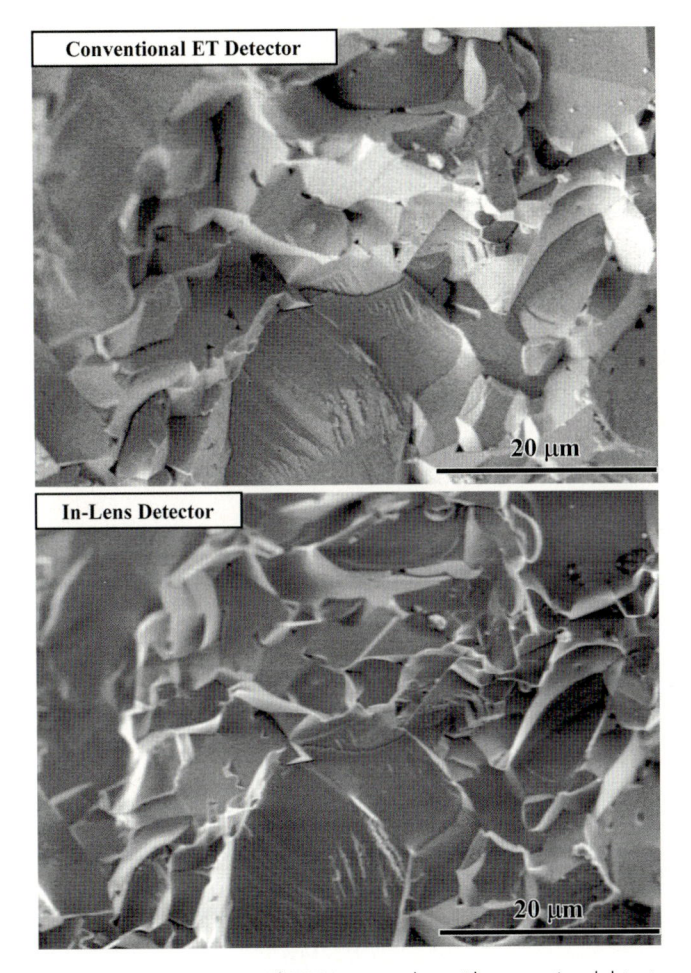

Figure 1.18 Comparison of SEM images taken with conventional detector and in-lens detectors. The SEM images show a fracture surface of alumina revealing cleavage facets and intergranular failure. Reproduced with permission from Ref. [4]; © 2008, John Wiley & Sons, 2008.)

surface topography is lost when using an in-lens detector located above the sample (Figure 1.18). In order to obtain topographic information with a segmented BSE detector, the signals of opposite sides of the sample must be subtracted, so as to enhance topographic contrast and reduce compositional contrast.

For high-magnification and small surface perturbations, SE images are superior in the topographic contrast and resolution compared to BSE images. This is due mainly to the escape depth, which is much larger for BSEs than for SEs, and results in a spatial resolution that is insufficient to differentiate between small surface topographic features in BSE images. When working with in-lens detectors registering only SE1 electrons, it is not only the depth resolution but also the lateral resolution that is strongly improved compared to BSE images. If a FEG is used in combination with an in-lens detector, then high-resolution imaging becomes possible (Figure 1.19). The resolution can be further improved to less than 1 nm when the microscope is operated at low accelerating voltages; this is a 10- to 20-fold improvement compared to the resolution obtained when using conventional SEM.

To summarize, the highest lateral resolution can be obtained by detecting SE1 using in-lens detectors and working at low incident accelerating voltages (low keV). Working at low keV can also be used to reduce charging. As can be seen in Figure 1.16, specific values exist (depending on the material) where the net current flow is zero; that is, the number of incident electrons is equal to the number of emitted electrons, which means that no net charge accumulates in the (nonconducting) sample, eliminating the need for conductive coatings (Figure 1.19).

Another point to mention here is the large depth-of-field in SEM, which allows for the imaging of large topographical changes, simultaneously. The large depth-of-field is due to the small angular aperture of the objective lens, and for large working distances is typically of the order of microns.

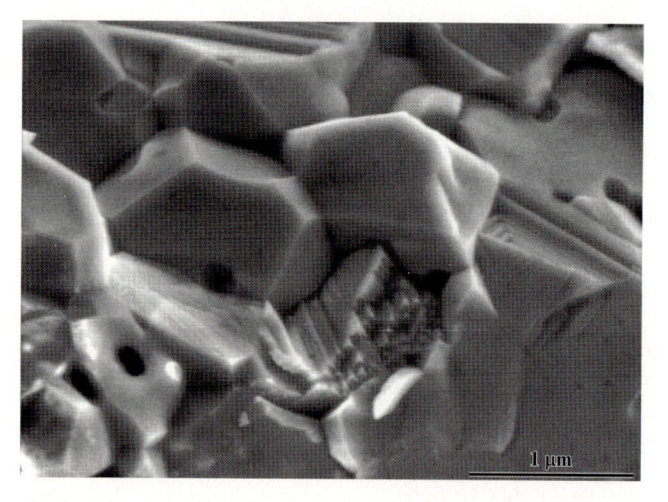

Figure 1.19 SEM image of a fracture surface of alumina taken with an in-lens detector at 5 keV. Reproduced with permission from Ref. [4]; © 2008, John Wiley & Sons.

1.3.2
Composition Contrast

As discussed above, the BSE signal shows a much stronger dependence on the atomic number than does the SE signal. While the signal from BSEs is higher for phases with a high mean atomic number, SE images are often dominated by surface topographic contrast, leading to a less straightforward interpretation of the data (Figure 1.20). Consequently, BSE images are more often used to display the phase distribution in composite materials. Any remaining contribution of topographic contrast can be removed from BSE images if a segmented BSE detector is used, where the signals from opposite sides of the sample are added. Compositional contrast in SE images is usually only visible if topographic contrast contributions are small and any contamination layers or conductive layers are of no concern. That is, if the sample is pretreated by, for example, plasma cleaning, and when the microscope is operated at high vacuum levels and low accelerating voltages. Under these circumstances, the resolution in the SE images is much better than that of BSE images, due to the reasons mentioned above.

1.3.3
Channeling Contrast

Until now, the ways in which crystal orientation can influence the SE and BSE efficiency, and hence the contrast in the image, have not been discussed. Primary electrons can penetrate more deeply into the crystal along certain crystallographic orientations ("channeling"), compared to a random grain orientation. Accordingly, the efficiency (η) of BSE generation becomes lower when crystals are oriented in specific crystallographic orientations; in addition, the secondary electron yield δ is also reduced. The BSE and SE signals of such grains will be much lower than that of randomly oriented grains, leading to the so-called "channeling contrast." Usually, channeling contrast is visible only for clean surfaces – that is, in the absence of any contamination layers that would obscure this effect.

1.4
Electron Backscattered Diffraction (EBSD)

In addition to detecting the number of BSEs, it is also possible to acquire their diffraction patterns by using an electron backscattered diffraction (EBSD) detector. The EBSD system consists basically of a phosphorescent screen, a charge-coupled device (CCD) camera, a control unit to operate the scanning electron microscope and EBSD detector, and software to evaluate the acquired data. To obtain a high-intensity diffraction pattern, the sample is usually oriented such that the incident electron beam is tilted 70° relative to the surface normal. As noted above, this leads to an asymmetric angular distribution for the BSEs, with the highest intensity of diffuse scattering in the forward direction. The diffraction patterns obtained contain

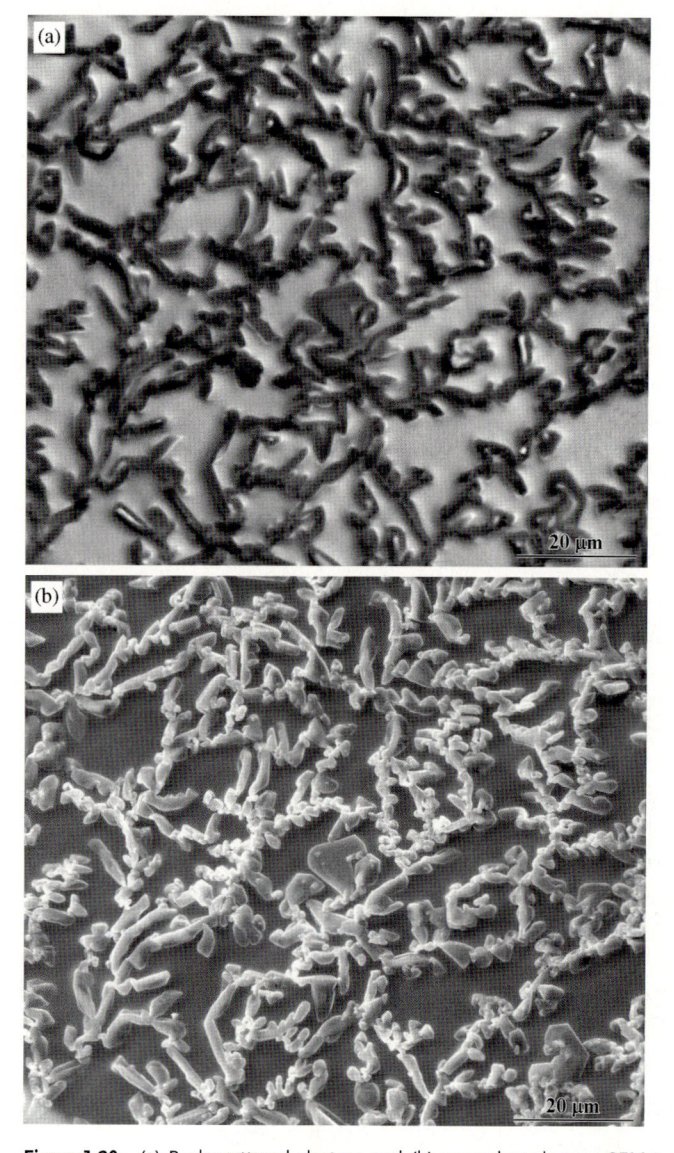

Figure 1.20 (a) Back-scattered electron and (b) secondary electron SEM images of small Al_2O_3 particles on the surface of a nickel substrate. Reproduced with permission from Ref. [4]; © 2008, John Wiley & Sons.

important information relating to the local crystal structure and orientation, thus providing a new aspect to the more conventional morphological and elemental information acquired when using SEM.

The origin of the EBSD pattern is similar to that of Kikuchi patterns in TEM. The intensity distribution of the inelastically scattered electrons is diffuse, but with a maximum in the forward direction. Some of these diffusely scattered electrons fulfill

Bragg's law for a given set of crystallographic planes and are diffracted into certain directions. This will add intensity to the darker, high-angle scattering regions and reduce intensity in the low-angle scattering region – that is, closer to the forward direction. The diffracted beams lie on so-called Kossel cones with a half angle of 90°- θ_B (where θ_B is the Bragg angle). The intersection of the Kossel cones with the plane of the detector are hyperbolae, but as all of the scattering angles are small the light and dark lines will appear straight and parallel. The distance between the lines is proportional to $2\theta_B$ and thus (for small angles) inversely proportional to the interplanar spacing. The EBSD pattern usually consists of several intersecting diffraction lines, each associated with a family of diffracting planes (Figure 1.21).

The EBSD pattern taken at one specific location can be used to obtain crystallographic information. For this, the EBSD patterns are calculated for various crystal structures and compared to the experimental data (Figure 1.21). In order to obtain

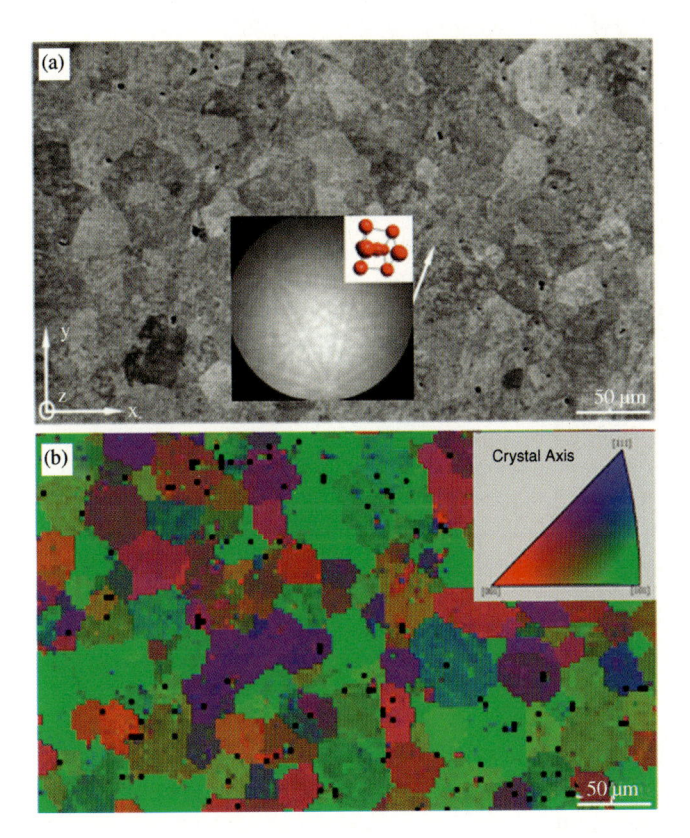

Figure 1.21 (a) Back-scattered electron image of a polycrystalline Tantalum wire with an EBSD pattern taken at the position marked by an arrow and corresponding crystal orientation; (b) Orientation map and corresponding color coding. Figures reproduced from S. Mayer, S. Pölzl, G. Hawranek, M. Bischof, C. Scheu, and H. Clemens (eds) (2006) Metallographic Preparation of Doped Tantalum for Microstructural Examinations using Optical and Scanning Electron Microscopy, in *Practical Metallography*, Vol. 43, No. 12, pp. 614–628.

orientation distribution maps, this procedure is repeated as a function of position across the sample surface during incident beam irradiation. After having evaluated the data, the results can be color-coded for selected ranges of crystal orientation (or phase). A prerequisite for the successful application of orientation imaging microscopy using SEM is a high-quality crystalline surface with minimal surface contamination, since both will affect the quality of the EBSD pattern.

1.5
Dispersive X-Ray Spectroscopy

The generated X-rays can be detected by using either an energy dispersive spectrometer (EDS) or a wavelength dispersive spectrometer (WDS). EDS is based on a solid-state detector that is a cryogenically cooled, negatively biased *p-i(intrinsic)-n* semiconductor diode. The incoming X-rays produce electron-hole pairs that are separated in the electrical field of the diode; this leads to a current pulse that can be converted to a voltage pulse which is then electronically amplified, digitized, and counted in a multichannel analyzer. The electrical charge generated is proportional to the energy of the incoming X-rays, and accordingly these are discriminated by their energies. The EDS detector is protected against contamination from the microscope by a thin window. Depending on the type of window used, elements with $Z > 5$ (boron) or $Z > 11$ (sodium) can be detected. Unfortunately, low-energy X-rays can be easily absorbed in the detector, leading to errors in quantitative analysis.

Consequently, for the analysis of light elements it is beneficial to use a WDS system. In this type of spectrometer, which is based on a curved crystal diffractometer, only those X-rays which fulfill Bragg's law for a specific orientation of the detector crystal are counted, and accordingly they are discriminated by their wavelengths. Due to the geometric discrimination, all characteristic X-ray peaks must be scanned sequentially. If several peaks are to be collected at a time, then several spectrometers must be used simultaneously (or extended acquisition times will be required). The detection of X-rays according to their wavelength has another major advantage compared to energy-dispersive systems, in that the energy resolution is at least an order of magnitude better than EDS. This is important if peak overlap inhibits identification and quantification. The disadvantage of the WDS is the much longer data collection time that is associated with the sequential acquisitions, compared to the parallel data collection mode used in the EDS.

As described above, in addition to the identification of elements present in a specimen, characteristic X-rays can also be used for chemical composition analysis. A semi-quantitative analysis can be performed by applying the following equation:

$$\frac{C_i}{C_{(i)}} = ZAF \frac{I_i}{I_{(i)}}. \qquad (1.20)$$

where C_i and $C_{(i)}$ are the concentration of species i in the unknown compound/specimen and in a pure standard, respectively, and I_i and $I_{(i)}$ are the intensities (number of counts) of the corresponding X-ray peaks after background subtraction.

The included *ZAF* correction represents the atomic number effects (*Z*), absorption (*A*) and fluorescence (*F*) which must be taken into account for bulk specimens. The background caused by the Bremsstrahlung can either be fitted by linear interpolation or calculated using a theoretical model (Kramers relation), and is then subtracted from the original data to obtain the net signal necessary for the quantification.

Several approaches may be employed to acquire X-ray data in SEM (Figure 1.22). In order to obtain the chemical composition of a specific location, the electron beam can be placed stationary on a sample position (point analysis). If contamination occurs, or if an average chemical composition analysis is of interest, the EDS spectrum can also be acquired while the electron beam is scanning over a certain area (area measurements). To determine the chemical composition across, for example, an interface, line-scans are often acquired (Figure 1.22). In this mode, the electron beam is scanned across a chosen line, and at specific points of this line either a complete spectrum is acquired or the intensity of individual, selected X-ray energies is measured. The latter measurement has the advantage of shorter acquisition times, with typical acquisition times of approximately 100 s for a complete spectrum being reduced by a factor of five, or more. However, this type of measurement permits only semi-quantitative analysis, and can only be used to display the intensity distribution

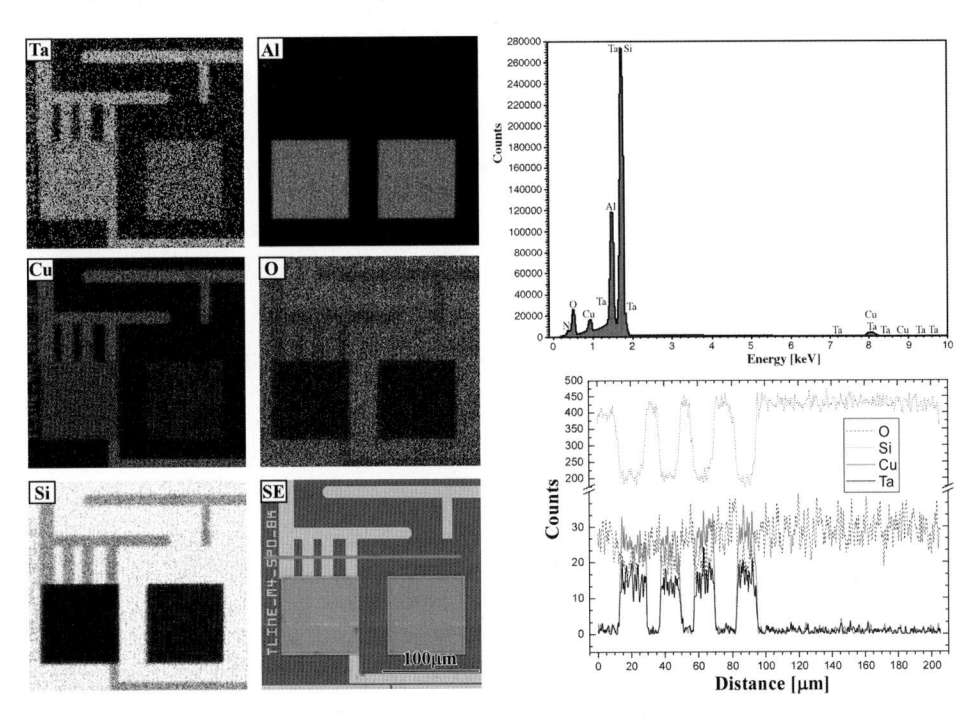

Figure 1.22 SEM image and corresponding X-ray dot images revealing the elemental distribution of Ta, Al, Cu, O and Si of a electronic device. At the right upper side is displayed an EDS spectrum taken at a specific point of the device. At the right lower side are shown the results obtained with a line scan measurement. Reproduced with permission from Ref. [4]; © 2008, John Wiley & Sons.

of the chosen elements. The acquisition of EDS data or chosen X-ray energy regions along a certain raster can be used to display the elemental distribution of specific sample areas (Figure 1.22). The corresponding X-ray dot images/elemental maps are then used to investigate the distribution of various phases within the sample (Figure 1.22). Together with the information obtained by BSE or SE images, these provide valuable information for materials scientists and engineers. However, from the above discussion it should be clear that the X-ray signal stems from an interaction volume that is much larger than the electron beam size used for the data acquisition.

1.6
Other Signals

As noted above, the interaction of the incident beam with an optically active material can lead to the emission of light. This cathodoluminescence (CL) can be detected by using a special detector that consists of a semi-ellipsoidal mirror for light collection and focusing, a light guide, and a photomultiplier. The wavelength of the emitted light depends on the composition and structure, and thus can be used to probe local changes of these microstructural features when the beam is scanned across the sample. At low temperatures, the wavelength distribution becomes very narrow and typically shows a pronounced maximum at the energy associated with the bandgap in semiconductors and isolators. Accordingly, local changes in bandgap and impurity states can be investigated. As CL is generated within the interaction volume, the spatial resolution is much worse than when acquiring BSE or SE images. Moreover, CL is very sensitive to low impurity levels (0.01 ppm) and, as such, much more sensitive than EDS.

Other alternative imaging modes are related to electron beam-induced current (EBIC) or electron beam-induced voltage (EBIV). These are useful when studying defects and structures in semiconductor devices, as the incident electron beam can be used to create electron-hole pairs in a semiconductor that can be separated with either an internal (e.g., pn- junction) or external electrical field. By measuring either the associated voltage (EBIV mode) or current (EBIC mode) while scanning the beam across the area of interest, it is possible to image local changes in the electronic properties.

1.7
Summary

In this chapter, the main components of the scanning electron microscope and the image formation process have been described. The important parameters that limit the resolution – primarily the incident electron beam diameter and the interaction volume in the sample – have been discussed in detail. The escape depth which governs the lateral and depth resolution depends on the type of signal to be used for imaging, and is different for SE1 and SE2 electrons, BSEs, X-rays, and light. How the contrast in SE and BSE images can be interpreted in terms of topology and composition variations has also been discussed, and the effect of beam tilt, detector position and channeling on the image contrast described. The analytical tools available for SEM to determine the chemical composition (via EDS and WDS) and the crystallographic orientation (EBSD), as well as the detection of other signals characteristic of the material being studied (e.g., CL, generated current and voltage) have also been outlined. In summary, modern SEM and its associated analytical tools provide unique possibilities for materials characterization. Notably, due to the simplicity of its use and its high resolution, SEM is also very well suited to *in-situ* studies.

References

1 Reimer, L. (1998) *Scanning Electron Microscopy: Physics of Image Formation and Microanalysis*, 2nd edn, Springer Verlag, Berlin, Heidelberg, New York; Germany and USA.
2 Goodhew, P.J., Humphreys, J. and Beanland, R. (2001) *Electron Microscopy and Analysis;*, 3rd edn, Taylor & Francis, London, New York.
3 Goldstein, J., Newbury, D., Joy, D., Lyman, C., Echlin, P., Lifshin, E., Sawyer, L., and Michael, J. (2003) *Scanning Electron Microscopy and X-Ray Microanalysis*, 3rd edn, Springer, New York.
4 Brandon, D. and Kaplan, W.D. (2008) *Microstructural Characterization of Materials*, 2nd edn, John Wiley & Sons, West Sussex, UK.
5 Joy, D.C., Romig, A.D. and Goldstein, J.I. (eds) (1986) *Principles of Analytical Electron Microscopy*, Plenum Press, New York, London.
6 Egerton, R.F. (1996) *Electron Energy-Loss Spectroscopy in the Electron Microscope*, 2nd edn, Plenum Press, New York.
7 Drouin, D., Couture, A.R., Joly, D., Tastet, X., Aimez, V., and Gauvin, R. (2007) CASINO V2.42 – A fast and easy-to-use modeling tool for scanning electron microscopy and microanalysis users. *Scanning*, **29** (3), 92–101.
8 Reimer, L. and Pfefferkorn, G. (1977) *Rasterelektronenmikroskopie*, Springer-Verlag, Berlin, Heidelberg, New York.

2
Conventional and Advanced Transmission Electron Microscopy
Christoph T. Koch

2.1
Introduction

2.1.1
Introductory Remarks

Despite its only moderate, 17.4-fold, level of magnification, the construction of the first conventional transmission electron microscope during the early 1930s by Ernst Ruska (who received the Nobel Prize in 1986) and Max Knoll [1, 2] laid the foundation for the development of instruments that today enable direct imaging of the atomic structure of matter. Although the first scanning transmission electron microscope was developed shortly afterwards, by Manfred von Ardenne [3], the results obtained at the time with this instrument could not compete with those acquired using transmission electron microscopy (TEM), until more coherent electron sources were developed during the late 1960s by Albert Crewe [4] and Akira Tonomura [5]. More than 30 years later, the introduction and commercialization of aberration-corrected electron optics by Ondrej Krivanek for the scanning transmission electron microscope [6] and by Max Haider for the transmission electron microscope [7–9] and later also the scanning transmission electron microscope, have set new standards in electron microscopy, allowing images and spectral information to be recorded with sub-Ångström resolution. Today, the strong elastic and inelastic interaction of fast electrons in the electron beam with the specimen allows the application of many different analytical methods to investigate structure, composition, and bonding within very small volumes, down to atomic resolution.

In this chapter, a brief introduction is provided to the basics of TEM, and also to a selection of imaging, diffraction and spectroscopic techniques that may be applied to extract information concerning the atomic structure, composition and electrostatic and magnetic fields of a specimen. The chapter does not include a detailed mathematical discussion of the underlying physics; rather, it is targeted at those who are not familiar with electron microscopy, providing the necessary theoretical background to understand the subsequent chapters in this book. If required, more

Table 2.1 De Broglie wavelength λ, relative velocity $\beta = v/c$ (c = speed of light in a vacuum), and electron specimen interaction constant (σ) for electrons at different accelerating voltages.

	Accelerating voltage (kV)					
	20	60	100	200	400	1250
λ (pm)	8.6	4.9	3.7	2.5	1.6	0.74
β	0.27	0.45	0.55	0.70	0.83	0.96
σ (V nm^{-1})	0.0186	0.0114	0.0092	0.0073	0.0061	0.0053

detailed discussions are available in several standard books on TEM (e.g., Refs [10–12]).

2.1.2
Instrumentation and Basic Electron Optics

The high spatial resolution achieved in TEM and scanning transmission electron microscopy (STEM) is made possible by the small de Broglie wavelength of the electrons that travel at a significant fraction of the speed of light (see Table 2.1). Even in the most advanced TEM systems the resolution is not limited by the diffraction limit (i.e., half the wavelength) but rather by aberrations, imperfections of the electron optical system, and instabilities, as well as the partial spatial and temporal coherence of the electron beam. An additional limitation is imposed by the sample itself – especially thermal vibration of the atoms. The resolution of modern transmission electron microscopes has already reached 0.05 nm, i.e. very close to the smearing out of the atomic positions due to room temperature vibrations of the atomic nuclei in most samples. This means that the sample itself imposes a practical resolution limit which cannot be overcome by any improvement of the instrumentation. However, there remains much potential and need to improve the other aspects of these instruments, which would be of great benefit for many applications. Increasing the brightness of electron sources, for example, would reduce acquisition times and enhance the signal-to-noise ratios (SNRs) of images, diffraction patterns, and spectra, and would be especially important in the context of *in-situ* microscopy. The complexity of the modern transmission electron microscope, especially if equipped with aberration correcting electron optics has reached a level that requires the operator to be very well trained. Another important area of improvement is, therefore, the ease of use of these instruments, including a larger degree of automation for routine tasks.

The principle of TEM and STEM imaging and diffraction is illustrated in Figure 2.1. Electrons extracted from a sharp tip (typically W, LaB$_6$, or LaB$_6$-coated W) by either thermionic emission (the work function of the metal is overcome mainly by heating the filament), cold field emission (the electrons are extracted by a high electric field at a sharp tip), or thermally assisted (Schottky) field emission (field emission at

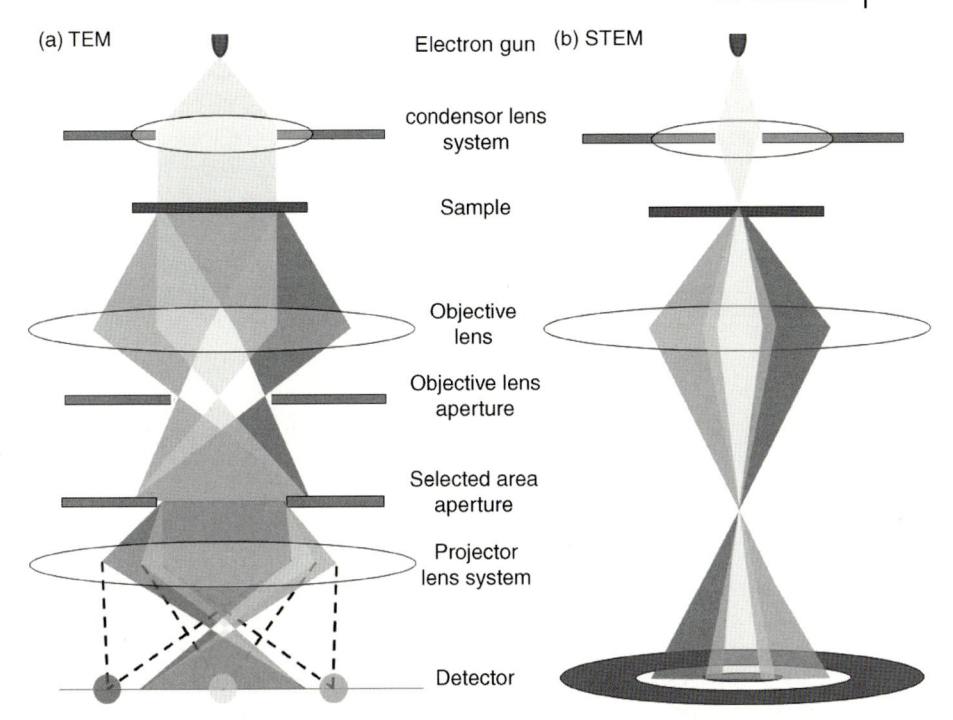

Figure 2.1 Simplified schematic diagrams of the principles of TEM imaging and diffraction and STEM imaging in modern TEM/STEM instruments. (a) In TEM mode, the condenser lens system produces a parallel electron beam on the sample, the size of which is determined by the size of the condenser aperture. The transmitted (light gray) and scattered (darker gray levels) electron partial waves are then spatially separated in the back focal plane of the objective lens, where the objective aperture may be used to select specific reflections. The selected area aperture which is located in the plane of the first intermediate image (enlarged usually by 30- to 70-fold) may be used to select electrons that have scattered from a specific region of the sample. The projector lens system is then used to produce either an image or a diffraction pattern (dashed lines) on the recording medium (film, CCD, image plate, or an energy filter with a subsequent detector); (b) In STEM mode, the electron beam is focused to a fine probe on the sample and scanned across it. The scattered electron can then be collected by either bright-field (small disc), or dark-field (larger ring) detectors to form bright-field (BF)-STEM or annular dark-field (ADF)-STEM images. The projector lens system in a scanning transmission electron microscope may be used to adjust the camera length on the detector, but some dedicated microscopes have no post-specimen lenses at all.

elevated temperature) are accelerated by an electrostatic potential ranging from 30 kV up to 3 MV, depending on the microscope. This beam of fast electrons is then shaped by the condenser and an objective pre-field lens system to produce a parallel beam (in TEM imaging mode) on the specimen, or a fine spot which may be positioned anywhere on the specimen surface or rastered laterally, as in STEM mode. Most modern transmission electron microscopes can be operated in either TEM or STEM mode; switching between the two modes is achieved relatively easily by changing the focal length of the condenser lens system and selecting a different size of the condenser aperture.

If the projector lens system of the microscope in Figure 2.1a is adjusted to image the back-focal plane of the objective lens, it will produce an electron diffraction pattern (the dashed lines in Figure 2.1a show that all electrons that have scattered by the same angle [indicated by same gray scale] are focused to the same pixel on the detector). The selected area (SA) aperture which limits the field of view in a TEM image is commonly used to select the area on the sample that will contribute to a diffraction pattern. In large-angle convergent beam electron diffraction (LACBED) mode, the intermediate image plane (this is where the SA aperture is located) contains a diffraction pattern, so that the SA aperture is used to select a certain diffraction spot (for further details, see Section 2.6).

A conventional or high-resolution TEM image is formed, if the intermediate image plane is projected on the detector by the projector lens system. This image is an interference pattern of all the partial electron waves which have scattered off the specimen and have passed the objective aperture. (For further details of the image-formation process, see Section 2.2).

In the case of a crystalline specimen, the periodicity of the atomic planes produces discrete diffraction spots in the back-focal plane of the objective lens. The objective aperture size may then be chosen to select either several or just one of these diffraction spots. One speaks of high-resolution (HR) TEM if several diffraction spots contribute to the image (see Section 2.2), and of conventional TEM (Section 2.3) otherwise. The experimental conditions required for HRTEM and three major conventional TEM imaging modes are illustrated in Figure 2.2, namely bright-field (BF); dark-field (DF), and weak-beam dark-field (WBDF) TEM. When applying the WBDF-TEM technique, a crystalline specimen must be oriented in a condition such that mainly one of the diffracted beams is strongly excited. The image is then recorded with the objective aperture selecting a second, diffracted beam that is only weakly excited. This technique is particularly useful for the investigation of strain contrast around dislocations (see Section 2.3).

2.1.3
Theory of Electron–Specimen Interaction

Before discussing the details of TEM image formation, it is important to examine more closely the situation that occurs when a fast beam electron passes through the sample. Electrons passing through a thin TEM sample are scattered by the atomic nuclei, while the electrons around them are scattered either elastically (when the kinetic energy of the beam electron is preserved) or inelastically (when the electron transfers some of its kinetic energy to the sample). The different signals that are produced by the incident electron beam are shown in Figure 2.3. To treat each of these processes in detail is beyond the scope of this chapter; however, in order to provide the necessary background for the remainder of this book, attention will be focused mainly on elastic scattering at this point, while electron energy loss and the emission of characteristic X-rays will be detailed in Section 2.8.

The elastic scattering of electrons by the sample occurs because of a slight acceleration of the fast beam electron by the (positive) electrostatic potential within

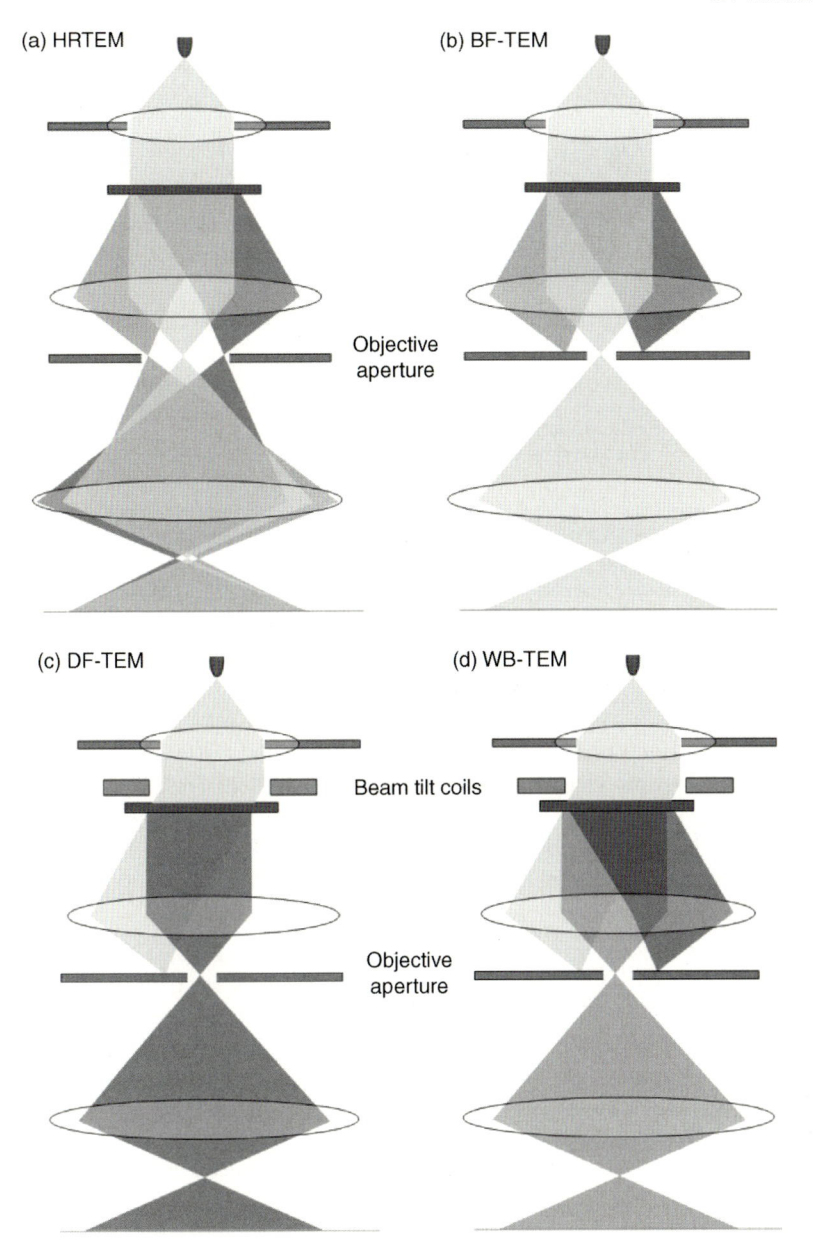

Figure 2.2 Illustration of experimental conditions for (a) high-resolution (HR) TEM, and the conventional TEM imaging modes (b) bright-field (BF), (c) dark-field (DF), and (d) weak-beam dark-field (WBDF) TEM.

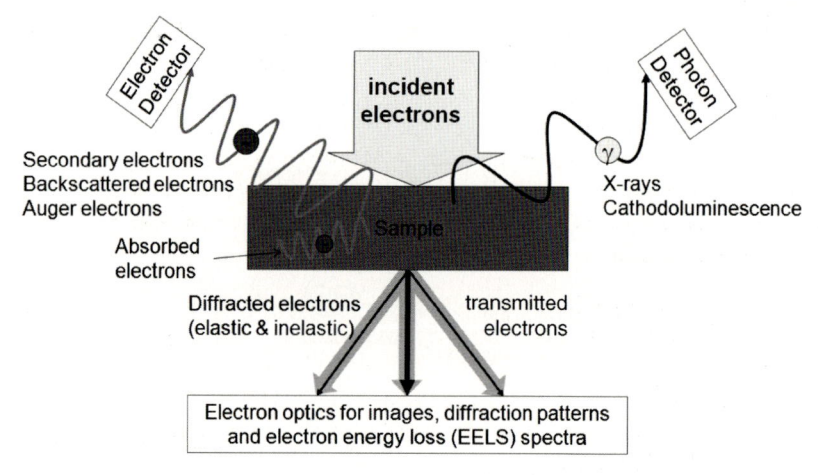

Figure 2.3 Diagram illustrating the result of various scattering processes of electrons within the sample. The incident electron beam can cause the emission of secondary electrons, Auger electrons, light (cathodoluminescence), or X-rays (element-specific characteristic X-rays, coherent and incoherent Bremsstrahlung, etc.). Transmitted electrons can be scattered either elastically (the kinetic energy of the electron remains unchanged, only its direction of propagation changes) or inelastically (the electron loses energy on its path through the sample). In thicker specimens there is a significant chance that an electron will be absorbed by the sample.

the sample; the latter is produced by the very localized positive charges of the atomic nuclei and the negative charge density "cloud" of the electrons around them. The scattering factor $f_{el}(s)$ for a single atom of atomic number Z can be obtained from the corresponding X-ray scattering factor $f_{X\text{-ray}}(s)$ (Fourier coefficients of the electron density distribution) by the Mott formula [13] (in units of Å) as:

$$f_{el}(s) = 0.0239 \, \text{Å}^{-1} \frac{Z - f_{X\text{-ray}}(s)}{s^2} \tag{2.1}$$

where $s = \sin(\theta)/\lambda$ is the scattering parameter, θ is half the scattering angle, and λ is the electron wavelength. Equation 2.1 implies that, at small scattering parameters, s, the elastic scattering strength of electrons is much stronger (in most cases 10^4- to 10^5-fold) than that of X-rays, so as to produce a strong scattering signal even if the illuminated volume is very small. These isotropic atomic scattering factors can be used to construct the electrostatic potential (in Volts) of a sample, consisting of atoms of atomic number Z_j at position r_j:

$$V(\vec{r}) = 47.878 \, \text{V} \, \text{Å}^2 \cdot \text{FT}^{-1} \left[\frac{1}{\Omega} \sum_j f_{el}^{Z_j} \left(\frac{|\vec{q}|}{2} \right) e^{2\pi i \vec{r}_j \cdot \vec{q}} \right] \tag{2.2}$$

Here, Ω is the volume containing all the atoms being summed over, and q ($|q| = 2s$) is the reciprocal space coordinate. This formulation is general, and may be applied to both crystalline and noncrystalline structures. The electrostatic potential of the specimen is the sole origin of elastic scattering of electrons in matter and, under ideal conditions, this is what will be seen when looking at a HRTEM image (see Equations 2.7 and Equation 2.8, and Section 2.2 for a more precise treatment) or even better, in an electron hologram (see Section 2.5).

For a crystalline specimen, it suffices to define the potential within a single unit cell, in which case Ω is the unit cell volume. Mathematically, this is equivalent to defining the Fourier components of the potential only at multiples of the reciprocal space lattice vectors $\mathbf{g} = h\mathbf{a}^* + k\mathbf{b}^* + l\mathbf{c}^*$, where h, k, and l are integer numbers. Conventionally, electron structure factors are defined as:

$$U_g = \frac{\gamma}{\Omega} \sum_j f_{el}^{Z_j}\left(\frac{|\vec{g}|}{2}\right) e^{2\pi i \vec{r}_j \cdot \vec{g}} \tag{2.3}$$

where $\gamma = 1/\sqrt{1 - v^2/c^2}$ is the Lorentz factor (remember that the electrons in a transmission electron microscope are usually relativistic, especially at energies >50 keV; see also Table 2.1).

At high scattering angles, inelastic scattering events are generally less likely than elastic events. Only for light atoms and small scattering angles will the inelastic scattering cross-section exceed the elastic scattering cross-section. On its path through the sample, the electron can lose energy in several ways: by exciting an electron of one of its atoms to a higher unoccupied orbital or even into vacuum (via ionization and the possible creation of secondary electrons); by exciting collective oscillations of the valence or conduction electrons in the sample (plasmons); by producing Cherenkov radiation (if the phase velocity of the beam electron exceeds the speed of light in the sample); by scattering with another electron in the sample (Compton scattering); by displacing atoms (knock on damage); or by exciting collective oscillations of several atoms (phonon excitation and heating of the sample). Some of these inelastic scattering mechanisms can be utilized to determine the local composition and other properties of the sample, and are discussed further in Section 2.8.

One common approximation to treat electrons that have scattered inelastically and do not contribute (apart from an inelastic background) to the elastic image contrast is to introduce an absorptive potential, $V'(r)$. This absorptive potential must be positive because the beam electrons do not gain kinetic energy by passing through the sample (see Ref. [14] for rare exceptions to this rule), and is often approximated as being proportional to the electrostatic potential $V(r)$, only by a factor of 10–20 smaller in amplitude:

$$V(r) \to V(r) + iV'(r) \approx V(r) + icV(r) \tag{2.4}$$

where the proportionality constant c is usually between 0.05 and 0.1. The corresponding general definition of the electron structure factor therefore includes also the Fourier components of the absorptive potential U'_g:

$$U_g \to U_g + iU'_g \tag{2.5}$$

Since, in a crystalline specimen, the atoms are located on a regular lattice in space, the electron structure factors are only nonzero on points in reciprocal space which lie on the *reciprocal lattice*. As illustrated in Figure 2.4, elastically scattered electrons

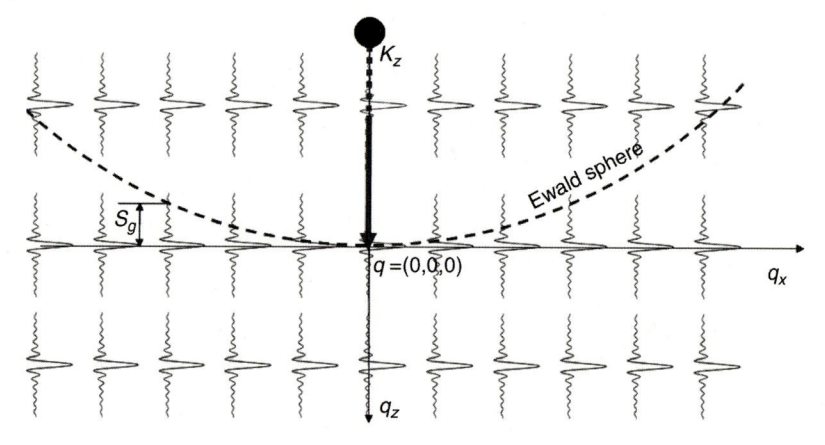

Figure 2.4 Diagram illustrating the Ewald sphere construction and the physical meaning of the excitation error s_g. The fast beam electron can only scatter elastically to reciprocal space vectors which lie on the Ewald sphere (dashed curve). The distance between a given point in reciprocal space and the Ewald sphere in the direction of the specimen's surface normal (assumed to be in the z-direction in this illustration) is called the "excitation error." For a thin specimen, the scattering amplitude oscillates with s_g according to Equation 2.9 or, in the kinematic approximation, Equation 2.17, as is indicated by the sinc curves on the position of the reciprocal lattice points.

maintain their momentum and may therefore scatter only to reciprocal lattice vectors which lie on a sphere (Ewald sphere). Reflections that are not intersected by the Ewald sphere may still be excited but, because of the nonvanishing excitation error s_g, in most cases they will have a reduced intensity (see Section 2.6.1 for a more detailed discussion of this point).

Kinematical scattering theory, which neglects the possibility that an electron scatters more than once on its path through the sample, predicts that the intensity in the diffraction pattern $I(q) = |\Psi(q)|^2$ (where Ψ is the complex wave function of the beam electron) at the reciprocal space vector q is proportional to $|U_q|^2$. While this simple approximation may hold for noncrystalline specimens (especially if they are thin and consist of light atoms), it breaks down for almost all crystalline samples because of the strong interaction of electrons with matter and the resulting high probability for an electron to scatter more than just once.

Dynamical scattering theory therefore defines the scattering matrix S in reciprocal space:

$$S = e^{i\pi\lambda t A} \tag{2.6}$$

which takes into account all possible multiple scattering events. The off-diagonal elements of the structure factor matrix A are the structure factors U_g and its diagonal elements are $2s_g/\lambda$. There exist a number of more or less approximate methods to compute this scattering matrix.

In the limit of a flat Ewald sphere which passes through all the reciprocal lattice points, the excitation errors are zero and the scattering matrix reduces to the phase object approximation which is most easily expressed in real space as

2.1 Introduction

$$\Psi(r) = e^{i\sigma V_{proj}}(r) \tag{2.7}$$

Here, $V_{proj}(r)$ is the electrostatic potential of the specimen projected along the direction of propagation of the incident electron beam (conventionally the z-direction), and σ is termed the electron–specimen interaction constant (see Table 2.1 for values of σ at different accelerating voltages).

If it is assumed that the exponent in Equation 2.7 is small, then a further approximation may be made:

$$\Psi(r) \approx 1 + i\sigma V_{proj}(r) \tag{2.8}$$

This weak phase object approximation is equivalent to the kinematical scattering approximation, and should only be applied to very thin and preferably noncrystalline specimen consisting of mainly light atoms; at low resolution, however, it may also be applied to somewhat thicker specimens. Although not very precise, this approximation provides a very intuitive approach to understand contrast in high-resolution images.

A much more accurate approximation to the exit-face wave function, and which also accounts for the curvature of the Ewald sphere, may be obtained by the multislice algorithm [15, 16]. This computes the dynamical wave function by slicing the sample into very thin slices, applying the phase object approximation for each one of them, and propagating the electron wave function through the empty space between them (Fresnel propagation). In the limit of very thin slices (i.e., ≤ 0.2 nm), this approximation becomes exact, even for crystalline samples in zone axis orientation. Consequently, this system is used in most simulation programs for simulating HRTEM images, especially if atomic-resolution images of defects are to be computed.

For the computation of electron diffraction patters of simple or moderately complex perfect crystals, it is most efficient to compute the scattering matrix in Equation 2.6 directly by a diagonalization of the structure factor matrix A (the Bloch wave method [17]), or alternatively by computing the matrix exponential. If the specimen is tilted in such a way that only one of the reflections in the diffraction pattern is strongly excited, it has been oriented in two-beam condition. The structure factor matrix can then be reduced to a 2×2 matrix containing only the structure factors for the central beam (the mean inner potential and the mean absorption) and reflections g and $-g$ as well as the excitation error s_g, allowing it to be diagonalized analytically. Neglecting absorption for a moment, the intensity I_g in the diffracted beam g in two-beam condition is then given by:

$$I_g |\Psi_g|^2 = \lambda^2 |U_g|^2 \frac{\sin^2\left(\pi t \sqrt{s_g^2 + \lambda^2 |U_g|^2}\right)}{s_g^2 + \lambda^2 |U_g|^2} \tag{2.9}$$

This expression shows that the intensity of the diffracted beam depends sensitively on the excitation error s_g and with that on the local crystal orientation. The intensity of the undiffracted beam I_0 shows a similar but inverted dependence on s_g because the sum of I_0 and I_g is assumed to be constant.

The more general case which includes absorption is treated by the Howie–Whelan differential equations [18, 19] which provide an expression for the change in the scattered (Ψ_g) and transmitted (Ψ_0) electron wave amplitude and phase as a function of depth z into the sample:

$$\frac{d\Psi_{\vec{g}}}{dz} = \pi\lambda\left(iU_{\vec{g}} - U'_{\vec{g}}\right)\Psi_0 + \pi\left[2is_{\vec{g}} - \lambda U'_0\right]\Psi_{\vec{g}} \quad (2.10a)$$

$$\frac{d\Psi_0}{dz} = \pi\lambda\left(iU_{\vec{g}} - U'_{\vec{g}}\right)\Psi_{\vec{g}} + \pi\lambda U'_0 \Psi_0 \quad (2.10b)$$

If the specimen is bent or strained due to the presence of defects, the local diffracting condition – and with that the excitation error s_g – becomes a function of lateral position (x,y) and depth (z). Thus, replacing the plain excitation error s_g in Equation 2.10a with

$$s_{\vec{g}} \rightarrow s_{\vec{g}} - \frac{d}{dz}\left[\vec{g} \cdot \vec{R}(x, y, z)\right] \quad (2.11)$$

allows both BF and DF strain contrast images to be computed. Here, $R(r)$ is the displacement of the unit cell at position (x, y, z) relative to the perfect lattice without defect. The Howie–Whelan Equations 2.10a and 2.10b form the basis for predicting and interpreting image contrast in conventional TEM imaging (see Section 2.3). They are particularly useful for understanding strain contrast around lattice defects, such as dislocations, stacking faults, and point defects.

2.2
High-Resolution Transmission Electron Microscopy

As shown in Figure 2.2a, the formation of a HRTEM (or lattice resolution) image requires the interference of the undiffracted beam with at least one Bragg-diffracted beam (or partial wave). For crystalline samples, the interference pattern between the central and the diffracted beams produces a magnified image of either one-dimensional lattice fringes (interference of the central spot with a single lattice reflection) or an image of the two-dimensional crystal lattice, if the central spot interferes with at least two noncollinear lattice spots. Although HRTEM images are often interpreted as if they displayed directly the projected atomic structure of the sample transmitted by the electron beam, in most cases this is not true. Only for very simple structures – and then only if these are oriented along a low-index zone axis and if the correct imaging conditions are met (including a very small specimen thickness) – might such a direct interpretation be considered.

An example of the complexity of the image-formation process in HRTEM is shown in Figure 2.5. This also demonstrates the enormous sensitivity of HRTEM contrast on local lattice orientation, and indicates that its direct interpretation can easily become impossible. Although the model structure shown in Figure 2.5a is only 2.7 nm thin, the simulated image shown in Figure 2.5b differs greatly from its

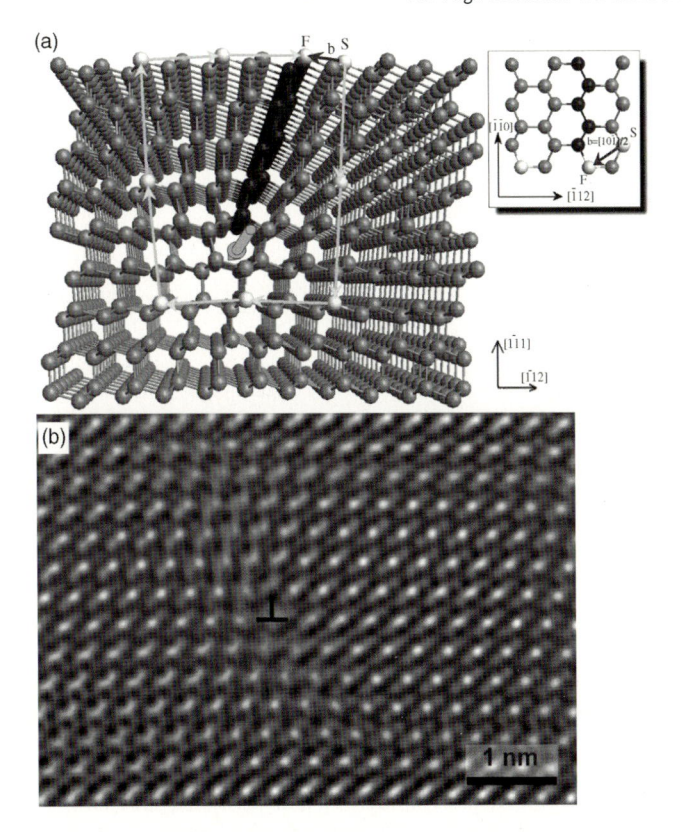

Figure 2.5 Model of an edge dislocation core in Si (a) and the HRTEM image simulated from it (b). Although the model was very thin ($t = 2.7$ nm), slight variations in the local lattice orientation caused by the strain field around the dislocation core change the HRTEM contrast substantially. Imaging conditions used for this simulation: $E_0 = 200$ kV, $C_s = 1.2$ mm, defocus $= -67.2$ nm (Scherzer defocus), illumination convergence angle $\alpha = 0.1$ mrad, focal spread $\Delta_f = 5$ nm. The model was constructed by J. F. Justo Filho, using an environment-dependent interatomic potential (EDIP) for silicon [20, 21].

projected potential. In most cases, extensive image simulations are required to interpret HRTEM contrast quantitatively; hence, the following text is devoted to explaining the image-formation process in a transmission electron microscope.

Within the quasi-coherent approximation, the intensity distribution in a TEM image is given in real space by:

$$I(\vec{r}) = \left| \Psi(\vec{r}) \otimes \text{CTF}(\vec{r}, \Delta f, C_s) \otimes FT^{-1} \left[E_{\text{temp}}(\vec{q}, \Delta_f) \cdot E_{\text{spat}}(\vec{q}, \Delta f, C_s, \alpha) \right] \right|^2 \quad (2.12)$$

where the vector r is the position in real space, $\text{CTF}(r, \Delta f, C_s) = FT^{-1}[\exp\{-i\chi(q, \Delta f, C_s)\}]$ is the contrast transfer function ($\chi(q, \Delta f, C_s)$ is called the phase distortion function), $E_{\text{temp}}(q, \Delta_f)$ and $E_{\text{temp}}(q, \Delta_f)$ are the temporal and spatial coherence

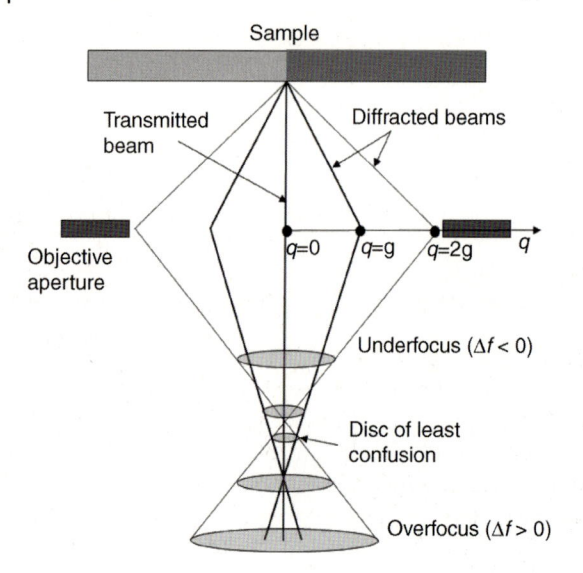

Figure 2.6 Simplified ray diagram of image HRTEM formation. The positive C_s of the objective lens (in the plane of the objective aperture) produces a shorter effective focal length for higher order diffracted beams. A C_s-corrector would be installed just below the objective lens to compensate this effect.

envelopes, respectively, and $q = |q|\cdot(\cos(\phi),\sin(\phi))$ is the two-dimensional reciprocal space coordinate.

Considering defocus Δf, astigmatism (amplitude A_1 and azimuthal direction ϕ_1), and spherical aberration C_s to be the dominant aberrations of the imaging system, the phase distortion function is given by

$$\chi(\vec{q}) = 2\pi \left[\frac{1}{2}\lambda(E)\{\Delta f(E,\phi) + A_1 \cos(2(\phi-\phi_1))\}q^2 + \frac{1}{4}\lambda^3(E)C_s q^4\right] \quad (2.13)$$

In 1936, Otto Scherzer [22] showed that all round electron lenses have an unavoidable positive spherical aberration coefficient C_s. The spherical aberration of objective lenses in a transmission electron microscope is typically between 0.5 mm and 3 mm. The effect of C_s on the image of, for example, a defect in a perfect crystal is shown in Figure 2.6. Spherical aberration prevents the diffracted beams that have scattered to large angles to be focused into the same image point as those having scattered to small angles only. This effect is termed "delocalization." Because of the small energy spread of electrons emitted by a field emission gun (FEG), transmission electron microscopes fitted with FEGs usually have a much higher information limit (maximum spatial frequency admitted by the temporal coherence envelope) than conventional systems equipped with a thermionic source. If no objective aperture is used, this higher-resolution information in the electron wave function results in a larger delocalization if the spherical aberration remains the same. The phase shift introduced by the spherical aberration of the objective lens (Equation 2.13) is proportional to q^4, while the defocus-induced phase shift is proportional to q^2. In order to achieve the highest possible directly interpretable resolution in a HRTEM image, the defocus and objective aperture size may be chosen to reduce the effects of C_s, but it is impossible to completely compensate for this. The defocus ($\Delta f_{Scherzer}$) and a corresponding objective aperture ($\theta_{Scherzer}$) size which optimize the point resolution of an image are given by:

$$\Delta f_{Scherzer} = -1.2\sqrt{C_s \lambda} \quad (2.14a)$$

$$\theta_{Scherzer} = 1.5\, C_s^{1/4}\, \lambda^{3/4} \quad (2.14b)$$

Because the actual information limit defined by the energy spread of a FEG usually exceeds this limitation in the point resolution (the resolution limited by an objective aperture of size $\theta_{Scherzer}$), many modern FEG microscopes are equipped with a C_s-corrector [7–9]. This is a highly complex electron optical device that features negative spherical aberration to compensate the positive spherical aberration of the objective lens. C_s-correctors allow for the use of much larger objective apertures, the size of which is now limited by higher-order aberrations, to record practically delocalization-free images at sub-Ångström resolution.

A perfect lens without any aberrations (such as defocus and C_s) would remove all phase contrast in the image. In order to regain phase contrast for at least the very small details (e.g., the difference in phase shift between two neighboring atomic columns), it is necessary to reintroduce at least some aberrations. In order to partially compensate the delocalization introduced by higher-order aberrations, which cannot be compensated by current aberration-correcting electron optics, a slightly negative value of C_s, combined with a small positive defocus, has been found to provide optimum conditions for recording directly interpretable images at very high resolution. This has an additional benefit of showing atomic columns being bright rather than dark, as is the case in images recorded with a large positive C_s and Scherzer conditions [23]. Negative C_s imaging (NCSI) conditions can be achieved by adjusting the C_s-corrector to overcompensate the spherical aberration of the objective lens.

At an image resolution close to 0.1 nm, changing the defocus of the objective lens by only a few nanometers will have a profound effect on the image contrast, and also on the pattern produced by a given atomic configuration. It is, therefore, very important to control the aberrations that lead to the formation of an image. While the spherical aberration is fixed by the mechanical design of the objective lens or is controlled by the software of the C_s-corrector (if the microscope is equipped with one), the objective lens defocus and astigmatism are much less stable and must be adjusted frequently. The diffractogram – the numerically obtained Fourier transform of the image intensity – provides a very straightforward way of correcting low-order aberrations of the microscope. The way in which the diffractogram varies with defocus, spherical aberration, and astigmatism is shown in Figure 2.7. For the case of vanishing spherical aberration ($C_s = 0$), ideal imaging conditions are met if the diffractogram resembles that shown at the top left corner. For $C_s = 1.2$ mm, the highest interpretable resolution is achieved if the diffractogram resembles that shown in the third column of the second row in Figure 2.7.

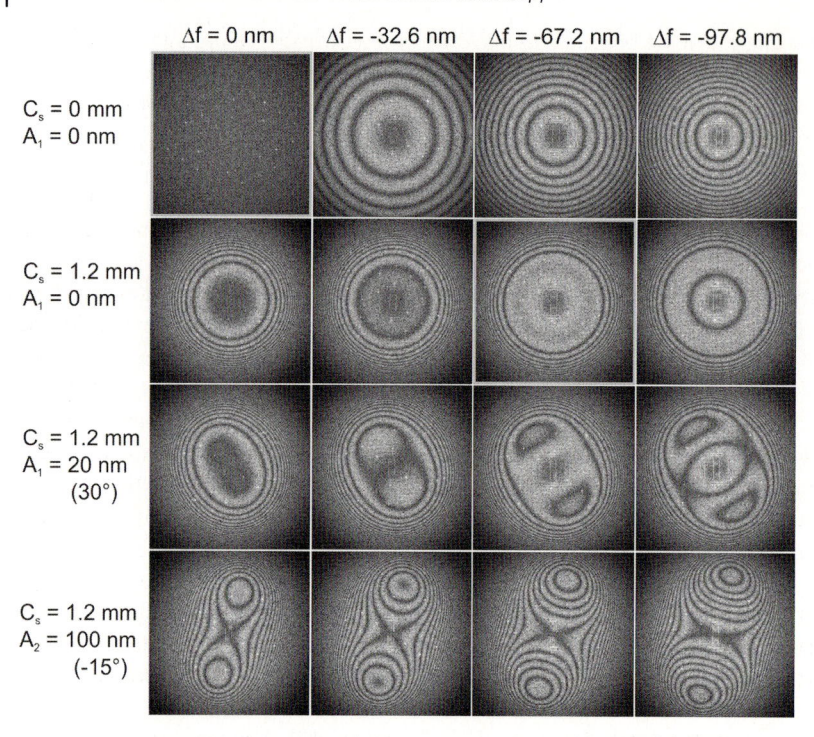

Figure 2.7 Simulated diffractograms for various imaging parameters. The imaging conditions were the same as those used in Figure 2.5. The third diffractogram in the second row has been simulated for Scherzer conditions; that is, for maximizing the range of spatial frequencies which are represented in the image without contrast reversal.

If several exposures can be afforded to record an image of the electron wave function for a number of different defocus values, it is possible to reconstruct both amplitude and phase of the exit-face electron wave function (see Section 2.5 for more details). Since the image aberrations introduced by the objective lens can be removed in this process, this procedure results in two delocalization-free images of the amplitude and phase of the exit face wave functions. This is especially important for nanostructures consisting of light atoms, that only produce a strong contrast for large defocus and/or C_s, in which case the delocalization is also quite severe. However, it must be borne in mind that the recording of such a focal series over only a small range of defocus values will not allow the reconstruction of phase differences across long distances. It is also very important that the sample remains unchanged throughout the whole focal series. Any damage produced by the electron beam, or changes in the sample that occur during *in-situ* experiments, may make a reliable reconstruction impossible.

Because of the finite energy spread of the electrons in an electron beam, and the fact that electrons are emitted from slightly different places on the emitting tip (finite effective source size), the experimental image recorded in any transmission electron microscope represents an average over single-electron images for a range of incident beam energies and directions. The resulting blurring of the HRTEM image is described by the temporal and spatial coherence envelope functions in Equation 2.12.

Recently, it has become possible to correct for the chromatic aberration (C_c) of an electron microscope [24]. Without such a C_c-corrector C_c is typically on the order of 1 mm. This means that in a 200 kV transmission electron microscope, electrons which differ in energy by 1 eV (relative energy difference $\Delta E/E = $ 5e-6) will produce images the defocus of which differs by the defocus spread $\Delta_f = C_c \times \Delta E/E = 5$ nm. Superimposing images of different defocus reduces the resolution. The full width at half-maximum (FWHM) of the energy distribution of electrons (temporal incoherence) emitted by thermionic guns is about 1.5 eV, which is twice that of Schottky emitters. The image blurring that results from superimposing the images of the corresponding difference in defocus is one of the main limitations in TEM image resolution. It is for this reason that a monochromator [25–27], which further reduces the energy spread in the electron beam, may also enhance the spatial resolution of the microscope, especially at lower accelerating voltages.

In addition to the elastically scattered electrons, electrons that have scattered inelastically will also contribute to the image, if they are not filtered away by an imaging energy filter [28–30]. Because of their largely different energy, these electrons produce a slowly varying background to the image produced by the elastically scattered electrons.

The quasi-coherent approximation (see Equation 2.12) is not exact, and may lead to (generally not very large) errors in some cases. A higher degree of accuracy in simulating the blurring of the TEM image contrast by a finite spread of energy (partial temporal coherence) and momentum (partial spatial coherence) of the incident electrons can be achieved by either a numerical Monte Carlo integration, which simulates incident electrons of slightly different energies and/or directions, or the use of the transfer cross-coefficient (TCC) [31]. Unfortunately, both of these approaches are computationally much more expensive and are, therefore, rarely used. The important point to keep in mind here is that the quasi-coherent approximation is not exact, and may in some cases lead to (albeit not very large) errors.

As discussed in Section 2.1.2, the parameters which limit the resolution in TEM images are incoherent aberrations, instabilities of the electron optics, and vibration of the sample. These define the information limit of any particular microscope. The insertion of an objective aperture limits the range of scattering angles of electrons, which in turn contribute to the image a maximum scattering angle, $\theta_{aperture}$. As electrons scattered to angles beyond the information limit produce an incoherent background to the image, the use of an objective aperture that is not much larger than the microscope's information limit will enhance the image contrast and should, therefore, always be used. If the objective aperture is smaller than the information limit, then the image resolution will be determined by the objective aperture that is being used. Details that can still be resolved in an image will have a size greater than $d_{min} = 1/q_{max}$, where $q_{max} = \sin(\theta_{aperture})/\lambda$ is the highest spatial frequency (or reciprocal space) component in the image.

2.3
Conventional TEM of Defects in Crystals

The observation of extended defects such as dislocations, using HRTEM, usually requires that the dislocation is straight and can be aligned with the optic axis of the microscope in order to view it "edge on" (as in Figure 2.5 or Figure 2.8a). From a high-quality HRTEM image of a dislocation which has been aligned in this way, it is then possible to determine the in-plane component of the Burgers vector from the gap in the Burgers circuit. The component of the Burgers vector which points along the dislocation (the screw component), however, cannot be determined using this method.

The screw and edge components of the Burgers vector of a dislocation are better determined using conventional BF, DF, or WBDF TEM at medium resolution. For an objective aperture which selects only a single reflection, and if the sample is aligned in a two-beam condition, the Howie–Whelan differential equations (Equations 2.10a and 2.10b) may be used to describe the variation of amplitude and phase of the transmitted and the diffracted beam for each position in the image. For a specimen of constant composition and thickness, Equations 2 and 2.10b predict that neither the BF nor the DF images will show any contrast variation, as long as the excitation error s_g does not vary. As shown in Figure 2.8, the presence of a lattice defect introduces a local displacement $R(x,y,z)$ of the unit cells from their perfect lattice positions. According to Equation 2.11, any displacement which is not perpendicular to the reciprocal lattice vector g of the diffracted beam will introduce variations in the excitation error s_g. This, in turn, will produce variations in the complex wave functions for the transmitted and diffracted beams $\Psi_0(r)$ and $\Psi_g(r)$. The edge dislocation used as an example in Figure 2.8 produces only distortions in the (horizontal) x-direction, while the lattice planes corresponding to a reciprocal lattice vector g that is normal to the Burgers vector will remain undistorted (Figure 2.8c). For the same reason, Equations 2.10a and 2.10b and Equation 2.11 predict that the dislocation contrast in a two-beam DF or BF image will vanish for dislocations where

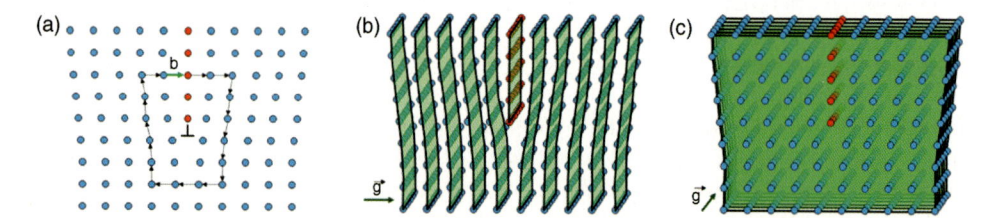

Figure 2.8 (a) Lattice distortion produced by the strain field around an edge dislocation, viewed end on. The dotted arrows indicate the Burgers circuit, and the solid arrow the Burgers vector b; (b) Lattice planes which correspond to a reflection g that is parallel to the Burgers vector are distorted around the dislocation core; (c) Lattice planes the reciprocal lattice vector of which is perpendicular to the direction of the Burgers vector (i.e., $g \cdot b = 0$) are undistorted. Both bright-field and dark-field images of the two configurations shown in (b) and (c) can be recorded if the electron beam is incident from the top.

the Burgers vector is normal to the reflection of the strongly excited beam (i.e., $g \cdot b = 0$). Dislocations with Burgers vectors that are not normal to g may produce contrast in the image. The differential Howie–Whelan equations (Equation 2.10a and 2.10b) must be integrated up to the thickness of the specimen, in order to quantitatively predict the contrast produced by a particular defect-induced strain field (a detailed discussion of this topic is available in Ref. [11]).

If neither the transmitted beam nor the strongly excited beam are used for imaging, but instead a third beam which is only weakly excited, but colinear with the strongly excited beam (see Figure 2.2d) is used, then the resolution and dynamic range of the strain contrast images may be greatly enhanced. The reason for this lies in the large excitation error of the weakly excited beam. Small changes in the excitation error will produce large differences in the diffracted intensity. Thus, instead of having to solve the Howie–Whelan equations, the images can be interpreted simply by looking for turning points in Equation 2.11 [32]. However, because of the low intensity of weak beam images, exposure times must generally be very long, which in turn imposes strict constraints on the specimen stability.

2.4
Lorentz Microscopy

In addition to the projected electrostatic potential of the specimen (Equation 2.7), the electron beam is also sensitive to magnetic fields. If the specimen contains domains where the magnetic induction B is not aligned with the optic axis of the microscope, then any electrons traveling with velocity v will be deflected according to the Lorentz force:

$$\vec{F}_{Lorentz} = -e\left(\vec{E} + \vec{v} \times \vec{B}\right) \qquad (2.15)$$

This implies that the electron beam may be used to map electrostatic as well as transverse (with respect to the microscope's optical axis) magnetic fields.

The magnetic field in the location of the specimen that is produced by the objective lens is usually of the order of 1 to 2 T, oriented along the optical axis of the microscope. In order to be able to image the unperturbed magnetic properties of the sample, it is necessary to keep the sample away from any external magnetic fields. This can either be achieved by constructing special low-field objective lenses, or by switching off the objective lens and using a lens that is located further down the optic axis of the microscope. Such lenses may be either special Lorentz lenses or transfer lenses which form part of the C_s corrector. Because of the increased distance from the specimen, the electron optical performance of these lenses is inferior to that of the objective lens (e.g., the spherical aberration is normally of the order of a few meters). However, many applications for Lorentz microscopy require rather large fields of view rather than an ultimate spatial resolution. In addition, if the microscope is equipped with a C_s-corrector the effective spherical aberration may be reduced to a few millimeters.

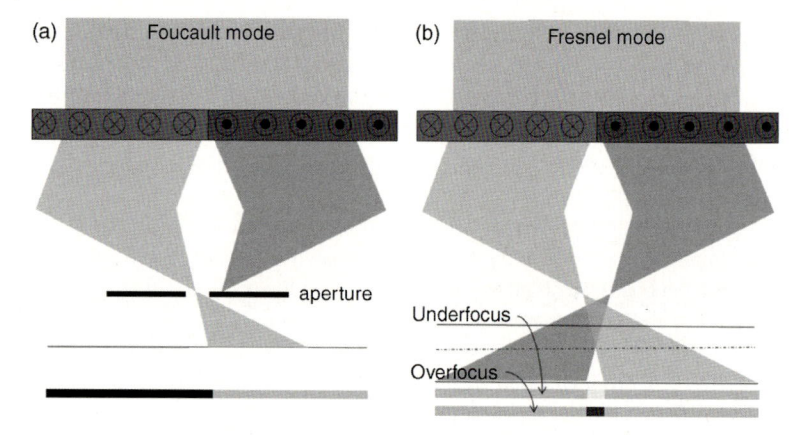

Figure 2.9 Diagram illustrating two possible Lorentz microscopy imaging modes for the investigation of magnetic structures. (a) In Foucault mode, domains with a selected magnetization will appear bright, and the rest of the sample dark. The selection is made using an aperture in the backfocal plane of the Lorentz lens; (b) In Fresnel mode, the domain walls and magnetization vortices will appear bright or dark in defocused images, depending on the magnetization gradient and defocus.

Two TEM imaging modes which are based on a deflection of the electron beam by the magnetic field in the sample are shown in Figure 2.9. In this case, a parallel beam of electrons traveling at speed v along the optical axis of the microscope is deflected perpendicular to the in-plane component of the magnetic induction according to the Lorentz force (Equation 2.15), while passing through the specimen. This adds a transverse component to the electron's momentum vector. In Foucault imaging mode (Figure 2.9a), the electrons of a specific offset in transverse momentum can be selected by using an aperture located in the back-focal plane of the (Lorentz) lens. A set of images may then be recorded for different positions of the momentum selecting aperture or (experimentally much easier) at different illumination tilt angles; these images can then be used to construct a map of the magnetic induction in the sample [33].

Based on the very same principle, the differential phase contrast (DPC) technique requires a scanning transmission electron microscope equipped with a partitioned BF detector [34]. As the electron beam is scanned across the specimen it is deflected by the local magnetic field in the sample, which changes the relative counts measured by the different segments of the detector. Again, this signal can be interpreted quantitatively in terms of the local magnetization.

In contrast to the Foucault imaging mode, which maps domains of equal magnetic induction, the Fresnel imaging mode can be used to map the position of domain boundaries. As illustrated in Figure 2.9b, the different transverse momentum component of parts of the electron beam which have passed through differently magnetized domains cause them to either overlap or separate at the domain boundaries, depending on the defocus and the relative magnetic induction between

2.5
Off-Axis and Inline Electron Holography

Holography in the context of electron microscopy aims to record the "whole image" – that is, the amplitude and phase of the complex electron wave function. Since it is physically impossible to directly detect the phase of an electron wave function, but only its probability density (intensity), the phase must be measured indirectly. Currently, two TEM techniques are commonly used to reconstruct the phase of the electron wave function that has been scattered by the object, namely off-axis and inline electron holography. Both of these methods incorporate a reconstruction of the phase of the exit face wave function from the interference pattern produced by interfering the electron wave function with a reference wave function.

In *off-axis holography*, an electrostatic biprism (a positively charged wire between two negatively charge plates) is used to split the electron wave function; this causes the part which has traveled through a vacuum to interfere with the part which has passed through the object. If the interference fringes are twice as fine as the smallest detail in the object wave function, then the reconstruction can be made either on an optical bench [35] or by solving a set of linear equations [36]. The need for very fine interference fringes can be overcome by recording several interference patterns with slightly different fringe positions [37]. An example of off-axis holography is shown in Figure 2.10. At this resolution, the phase contrast is well described by the phase object approximation which states that, in the absence of electrostatic fields, the phase shift is directly proportional to the local projected electrostatic potential (Equation 2.7). In general, the relative phase shift $\Delta\phi$ between two points $\mathbf{r}$ and $\mathbf{r}_0$ in the electron wave function depends on the electrostatic potential and also on the transverse component of the magnetic induction $\mathbf{B}_\perp(\mathbf{r})$ in the sample, and is given by

$$\Delta\Phi(\vec{r}) = \Phi(\vec{r}) - \Phi(\vec{r}_0) = \int_{-\infty}^{\infty} \left(\sigma(E)[V(\vec{r}) - V(\vec{r}_0)] - \frac{e}{\hbar} \int_{\vec{r}_0}^{\vec{r}} \vec{B}_\perp(\vec{r}')d r' \right) dz \tag{2.16}$$

where $\sigma(E)$ is the electron interaction constant given in Table 2.1, and the outer integral over $\mathbf{B}_\perp(\mathbf{r})ds$ is along the path $r \to r_0$. Under the assumption that the reference wave has traveled through field-free space, the electrostatic potential and magnetic induction at r_0 are zero, which makes these measurements absolute.

The original proposal of Dennis Gabor (who received the Nobel Prize for inventing holography in 1971) did not involve an off-axial reference wave; rather, the transmitted wave itself served as the reference wave [38]. For plane wave illumination this results in two superposed holograms which cannot be separated by the linear algebra methods applied in the reconstruction of off-axis holograms (twin image problem).

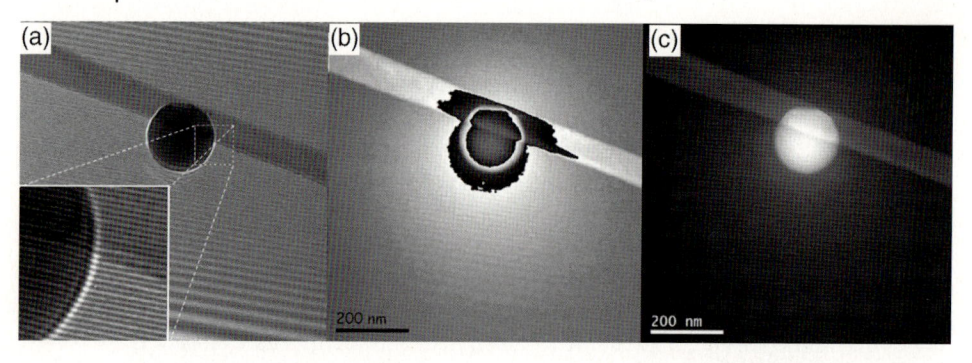

Figure 2.10 (a) Off-axis hologram of a latex sphere supported by lacey carbon. The fine, almost horizontal, fringes (see inset) are due to interference between the object wave which has passed through the region of interest, and the reference wave which has passed through vacuum only. Reconstruction of the off-axis hologram yields a complex wave function, the phase of which is displayed in (b). The abrupt transitions between white and black contrast are due to the fact that the phase of a complex number can only assume values between π and $-\pi$. "Unwrapping" these phase jumps to produce a continuous phase map results in the phase image displayed in (c), the gray levels of which correspond to values of the phase between 0 and 14 rad. Images kindly provided by Dr Petr Formanek, Leibniz Institute for Polymer Research Dresden, Germany.

The recording of several such *inline holograms* (two or more), however, allows the twin image problem to be solved, and the complex object wave function to be reconstructed. Because the system of equations that must be solved to reconstruct inline holograms is nonlinear [39, 40], the reconstruction is usually more time-consuming than that of off-axis holograms, unless linear approximations to the problem are used [41, 42]. Some inline holograms of the same sample used for the off-axis holograms in Figure 2.10, and the phase shift reconstructed from them, are shown in Figure 2.11.

The amount of defocus required to obtain interference between partial waves that have scattered from two points on the sample that are a distance d apart requires a defocus of at least $d/\tan(\theta_{aperture})$. However, the spatial coherence of the illuminating electron beam must be good enough to produce sufficient fringe contrast across the desired length scale d. As in off-axis holography, for some reconstruction algorithms the spatial coherence of the illuminating electron beam places an upper limit on the lengths scales across which phase differences can be measured. One of the main advantages of inline holography is that the object wave serves as its own reference wave, and that it does not require any part of this coherent wave function to travel through vacuum. This allows it to be applied anywhere in the electron-transparent region of the sample, and also far away from the specimen edge. Likewise, as for the off-axis fringe shifting method [37], for the same resolution in the reconstructed phase the magnification can be at least twice as low as for off-axis holography, which results in principle in larger field of view. However, just as for high-resolution focal series reconstruction, multiple (at least three [44]) images produced by the same wave function must be recorded, which requires that the specimen remains unchanged between exposures.

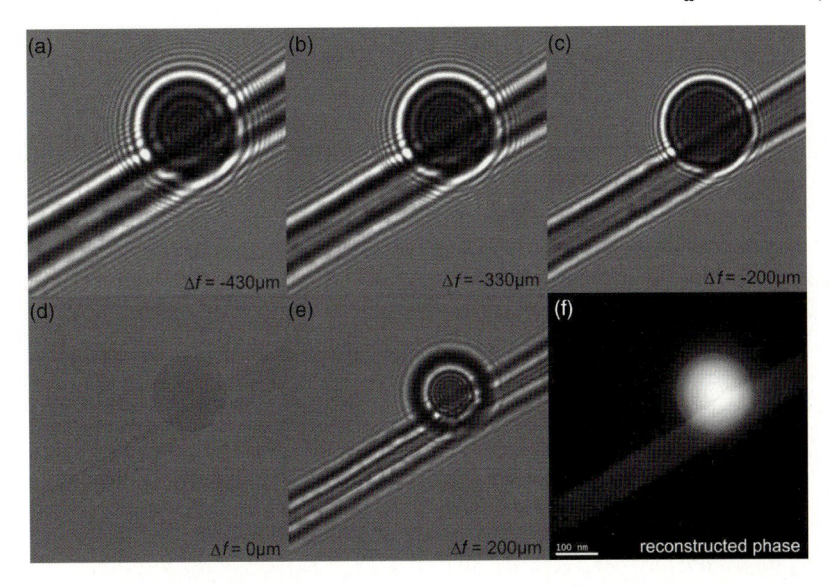

Figure 2.11 (a–e) Images of the same latex sphere shown in Figure 2.10, on a strand of lacy carbon and recorded for different values of the objective lens defocus Δ_f. These data were kindly provided by Dr Petr Formanek, Leibniz Institute for Polymer Research Dresden, Germany; (f) Phase shift reconstructed from images (a) to (e). The gray scale in (f) ranges from 0 to 14 rad. The reconstruction algorithm used here directly reconstructs a continuous phase map – there is no need to "unwrap" the phase afterwards. Changes in magnification and image rotation and distortions (panels (a)–(e) show the corrected images) have been fitted and corrected for automatically by the reconstruction algorithm [43].

2.6
Electron Diffraction Techniques

2.6.1
Fundamentals of Electron Diffraction

As described in Section 2.1, in a transmission electron microscope a diffraction pattern can be obtained by adjusting the microscope's projector lens system to produce an image of the back-focal plane of the objective lens. If the condenser lens system produces a parallel (plane wave) beam that illuminates a single crystal, the diffraction pattern will consist of discrete diffraction spots on the detector at positions $L\lambda \mathbf{g}$, where L is the camera length. A polycrystalline specimen will produce ring patterns, because of the superposition of diffraction patterns of many differently oriented grains, whereas amorphous structures will produce speckle patterns that transition smoothly into diffuse ring patterns with decreasing spatial coherence of the illumination. The diffracted intensity very close to the origin of the diffraction pattern may be used to determine the size of the scattering object, if this is smaller than the coherence length of the electron beam and the area selected for diffraction (e.g., see the one-dimensional shape transform of an interface in Figure 2.12).

Figure 2.12 Energy-filtered diffraction pattern obtained by illuminating the interface between two grains in a Si$_3$N$_4$ ceramic with a narrow parallel beam. The continuous horizontal streak through the center of the diffraction pattern represents the shape transform of the interface between the two grains which contains a 1 nm-wide (nonperiodic) inter-granular glassy film. The spacing of the diffraction spots corresponds to the basal plane of Si$_3$N$_4$. The diffuse background is produced by a thin layer of amorphous carbon on the sample. Illustration courtesy of S. Bhattacharyya; recorded on a Zeiss EM912 on imaging plate.

In the kinematical approximation, the intensity in a diffraction pattern is proportional to $|U_q|^2$, the modulus square of the Fourier components of the potential of the specimen at the three-dimensional reciprocal lattice points that lie on the Ewald sphere. For crystalline samples, U_q vanishes for all reciprocal space vectors that do not correspond to reciprocal space lattice vectors **g**. The kinematical diffraction intensity for reflection g is then given by

$$I_{\vec{g}} = |U_{\vec{g}}|^2 \frac{\sin^2(\pi \lambda t s_{\vec{g}})}{s_{\vec{g}}^2} \qquad (2.17)$$

where U_g is the structure factor defined by expression Equation 2.3 (and Equation 2.5) and, as shown in Figure 2.4, s_g is the excitation error given by

$$s_{\vec{g}} = \frac{\lambda}{2}\left(-|\vec{g}|^2 + \vec{k}\cdot\vec{g}\right) \qquad (2.18)$$

For reflections in the zero-order Laue zone (ZOLZ), this reduces to

$$s_{\vec{g}} = \frac{\lambda}{2}\left(-|\vec{g}|^2 + \vec{k}_t\cdot\vec{g}\right) \qquad (2.19)$$

splitting the incident wave vector $\mathbf{k} = (\mathbf{k}_t, k_z)$ into its vertical component $k_z \approx 1/\lambda$ and its transverse component $\mathbf{k}_t = (k_x, k_y)$.

As mentioned in Section 2.1.3, the Ewald sphere construction illustrated in Figure 2.4 is very useful for understanding the scattering geometry of elastic and inelastic scattering. All scattering events with the same final momentum of the scattered electron must have wave vectors that lie on a sphere that intersects the origin of reciprocal space. For elastic scattering events, the radius of the Ewald sphere is the magnitude of the wave vector of the incident electrons inside the sample ($|\mathbf{k}_0| = 1/\lambda$), whereas for inelastic scattering events the radius of the Ewald sphere is reduced because of the loss of energy of the scattered electron and the corresponding slight increase in wavelength.

2.7
Convergent Beam Electron Diffraction

Varying the direction of the incident electron beam relative to the zone axis of a crystal by small amounts leads to a change in the transverse component of the incident wave vector and, with that, the excitation error of those reflections which do not lie perpendicular to the tilt direction. The variation in the intensity of a diffracted beam with the incident beam direction is referred to as a "rocking curve." Convergent beam electron diffraction (CBED) represents a very elegant means of recording two-dimensional rocking curves of all reflections within a single exposure. A convergent electron beam, which may be focused to a very small spot on the specimen surface, consists of a range of plane waves described by all wave vectors within a cone. The angle at the tip of this cone, and with that the range of transverse components of the incident wave vector, is defined by the size of the condenser aperture.

Each of these incident plane waves produces its own diffraction pattern on the detector. Limiting the convergence angle by a round condenser aperture results in round diffraction discs instead of spots, as illustrated in Figure 2.13. Each point within a CBED disc corresponds to a specific $\mathbf{k}_t$ and with that, to a specific excitation error for that reflection. In the direction perpendicular to the reciprocal lattice vector $\mathbf{g}$ of a reflection the value of s_g is constant; this leads, for example, to the formation of dark lines in the central and diffracted discs, termed HOLZ lines. It can be seen in Figure 2.14 that, because of the steep angle at which the Ewald sphere intersects higher-order Laue planes, the exact Laue condition for reciprocal lattice points in these planes will only be met for incident beam directions that lie on sharp lines. The

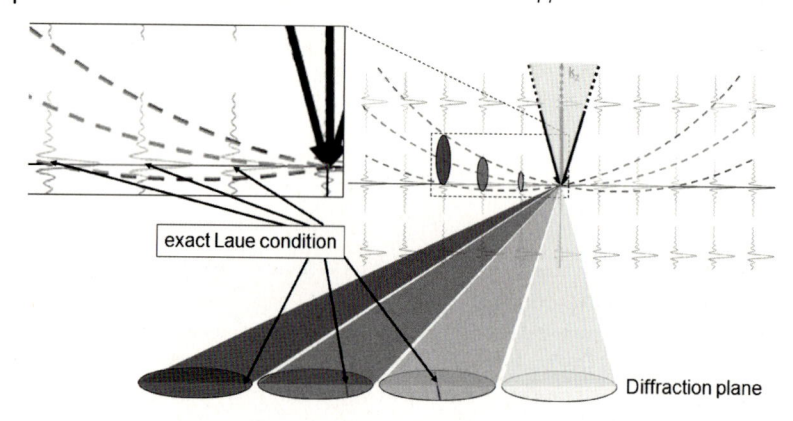

Figure 2.13 Illustration of the scattering geometry in convergent beam electron diffraction. The convergent incident beam contains a range of wave vectors (representing plane waves), each of which points from the origin of a different Ewald sphere to the origin of reciprocal space. The curvature of the Ewald sphere causes none of the incident plane electron waves to satisfy the Laue condition for all reflections simultaneously. In the two-beam case, bright lines indicate the exact Laue condition in the diffracted discs and corresponding deficiency lines in the central disc.

strong excitation of many HOLZ reflections produces bright sharp HOLZ rings consisting of many HOLZ line segments. The radius R_n of the nth order HOLZ ring for a crystal oriented in the zone axis defined by g_{zone} is given by

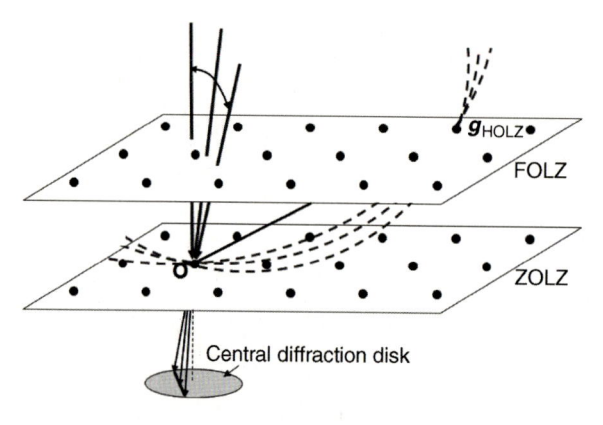

$$R_n = \cos^{-1}\left(1 - \lambda n |\vec{g}_{zone}|\right) \tag{2.20}$$

Figure 2.14 Illustration of the scattering geometry that leads to the formation of higher order Laue zone (HOLZ) lines in diffraction discs lying in the zero-order Laue zone (ZOLZ). Incident beam directions for which the Ewald sphere intersects a particular reciprocal lattice vector in a higher-order Laue plane lie on a line. The strong excitation of reflection g_{HOLZ} leads to missing intensity and thus a dark (HOLZ) line in the diffraction discs of the transmitted beam and other ZOLZ reflections.

and dark HOLZ deficiency lines in the central disc and other ZOLZ reflections. The sharp nature of HOLZ lines and rings leads to these features of the diffraction pattern being well suited for measuring lattice parameters or the precise calibration of the high voltage of the microscope.

2.7.1
Large-Angle Convergent Beam Electron Diffraction

In addition to providing two-dimensional rocking curves, the diffraction data contained in CBED patterns is quite local, because they are produced by focusing a fine convergent probe on the specimen. Keeping the illumination conditions constant and simply raising the specimen produces a set of diffraction spots in the first intermediate image plane. The selection of one of these diffraction spots with the SA aperture will remove all diffraction discs from the diffraction pattern, except for the one being selected. The convergence angle can then be increased without producing any overlap between neighboring discs. Because the specimen is illuminated by a convergent but defocused electron beam, each position on the sample will be illuminated from a different angle. This so-called large-angle convergent beam electron diffraction (LACBED) technique [45] is, therefore, especially useful for examining the Burgers vector of dislocations, because the diffracting conditions change in a very well-defined manner along the length of the dislocation line.

2.7.2
Characterization of Amorphous Structures by Diffraction

Although much less ordered than crystals, the structure of amorphous materials may also be characterized by electron diffraction. A major characteristic of an amorphous structure is its partial pair distribution functions; that is, the probability of finding an atom pair with atomic numbers Z_1 and Z_2 at a distance r_{ij} apart. An electron diffraction pattern of an amorphous structure allows its reduced density function (RDF), which is the weighted sum of partial pair distribution functions, to be determined [46, 47]. The RDF of amorphous structures varies rather smoothly compared to that of crystals, which features very sharp peaks. Molecular dynamics (MD) or reverse Monte Carlo (RMC) simulations [48] are required to extract structures of atomic clusters from such data.

While the RDF provides information about inter-atomic distances, fluctuation electron microscopy (FEM) [49, 50], which may be implemented as a DF TEM [49] imaging or STEM [51] diffraction technique, provides information about four-body correlations, and is therefore very sensitive to medium range order.

2.8
Scanning Transmission Electron Microscopy and Z-Contrast

As shown in Figure 2.1b, in STEM the electron beam is focused to a fine spot which is scanned across the sample; the image is then constructed by assigning the signal that has been collected at a given beam position to the corresponding pixel position in the image. In contrast to TEM, the resolution of STEM images is determined mainly by the electron optics in front of the specimen, which produces the fine spot. There exist two major STEM imaging modes: (i) bright-field STEM (BF-STEM), in which scattered and transmitted electrons are collected; and (ii) annular dark-field STEM (ADF-STEM) where only scattered electrons are collected. The commonly used detector geometries used for these two modes are shown in Figure 2.1b, where the small disc in the center is the BF detector and the ring around it is the ADF detector. If the inner radius of the ADF detector is so large that most of the Bragg diffracted electrons pass through the hole in the detector (usually about 60 mrad in a 200 kV microscope) it is termed a high-angle annular dark-field (HAADF) detector. Typically, HAADF detectors collect mainly electrons that have been scattered very close to the nucleus of an atom, and are therefore well described by the Rutherford scattering cross-section which is proportional to Z^2 (where Z is the atomic number). It is for this reason that HAADF-STEM is often also referred to as Z-contrast imaging. Because the scattering signal collected by the HAADF detector is not pure Rutherford scattering and depends on a precise detector geometry, the image contrast will scale approximately as $Z^{1.7}$ [52]. This sensitivity to the atomic number may be used for fast elemental mapping, if only very few different atomic species are present within the sample, and if these show large differences in mean Z-value (e.g., see Ref. [53] for atomic resolution mapping of individual Sb atoms in silicon). Thermal vibration of the atoms in the sample causes each incoming beam electron to probe slightly different inter-atomic distances, which in turn produces a diffuse incoherent scattering background known as thermal diffuse scattering (TDS). In crystalline specimens this TDS background is not homogeneous, but produces so-called Kikuchi bands [54] that are normal to the atomic planes. HAADF-STEM image contrast is mainly produced by TDS and is therefore, to some degree, an incoherent imaging method. Figure 2.15 shows the integrated intensity on the ADF detector as a function of specimen thickness. Plots for various inner detector collection angles and a fixed outer detector angle are shown. When comparing these, it becomes clear that the thickness dependence of the ADF signal becomes more monotonous with increasing inner collection angle of the ADF detector. The reduction of these thickness oscillations may also be interpreted as an increase in the degree of incoherent nature of the image contrast. It also shows that cooling a specimen will result in a reduction of TDS and thus a reduced image signal, particularly for very high-angle ADF-STEM. The enhanced danger of contamination during STEM imaging, caused by the localized electron probe in some microscopes, may be overcome by cooling the specimen. However, it must be borne in mind that this also reduces the TDS.

In BF-STEM, it is the central portion of the diffraction pattern that is collected. Whereas, ADF-STEM images resemble the local scattering strength of the material, at medium resolution BF-STEM images provide complimentary information, i.e. count of those electrons which have not scattered to high angles. According to the reciprocity principle, which states that for elastic scattering the source and detector

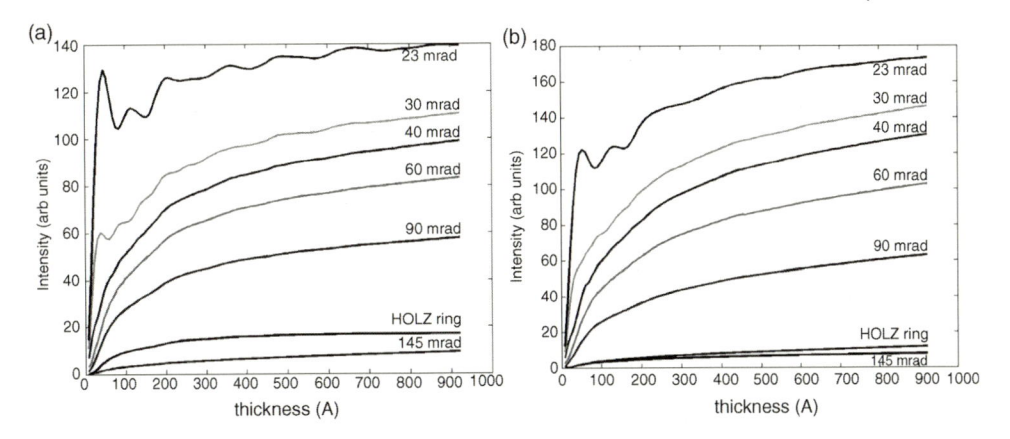

Figure 2.15 Thickness and temperature dependence of the ADF-STEM signal collected for a small 100 kV probe positioned over an atomic column in Si[110] for various inner detector cut-off angles ranging from 23 to 145 mrad, and a large outer cut-off angle (200 mrad). The Laue ring detector is a narrow annulus which only collects the scattering within the first-order Laue ring. These data were obtained by multislice simulations which include TDS by the frozen lattice approximation [52]. The root mean square atomic displacement was (a) 0.0044 nm (sample temperature T = 20 K) and (b) 0.0076 nm (room temperature).

can be interchanged, BF-STEM imaging contrast is equivalent to BF-TEM contrast. The angular range collected by the BF-detector corresponds to the beam convergence angle in BF-TEM, and the condenser aperture in BF-STEM corresponds to the objective aperture in TEM. This equivalence between BF-TEM and BF-TEM explains the strong sensitivity of BF-STEM images to local diffraction conditions, defocus, and strain induced contrast at, for example, dislocations or other defects. As in HRTEM imaging, where only those crystal lattice planes can be resolved for which the corresponding reflections pass the objective aperture, in BF-STEM the illumination convergence angle determined by the size of the condenser aperture must be sufficiently large to make the diffraction discs corresponding to these planes overlap with the central disc.

One major advantage of STEM imaging is that several signals can be collected simultaneously. In addition to a BF- and an ADF-detector (as shown in Figure 2.1b), it is possible – at least in principle – to configure additional (HA)ADF detectors, or an energy-dispersive X-ray (EDX) spectrometer; the BF-detector may also be replaced with an electron energy loss (EEL) spectrometer. However, it must be borne in mind that STEM images are acquired sequentially, which means that the acquisition times are normally longer for STEM images than for TEM images, where all pixels of the image are acquired simultaneously. This may be important for *in-situ* investigations, because a different acquisition time must be assigned to each pixel in the image. While specimen drift produces a smearing of contrast in TEM images, it produces distortions in STEM images.

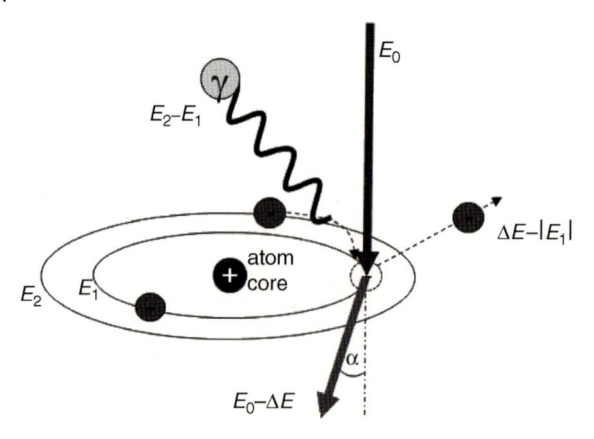

Figure 2.16 Illustration of the origin of the signal on core-loss EELS and EDXS. The incident beam of electrons of energy E_0 may transfer energy ΔE to, for example, a K-shell electron which was bound to the atom with a binding energy E_1. This electron is then excited to either an unoccupied orbital of higher energy ($\Delta E - |E_1| < 0$) or even to outside the sample, where it can then be detected as a secondary electron ($\Delta E - |E_1| > 0$). The reduction of kinetic energy ΔE in the electron beam can be detected using EELS. If an electron from a higher occupied energy level E_2 relaxes to fill in the empty K-shell position, the excess energy $E_2 - E_1$ will lead to the emission of X-rays or Auger electrons.

2.9
Analytical TEM

While elastic electron scattering provides very valuable information relating to interatomic distances and electrostatic and magnetic fields on the nanoscale, inelastic scattering may be more sensitive to the chemical composition of a sample. As illustrated in Figure 2.16, a fast beam electron which ionizes an atom will transfer part of its energy to the electron that it has excited. This energy loss can then be detected using electron energy-loss spectroscopy (EELS) which makes use of the velocity-dependence of the Lorentz force acting on a fast electron in a transverse magnetic field (Equation 2.15), or photons of energy no greater than this energy loss by energy-dispersive X-ray spectroscopy (EDXS).

When operating the microscope in STEM mode and scanning a focused beam over the sample, the EELS signal may be recorded at each pixel position, so as to produce EELS maps with a high spatial resolution [55]. Since electrons cannot be excited into orbitals that are already occupied, there is a minimum energy threshold below which electrons from a given shell cannot be excited. This leads to features in the EEL spectrum that can be assigned to specific elements. Consequently, STEM-EELS maps may be in turn used to construct maps of the elemental distribution, the variation in EELS fine structure, or the distribution of localized plasmons [56]. Alternatively, if the microscope is equipped with an imaging energy filter, the images may be recorded at a constant energy loss ΔE. Elemental maps extracted from only three such energy-filtered TEM (EFTEM) images have the advantage over STEM-EELS, in that large images (containing up to millions of pixels) may be constructed over comparatively short acquisition times. However, in most cases STEM-EELS has a higher energy and spatial resolution than EFTEM.

An additional advantage of STEM-EELS is that the high-angle (Z-contrast) ADF-STEM signal can be recorded parallel to the EEL spectrum. It is also possible to simultaneously record the emitted X-rays using an EDX spectrometer. As shown in Figure 2.16, when the electron beam has excited an electron from one of the inner shells, the atom will relax by filling the empty position with an electron from a higher atomic orbital. The energy gained by this electron will then be emitted as X-rays, which can be detected using either energy-dispersive or wavelength-dispersive X-ray spectrometry. Because of their much simpler operation, EDX spectrometers are more common. Whereas, EEL spectra provide information concerning unoccupied electronic levels, EDX spectra provide information concerning the electronic levels that are already occupied.

References

1 Knoll, M. and Ruska, E. (1938) Das Elektronenmikroskop. *Z. Phys.*, **78**, 318.
2 Ruska, E. (1987) The development of the electron microscope and of electron microscopy. *Rev. Mod. Phys.*, **59**, 627–638.
3 von Ardenne, M. (1938) Das elektronen-rastermikroskop. Praktische ausführung. *Z. Tech. Phys.*, **19**, 407–416.
4 Crewe, A.V., Isaacson, M., and Johnson, D. (1969) A simple scanning electron microscope. *Rev. Sci. Instrum.*, **40**, 241–246.
5 Tonomura, A., Matsuda, T., Endo, J., Todokoro, H., and Komoda, T. (1979) Development of a field-emission electron microscope. *J. Electron Microsc. (Tokyo)*, **28**, 1–11.
6 Krivanek, O.L., Dellby, N., Spence, A.J., Camps, A., and Brown, L.M. (1997) Aberration correction in the STEM, in *Proceedings of the 1997 Biennial Meeting of the Electron Microscopy and Analysis Group of the Institute of Physics (EMAG 97), Cambridge, UK* (ed. J.M. Rodenburg), Institute of Physics, Bristol and Philadelphia, pp. 35–39.
7 Haider, M., Braunshausen, G., and Schwan, E. (1995) Correction of the spherical aberration of a 200kV TEM by means of a hexapole-corrector. *Optik*, **99**, 167–179.
8 Haider, M., Uhlemann, S., Schwan, E., Rose, H., Kabius, B., and Urban, K. (1998) Electron microscopy image enhanced. *Nature*, **392**, 768–769.
9 Haider, M., Rose, H., Uhlemann, S., Kabius, B., and Urban, K. (1998) Towards 0.1nm resolution with the first spherically corrected transmission electron microscope. *J. Electron Microsc. (Tokyo)*, **47**, 395–405.
10 Spence, J.C.H. (1981) *Experimental High Resolution Electron Microscopy*, Clarendon Press, Oxford.
11 Hirsch, P.B., Howie, A., Nicholson, R.B., Pashley, D.W., and Whelan, M.J. (1965) *Electron Microscopy of Thin Films*, Krieger, Malabar, FL.
12 Williams, D.B. and Carter, C.B. (1996) *Transmission Electron Microscopy: A Textbook for Materials Science*, Plenum, New York.
13 Mott, N. and Massey, H. (1965) *The Theory of Atomic Collisions*, Clarendon Press, Oxford.
14 García de Abajo, F.J. and Kociak, M. (2008) Electron energy-gain spectroscopy. *New J. Phys.*, **10**, 073035.
15 Cowley, J.M. and Moodie, A.F. (1957) The scattering of electrons by atoms and crystals. 1. A new theoretical approach. *Acta Crystallogr.*, **10**, 609.

16. Ishizuka, K. and Uyeda, N. (1977) A new theoretical and practical approach to the multislice method. *Acta Crystallogr. A*, **33**, 740.
17. Bethe, H. (1928) Theorie der Beugung von Elektronen an Kristallen. *Annalen der Physik*, **392**, 55–129.
18. Howie, A. and Whelan, M.J. (1961) Diffraction contrast of electron microscope images of crystal lattice defects. 2. Development of a dynamical theory. *Proc. Royal Soc. A*, **263**, 217–237.
19. Howie, A. and Whelan, M.J. (1962) Diffraction contrast of electron microscope images of crystal lattice defects. 3. Results and experimental confirmation of dynamical theory of dislocation contrast. *Proc. Royal Soc. A*, **267**, 206–230.
20. Justo, J.F., Bazant, M.Z., Kaxiras, E., Bulatov, V.V., and Yip, S. (1998) Interatomic potential for silicon defects and disordered phases. *Phys. Rev. B*, **58**, 2540.
21. Koch, C.T. (2002) Determination of core structure periodicity and defect density along dislocations. PhD thesis, Arizona State University.
22. Scherzer, O. (1936) Über einige Fehler von Elektronenlinsen. *Z. Phys.*, **101**, 593–603.
23. Lentzen, M., Jahnen, B., Jia, C.L., Thust, A., Tillmann, K., and Urban, K. (1998) High-resolution imaging with an aberration-corrected transmission electron microscope. *Ultramicroscopy*, **92**, 233–242.
24. Haider, M., Müller, H., Uhlemann, S., Hartel, P., and Zach, J. (2008) Developments of aberration correction systems for current and future requirements, EMC2008, in *Proceedings of the European Microscopy Congress 2008* (eds M. Luysberg, K. Tillmann, and T. Weirich), Springer, Berlin, p. 9.
25. Batson, P.E., Mook, H.W., and Kruit, P. (2000) High brightness monochromator for STEM. Proceedings of the Second Conference of the International Union of Microbeam Analysis Societies (Microbeam Analysis 2000), Kailua-Kona, Hawaii, 9–14 July 2000, vol. 165, pp. 213–214.
26. Su, D.S., Zandbergen, H.W., Tiemeijer, P.C., Kothleitner, G., Hävecker, M., Hébert, C., Knop-Gericke, A., Freitag, B.H., Hofer, F., and Schlögl, R. (2003) High resolution EELS using monochromator and high performance spectrometer: comparison of V_2O_5 ELNES with NEXAFS and band structure calculations. *Micron*, **34**, 235–238.
27. Kahl, F. and Rose, H. (1996) Design of an electron monochromator with small Boersch effect, in *Proceedings of the 11th European Congress on Electron Microscopy (EUREM)* (ed. U.C. Belfield), EUREM, Dublin, Ireland, pp. 478–479.
28. Rose, H. (1994) Correction of aberrations, a promising means for improving the spatial and energy resolution of energy-filtering electron microscopes. *Ultramicroscopy*, **56**, 11–25.
29. Krivanek, O.L., Gubbens, A.J., Delby, N., and Meyer, C.E. (1992) Design and first applications of a post-column imaging filter. *Microsc. Microanal. Microstruct.*, **3**, 187–189.
30. Uhlemann, S. and Rose, H. (1994) The MANDOLINE filter – a high-performance imaging energy filter for sub-eV EFTEM. *Optik*, **96**, 163–178.
31. Ishizuka, K. (1980) Contrast transfer of crystal images in TEM. *Ultramicroscopy*, **5**, 55.
32. Cockayne, D.J., Ray, I.L.F., and Whelan, M.J. (1969) Investigations of dislocation strain fields using weak beams. *Philos. Mag.*, **20**, 1265–1270.
33. Daykin, A.C. and Petford-Long, A.K. (1995) Quantitative mapping of the magnetic induction distribution using Foucault images formed in a transmission electron microscope. *Ultramicroscopy*, **58**, 365.
34. Chapman, J.N., Ploessl, R., and Donnet, D.M. (1992) Differential phase contrast microscopy of magnetic materials. *Ultramicroscopy*, **47**, 331.
35. Tonomura, A., Fukuhara, A., Watanabe, H., and Komoda, T. (1968) Optical reconstruction of image from Fraunhofer electron-hologram. *Jpn. J. Appl. Phys.*, **7**, 295.
36. Takeda, M. and Ru, Q. (1985) Computer-based highly sensitive electron-wave interferometry. *Appl. Opt.*, **24**, 3068–3071.

37 Ru, Q., Endo, J., Tanji, T., and Tonomura, A. (1991) Phase-shifting electron holography by beam tilting. *Appl. Phys. Lett.*, **59**, 2372–2374.

38 Gabor, D. (1949) Microscopy by reconstructed wave-fronts. *Proc. R. Soc. Lond. A*, **197**, 454–487.

39 Kirkland, E.J. (1982) Non-linear high-resolution image processing of conventional transmission electron micrographs. 1. Theory. *Ultramicroscopy*, **9**, 45–64.

40 Kirkland, E.J. (1984) Improved high resolution image processing of bright field electron micrographs. 1. Theory. *Ultramicroscopy*, **15**, 151–172.

41 Teague, M.R. (1983) Deterministic phase retrieval: a Green's function solution. *J. Opt. Soc. Am.*, **73**, 1434–1441.

42 Kirkland, E.J., Siegel, B.M., Uyeda, N., and Fujiyoshi, Y. (1980) Digital reconstruction of bright field phase contrast images from high resolution electron micrographs. *Ultramicroscopy*, **5**, 479–503.

43 Koch, C.T. (2008) A flux preserving inline holography reconstruction algorithm. *Ultramicroscopy*, **108**, 141–150.

44 Allen, L.J., Faulkner, H.M.L., Oxley, M.P., and Paganin, D. (2001) Phase retrieval and aberration correction in the presence of vortices in high-resolution transmission electron microscopy. *Ultramicroscopy*, **88**, 85–97.

45 Tanaka, M., Saito, R., Ueno, K., and Harada, Y. (1980) Large-angle convergent beam electron diffraction. *J. Electron Microsc.*, **29**, 408–412.

46 Cockayne, D.J.H. and McKenzie, D.R. (1988) Electron-diffraction analysis of polycrystalline and amorphous thin films. *Acta Crystallogr. A*, **44**, 870–878.

47 Cockayne, D.J.H. (2007) The study of nanovolumes of amorphous materials using electron scattering. *Annu. Rev. Mater. Res.*, **37**, 159–187.

48 McCulloch, D.G., McKenzie, D.R., Goringe, C.M., Cockayne, D.J.H., and McBride, W. (1999) Experimental and theoretical characterization of structure in thin disordered films. *Acta Crystallogr. A*, **55**, 178–187.

49 Treacy, M.M.J. and Gibson, J.M. (1996) Variable coherence microscopy: a rich source of structural information from disordered materials. *Acta Crystallogr. A*, **52**, 212–220.

50 Treacy, M.M.J., Gibson, J.M., Fan, L., Paterson, D.J., and McNulty, I. (2005) Fluctuation microscopy: a probe of medium range order. *Rep. Prog. Phys.*, **68**, 2899–2944.

51 Cowley, J.M. (2002) Electron nanodiffraction methods for measuring medium-range order. *Ultramicroscopy*, **90**, 197–206.

52 Kirkland, E.J., Loane, R.F., and Silcox, J. (1987) Simulation of annular dark field STEM images using a modified multislice method. *Ultramicroscopy*, **23**, 77.

53 Voyles, P.M., Muller, D.A., Grazul, J.L., Citrin, P.H., and Gossmann, H.-J.L. (2002) Atomic-scale imaging of individual dopant atoms and clusters in highly *n*-type bulk Si. *Nature*, **416**, 826.

54 Kikuchi, S. (1928) Diffraction of cathode rays by mica. *Nature*, **121**, 1019–1020.

55 Browning, N.D., Chisholm, M.F., and Pennycook, S.J. (1993) Atomic-resolution chemical analysis using a scanning transmission electron microscope. *Nature*, **366**, 143–146.

56 Nelayah, J., Kociak, M., Stéphan, O., de Abajo, F.J.G., Tencé, M., Henrard, L., Taverna, D., Pastoriza-Santos, I., Liz-Marzán, L.M., and Colliex, C. (2007) Mapping surface plasmons on a single metallic nanoparticle. *Nature Phys.*, **3**, 348–353.

3
Dynamic Transmission Electron Microscopy

Thomas LaGrange, Bryan W. Reed, Wayne E. King, Judy S. Kim, and Geoffrey H. Campbell

3.1
Introduction

Recent electron optic and instrumentation developments have pushed the spatial resolution of transmission electron microscopy (TEM) to sub-Ångstrom levels [1–3]. Very little effort has been made, however, to improve the temporal resolution of *in-situ* TEM observations which is, for most *in-situ* techniques, limited to the video rate acquisition times. In materials, most transient processes – such as rapid dislocation nucleation and dynamics, diffusionless and rapid diffusional phase transformations, and catalysis – occur at rates that are much faster than video rate (typically 33 ms per frame) [4, 5]. As multiple reactions can occur within this interframe time period, the experimenter is left to infer the mechanistic path of the material process. Hence, if the fundamental understanding of rapid material processes is to be broadened, then the spatiotemporal resolution of *in-situ* TEM measurements must be improved, in order to observe the salient features of these processes at their relevant length and time scales.

One way to improve such temporal resolution would be to control the electron emission, such that it is correlated in time with the transient process being studied [6–16]; this is analogous to optical and X-ray pump-probe techniques. In the pump-probe approach, a transient state is first driven in the material (e.g., by an initiating laser pulse), and later examined using an analytical pulse of electrons at a set time delay to the initiating pulse. In order to observe an irreversible process, it is essential to generate sufficient electrons within a single pulse to acquire a snapshot of the transient events in that process (the "single-shot approach"). Alternatively, if the process is reversible, the specimen can be pumped and probed by using multiple, lower-intensity pulses that can be summed to the point where the signal-to-noise ratio (SNR) allows an interpretable image to be formed (the "stroboscopic approach").

Previously, Zewail *et al.* [17, 18] have refined the stroboscopic approach [19] to subpicosecond time resolution by employing femtosecond lasers. Each femtosecond laser pulse produces, on average, only a few photoemitted electrons, thereby eliminating any spatial resolution-limiting space charge effects and enabling an

---

In-situ Electron Microscopy: Applications in Physics, Chemistry and Materials Science, First Edition.
Edited by Gerhard Dehm, James M. Howe, and Josef Zweck.
© 2012 Wiley-VCH Verlag GmbH & Co. KGaA. Published 2012 by Wiley-VCH Verlag GmbH & Co. KGaA.

imaging resolution equivalent to that of conventional TEM in normal operation. This approach requires that the images are built up from millions of single-electron pulses that have been precisely correlated with the specimen drive laser. Although this method allows temporal and spatial resolution to be maintained at optimum levels, the fact that the specimen must be laser-pumped millions of times means that the process being studied must be highly repeatable, or that the sample must completely recover between specimen drive laser shots.

The single-shot approach was pioneered by Bostanjoglo and coworkers [20–25], who demonstrated a spatial resolution of approximately 200 nm with a pulse duration <10 ns. Since all the information is obtained from a single specimen drive event, this technique can be used to measure irreversible and unique material events that cannot be studied stroboscopically. The obvious caveats to the single-shot approach are that it requires a high brightness source ($>10^9$ A cm^{-2} steradian), and that electron–electron interactions in high current electron pulses will limit the spatial resolution. In addition, modern transmission electron microscopes are not designed for the high-current operation necessary for a single-shot technique; consequently, the instrument must be modified and optimized to accommodate high-current pulsed electron probes, which makes it a much more technically challenging approach. Despite these difficulties, the single-shot approach is worth pursuing because it dramatically widens the range of phenomena that may be studied, as most processes of interest in the materials sciences are irreversible. Recently, an ongoing effort at Lawrence Livermore National Laboratory (LLNL) has focused on the single-shot approach, developing the technique of dynamic transmission electron microscopy (DTEM) with this capability [4, 26–28].

In this chapter, the design and construction of the LLNL dynamic transmission electron microscope for single-shot operation is described, after which details will be provided as to how the current microscope's capabilities (spatial resolution better than 10 nm using 15 ns electron pulses) have been employed to study phase evolution in solid-state reactions in NiAl reactive multilayer foils, and the martensitic nucleation behavior and α to β transformation rates in nanocrystalline Ti. A brief summary of the main results of these investigations will be provided, followed by a description of new holder designs that will enable unique *in-situ* capabilities for studying chemical and biological reactions in controlled gaseous and aqueous environments. Finally, the technical challenges and limitations of the single-shot approach will be described, and consideration given to the future of DTEM instrumentation.

3.2
How Does Single-Shot DTEM Work?

The dynamic transmission electron microscope is based on the JEOL 2000FX microscope platform (see Figure 3.1), with the electron optical column having been modified to provide laser access to the photocathode and specimen. A brass drift section has been added between the gun alignment coils and condenser optic that contains a 2.5 cm laser port and 45° Mo mirror, which directs an on-axis 211 nm laser

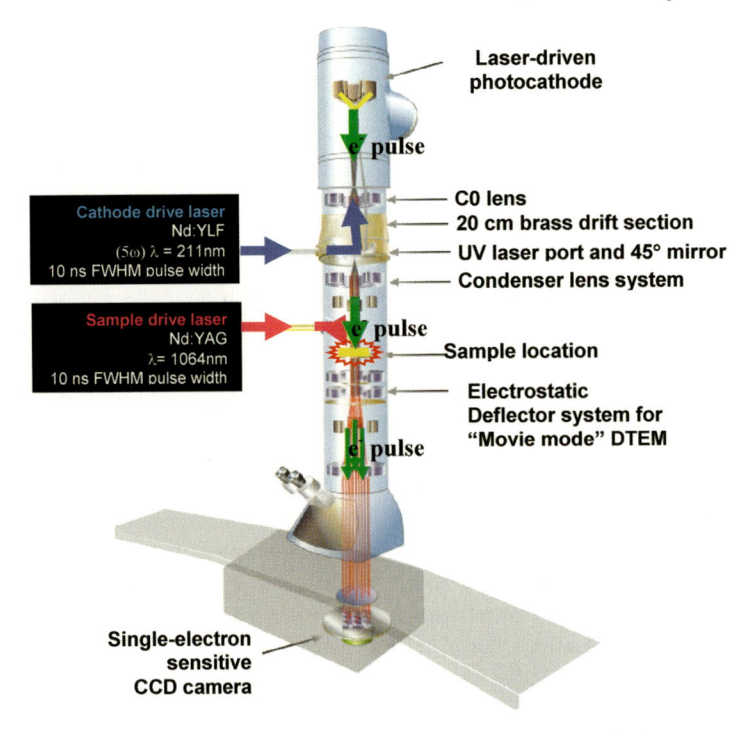

Figure 3.1 Schematic diagram of the Lawrence Livermore National Laboratory (LLNL) dynamic transmission electron microscope.

pulse to a 825 μm Ta disk photocathode [29].[1] The 10 ns UV laser pulse photoexcites a ~15 ns full-width at half-maximum (FWHM) electron pulse from the cathode, which is accelerated through the electron gun and passes through a hole in the Mo[2] laser mirror into the electron optics of the transmission electron microscope column. The electron pulse is aligned and illuminates specimens as in standard transmission electron microscope operation; in this way, all imaging modes can be utilized, including bright-field (BF) and dark-field (DF) selected area electron diffraction (SAED).

To better couple the photoemitted electron pulse into the condenser electron optics, a weak lens has been installed above the Mo laser mirror and brass drift section. This lens provides an increased current by focusing the spatially broad pulsed electron beam through the hole in the laser mirror and condenser system entrance apertures. In prior configurations without the coupling lens, the mirror was found to be the limiting aperture and to reduce the electron current by a factor of 20. The coupling lens, combined with appropriate condenser lens settings and imaging

1) The Ta disk photocathode can also be used as a thermionic source in nominal TEM operational mode.
2) Solid molybdenum mirrors have a high reflectivity (>60%) in the UV range, and molybdenum can be easily polished to produce high-quality, uniform surfaces (low wavefront distortion) needed for laser applications.

conditions, preserves the brightness of the photogun and improves the beam quality by reducing the aberrations that can result from a spatially broad electron pulse and high-angle, off-axis electrons. More details on the coupling lens design are provided in Section 3.2.2.

Time-resolved experiments in the dynamic transmission electron microscope are conducted by first initiating a transient state in the sample, and then taking a snap-shot of the transient process with the 15 ns electron pulse at a preferred time delay after the initiation. In most DTEM experiments, the transient process is initiated with a second laser pulse, which enters the transmission electron microscope column through a modified high-angle X-ray port. For nanosecond timescale experiments, neodymium-doped yttrium aluminum garnet (YAG) lasers with a pulse duration ranging from 3 to 25 ns are used that can produce fluences up to 1500 J cm^{-2} on the specimen. The fundamental wavelength (1064 nm) of these lasers can be frequency-converted using nonlinear harmonic generation crystals, for example, doubled (532 nm) or tripled (355 nm), as dictated by the absorption characteristics of the sample and the desired experimental conditions. For instance, as metals have broadband absorption, all of these wavelengths can be used, whereas certain semiconductors absorb only sufficient amounts of laser energy in the UV range and may require frequency-tripled laser pulses.

In a typical DTEM experiment, the instrument and specimen are first aligned in standard, thermionic mode, after which the alignment is optimized further in pulsed mode. As DTEM pump-probe experiments usually cause a permanent alteration to the sample, the latter is moved to a new location between shots, or an entirely new sample may need to be introduced into the microscope. The evolution of an irreversible materials process is studied through a series of pump-probe experiments with different time delays. For ultrafast optical pump-probe experiments, the time delay is often generated using an optical delay line. This is impractical for the wide range of time delays used in DTEM, since the technique requires 300 m of distance per microsecond of delay time. This is impractical given the microsecond delays required for some experiments with the dynamic transmission electron microscope. Instead, the microscope is used to control the time delays electronically, triggering each of the two lasers independently, and enabling a precision of approximately 1 ns, even with time delays of many microseconds. In order to permit a precise calibration of the pulse arrival times, the laser focus, and spot sizes, each laser is passed through a beam splitter shortly before it enters the transmission electron microscope column. The split-off beam is then directed to a camera placed at a position that is optically equivalent to the sample position (for the sample drive laser) or the cathode (for the cathode laser).

3.2.1
Current Performance

In its current configuration, the dynamic transmission electron microscope can be used to acquire images with better than 10 nm spatial resolution with 15 ns electron pulses. An example of the current spatial resolution and image capabilities is shown

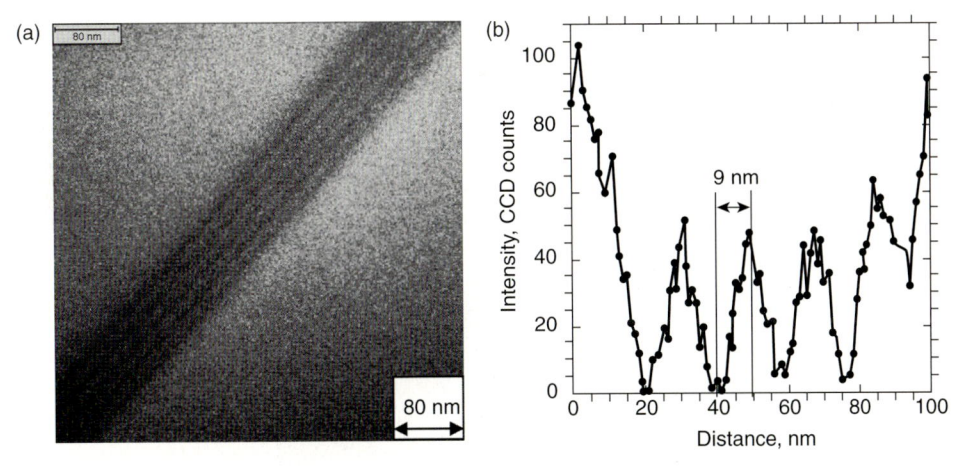

Figure 3.2 (a) Single-shot 15 ns image of a Au/C grating; (b) A plot of the image intensity across the Au and C layers, showing that the spatial resolution is less than 10 nm.

in Figure 3.2, which provides a cross-sectional view of a gold (dark layers) and carbon (light contrast layers) multilayer foil. In this case the individual layer thicknesses are less than 10 nm, and the layers are clearly resolved in the single-shot image (Figure 3.2a). To better illustrate the resolution limits in the single-shot image, the pixel intensity across the multilayer was measured and plotted (Figure 3.2b). The intensity from 9 nm thick layers was clearly visible above the background, which indicated that the resolution was at least 9 nm, if not better.

One critical advancement of the LLNL dynamic transmission electron microscope over previous instruments developed by the Bostanjoglo group is the ability to capture dynamical contrast images [30]. The unique electron source and lens system of the LLNL microscope allows a higher electron beam current and coherence, which in turn enables the use of objective apertures to obtain BF and DF images of defects such as grain boundaries, dislocations, and stacking faults. An example of a single-shot, 15-ns exposure DTEM image that clearly shows the line defects (dislocations) and planar defects (stacking faults) in a stainless steel material is shown in Figure 3.3. Dynamical image contrast formation is an important capability, as these defects play an important role in the mechanical properties of materials. Capturing their dynamics under a transient thermal or mechanical load can provide a better understanding of how the mechanical behavior of the material evolves.

3.2.2
Electron Sources and Optics

The standard electron optics of modern transmission electron microscopes are designed to operate with electron beam currents <100 nA. Most of the electrons generated at the source (several microamperes of current in the case of a thermionic source) are blocked by apertures in order to increase beam coherence and spatial

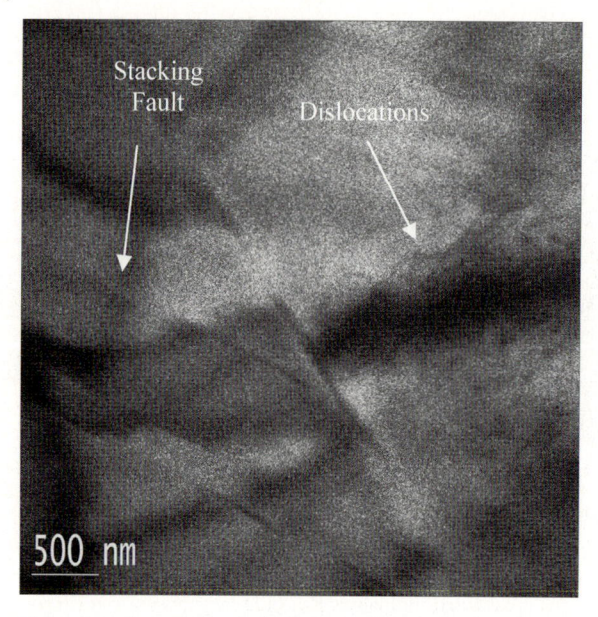

Figure 3.3 A single-shot pulsed image of a stainless steel sample. This shows clearly that DTEM can be used to image material defects such as dislocations and stacking faults with a single 15 ns exposure.

resolution, and typically, only 0.1% of the generated electrons will reach the sample. This is a practical trade-off in a conventional transmission electron microscope, as there is sufficient dose in a 1 s exposure of a nanoampere electron beam to acquire high-quality images. However, for the single-shot operation in the dynamic transmission electron microscope, the number of electrons per pulse (10^8–10^9) and the signal are low, requiring the use of large apertures and to make a trade-off between beam coherence and signal.

The number of electrons required for a single-shot DTEM image can be estimated through Rose criterion arguments [31] which stipulate that, for adequate imaging, the signal must be fivefold the shot noise in the image.

If the BF image contrast at a point is defined as c, and the number incident electrons per unit area is n, then the signal for a defined resolution element (d), for example, image pixel size, can be described as:

$$\text{signal} = cnd^2 \tag{3.1}$$

Accordingly, the shot noise is defined as

$$\text{shot noise} = d\sqrt{n} \tag{3.2}$$

For adequate image recognition according to the Rose criteria, the required electron dose would be,

$$n \geq 25/c^2 d^2 \tag{3.3}$$

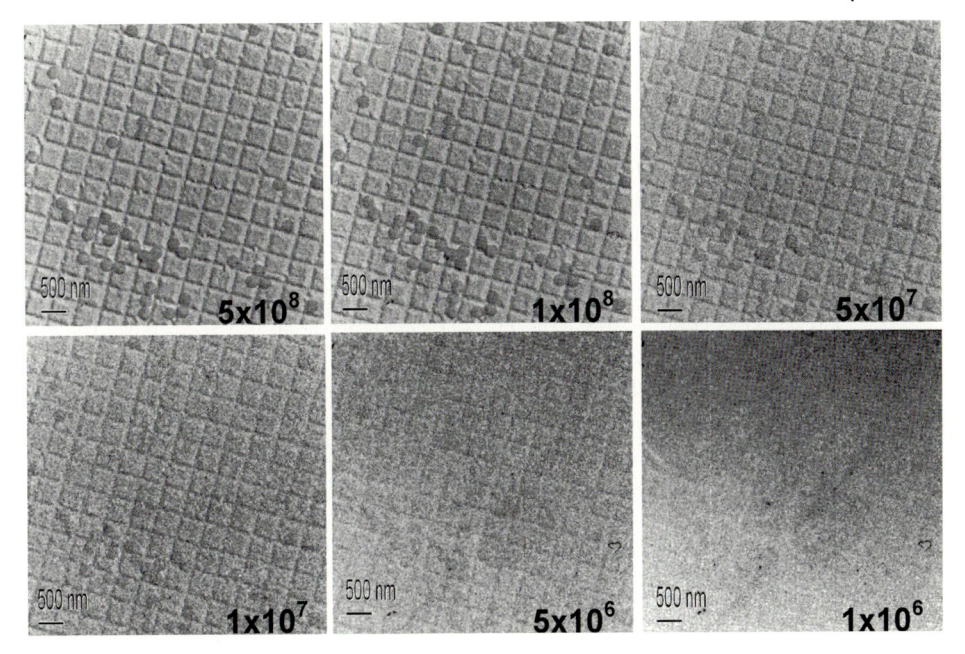

Figure 3.4 Collection of bright-field images of a carbon replica grating (463 nm grid squares), showing the variation in contrast and resolution as a function of the electron dose (shown at lower right-hand corners of panels).

To obtain an image contrast of 50%, the noise would have to be below 10% and the required dose would have to be at least 100 electrons per pixel, or a total of 10^8 electrons in a one-megapixel image. An example is given in Figure 3.4, which shows the image contrast and quality as function electron dose on the charge-coupled device (CCD) camera.

Of course, this analysis does not take into account the imperfections of the imaging system (lens aberrations) nor any space charge effects (discussed later) which would reduce this resolution [28, 32]. The factor of 5 criterion between signal and noise is also quite conservative, and might be relaxed for some conditions; for example, SAED requires 1% of the dose needed for BF imaging, while the contrast in the 10^7 electron dose image in Figure 3.4 is sufficient to observe the salient features. Ultimately, however, the required dose will depend on the sample and the desired observations. Although conservative, the Rose criterion analysis sets a lower bound on the required pulse current and source brightness to achieve high contrast and a spatial resolution of about 1 nm.

Having such a large number of electrons (10^8) in a 15 ns pulse results in a very large peak current (1 mA) as compared to the nanoampere currents of conventional TEM images. In order to achieve these currents, the electron source must be bright ($>10^7 \, \text{A cm}^{-2} \, \text{steradian}^{-1}$), and the imaging system and electron optics must be designed in such a way as to preserve the brightness of the source and to minimize any space charge effects that would reduce gun brightness and spatial resolution.

The only way to generate such high currents in a nanosecond pulse is by photoemission and the use of a high-quantum efficiency photocathode. Common photocathodes, such as cesiated transition metals, have quantum efficiencies of $>10^{-4}$, but require special preparation and high vacuum pressures to limit the surface contamination and oxidation that reduce their efficiency. In the case of the LLNL dynamic transmission electron microscope, the choice of cathode material has been based on practical considerations. Typically, cathode materials are selected that can be used as both photocathodes and thermionic sources, and which are commercially available in configurations that require little or no alteration of the electron gun and have moderate quantum efficiencies for the available laser wavelengths.

In the past, Ta disk cathodes have been an ideal choice as they have a quantum efficiency ranging from 10^{-5} to 10^{-4} at 211 nm, and similar properties to W hairpin thermionic emitters. They also have a large size that facilitates laser alignment. The Ta disk cathodes are illuminated with between 100 µJ and 1 mJ of 211 nm light (fifth harmonic) and a $1/e^2$ diameter spot of approximately 350 µm, which generates about 10^9 to 10^{10} electrons per pulse at the cathode surface. As the initial source size is large (ca. 350 µm diameter) and the electron pulse is divergent, most of these electrons will be blocked by the laser mirror and the entrance aperture of condenser system (ca. 500 µm). For this reason, a coupling lens and drift section have been installed between the accelerator section and the brass section containing the cathode laser mirror.

The LLNL dynamic transmission electron microscope condenser lens system has been modified to reoptimize the electron-optical coupling between the accelerator and the first condenser lens (Figure 3.5). This new coupling lens is necessitated by the very different requirements of single-shot DTEM versus conventional continuous-wave (CW) TEM. In CW TEM, the image quality and resolution are determined by an optimized interplay of lens aberrations and beam coherence. In this case, the easiest way to improve the performance is to use small apertures with lens conditions that are designed to throw away all but a tiny fraction of the current. Especially for thermionic sources, it is very common to deliberately block over 99% of the electrons in order to obtain high-resolution images or nanometer probes (Figure 3.5a).

Single-shot DTEM does not have this luxury, as the source brightness is limited and the required current density at the sample is strictly dictated by temporal resolution and SNR requirements (Rose criteria). The inevitable conclusion is that some sacrifice of spatial coherence is required, which means that microscope's condenser system must be able to handle much wider (or, more precisely, higher-emittance) beams than does a CW transmission electron microscope. This, in turn, means that high-coherence modes such as phase-contrast atomic resolution imaging will be extremely challenging on the nanosecond scale, although other modes with less stringent coherence requirements (e.g., mass-thickness and diffraction contrast imaging) would be possible. While some gains can be made by removing fixed apertures and weakening the C1 lens (see Figure 3.5b), this can still leave well over 90% of the electrons unused, especially when C1 aberrations are

3.2 How Does Single-Shot DTEM Work?

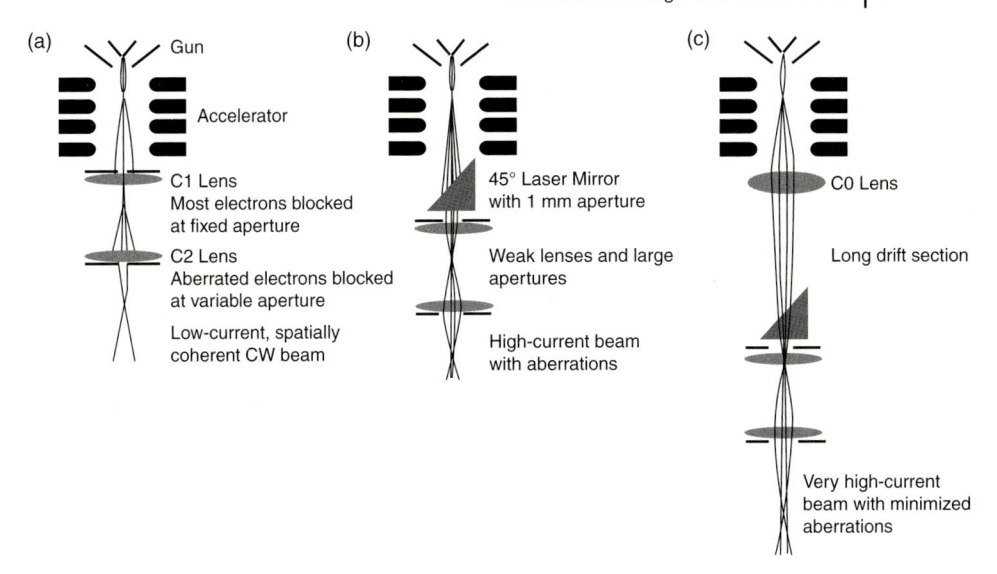

Figure 3.5 Schematic beam paths in the gun, accelerator, and condenser lenses for (a) conventional continuous wave (CW) TEM, (b) the first version dynamic transmission electron microscope, and (c) the current microscope with the added C0 lens. While the conventional design is an excellent optimization for few-second exposure times, many DTEM experiments call for a broader ability to sacrifice spatial coherence for an increased signal without introducing excessive aberrations. The weak C0 lens solves this problem.

Labels in figure:
(a) Gun; Accelerator; C1 Lens – Most electrons blocked at fixed aperture; C2 Lens – Aberrated electrons blocked at variable aperture; Low-current, spatially coherent CW beam.
(b) 45° Laser Mirror with 1 mm aperture; Weak lenses and large apertures; High-current beam with aberrations.
(c) C0 Lens; Long drift section; Very high-current beam with minimized aberrations.

considered. This situation is far from optimal, as the estimated spatial resolution limit due to spatial incoherence (ca. 1 nm) is still far better than the resolution limit due to the SNR (>10 nm).

To overcome this problem, a very weak, very large-bore lens (dubbed the C0 lens, as it precedes C1) has been introduced below the accelerator, followed by a long (ca. 20 cm) drift space to allow the beam to reconverge (see Figure 3.5c) [29]. The geometry allows a 3 mm-diameter portion of the beam to be focused through the 1 mm-diameter hole in the mirror and into the center of the C1 lens, without introducing any excessive aberrations. Simulations of the modified condenser lens system have shown that the thermal emittance (i.e., the spatial incoherence deriving from the finite source brightness) dominates the effects of condenser lens aberrations when the dynamic transmission electron microscope is operated in this mode. In other words, the coupling is essentially as good as it can get since, even with the large apertures in use the aberration-induced emittance growth from source to sample is negligible. When these simulations are coupled with the measured characteristics of the beam at the exit of the accelerator, it is found that a conservatively estimated 10^9 electrons should become available in a 10 ns pulse, this being a 20-fold improvement in beam current compared to the dynamic transmission electron microscope without the C0 lens. At this point, the system will be limited by the source rather than (as at present) by the lens and aperture designs, and the signal levels and coherence can be optimized against each other across the entire gamut of foreseeable experiments.

3.2.3
Arbitrary Waveform Generation Laser System

For most Q-switched, pulsed laser systems, the resonator cavity, lasing medium and excitation source determine the pulse duration and repetition rate, such that these systems will have a fixed pulse width, spatial mode, and repetition rate. As the cathode laser pulse and electron pulse durations are similar, the time resolution is fixed by the type of cathode laser system used. Since not all experiments require nanosecond time resolution, however, a compromise can be made between the temporal and spatial resolutions.[3] For instance, microsecond time resolution is sufficient for some catalytic reactions and dislocation dynamic studies, and these studies can benefit from the added electron dose and increased spatial resolution. To enable a flexible laser system and tailor the laser parameters for a given experiment, an arbitrary waveform generation (AWG) laser has been designed and constructed which can temporally shape laser pulses, thus allowing easy changes to be made to the pulse duration.

The AWG is composed of a waveform generator which drives a fiber-based electro-optical modulator and temporally shape a CW fiber laser seed pulse. The modulated waveforms are then amplified, frequency-converted, and delivered to the dynamic transmission electron microscope at an appropriate energy for use as either cathode or specimen drives. The AWG cathode drive laser allows for continuously variable and controlled electron pulse duration, from 250 μs down to 10 ns. The AWG sample drive laser enables the precise control of pulse time and shape, so that a better control of the drive conditions can be achieved at the specimen, including the heating rate, isothermal holds and cooling rate, so as to provide a wide variety of temperature–time profiles on the sample. At present, the sample is rapidly heated with a pulse of about 10 ns duration (there is very little ability to change this value at present), and then allowed to cool according to its own thermal conductivity, usually on a time scale of between 10 and 100 μs. By compensating the thermal conduction heat loss from the sample region of interest with an appropriately shaped pulse, it should be possible to sustain the sample temperature at a constant level for some microseconds before allowing it to cool. The AWG capability would enable a broad class of experiments in catalysis and surface sciences, and also permit the production of electron pulse trains, which is an essential element of the planned DTEM "movie mode" of operation (see Section 3.2.4). Multiple pulses with user-set time delays between the pulses are also used for double-exposure images to measure precise displacements of features during dynamic events

---

3) Longer pulses have reduced space charge effects and, thus, have a higher electron dose that enables the acquisition of images with a higher spatial resolution. For example, with a 1 μs electron pulse, an image with subnanometer resolution can be acquired.

3.2.4
Acquiring High Time Resolution Movies

The ability to acquire high time resolution movies – termed "movie mode DTEM" – expands the scientific capabilities of the dynamic transmission electron microscope in single-shot mode by providing detailed histories of unique material events on the nanometer and nanosecond scales. Previously, the microscope's hardware allowed only single-pump/single-probe operation, building up the typical time history of a process by repeating an experiment with varying time delays at different sample locations. In contrast, the movie mode DTEM upgrade enables single-pump/multi-probe operation, providing not only an ability to track the creation, motion, and interaction of individual defects, phase fronts, and chemical reaction fronts, but also invaluable information on the chemical, microstructural and atomic level features that influence the dynamics and kinetics of rapid material processes. For example, the potency of a nucleation site is governed by many factors related to defects and local chemistry. Whereas, a single pump-probe snapshot provides statistical data related to these factors, a multi-frame movie of a unique event would allow all of these factors to be identified, and the progress of nucleation and growth processes to be explored in detail. This would provide an unprecedented insight into the physics of rapid material processes from their early stages (e.g., nucleation) to their completion, providing direct, unambiguous information with regards to the dynamics of complex processes.

The two core components of the movie mode technology (Figure 3.6) are the arbitrary waveform generator (AWG) cathode laser system and a high-speed

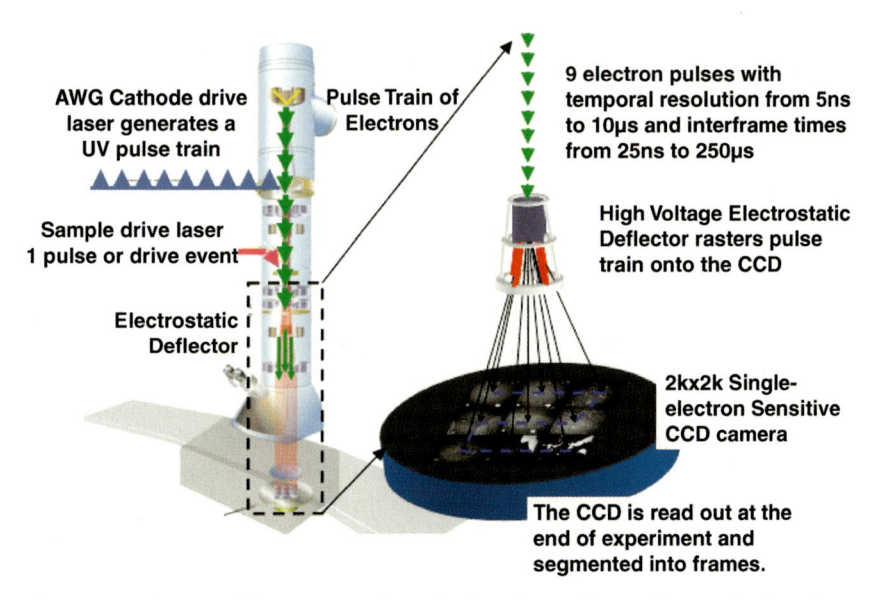

Figure 3.6 Schematic of the "movie mode" technology that enables single-pump/multi-probe operation and true, *in-situ* microscopy capabilities in the dynamic transmission electron microscope in which multiframe movies of ultrafast material dynamics can be acquired.

electrostatic deflector array. The AWG cathode drive laser enables continuously variable and controlled electron pulse durations from 250 μs down to 10 ns, in which a series of laser pulses is produced with user-defined pulse durations and delays that stimulates a defined photoemitted electron pulse train for a single sample drive event. Each pulse captures an image of the sample at a specific time. A fast-switching electrostatic deflector located below the sample directs each image to a separate patch on a large, high-resolution CCD camera. At the end of the experiment, the entire CCD image is read out and segmented into a time-ordered series of images – that is, a movie. The current technology produces nine-frame movies, but near-term modification to the system should enable up to 25-frame movies with interframe times as low as 25 ns. This frame rate is six orders of magnitude faster than modern video-rate *in-situ* TEM. Future versions of movie mode may also include fast-framing CCD technology, which can capture hundreds of frames within a few microseconds. The operating principle of these devices is that the photoelectron CCD data from multiple frames is stored in on-chip buffers that are read out at the end of the acquisition.

3.3
Experimental Applications of DTEM

3.3.1
Diffusionless First-Order Phase Transformations

One key application of DTEM is the investigation of rapid solid–solid-phase transformations, such as diffusionless, martensitic (shear-dominated) transformations. The latter represent a fundamental and much-studied process in the materials sciences, where the time resolution of DTEM can provide new information on the nucleation and growth behavior associated with this solid-state transition. Such transformation occurs by a coordinated atomic shuffle and shear displacement (shear-dominated lattice distortion). As this process does not involve any long-range diffusion, the transformation rates and interface velocities may be very high, approaching one-third the speed of sound [33]. Phenomenological crystallographic theories of martensitic phase transformation suggest that, in order for a structural transition to occur, there must be an undistorted (invariant) habit plane for the macroscopic shape deformation, which forms a highly coherent interface between the parent and martensite phase. It has been postulated, in theoretical models, that the parent–martensite interface consists of pairs of dislocations that develop the necessary lattice strains for the structural transition, while maintaining coherency across the interface, and that they must also be highly glissile to account for the high growth velocities. Due to the innate speed of these interfaces, however, the experimental observation of the interface structure, and even of the overall transformations kinetics, has been difficult. Nonetheless, the high time resolution of DTEM can be used to bring new insights into the interface dynamics and nucleation mechanisms in martensitic phase transformations.

For the first experiments, the high-temperature phase transformation in titanium (Ti) was studied. Titanium is a dimorphic element that undergoes a crystal structure change from hexagonal close packing (HCP; α-phase) to body-centered cubic (BCC; β-phase) atomic coordination upon heating above 1155 K, which is accomplished by a simple shear strain of the HCP unit cell [34]. The fast kinetics of the α → β phase transformation in nanocrystalline Ti films were investigated using single-shot electron diffraction and BF TEM images [5, 35–37]. Subsequently, the selected area electron diffraction patterns (SAEDPs) were monitored at different delays between the heating, drive laser pulse and electron pulse and drive laser energies. The laser-induced temperature change was determined by correlating the observed onset of melting in the SAEDP with the laser energy; this was then used as a one-point temperature calibration for subsequent experiments. The SAEDP were rotationally averaged to increased the SNR, and the variation in radial intensity distribution as a function of scattering vector was plotted and compared to modeled intensities and distributions for the α- and β-phases, as shown in Figure 3.7.

By using Rietveld [38] fitting routines, the phase fractions and corresponding α → β transformation rates were determined for temperatures between the transition start (1155 K) and the melt temperature (1943 K). The experimental data

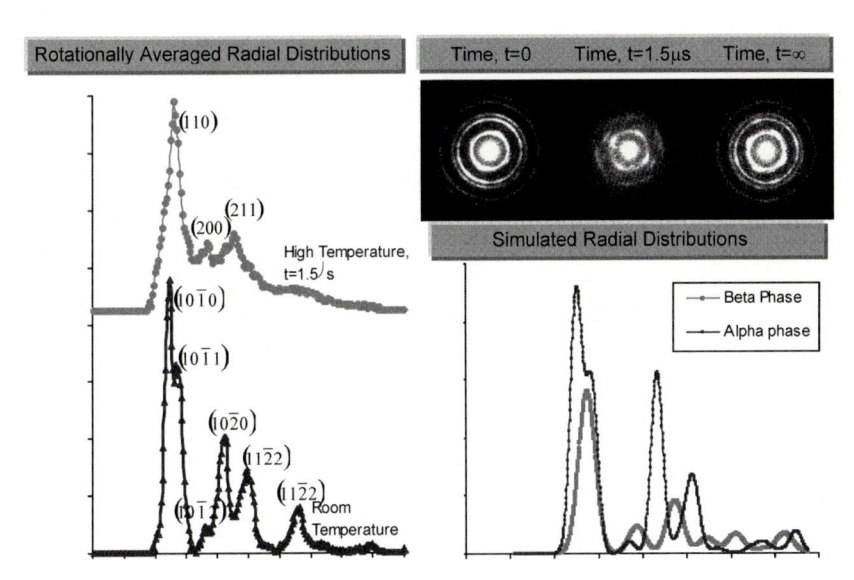

Figure 3.7 Single-shot electron diffraction data (15 ns exposure) of the α → β transformation in nanocrystalline Ti, occurring as the result of a laser-induced temperature rise to 1300 K. The upper-right diffraction patterns show the transformation occurring from α-phase to β-phase at 1300 K and a time delay of 1.5 μs which, after cooling back through the transition temperature, transforms back to α-phase ($t = \infty$). Comparison of the rotationally averaged radial intensity distributions (plot at left-hand side) with simulated diffraction data (plot at lower right-hand side) indicate that transformation to the β-phase was complete at 1.5 μs.

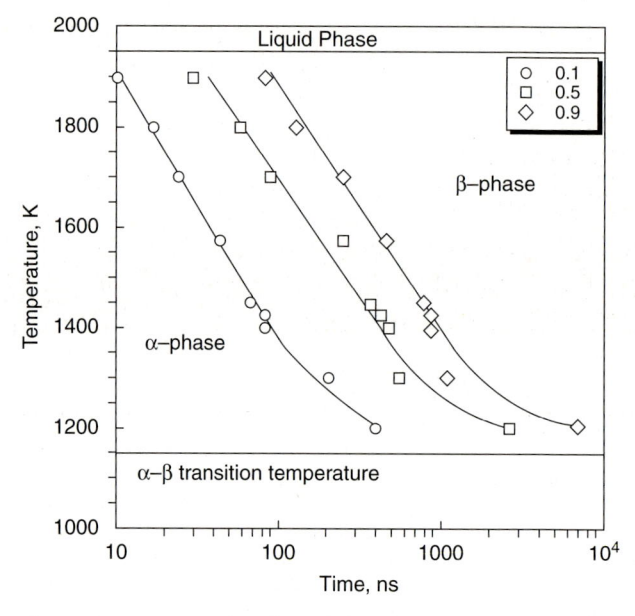

Figure 3.8 Experimental isothermal phase diagram for nanocrystalline Ti film constructed from quantitative analysis of the pulsed electron selected area diffraction data. Lines through are a smoothing curve fit through the data set [5].

were summarized in a time-temperature-transformation (TTT) curve with nanosecond time resolution (see Figure 3.8). Theoretical TTT curves were calculated using analytical models for isothermal martensite, available thermodynamic data, and experimental data gathered from pulsed electron BF TEM images (see Figure 3.9). Above 1300 K, there was an excellent agreement between the experiment and the discrete-obstacle interaction model [39]; this suggested that the nucleation rate and thermally assisted motion of the martensite interface were controlled by the interface–solute atom interactions. However, theory predicts much slower transformation rates near the transition temperature than does experiment. Experimental data fits using the Pati–Cohen model [40, 41] suggested that an increase in autocatalytic nucleation might partially account for the fast transformation rates at lower temperatures. It has also been speculated that the transition temperature may be lower for nanocrystalline material, which may also explain the discrepancy.

Another important point identified in these studies was that the incubation times for nucleation were heavily dependent on the grain size. For example, a Ti foil with an average grain size of 40–75 nm had incubation times of approximately 100 ns, whereas materials with 100 μm grain sizes had incubation times on the order of 500 ns. This suggests that numerous grain boundaries in nanocrystalline materials act as strong defects and heterogeneous sites for nucleation, thus reducing the nucleation barrier energy and increasing the nucleation and transformation rate.

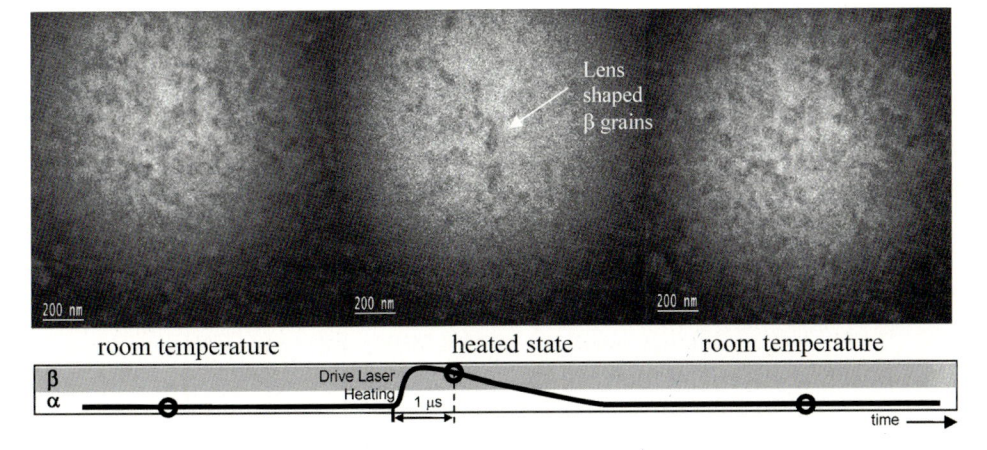

Figure 3.9 A series of single-shot bright-field, time-resolved TEM images showing the change in grain morphologies induced by the $\alpha \rightarrow \beta$ phase transformation. The left image was taken at room temperature, before the laser struck the film; the center image was taken 1 μs after the laser struck the film, and shows the specimen in the high-temperature state (estimated temperature rise ~1700 K); the right image was taken after the specimen had cooled to room temperature. The schematic temperature plot below is qualitative, and for illustration purposes only [5].

3.3.2
Observing Transient Phenomena in Reactive Multilayer Foils

Another interesting topic that can employ the high time resolution capabilities of DTEM is the observation of transient phases and morphologies that result from rapid solid-state chemical reactions in reactive multilayer foils (RMLFs) [28, 42, 43]. These foils, which are also referred to as "nanostructured metastable intermolecular composites," are layers of polycrystalline reactant materials that pass through exothermic, self-propagating reactions when the layer mixing is driven by an external stimulus [44, 45]. Depending on the composition, the amount of bilayers and the layer thickness, the exothermic reaction front may reach temperatures well above 1100 K and travel at a velocity of about $10\,\mathrm{m\,s^{-1}}$, governed by interface diffusion [46, 47]. As the RMLFs produce immense heat over a small surface area, they are often applied as localized heat sources for material bonding or biological sterilization [48, 49]. In addition, the periodic nanoconstruction renders RMLFs relevant to an examination of the *in-situ* progression of interface-controlled diffusion and transient phase evolution.

Conventional methods lack the necessary combined spatial and time resolution to observe nanolayer mixing directly, which may leave some details of the reaction unclear. However, by using the time resolution in DTEM, the transient states may be observed *in-situ*. In this case, the transient states of the reaction front were observed in RMLFs by using time-resolved diffraction and the imaging of plan-view TEM samples that were ignited with an infrared laser pulse. Studies of the transient states of this dynamic material, and of the mechanisms that govern the rate of heat generation and transport, may lead to a better understanding of atomic diffusion between thin films and phase boundary motion for optimized engineering applications in the future.

For such studies the Al/Ni-7 wt% V multilayer system was selected, with phase evolution and metastable phase formation being investigated using a 15 ns exposure, single-shot electron diffraction, and BF imaging. The 15 ns "snap shot" diffraction data showing the phase evolution from discrete Al and NiV multilayers to the final intermetallic phase is illustrated in Figure 3.10. Here, single-shot diffraction has been acquired in front of ($t = 0$), in line with ($t = 300$ ns), and beyond ($t = \infty$) the reaction front region to detect the evolution of phase formation and metastable states. Thermal effects of the exothermic behavior in the material are evident by the diffuse background and the reduction in higher-order reflections in the time-resolved 300 ns diffraction pattern (Figure 3.10). When this hot NiAL structure was allowed to cool to

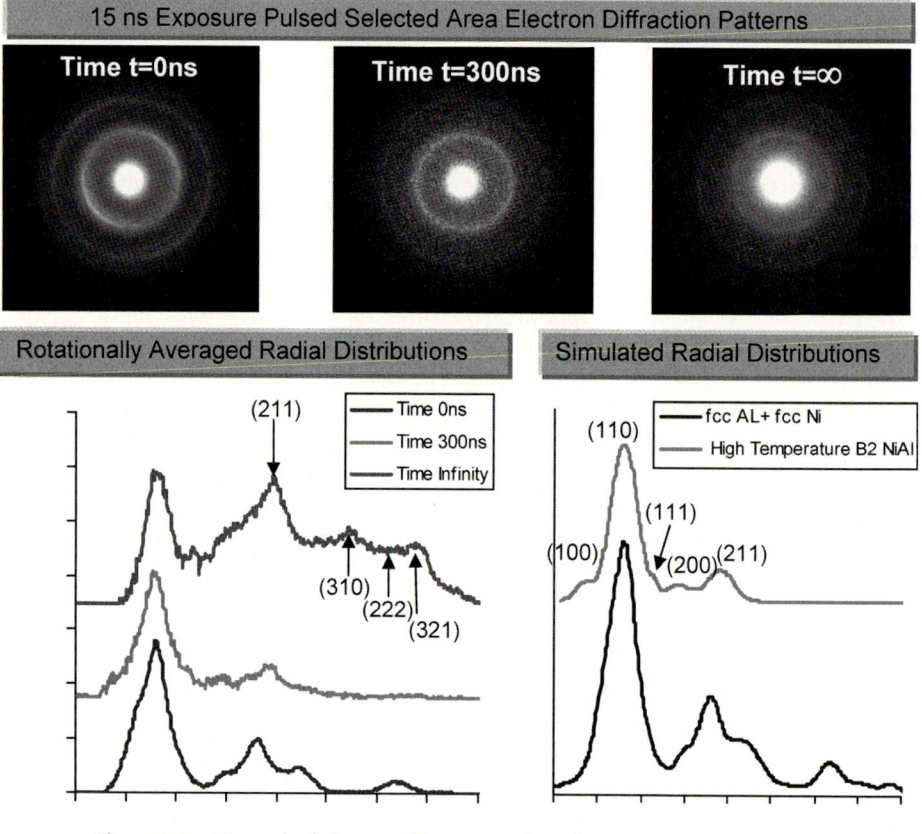

Figure 3.10 15 ns pulsed electron diffraction data acquired in front ($t=0$), in line with ($t=300$ ns), and behind ($t=\infty$) the reaction front. The rotationally averaged distributions show that at $t=300$ ns, an order B2 NiAl structure is formed that becomes highly textured when cooled to room temperature ($t=\infty$).

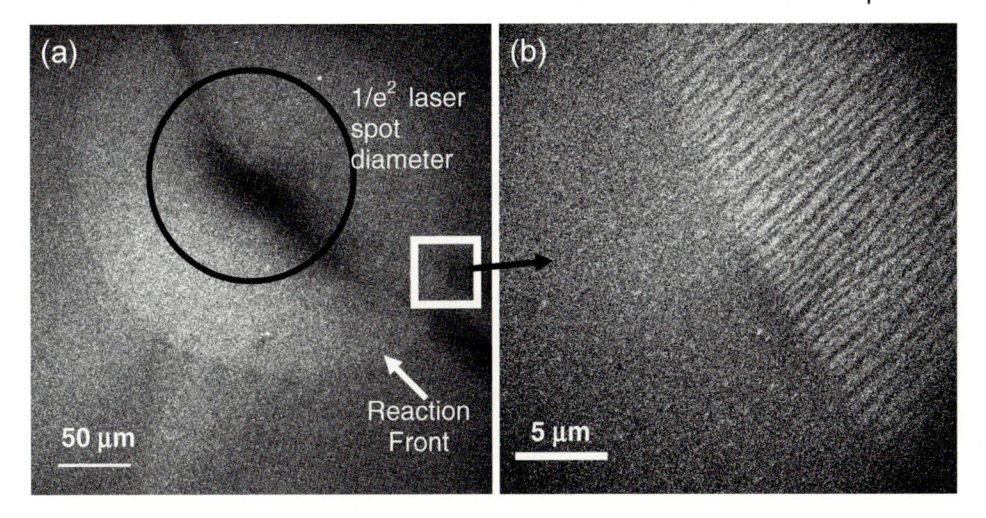

Figure 3.11 15 ns exposure "snap-shot" images of the reaction front morphology. (a) A low-magnification image of the reacted zone (light contrast). Note the sharp transition between unreacted (dark) and reacted (light) zones, indicating the position of the reaction front; (b) A higher-magnification pulsed image of the cellular microstructure which forms behind the reaction front.

room temperature, the concomitant grain growth processes led to a preferential texturing of the films, as indicated by the change in relative peak intensities. The primary conclusion here was that an ordered, B2 intermetallic NiAl phase had been formed within an extraordinarily short time period (<300 ns); this was to be expected due to the short (a few nanometers) atomic diffusion distances, the high temperatures, and the high phase stability of the B2 structure at near equi-atomic compositions.

In addition, the reaction front appeared as a sharp ring that moved outwards, radially from the central laser initiation point (as shown in Figure 3.11a). The sharp transition between the unreacted and reacted material was consistent with electron diffraction, and indicated that the reactants had formed rapidly, on the order of the 15 ns electron pulse. A closer examination of the microstructure (Figure 3.11b) revealed that a cellular structure had formed directly behind the reaction front. Simple calculations of the heat released by the mixing of Al and Ni indicated that temperature rises on the order of 1700 K would have been expected ($\Delta H \approx 77$ kJ mol^{-1} and $Cp^{Ni/Al} = 24$ J mol^{-1} K^{-1}). At such temperatures and 60 atom % Ni, the material could exist in a two-phase state, as solid-NiAl + liquid-NiAl; this might explain the dark contrast line, which may be related to the liquid structure,[4] and a well-defined periodicity (ca. 750 nm) that is analogous to cellular liquid–solid structures in alloy solidification [50]. These cellular microstructures persist only for 2–3 μs after the reaction front has passed, and are not present in the post-mortem microstructure. Although the mechanism for their formation remains unclear, their

---

4) The dark contrast arises from the diffuse, incoherent scattering that occurs in liquid phase, making it more opaque than the crystalline solid phase.

3.4
Crystallization Under Far-from-Equilibrium Conditions

The crystallization processes of as-deposited, amorphous NiTi thin films have been studied in detail using techniques such as differential scanning calorimetry (DSC) and *in-situ* TEM [51–54]. The kinetic data have been analyzed in terms of the Johnson–Mehl–Avrami–Kolomogrov (JMAK) semi-empirical formulae [55, 56], and the kinetic parameters determined have proved beneficial when defining process control parameters for tailoring microstructural features and shape memory properties. Based on the commercial urgency to shrink thin film-based devices, a series of unique processing techniques has been developed using laser-based annealing to spatially control the microstructure evolution down to submicron levels. In this respect, nanosecond, pulse laser annealing is particularly attractive as it limits the amount of peripheral heating of, and undesirable microstructural changes to, the underlying or surrounding material. Nonetheless, crystallization under pulsed laser irradiation can differ significantly from conventional thermal annealing, for example, when slow heating in a furnace. This is especially true for amorphous NiTi materials and relevant to shape memory thin film-based micro-electromechanical system (MEMS) applications.

At present, little to no data exists on the crystallization kinetics of NiTi under pulsed laser irradiation. This is due primarily to the high crystallization rates intrinsic to high-temperature annealing, and also to the spatial and temporal resolution limits of standard techniques. However, with the high time- and spatial-resolution capabilities of DTEM, the rapid nucleation events that occur from pulsed laser irradiation can be observed directly and the nucleation rates quantified [57]. An example of a series of nanosecond time-resolved, DTEM images of the pulsed laser-induced crystallization process in NiTi is shown in Figure 3.12. Here, the 15 ns delay image is taken at 1.5 μs

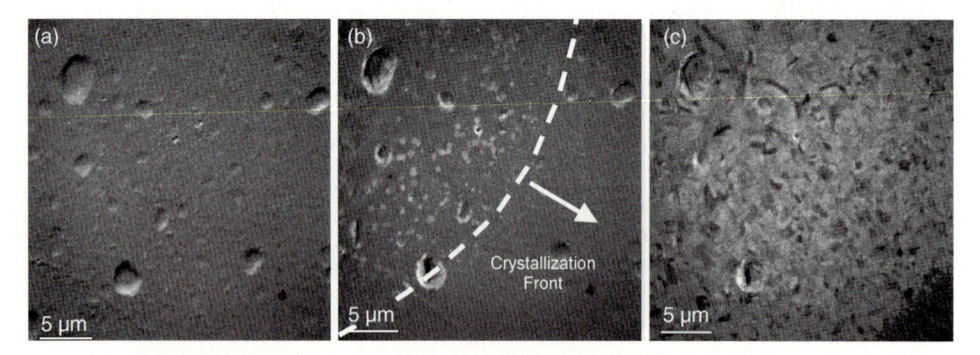

Figure 3.12 Series of 15 ns exposure pulsed electron images. (a) Image taken before laser heating; (b) Image taken at 1.5 μs after the laser struck the foil; (c) Image taken after the foil had cooled to 300 K.

after the pump laser has irradiated the sample; the center of the pulsed laser spot (120 μm $1/e^2$ diameter) is located in the upper left-hand corner of the image. Note the semicircular pattern of the newly formed crystallites (light contrast) in the amorphous matrix (dark background) that radiates outward from the center of laser-irradiated zone. The pulsed laser crystallization exhibits a radially propagating crystallization front that moves both inwards and outwards, with varying nucleation rates and crystallization times that are proportional to the temperature gradient across the gaussian laser-heated zone.

Variations in the number density of nuclei and their size are shown in Figure 3.13. In this case, at radial distances close to the center of the gaussian heated zone, where the temperatures approach an estimated 1500 K, the density of nuclei and nucleation rates are quite low (>5 × 10^4 nuclei μm^{-3} s^{-1}), but the grain sizes are large (>1 μm) due to high growth rates. At distances of 25 μm from the center, however, where the temperature are estimated to be about 1200 K, the nucleation rates are quite high (10^6 nuclei μm^{-3} s^{-1}), while at temperatures below 900 K no nucleation is observed. This crystallization behavior can be described under classical phase transformation theory as the competition between thermodynamic driving force and kinetics, which is further illustrated in Figure 3.14a, which shows the curve for 90% crystallization as a function of time for pulsed laser annealing. It should be noted that the high rates of crystallization occur at temperatures around 1200 K (the nose of the C-curve). At temperatures above 1200 K, the thermodynamic driving force for nucleation decreases, but growth rates are kinetically enhanced at the higher temperature, leading to large grains. At temperatures below 1200 K, the nucleation rates are high due to higher driving forces, but the crystallite growth is kinetically limited, leading to fine-grained microstructures. The time-resolved DTEM image in Figure 3.14b, taken at a delay of 6 μs, displays a direct measurement of this C-curve behavior, where the crystallization fronts are propagating inwards and outwards and there is radial

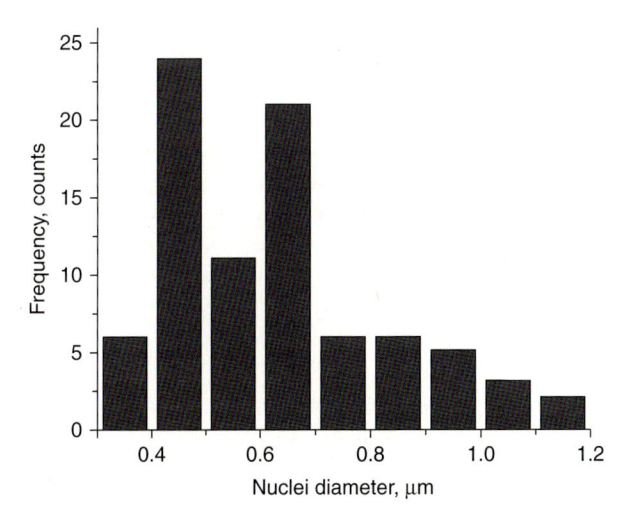

Figure 3.13 Plot of the number density of nuclei as a function of the nuclei size.

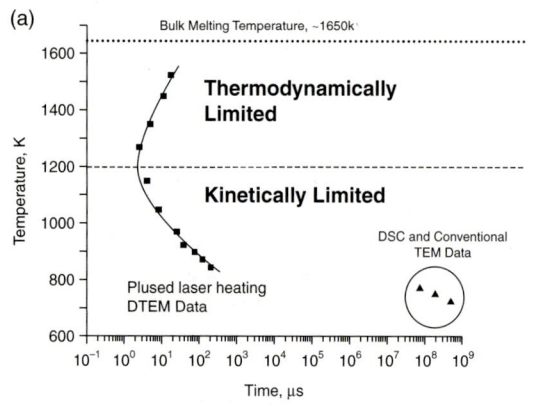

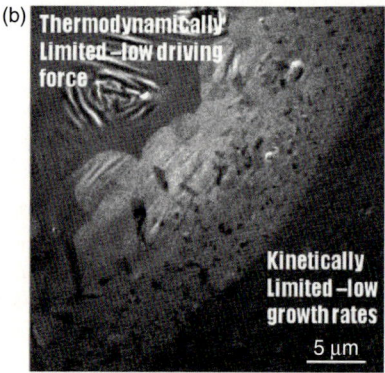

Figure 3.14 (a) Plot of the 90% crystallized volume as a function of temperature and time. Note the classical C-curve behavior for diffusional phase transformation; (b) A 15 ns exposure of the crystallization taken 6 μs after the laser pulse had struck the foil. The center of the gaussian laser spot is located in the upper left-hand corner.

variation in grain size. The direct measurement of nucleation and growth rates as a function of temperature allows the prediction and control of grain size, which is necessary when fabricating devices. By using DTEM, rapid nucleation and growth phenomena can be observed directly and quantified with nanosecond time resolution. Moreover, with this high time resolution it is possible to gain insight into the subtle details of materials processes in previously unexplored regimes, such as crystallization processes at high temperatures.

3.5
Space Charge Effects in Single-Shot DTEM

As described in Section 3.2.2, the large number of electrons ($>10^9$) generated in a 15 ns pulse results in very large peak currents (1 mA) and charge densities, especially at strong lens cross-overs and at low accelerating voltages. The close proximity of electrons in space and time results in coulombic interactions that can greatly affect both the spatial and temporal resolution. The effects of coulombic repulsion can be divided into two broad categories, namely global space charge and stochastic blurring.

3.5.1
Global Space Charge

Global space charge (GSC) involves an overall expansion of the electron bunch although, as GSC expansion is a large-scale effect it is more easily modeled than stochastic blurring, and hence is used widely among the particle accelerator community [58–60]. The expansion of the electron pulse lengthens the pulse as it propagates through the system, which in turn will limit time resolution to the picosecond range for a 200 kV pulse. The lateral broadening due to GSC serves as a perfect negative electron lens, and may be compensated – at least in principle – by refocusing the beam with standard electron optics. Nonetheless, as the pulse distribution will generally be nonuniform, the effective GSC lens will be aberrated, although coupling such aberrations with the convergence angles required to achieve a high dose on the specimen may result in a reduced spatial resolution [4, 26, 28, 32, 60]. The effects of global space charge are less severe for pulses of nanoseconds and are longer, but can be compensated with an appropriate gun and condenser system design that will also optimize the electron generation and collection. Accelerating voltages of MeV magnitude may also reduce such effects, as the electrons will be more spread out in space. GSC is not considered to be a limiting factor in the spatial resolution; rather, randomization of the specimen information due to a stochastic scattering of the electrons would be the dominant effect limiting spatial resolution at high time resolution.

3.5.2
Stochastic Blurring

Stochastic blurring is the irreversible loss of high-resolution spatial information encoded in an electron pulse as a result of random electron scattering events [28, 32, 60, 61]. These effects are most prominent at the objective and projector lens cross-overs. By using multiparticle simulation codes to model the coulombic forces and electron scattering in a TEM lens system, Armstrong *et al.* [32] suggested that MeV electron voltages would be required to obtain both high temporal and spatial resolution. For example, with a 10 ps electron pulse the spatial resolution would be limited to about 10 nm, which would still provide sufficient resolution to image defects such as dislocations. In addition to using relativistic electron beams (MeV range), the use of techniques such as annular dark field (ADF) imaging can also mitigate stochastic blurring effects. When employing ADF imaging, only the diffracted beams at a particular angular range (as defined by the annulus diameter and the width of the objective aperture) are used to form the image. The strong, low-angle scatter beams are omitted, which in DF imaging, do not add to the contrast. This reduces not only the charge densities at the strong objective and projector lenses but also the resulting stochastic effects, thus improving the spatial resolution. Ultimately, techniques such as ADF and relativistic electron beams will be required to increase single-shot DTEM time resolution to the picosecond range.

3.6
Next-Generation DTEM

3.6.1
Novel Electron Sources

The brightness of the Ta photocathode source is $\sim 10^8 \, \text{A cm}^{-2} \, \text{steradian}^{-1}$, although to achieve a higher spatial resolution than 10 nm and produce high-quality BF images the gun brightness must be improved. Photoemission can achieve a gun brightness in excess of 10^9 A cm^{-2} steradian^{-1}, while the photocathode brightness can be substantially improved by optimizing the source material, the laser parameters and the high-voltage accelerator design, the aim being to reduce thermal emittance growth and increase the photoemitted electron current. Another essential element in the source design is to appropriately match the laser photon energy to the cathode work function; this would minimize the thermal energy of the photoemitted electrons, and also the associated thermal emittance growth that reduces gun brightness [4, 28, 32, 58, 60]. Thus, the optimal source material and laser parameters must be selected based on the best compromise between quantum efficiency and thermal emittance growth effects. In practice, the range of available photocathode materials and laser wavelengths facilitates this optimization, but to fully optimize the system and the electron currents both the gun and accelerator section must be modified to increase the electric field at the cathode surface.

The Child–Langmuir effect limits the current that may be emitted from a cathode, as a function of the cathode–anode separation and the potential difference. In general, the main approaches for increasing current in future DTEM designs are the same as have been developed for other high-brightness electron sources [62–65], namely to increase the electric field at the surface and to decrease the acceleration gap. The electron accelerator of modern TEM systems are designed in such a way as to provide for a smooth acceleration of the beam; this will reduce the electron energy spread and increase the spatial and spectroscopic imaging resolution, such that the electric fields are small. Furthermore, by design the Wehnelt assembly shields the cathode from high fields, which means the electron gun must be shortened and modified in order to produce higher extraction fields close to the cathode, such as extraction electrodes used by field emission gun sources. An alternate solution would be to use radiofrequency (RF)-based photoguns that generate high electric fields (GeV cm^{-1}) and have a high brightness [62, 66–69]. These guns also produce MeV electrons, which is necessary for the reduction of space charge effects; other suggested gun designs might include the use of a cooled gas to emit an ultracold gas of electrons [59, 70]. Ultimately, however, the electron gun must generate both a high peak current and a high beam quality to obtain the brightness required for sub-nanosecond, single-shot images [60, 71].

3.6.2
Relativistic Beams

An alternate method for achieving a high time resolution and to reduce any space charge effects would be to increase the accelerating voltage to the MeV range. In fact, by operating in the MeV range and employing large accelerating fields (ca. 10^8 V m^{-1}) in an RF cavity, a sub-picosecond temporal resolution can be achieved with 10^8 to 10^{10} electrons per pulse (see Refs [32, 60] for more detail). This capability would enable the observation of very fast processes in materials, an example being the irreversible phase transformation induced by high-pressure impulse loading. To operate in the MeV range would also have the benefit of greatly increasing the maximum sample thickness (in transmission), from the tens of nanometers attainable with current instruments to the micron scale.

Pulsed RF guns have been used for many years as electron sources for synchrotrons, and are able to generate extremely bright, pulsed electron beams [62, 66–69]. Such bright beams are generated due to the high RF fields (typically 100 MV m^{-1} or more) that are used to accelerate them over short distances [68]. Subsequently, the beams become relativistic within a few centimeters, such that the space-charge force repulsion is quickly reduced and the electrons are typically accelerated to about 5–6 MeV ranges, depending on the RF amplitude employed and the electron beam launch phase. In principle, electron optics can be engineered to operate with such a gun and electron energy ranges; however, MeV DTEM may more closely resemble an accelerator than a high-voltage electron microscope.

3.6.3
Pulse Compression

At present, DTEM time resolution is set by the electron pulse duration, which is essentially equal to the cathode laser pulse duration, and is dictated by the experiment's fluence requirements and the electron gun brightness. The size, shape, and temporal resolution of the electron pulse impacting the specimen is a function of the broadening and de-coherence effects in the column. In next-generation DTEM, an RF-cavity will conceivably be incorporated into the column and used to shape the pulse by applying a field; this might speed up the electrons at the back of the pulse but slow down those at the front, thus shortening the overall pulse duration. It would also provide the flexibility to shape the pulse in more ways than simply longitudinal compression. Today, RF cavity compression devices are often used in acceleration facilities to compress multi-MeV pulses of nano-Coulomb magnitude down to picosecond durations while maintaining normalized emittance values of about 1 μm-radian and an energy spread less than about 1% [72–76]. The RF cavity is most likely to be located between the condenser lens and the objective lens; this would allow for the shortest distance (time) after the pulse shaping until interaction with the specimen, while still permitting control of the pulse with a final lens. In fact, pulse compression may be the sole means of digging deeply into the sub-nanosecond regime and, combined with MeV accelerating voltages, may ultimately enable picosecond-nanometer resolution single-shot imaging. Without increasing the effective brightness of the electron source by crowding the electrons into a shorter pulse, however, many measurements that require a combination of high fluence, small convergence angle, and high time resolution will simply be impossible.

3.6.4
Aberration Correction

To further maximize the current and the spatial coherence in the illumination on the specimen, the pre-specimen objective lens should be spherical aberration-corrected.

Today, aberration correctors are used several commercial instruments, and studies are also under way to evaluate their use with short electron pulses. Correcting for spherical aberration appears to reduce any incoherent broadening of the contrast transfer function, and to accentuate phase contrast from small signals, which makes them attractive for liquid cell biological imaging. It would also allow the use of larger aperture and convergence angles at the specimen, thereby increasing the dose without any significant loss in beam coherence.

To ensure that the maximum contrast is maintained when the beam has interacted with the specimen, the post-specimen objective lens should be aberration-corrected. In this case, the lens would be corrected for both spherical and chromatic aberrations, with the spherical aberration correction having the same advantages as for the pre-specimen lens. Additionally, the chromatic aberration correction should reduce the energy spread associated with photoelectron generation and the global space charge effect, which would in turn enable high-resolution images to be recorded with even a large, initial energy spread. Currently, dynamic transmission electron microscopes have an energy spread in the beam of about 8 to 10 eV although, by using the RF cavity for pulse compression, this energy spread could be increased by a factor of ten. Although chromatic aberration correctors have been developed for conventional TEM use, the current designs should be able to cope with energy spreads of up to $\sim$100 eV, which would make them viable for all DTEM applications.

3.7
Conclusions

The dynamic transmission electron microscope provides a fundamentally new capability for the characterization of dynamics at the nanoscale, notably by offering a many orders of magnitude increase in time resolution over conventional electron microscopy. During its development and subsequent operation, DTEM has made fundamental contributions to the fields of martensitic transformations, rapid chemical reactions in multilayer thin films, and crystallization processes occurring under extreme driving forces. Clearly, the technique holds great promise for performing "*in-situ*" experiments, and many more exciting developments are anticipated in this field as the technique comes to maturity.

Acknowledgments

The authors thank Richard Shuttlesworth, Glenn Huete and Benjamin Pyke for their excellent technical assistance. These studies were conducted under the auspices of the US Department of Energy, Office of Basic Energy Sciences, Division of Materials Sciences and Engineering by Lawrence Livermore National Laboratory under contract DE-AC52-07NA27344.

References

1. Batson, P.E., Dellby, N., and Krivanek, O.L. (2002) *Nature*, **419**, 94–94.
2. Urban, K.W. (2007) *MRS Bull.*, **32**, 946–952.
3. Walther, T., Quandt, E., Stegmann, H., Thesen, A., and Benner, G. (2006) *Ultramicroscopy*, **106**, 963–969.
4. King, W.E., Campbell, G.H., Frank, A., Reed, B., Schmerge, J.F., Siwick, B.J., Stuart, B.C., and Weber, P.M. (2005) *J. Appl. Phys.*, **97**, 111101.
5. LaGrange, T., Campbell, G.H., Turchi, P.E.A., and King, W.E. (2007) *Acta Mater.*, **55**, 5211–5224.
6. Miller, R.J.D. (2004) Femtosecond Electron Diffraction: Making the Molecular Movie, in First National Lab and University Alliance Workshop on Ultrafast Electron Microscopies, Pleasanton, CA.
7. Williamson, S. and Mourou, G. (1982) *Appl. Physics B*, **28**, 249–250.
8. Mourou, G. and Williamson, S. (1982) *Appl. Phys. Lett.*, **41**, 44–45.
9. Ruan, C.Y., Vigliotti, F., Lobastov, V.A., Chen, S.Y., and Zewail, A.H. (2004) *Proc. Natl Acad. Sci. USA*, **101**, 1123–1128.
10. Weber, P. (2004) Ultrafast electron diffraction in Chemistry, in First National Lab and University Alliance Workshop on Ultrafast Electron Microscopies, Pleasanton, CA.
11. Elsayed-Ali, H.E. and Herman, J.W. (1990) *Appl. Phys. Lett.*, **57**, 1508–1510.
12. Zewail, A.H. (2000) *J. Phys. Chem. A*, **104**, 5660–5694.
13. Ischenko, A.A., Ewbank, J.D., Lobastov, V.A., and Schäfer, L. (1995) Stroboscopic gas electron diffraction: a tool for structural kinetic studies of laser excited molecules, in: *Time-resolved electron and X-ray diffraction: 13–14 July, 1995, San Diego, California* (ed. P.M. Rentzepis), SPIE-Society of Photo-Optical Instrumentation Engineers, Bellingham, Washington, USA, pp. ix.
14. Williamson, J.C., Dantus, M., Kim, S.B., and Zewail, A.H. (1992) *Chem. Phys. Lett.*, **196**, 529–534.
15. Cao, J. (2004) Electrons as an ultrafast probe: femtosecond electron diffraction, in First National Lab and University Alliance Workshop on Ultrafast Electron Microscopies, 2003, Pleasanton, CA.
16. Fleming, G.R. (1986) *Chemical Applications of Ultrafast Spectroscopy*, Oxford University Press; Clarendon Press, New York; Oxford, p. ix.
17. Lobastov, V.A., Srinivasan, R., and Zewail, A.H. (2005) *Proc. Natl Acad. Sci. USA*, **102**, 7069–7073.
18. Zewail, A.H. (2005) *Philos. Trans. R. Soc. A*, **363**, 315–329.
19. Spivak, G.V., Pavlyuchenko, O.P., and Petrov, V.I. (1966) *Bull. Acad. Sciences, USSR, Phys. Series*, **30**, 822–826.
20. Bostanjoglo, O. (2002) *Adv. Imag. Elect. Phys.*, **121**, 1–51.
21. Bostanjoglo, O., Elschner, R., Mao, Z., Nink, T., and Weingartner, M. (2000) *Ultramicroscopy*, **81**, 141–147.
22. Bostanjoglo, O., Endruschat, E., Heinricht, F., Tornow, R.P., and Tornow, W. (1987) *Eur. J. Cell Biol.*, **44**, 10–10.
23. Bostanjoglo, O., and Kornitzky, J. (1990) Nanoseconds double-frame and streak transmission electron microscopy. in: Proceedings, 12th International Congress of Electron Microscopy, San Francisco, CA, p. 180.
24. Bostanjoglo, O. and Rosin, T. (1976) *Mikroskopie*, **32**, 190–190.
25. Bostanjoglo, O., Tornow, R.P., and Tornow, W. (1987) *J. Phys. E. Sci. Instrum.*, **20**, 556–557.
26. LaGrange, T., Armstrong, M.R., Boyden, K., Brown, C.G., Campbell, G.H., Colvin, J.D., DeHope, W.J., Frank, A.M., Gibson, D.J., Hartemann, F.V., Kim, J.S., King, W.E., Pyke, B.J., Reed, B.W., Shirk, M.D., Shuttlesworth, R.M., Stuart, B.C., Torralva, B.R., and Browning, N.D. (2006) *Appl. Phys. Lett.*, **89**, 044105.
27. King, W.E., Armstrong, M., Malka, V., Reed, B.W., and Rousse, A. (2006) *MRS Bull.*, **31**, 614–619.

28 Armstrong, M.R., Boyden, K., Browning, N.D., Campbell, G.H., Colvin, J.D., DeHope, W.J., Frank, A.M., Gibson, D.J., Hartemann, F., Kim, J.S., King, W.E., LaGrange, T.B., Pyke, B.J., Reed, B.W., Shuttlesworth, R.M., Stuart, B.C., and Torralva, B.R. (2007) *Ultramicroscopy*, **107**, 356–367.

29 Reed, B.W., LaGrange, T., Shuttlesworth, R.M., Gibson, D.J., Campbell, G.H., and Browning, N.D. (2010) *Rev. Sci. Instrum.*, **81**, 053706.

30 Domer, H. and Bostanjoglo, O. (2003) *Rev. Sci. Instrum.*, **74**, 4369–4372.

31 Rose, A. (1948) Television pickup tubes and the problem of vision, in *Advances in Electronics and Electron Physics*, Academic Press, New York, pp. 131–166.

32 Armstrong, M.R., Reed, B.W., Torralva, B.R., and Browning, N.D. (2007) *Appl. Phys. Lett.*, **90**, 114101.

33 Raghaven, V. (1992) *Martensite: A Tribute to Morris Cohen*, AMS International, Materials Park.

34 Lütjering, G. and Williams, J.C. (2003) *Titanium*, Springer-Verlag, Berlin, Heidelberg, New York.

35 LaGrange, T., Campbell, G.H., Colvin, J.D., King, W.E., Browning, N.D., Armstrong, M.A., Reed, B.W., Kim, J.S., and Stuart, B.C. (2005) In-*situ* studies of the martensitic transformation in pure Ti thin films using the dynamic transmission electron microscope (DTEM), in: *Materials Research Society Fall Meeting 2005*, vol. **907E** (eds P.J. Ferreira, I.M. Robertson, G. Dehm, and H. Saka), MRS, Boston, pp. 0907-MM05-02 to 0907-MM05-10.

36 LaGrange, T., Campbell, G.H., Colvin, J.D., Reed, B., and King, W.E. (2006) *J. Mater. Sci.*, **41**, 4440–4444.

37 Campbell, G.H., LaGrange, T.B., King, W.E., Colvin, J.D., Ziegler, A., Browning, N.D., Kleinschmidt, H., and Bostanjoglo, O. (2005) The HCP to BCC phase transformation in Ti characterized by nanosecond electron microscopy, in *Solid-Solid Phase Transformations in Inorganic Material*, The Minerals, Metals and Materials Society, vol. **2**, pp. 443–448.

38 Rietveld, H.M. (1969) *J. Appl. Crystallogr.*, **2**, 65.

39 Grujicic, M., Olson, G.B., and Owen, W.S. (1985) *Metall. Trans. A*, **16**, 1713–1722.

40 Pati, S.R. and Cohen, M. (1969) *Acta Metall.*, **17**, 189.

41 Pati, S.R. and Cohen, M. (1971) *Acta Metall.*, **19**, 1327.

42 Kim, J.S., LaGrange, T., Reed, B.W., Taheri, M.L., Armstrong, M.R., King, W.E., Browning, N.D., and Campbell, G.H. (2008) *Science*, **321**, 1472–1475.

43 Kim, J.S., Reed, B.W., Browning, N.D., and Campbell, G.H. (2006) *Microsc. Microanal.*, **12** (Suppl. 2), 148–149.

44 Besnoin, E., Cerutti, S., Knio, O.M., and Weihs, T.P. (2002) *J. Appl. Phys.*, **92**, 5474–5481.

45 Michaelsen, C., Barmak, K., and Weihs, T.P. (1997) *J. Phys. D Appl. Phys.*, **30**, 3167–3186.

46 Gavens, A.J., Van Heerden, D., Mann, A.B., Reiss, M.E., and Weihs, T.P. (2000) *J. Appl. Phys.*, **87**, 1255–1263.

47 Mann, A.B., Gavens, A.J., Reiss, M.E., Van Heerden, D., Bao, G., and Weihs, T.P. (1997) *J. Appl. Phys.*, **82**, 1178–1188.

48 Wang, J., Besnoin, E., Knio, O.M., and Weihs, T.P. (2005) *J. Appl. Phys.*, **97**, 114307.

49 Wang, J., Besnoin, E., Duckham, A., Spey, S.J., Reiss, M.E., Knio, O.M., and Weihs, T.P. (2004) *J. Appl. Phys.*, **95**, 248–256.

50 Porter, D.A., and Easterling, K.E. (1981) *Phase Transformations in Metals and Alloys*, Van Nostrand Reinhold Co., New York.

51 Chen, J.Z. and Wu, S.K. (2001) *J. Non.-Cryst. Solids*, **288**, 159–165.

52 Kim, J.J., Moine, P., and Stevenson, D.A. (1986) *Scr. Math.*, **20**, 243–248.

53 Vestel, M.J., Grummon, D.S., Gronsky, R., and Pisano, A.P. (2003) *Acta Mater.*, **51**, 5309–5318.

54 Wang, X. and Vlassak, J.J. (2006) *Scr. Math.*, **54**, 925–930.

55 Avrami, M. (1939) *J. Chem. Phys.*, **7**, 1103–1112.

56 Johnson, W.A. and Mehl, R.F. (1939) *Trans. Am. Inst. Min. Metall. Eng.*, **135**, 416–442.

57 LaGrange, T., Grummon, D.S., Reed, B.W., Browning, N.D., King, W.E., and Campbell, G.H. (2009) *Appl. Phys. Lett.*, **94**, 184101.

58 Reed, B. (2004) Propagation dynamics of femtosecond electron packets and relativistic effects in ultrafast electron microscopy (UEM), in First National Lab and University Alliance Workshop on Ultrafast Electron Microscopies, Pleasanton, CA.

59 Luiten, O.J., van der Geer, S.B., de Loos, M.J., Kiewiet, F.B., and van der Wiel, M.J. (2004) *Phys. Rev. Lett.*, **93**, 094802.

60 Reed, B.W. (2006) *J. Appl. Phys.*, **100**, 034916.

61 Jansen, G.H. (1990) *Coulomb Interactions in Particle Beams*, Academic, San Diego, CA, p. 546.

62 Bluem, H., Todd, A.M.M., Cole, M.D., Rathke, J., and Schultheiss, T. (2003) *Nucl. Instrum. Methods Phys. Res., Sect. A*, **507**, 215–219.

63 Dowell, D.H., Bolton, P.R., Clendenin, J.E., Emma, P., Gierman, S.M., Graves, W.S., Limborg, C.G., Murphy, B.F., and Schmerge, J.F. (2003) *Nucl. Instrum. Methods Phys. Res., Sect. A*, **507**, 327–330.

64 Merano, M., Collin, S., Renucci, P., Gatri, M., Sonderegger, S., Crottini, A., Ganiere, J.D., and Deveaud, B. (2005) *Rev. Sci. Instrum.*, **76**, 085108.

65 Wishart, J.F., Cook, A.R., and Miller, J.R. (2004) *Rev. Sci. Instrum.*, **75**, 4359–4366.

66 Bolton, P.R., Clendenin, J.E., Dowell, D.H., Ferrario, M., Fisher, A.S., Gierman, S.M., Kirby, R.E., Krejcik, P., Limborg, C.G., Mulhollan, G.A., Nguyen, D., Palmer, D.T., Rosenzweig, J.B., Schmerge, J.F., Serafini, L., and Wang, X.J. (2002) *Nucl. Instrum. Methods Phys. Res., Sect. A*, **483**, 296–300.

67 Serafini, L. (1996) *IEEE Trans. Plasma Sci.*, **24**, 421–427.

68 Travier, C. (1994) *Nucl. Instrum. Methods Phys. Res., Sect. A*, **340**, 26–39.

69 Schmerge, J., Dowell, D., and Hastings, J. (2004) Measured RF gun parameters at the SLAC Gun Test Facility, in First National Lab and University Alliance Workshop on Ultrafast Electron Microscopies, Pleasanton, CA.

70 Claessens, B.J., van der Geer, S.B., Taban, G., Vredenbregt, E.J.D., and Luiten, O.J. (2005) *Phys. Rev. Lett.*, **95**, 164801.

71 Lewellen, J.W. (2004) Proceedings of LINAC 2004, Lübeck, Germany, pp. 842–846.

72 Anderson, S.G., Rosenzweig, J.B., Musumeci, P., and Thompson, M.C. (2003) *Phys. Rev. Lett.*, **91**, 072302.

73 Qian, B.L. and Elsayed-Ali, H.E. (2002) *Phys. Rev. E*, **65**, 046502.

74 Qin, H. and Davidson, R.C. (2002) *Phys. Rev. Spec. Top. Accel. Beams*, **5**, 034401.

75 Dowell, D.H., Adamski, J.L., Hayward, T.D., Johnson, P.E., Parazzoli, C.D., and Vetter, A.M. (1997) *Nucl. Instrum. Methods Phys. Res., Sect. A*, **393**, 184–187.

76 Schroeder, W.A. (2004) Pulse Compression for UEM. in: First National Lab and University Alliance Workshop on Ultrafast Electron Microscopies, Pleasanton, CA.

4
Formation of Surface Patterns Observed with Reflection Electron Microscopy
Alexander V. Latyshev

4.1
Introduction

It is generally accepted that the evolution of a crystal surface is determined by the migration of adsorbed atoms onto its surface. This explains the increasing urge to acquire an understanding of the elementary acts that occur on a crystal surface during sublimation and epitaxial growth. Recently, much effort has been expended on studies of the kinetics of adsorbed atoms and the mechanisms of their interaction with the surface, mainly because of the associated technological aspects of crystal growth. Today, a need to develop the technology of nanoelectronics demands the fabrication of thin films and of both homointerfaces and heterointerfaces with well-defined compositions, structures, levels of impurity doping, and perfect qualitative properties for the structures of the interface. This quest has been stimulated by a tendency for solid-state electronics to be directed towards an increased integration, and a transition to quantum size phenomena. The optimal conditions for the growth of two-dimensional (2-D) crystals are provided by molecular beam epitaxy (MBE) methods, which include film growth from exceptionally clean matter under ultrahigh-vacuum (UHV) conditions with a high degree of control of not only the substrate temperature but also the rate of deposition and the composition of the growing films. This allows both sublayer and monolayer coverage of the substrate to be obtained, under full automatic control of the growth processes. The main problems encountered with MBE relate to an understanding of the elementary processes that occur on the crystal surface during sublimation, epitaxial growth, and phase transitions. Despite extensive structural investigations of atomically clean semiconductor surfaces having been conducted, there remains a distinct lack of detailed information regarding the micromorphology, the kinetics of adatoms, and the mechanisms of their interaction with surface sinks. This is due mainly to problems of visualizing the elementary (atomic) acts with a high spatial resolution at elevated temperatures, despite vast progress having been made in this field during the past years. Difficulties persist in the solution of problems in application, including the preparation of a perfect surface-substrate with special required surface profiles, and an evaluation of

In-situ Electron Microscopy: Applications in Physics, Chemistry and Materials Science, First Edition.
Edited by Gerhard Dehm, James M. Howe, and Josef Zweck.
© 2012 Wiley-VCH Verlag GmbH & Co. KGaA. Published 2012 by Wiley-VCH Verlag GmbH & Co. KGaA.

the data related to the kinetics of surface rearrangements that is essential in order to control the density of the surface sinks for adatoms, and to determine the mechanism of thin film growth. In addition, problems persist with regards to the nature of surface reconstruction and its role in the atomic processes at the surface.

Until about 10 years ago, by far the best means of investigating surface structures and their structural reorganizations was that of electron microscopy [1]. Indeed, the development of numerous modifications of the technique, including scanning tunnel microscopy (STM) and atomic force microscopy (AFM), has provided – at least potentially – an optimum approach to the study of surface structures [2, 3]. It is important to appreciate, however, that neither STM nor AFM can provide a high spatial resolution; rather, alternative methods must be sought that demonstrate a much greater ability to perform experiments *in-situ* [4]. In fact, when analyzing the benefits of structural methods used for the direct visualization of the surface morphology of crystals, the likelihood is that electron microscopy techniques employing UHV conditions will be required, in addition to atomically clean surfaces.

The development of methods that allow the visualization of surface structures will, in turn, also permit the study of the kinetics of surface processes with a high spatial resolution, sufficient to image single monatomic steps and superstructure domains. One essential condition that must be maintained when studying an atomically clean surface is the ability to clean that surface, and to preserve such cleanliness while the experiments are conducted. This condition is necessary because, when exposed to a gaseous atmosphere, a surface will adsorb extrinsic atoms that will, in turn, change not only its composition but also its atomic surface structure. However, the influence that an adsorbed layer can exert on atomic transformations will be reduced if the pressure of the remaining gas atmosphere can be minimized. Consequently, the analysis of atomically clean surfaces has only become possible following the development of pumps that allow a reduction in the pressure of the residual atmosphere, to less than 10^{-8} Pa.

The structural diagnostic methods employed should meet the following requirements:

- A high spatial resolution (equal to or less than the size of the interplane spacings).
- An UHV condition around the sample (a residual atmosphere below 10^{-8} Pa).
- The possibility to perform *in-situ* experiments on heating, cooling, epitaxy, adsorption, deformation, gas exposure, and so on.
- The real-time registration of dynamic processes.

In addition, the methods applied should whenever possible be nondestructive; in particular, any influence that the probe beam might have on any surface processes should, ideally, be negligible.

From this point of view, the methods of structural analysis that employ an electron beam as the analyzing probe have demonstrated the most promise for studying structural phenomena on a surface. Among such well-known techniques can be included reflection high-energy electron diffraction (RHEED) and low-energy electron diffraction (LEED), both of which allow the study of elements located periodically on a surface.

Unfortunately, the field of application of both RHEED and LEED is limited by several factors:

- The information acquired using these methods is limited by the lateral coherence length of the electron beam; in other words, any periodicity which extends over more than some hundreds of Ångstroms will not be displayed on diffraction patterns.
- The data provided from diffraction patterns is averaged over a large area of the sample surface, so that the spatial resolution – which is defined by the size of the electron beam (ca. 1 μm diameter) – does not permit the study of local deviations in periodicity.
- The absence of an adequate dynamic theory of diffraction of electrons does not allow an unequivocal comparison between the calculated and experimentally measured intensities of diffracted electron beams.
- Neither RHEED nor LEED allow images of single surface features to be obtained directly with a sufficiently high spatial resolution; this is linked to an absence of any optical system to form an image.

However, for the first time a direct visualization of the structure of a crystal surface containing a system of atomic steps, became possible by the application of transmission electron microscopy (TEM), which allowed surfaces to be "decorated" by noble metals, with the subsequent preparation of a carbon replica [5]. The high sensitivity of the method used to decorate the structural heterogeneities of a surface allowed the direct imaging of such features on an ionic crystal surface that included atomic steps that were only one or two interplanar distances high, as well as dislocations with a screw component that emerged on a surface. In addition, the technique allowed data to be obtained on the behavior of atomic steps on a surface of sodium chloride during sublimation, dissolution, and epitaxial growth [6]. A system of steps on a silicon(111) surface, presumably with a height of only one interplanar distance, was revealed using low-voltage scanning electron microscopy (SEM), following decoration of the surface with metals [7] and particles of contamination from the residual gas atmosphere within the electron microscope [8]. The use of forbidden reflections or weak beams has allowed the imaging of atomic steps in the electron microscope [9, 10]. Recent progress with TEM has led to the possibility of resolving separate atomic columns [11], while its application to studies on cross-sections has allowed the analysis of images of atomic structures on a surface [12–14]. Problems in creating UHV conditions within the column of the electron microscope has limited the application of methods for studying structural processes on the atomically clean surfaces of semiconductors [15]. However, an improvement in UHV conditions around a sample in the column of a transmission electron microscope, by using differential pumping devices, was reported by various groups, including Takayanagi et al. [16], Poppa et al. [17, 18], Gibson et al. [19], Lehmpfuhl [20], Yagi [21], Metois [22], and others [23–25]. Visualization of the structure of atomic steps of silicon and germanium under a layer of the elements' own amorphous oxide, has also been demonstrated using high-resolution TEM (HRTEM) [26].

To investigate the atomically clean surfaces of metals and semiconductors, the most effective method is ultrahigh-vacuum reflection electron microscopy (UHV-REM), especially when the experiments are to be conducted *in-situ*. Moreover, the UHV-REM system can be built by using a commercially available TEM set-up as a base, with modifications to the vacuum system and other column components, and using an additional set of electromagnetic coils to tilt the electron beam. In UHV-REM, the contrast is created by the electrons being diffracted from a surface; hence, the technique will also allow studies to be made of the periodic distribution of atoms and crystallographic parameters of a surface using RHEED. However, the main advantage of UHV-REM over other surface-sensitive techniques is an ability to visualize surface elements such as monatomic steps, dislocations, 2-D islands and other structural defects [27, 28]. Another important point is that UHV-REM allows the investigation of the kinetics of structural reorganizations on surfaces, but with a variety of external influences (e.g., thermal annealing, epitaxial growth, adsorption, sublimation).

4.2
Reflection Electron Microscopy

The concept of REM was first proposed by Ernst Ruska, who had earlier received the Nobel prize for inventing the electron microscope in 1933 [29]. Unfortunately, however, the development of scanning (raster) electron microscopy as a replica of TEM (but with similar or better spatial resolution) meant that REM was not adopted worldwide. This was because, at the time, the basic type of contrast in REM was simply a "shadow" caused by the roughness of a surface that usually was covered by a layer of natural oxide and contamination, and which led to a much lower capacity of REM compared to other methods. Subsequently, REM was further developed by applying a Bragg (diffraction) contrast for image formation from a crystal surface, rather than using diffusively scattered electrons [30]. The basic description of contrast formation from atomic steps and dislocations was based on the fundamental theory of dynamic scattering of electrons on a surface atomic lattice. However, significant progress in REM was made only when UHV conditions were applied within the microscope [31].

Consideration should be given to the differences between electron beam paths in TEM and REM. Whereas, in regular TEM the electron beam strikes the specimen at normal incidence, the situation is different in REM. In the latter case the electron beam generated by the microscope's optical system is at an angle $\theta + \varphi$ with respect to the optical axis; this is achieved by using electromagnetic coils that cause the beam to strike the crystal surface at angle φ with respect to the optical axis (Figure 4.1). Typically, the size of the incident angle θ is only a few degrees, in order to restrict the depth of electron penetration into the bulk of the crystal, whereas angle φ has the magnitude of a Bragg angle.

As in TEM, one or several diffracted electron beams $n(hkl)$ are used to form the image in REM, where n is the order of the reflections transmitted through an aperture

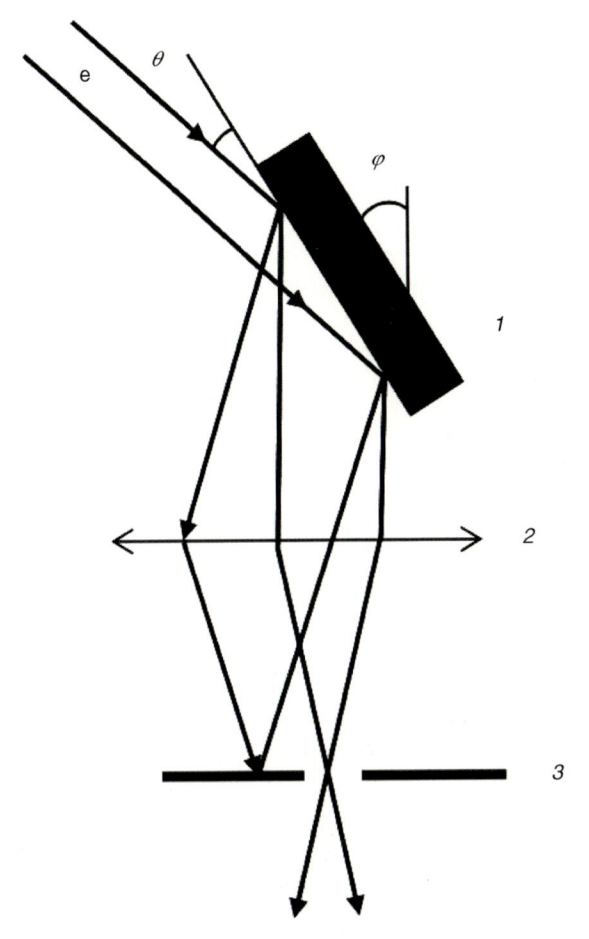

Figure 4.1 The schematic representation of rays of electron beam in reflection electron microscopy. 1, crystal sample; 2, electromagnetic lens; 3, diaphragm.

of the microscope. The intensity of the diffracted beams is defined by the structure of few atomic planes (hkl) close to the surface of the crystal under investigation. This provides an explanation for the high sensitivity of REM when determining the structure of the surface layers. In REM mode, the images of a surface will always be dark-field, albeit with a high contrast.

The REM images demonstrate three main features:

- The magnification scale in a direction parallel to the optical axis of the microscope is reduced compared to the scale perpendicular to the axis: $M_{\parallel} = M_{\perp} \sin \theta$. The value of $M_{\perp}$ is defined by parameters of the electron-optical system of the microscope, while the angle θ depends on the order of the reflection used for image formation. Due to these different directionally dependent magnifications in the REM image, the true geometric shape of a surface will be distorted by a

"foreshortening" effect along the direction of the incident electron beam by a factor of $M_{||}/M_{\perp} = 1/30 \sim 1/50$.
- Only a narrow part of the image along a line perpendicular to the optical axis will be precisely focused; regions farther away from this line will be increasingly out-of-focus.
- The extreme sensitivity of REM to the structure of thin surface layers leads to an absolute requirement to prepare and preserve of an atomically clean surface of the semiconductors to be investigated in the electron microscope column. Failure to do this will cause the contrast in the REM image to be defined almost completely by a film of natural oxide and the contamination that is always present on a real crystal surface. This requires the pressure within the microscope column to be in the region of $10^{-8} \sim 10^{-6}$ Pa.

Following an initial report by the Tokyo Institute of Technology group of exceptional results obtained with REM [16], a commercially available transmission electron microscope (JEM-7A) with an accelerating voltage of up to 100 kV was used to create a UHV-REM system at the Institute of Semiconductor Physics, Russia (Figure 4.2). Typically, a reduction of the residual pressure in the electron microscope column is complicated by a need to include multiple mobile O-rings that are lubricated with an oil. In order to improve the vacuum conditions in the column, an additional nitrogen trap was inserted into the central vacuum pipe of the microscope, while a technical

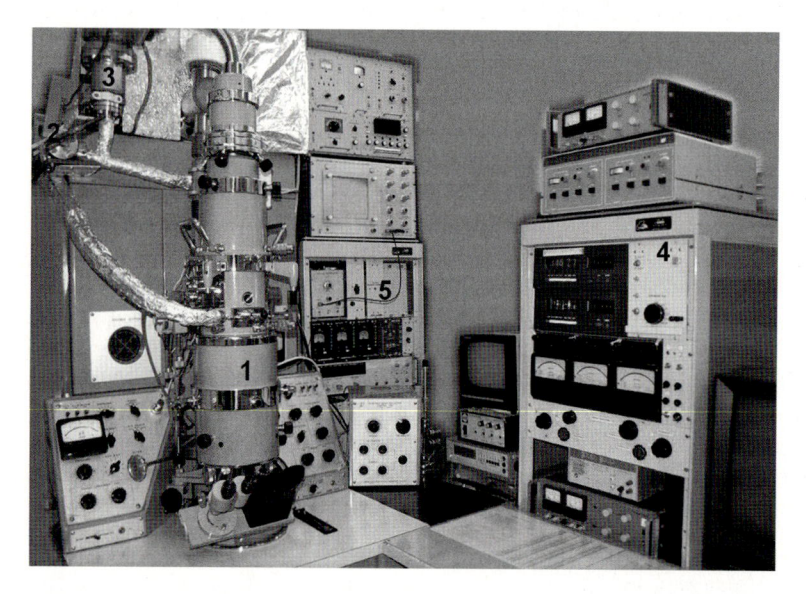

Figure 4.2 An ultrahigh vacuum reflection electron microscope, based on a commercial JEM-7A (JEOL) instrument, that has been home-modified for the *in-situ* investigation of structural transformations on atomically clean surfaces. 1, microscope column; 2, adsorption pump; 3, mass spectrometer; 4, control cabinet for sample and evaporators; 5, electric vault for vacuum control and intensity electron beam units.

upgrade of the standard pumping system provided much-improved vacuum conditions and helped to stabilize the microscope's crucial working parameters. These new vacuum conditions not only improved the stability of the accelerating high voltage but also prolonged (several-fold) the lifetime of the tungsten V-shaped cathode of the electron gun, despite the microscope being used intensively. Despite the above-described modifications, however, UHV conditions could still not be achieved in the sample chamber.

In order to achieve a vacuum of 10^{-6} Pa and better, a differential cryogenic pumping device was applied to the quasi-closed chamber around the sample (as described in Refs [32–34]). In this case, the device was developed to fit a transmission electron microscope with a top-entry goniometer of the sample [35, 36]. The main advantage of a device with differential cryogenic pumping is the high rate of gas evacuation and the avoidance of any magnetic and/or electric fields that might adversely affect the electron beam of the microscope. For a conventional UHV chamber, the system would involve thermal annealing of the internal armatures and the chamber housing at temperatures of 100–200 °C over several tens of hours, in order to remove any layers of adsorbed water vapor and other gases. However, such thermal annealing would be detrimental to the column of an electron microscope that contained sets of rubber gaskets and alignment coils made from copper wire. Consequently, cooling the specimen chamber walls down to cryogenic temperatures will allow an adsorption layer to be "frozen" onto the chamber walls that, in turn, will avoid thermal annealing of the chamber and permit the provision of UHV conditions.

The design of the differential pumping device is shown schematically in Figure 4.3. The device represents a cylindrical cartridge made from brass (1), with top and

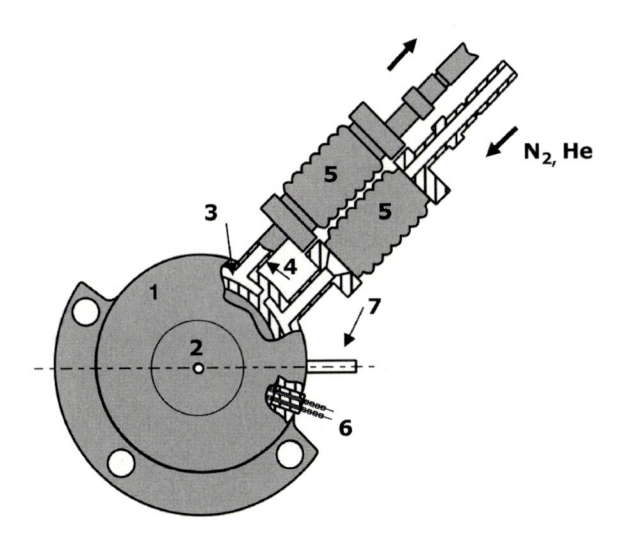

Figure 4.3 Schematic drawing of the differential cryogenic pumping device for a quasi-closed chamber around of the sample. 1, brass cylindrical cartridge; 2, cover; 3, coolant channel; 4, fitting with rolling bellows (5) for circulating of a coolant; 6, electric current inputs; 7, axis of sample tilting. The arrows indicate the directions of coolant flow.

bottom covers (2) which are fixed by screws and have apertures that allow an electron beam to enter and leave the device. Inside the cylinder wall is a channel (3) to which two fittings (4) are attached and connected to bellows that allow a coolant (liquid nitrogen or evaporating liquid helium) to circulate within the device. The nonmagnetic bellows allow the cylindrical cartridge, which is fixed by screws with Teflon rings onto the sample stage of the microscope, to move. The coolant flow into and out of the microscope column occurs via a Teflon aperture (3). The pumping of liquid nitrogen or the evaporation of liquid helium through the cryogenic differential device is achieved by using an additional vacuum pre-pump, through a ballast balloon. During cooling, the temperature of the cylindrical cartridge is monitored with a thermocouple that is attached physically to the cartridge's external surface. While the liquid nitrogen is being pumped the temperature of the cylinder should not exceed 80–90 K (30 K for helium). Although UHV conditions are known to exist inside the differential cryogenic device, any direct measurement of the vacuum within the chamber is difficult due to the small volume of the latter compared to the vacuum measurement tube.

After having installed the crystal specimen into the UHV chamber of the reflection microscope, it is heated by passing through it either a direct current (DC) or an alternating current (AC). This simple method of sample heating has several benefits in electron microscopy:

- The geometric size of the heater within the microscope column is minimal.
- Contamination of the sample from the heater's hot regions is avoided, as the maximally heated object is the sample itself.
- Resistive heating of the sample has a minimal electromagnetic influence on the electron beams, which otherwise might reduce the resolution of the microscope.

The temperature of the sample (T) is controlled by the applied electric current (I) that is passed through the heated sample, and is defined from the equation of thermal balance:

$$\varepsilon \sigma_0 T^4 S = \varrho(T) I^2 \frac{l_a}{s},$$

where $S \approx 2l_a(l_b + l_c)$ is the surface area of heat radiation, $s = l_b \times l_c$ is the cross-section of the heated crystal, l_a, l_b, and l_c are the length, width, and thickness of the sample, respectively, $\varepsilon (= 0.65)$ is the coefficient of emissivity of silicon, $\varrho(T)$ is the temperature dependence of the resistance for silicon, and σ_0 is the Stefan–Boltzmann constant. The temperature of the sample is calibrated preliminarily by means of a thermocouple at low temperatures ($T < 800\,°C$) and with an optical pyrometer at high temperatures ($T > 800\,°C$). An additional reference point is the temperature of the phase transition $(1 \times 1) \Leftrightarrow (7 \times 7)$ on the silicon(111) surface, which occurs at $830\,°C$ (as reported in Ref. [37]). Based on current estimations, the accuracy of relative temperature measurements in the experiments was $\pm 5\,°C$, and for absolute temperature measurements was $\pm 30\,°C$.

When investigating the initial stages of epitaxial growth, various types of small-scale evaporators were developed and adopted in the UHV chamber of the microscope. A silicon evaporator was prepared from a silicon plate similar to the sample studied, and mounted opposite the sample at a distance of 15 mm. A germanium evaporator was employed that consisted of a home-made crucible of tungsten wire. The Si, Ge, Au, and Ca atoms were deposited onto the silicon substrates at various deposition rates, measured precisely via RHEED oscillation analysis or by surface reconstruction diagrams at appropriate temperatures. During the *in-situ* dynamical experiments, the REM images were either exposed directly to photographic films or recorded using a Gatan TV system with a tape recorder, and later with a CCD camera.

4.3
Silicon Substrate Preparation

The samples, which were $0.3 \times 1 \times 8 \, mm^3$ in size, were cut from a silicon wafer (resistivity of a few $\Omega \, cm^{-1}$) with a misorientation less than $1°$ from the (111) or (100) planes. The incident electron beam was almost normal to the longer side of the specimen, and was directed approximately parallel to the steps (the <110> directions). Before installation into the microscope, the specimen was cleaned and chemically etched using ordinary treatments, and then outgassed in the UHV microscope chamber at approximately $900 \, °C$ for several minutes. As the most convenient means of cleaning the silicon surface is thermal annealing, the clean Si(111) surfaces typically were prepared by flashing to about $1250 \, °C$ in the UHV chamber of the microscope. Such flashing removed the natural oxide films and various types of contaminations (which mostly was silicon carbide resulting from carbon–silicon reactions). In order to reduce thermal drift, the specimen was maintained at $900–1050 \, °C$ for about 30 min. Following such final cleaning treatments, atomic steps were observed reproducibly in the REM images, without any pinning points being associated with particles of contamination [38]. For germanium crystals, a method that employed a protective coating of a sulfidic film evaporated in vacuum was used [39].

The image and microdiffraction (on an insert) of a Si(111) surface after transfer into the REM column is shown in Figure 4.4. A wavy contrast on the REM image is caused by the delamination of a natural oxide film, 1–2 nm in height, creating "shadows." Heating the sample to $900–1000 \, °C$ leads to removal of the natural oxide, as confirmed by the disappearance of the inelastic diffuse amorphous background on the diffraction pattern, and the formation of particles of epitaxial silicon carbide with sizes ranging from 5 to 10 nm on the surface. The SiC particles are clearly visible on the REM image, and also produce additional reflections on the diffraction image (Figure 4.4b).

The consecutive stages of the removal of SiC particles during thermal annealing at a temperature above $1000 \, °C$ are illustrated in Figure 4.4c and d, which confirm a reduction in the density of the SiC particles. In Figure 4.4d, the thin lines corresponding to monatomic steps can be clearly seen; these limit the singular terraces (111) and have a height of one interplanar distance, as shown below [40]. In addition,

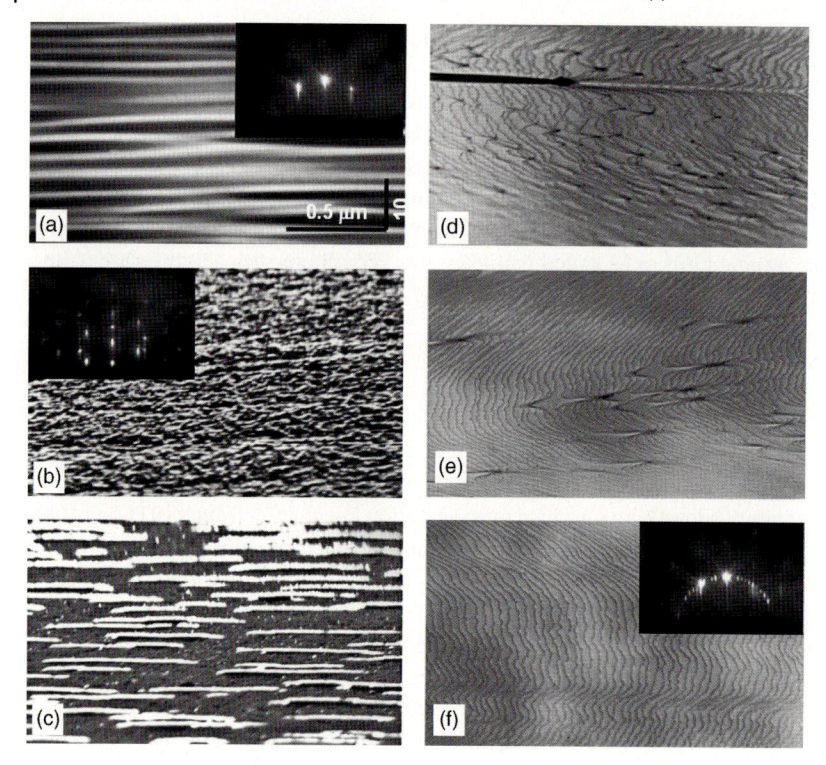

Figure 4.4 images of silicon(111) during consecutive stages of surface cleaning at thermal annealing. (a) The initial surface; (b) Surface of silicon completely covered by silicon carbide after 2-min warming at $T = 950\,°C$ (the image is received in reflection SiC); (c–e) Consecutive stages of removal of SiC particles from a surface at 1250 °C; (f) Atomically clean silicon(111) surface. The inserts show the corresponding patterns of microdiffraction.

the residual particles of SiC are visible as black points that are the pinning centers of the atomic steps. During annealing at a temperature above 1000 °C, there is a reduction in both the geometric sizes and the density of the SiC particles (Figure 4.4e). Unfortunately, when using the UHV-REM technique it is not possible to define concisely what happens to the SiC crystallites – whether they are sublimated from the surface or they dissolve into the bulk of the silicon substrate although, based on current knowledge, the diffusion of SiC into the bulk of silicon is the most likely occurrence. A complete cleaning of a silicon surface results in the removal of any residual SiC, an effect which is visible on both the diffraction pattern and on the REM images (Figure 4.4f). Notably, the time required for a total removal of SiC does not exceed 30 s at a temperature of 1200 °C.

With regards to the diffraction pattern from the clean silicon surface, no reflections were observed that could be assigned to crystalline contamination. The typical sensitivity for reflection high-energy electron diffraction can be estimated as approximately 0.05 monolayers of SiC. In contrast, at a temperature below 830 °C, it is possible to observe diffraction patterns of clean silicon surfaces, in addition to superstructural reflections corresponding to the reconstructed surface (7 × 7) (see insert in Figure 4.4f).

When using the UHV-REM technique, it is possible to refer to a contamination-free silicon surface by the following criteria [41]: (i) a complete absence of additional diffraction spots or an amorphous background on the diffraction pattern; (ii) the occurrence of reversible phase transitions of (7 × 7) ⇔ (1 × 1) for Si(111) and (1 × 2) ⇔ (1 × 1) for Si(001) surfaces observed for a sample temperature variation around 830 and 1190 °C, correspondingly; and (iii) the observation of atomic steps without pinning centers on REM images. These surface-cleaning procedures have also been confirmed in UHV chambers used for molecular beam epitaxy, with good results.

4.4
Monatomic Steps

A typical REM image of the vicinal silicon(111) surface after high-temperature cleaning in the UHV chamber of the microscope is shown in Figure 4.4f. In this image, the dark lines are monatomic steps, while the uniform areas between them are silicon crystallographic planes known as "terraces" with (111) orientation. The average width of the terraces depends on the surface misorientation from a (111) plane. The atomic step contrast in REM images is a superposition of diffraction and phase contrasts [42]. The former condition (which is also known as the Bragg contrast) is caused by a monatomic step introducing a deformation of the atomic lattice in its vicinity, which in turn leads to a modification of the diffraction conditions locally at the steps. In contrast, the phase (or Fresnel) contrast relates to the phase shift of the electrons reflected from the neighboring terraces of the step. Due to such superposition, the intensity of the diffracted electron beams will be extremely sensitive to the structural perfection of the crystal lattice close to the sample surface. Subsequently, the conclusion of several approaches was that the height of atomic steps on Si(111) surface was equal to 0.31 nm [43–45].

The monatomic step moves in the step-up direction during sublimation due to the detachment of atoms from step kinks, the diffusion along the surface, and desorption from the terrace. This step motion can be observed, using REM, up to a sample temperature close to the melting point (Figure 4.5). According to the classical Burton–Cabrera–Frank (BCF) theory [46], the step motion depends on the temperature and the distance between the neighboring steps. A direct REM measurement of step shifting during sublimation has revealed that the rate of step motion is proportional to the distance between the neighboring steps [47, 48]. This means that an atom detached from a step kink has the probability of reaching the neighboring steps until it evaporates from the surface. Measurements of the step motion rates at different crystal temperatures (Figure 4.5d) revealed an Arrhenius dependence with an activation energy of 4.2 ± 0.2 eV for a silicon atom to detach from a step kink and to evaporate from a terrace of the Si(111) [38].

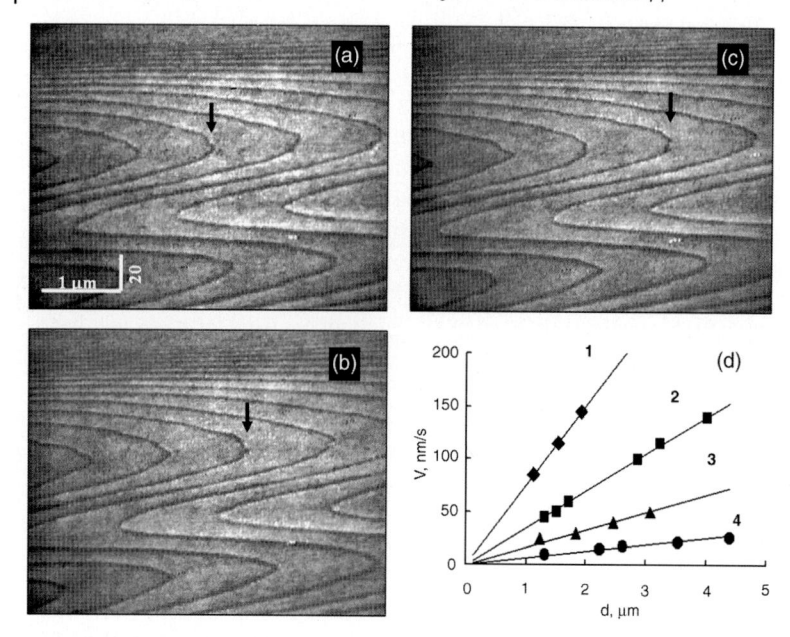

Figure 4.5 Consecutive reflection electron microscopy images of the same surface area of vicinal silicon(111) surface at a temperature of 1380 °C at: (a) t = 0; (b) t = 0.44; and (c) t = 0.92. The arrows indicate the position of the same step at various times; (d) Dependence of step rate on terrace width during sublimation at 1200 °C (1), 1170 °C (2), 1130 °C (3), and 1090 °C (4).

According to the competition between two tendencies – namely, the free energy minimization and the increase in the system's entropy – a straight monatomic step is unstable and leads to "step shape meandering." The temporal and amplitude analyses of these fluctuations, at thermodynamic equilibrium, has allowed a number of important parameters of adatoms and steps to be deduced [49, 50]. In practice, however, it may prove to be more interesting to analyze step fluctuation under conditions far from equilibrium, perhaps during sublimation or epitaxial growth. Despite much interest having been shown, however, step fluctuations occurring during sublimation have not been investigated previously, mainly because problems of visualization and registration of the step fluctuations at sublimation temperatures have been identified. The linear tension of monatomic steps on a Si(111) surface at nonequilibrium has been examined experimentally by observing the interaction between a single step and the dislocation emerging at the surface [51]. It has been shown that, during sublimation, the motion of the steps is suppressed in the strain field of a dislocation core [52]. Subsequently, dislocations were introduced into the silicon substrate which was fixed inflexibly in the sample holder by means of uncontrollable mechanical and thermal stresses, initiated during the heating treatments. As a result of the step–dislocation interactions, the step shape was changed by a pinning phenomenon in the dislocation core; the activation energy of the step–dislocation interactions was then estimated [53]. Step fluctuation analysis using REM allows the dynamical step edge stiffness (10^{-9} J cm^{-1} at 1350 K) to be deduced, and this was drastically larger in comparison with the equilibrium case (10^{-12} J cm^{-1}). In an attempt to explain this difference, it was speculated that various probabilities existed for atoms to be incorporated into the step from the upper and lower terraces (the "Schwoebel effect") [54]. As a result, a theoretical explanation of the larger effective step stiffness for sublimation, and of the smaller stiffness for growth, was developed by Uwaha and Saito [55, 56] and Pimpinelli et al. [57]. Moreover, a smoothing effect of the step shape at sublimation temperatures was claimed by Alfonso et al. [58].

4.5
Step Bunching

The system of diffusion-linked steps was found to be unstable with respect to step fluctuations, which is the reason for a rearrangement of regular monatomic steps into step bunches on the Si(111) surface during sublimation (Figure 4.6a). It should be noted that the phenomenon of step bunching is completely reversible, and depends heavily on the direction of the electric current being used for specimen heating [47] (Figure 4.6c). The step bunching suggests an electromigration effect of silicon adatoms with an effective charge in the applied electric field. The first theoretical explanation of step bunching under DC heating was proposed by Stoyanov [59].; subsequently, step bunching and the electromigration of adatoms were discussed in a variety of experimental [48, 60–64] and theoretical [65–70] reports.

Additionally, step anti-bunch formation on the Si(111) surface was identified that represented bunches of the step-up steps, instead of the original step-down steps [71]. Typically, anti-bunches are observed near step bunches, while the distance between bunch and anti-bunch depends on parameters such as the adatom concentration, the density of inclined steps in the area between bunch and anti-bunch, and the direction of the electric heating current [72]. Similar to step bunching, the anti-bunch formation is a reversible process; hence, step bunching and anti-bunching allow the surface morphology of the substrates to be controlled during high-temperature treatments.

An additional surface instability was discovered on silicon(111) surfaces that occurred due to step rearrangement during phase transitions at polynucleation of the reconstructed domains on the surface [73]. The step groupings in macrosteps under surface phase transitions were named by different groups as step clustering [73], sub-bunch formation [74], step bunching [75], and faceting [76]. The influence of phase transitions on step redistribution was demonstrated during the formation of superstructural domains: (7×7) and (5×5) at Ge/Si(111) [73], (5×2) at Au/Si(111) [38], and (3×1) at Ca/Si(111) [77]. The above-described morphological transition is reversible, which means that the clustered steps will break into regular monatomic steps when a reverse superstructure transition occurs. During this step rearrangement, however, an electromigration phenomenon has also been identified [77]. At this point, it should be noted that the physical driving force of this step

4 Formation of Surface Patterns Observed with Reflection Electron Microscopy

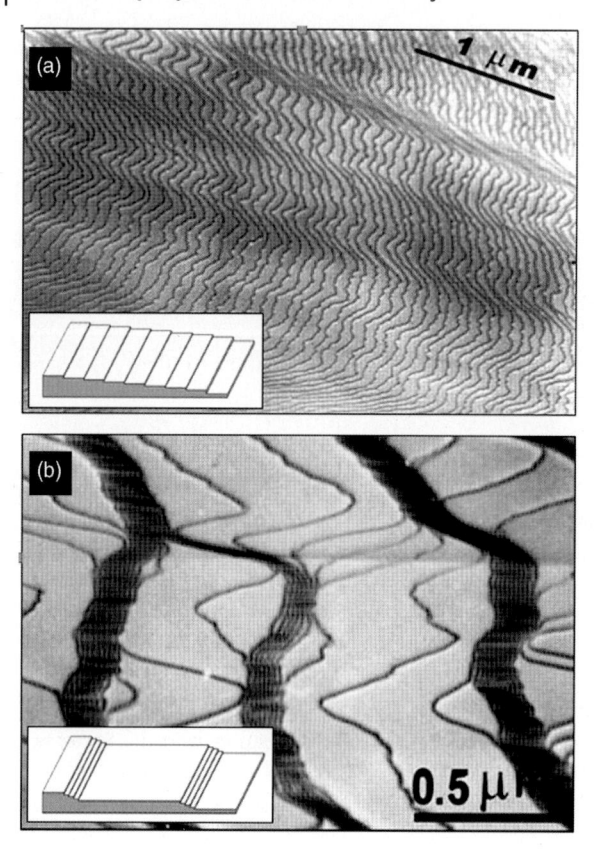

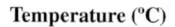

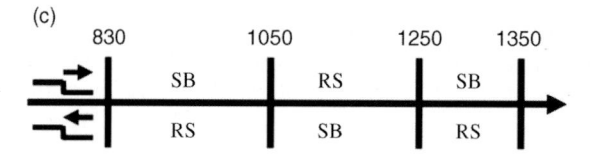

Figure 4.6 Reflection electron microscopy images of stepped silicon(111) surfaces at temperatures of (a) 1270 °C and (b) 1180 °C. The direction of direct electric current for sample heating was step-up one. The inserts show the corresponding schematic drawing of the step distribution; (c) Dependence of surface relief [regular steps (RS) or step bunches (SB)] on sample temperature and direction of heating current.

rearrangement is a phase transition which occurred on the surface, rather than sublimation as for step bunching.

In an attempt to display the role of the effective charge of adatoms in step bunching, a small amount of gold atoms was deposited onto the Si(111) surface at bunching temperatures. It was found that, in the case of sample heating with a DC in a step-up direction, the regular steps on the clean surface were unstable following sub-monolayer gold adsorption, whereas step bunches were transformed to the regular steps on the gold-adsorbed surface during heating by DC in a step-down direction [78]. It could be proposed that the adsorption of gold changes the sign of the effective charge of silicon adatoms, from positive to negative. At this temperature range, the value of the adatom effective charge was evaluated as 0.004 ± 0.001 (in units of electron charge) [79].

Recently, UHV-REM has been applied *in-situ* to investigate the impact of adsorbed gold atoms on Si(111) surface morphology stability at elevated temperatures (830–1260 °C), with DC heating of the sample [80]. Consequently, a new phenomenon of surface morphology instability was observed on the gold-covered surface by means of a periodic redistribution of the regular atomic steps (RS) into the step bunches (SB), and *vice versa* as the gold coverage decreased. A series of REM images illustrating the transformations in Si(111) surface morphology following the deposition of 0.75 monolayer of gold, and subsequent annealing at 900 °C and with a step-up current direction, are shown in Figure 4.7. No additional reflections corresponding to surface reconstruction were observed immediately at the RHEED pattern at this temperature; however, during high-temperature annealing a decrease in the gold concentration occurred, most likely due the gold evaporating and dissolving into the bulk. Following a 16-min period of annealing, the step-bunched structure (Figure 4.7a) was fluently transformed into an array of regular steps (Figure 4.7b). Subsequent isothermal annealing at 900 °C in a UHV chamber for 23 min caused step bunching (Figure 4.7c), while annealing for a further 27 min led to the formation of regular steps on the surface (Figure 4.7d). Thus, isothermal annealing at 900 °C is accompanied by the following transitions on the silicon surface with a predeposited 0.75 monolayer gold coverage: RS (0.72) → SB (0.42) → RS (0.24) → SB (0.07) → RS(0). Here, the numbers in parentheses are estimated values of the critical gold coverage (in monolayer units) at which the morphological transitions are observed. A decrease in gold atom concentration during annealing nay be due to both evaporation and dissolution into the silicon bulk.

Switching the direction of the direct electric current used for sample heating leads to the reversible changes RS → SB and SB → RS at the same values of gold coverage.

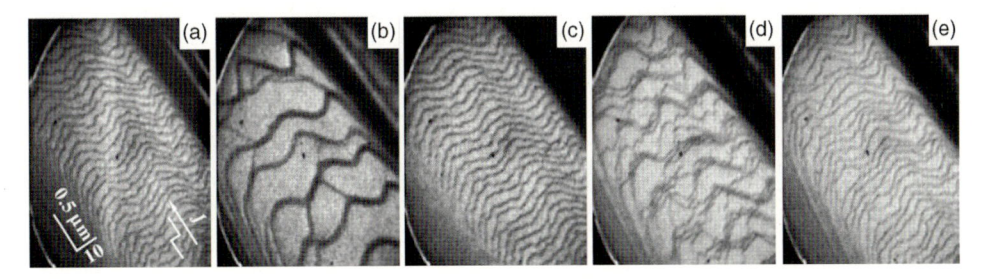

Figure 4.7 Reflection electron microscopy images of step rearrangement on silicon(111) surface after deposition of 0.7 monolayer of gold during annealing at 900 °C for 4, 20, 43, and 50 min, respectively. The arrow in panel (a) shows the direction of an electric current using for sample heating.

The dependence of silicon surface morphology on gold coverage for opposite directions (i.e., step-up and step-down) of the heating electrical current at a sample temperature of 900 °C is shown schematically in Figure 4.8. Here, the dark areas correspond to the step bunched morphology, whereas white areas correspond to the regular distributed atomic steps. The critical gold concentrations required to change the surface relief were determined at different temperatures and directions of DC passing through the sample. Within the investigated range of gold coverage (0–0.71 monolayer) the silicon(111) surface showed only regular steps when the sample was heated using AC.

The analysis of REM images of three-dimensional (3-D) gold islands nucleated at a sufficiently high gold supersaturation showed a dominating migration of the islands on the silicon(111) surface in the step-up direction [81]. This migration was independent of the direction of the DC heating of the crystal. It should be noted here that the islands moved faster when the heating current was flowing in a step-down direction, with smaller gold islands moving faster than their larger counterparts due to a smaller interaction energy with the silicon substrate. A continued coalescence of small gold particles into a larger particle led to an increase in the average size of the 3-D islands, while the total amount of the islands was reduced. Although the 3-D islands showed a tendency to orient along the step bunch [82], no interaction between the gold islands and single monatomic steps was observed at high temperatures.

4.6
Surface Reconstructions

Previously, *in-situ* UHV-REM studies have revealed that the reconstructed (7 × 7) domains on the silicon(111) surface would be nucleated preferentially close to the upper edges of a step at the (1 × 1) ↔ (7 × 7) phase transition. Yet, by monitoring the positions of individual monatomic steps (using REM), a step shifting at the (1 × 1) to (7 × 7) transition was found that was below the transition temperature (ca. 830 °C); in this case, the steps moved in a step-down direction, whereas for the reverse (7 × 7) to (1 × 1) transition they moved along the step-up [83, 84]. On measurement, the step displacement was found to be 20–30% of the terrace width. Thus, the step generates surface adatoms at the (7 × 7) domain formation, but adsorbs them during the

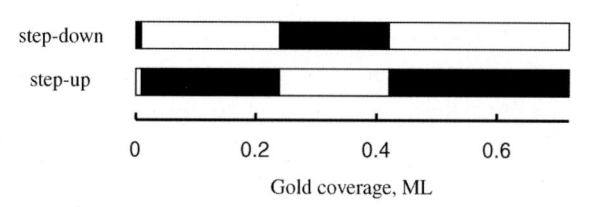

Figure 4.8 Schematic representation of the silicon(111) surface morphology at 900 °C for step-down and step-up directions of heating current at different gold coverages. The dark areas correspond to the step-bunched morphology; the white areas correspond to the regular distributed steps.

reversal process and formation of the unreconstructed (1 × 1) surface. An explanation of the above-described step shifting at phase transition might involve the existence of an anomalously high concentration of adsorbed atoms on the silicon (1 × 1) surface, corresponding to a monolayer coverage of approximately 0.25 ± 0.05. The same conclusion regarding a high density of adatoms was drawn at the base of the electron diffraction intensity calculations for unreconstructed Si(111) [85]. Direct evidence for the existence of a disordered layer of mobile adatoms on the unreconstructed Si(111) also was evaluated from high-temperature STM studies of the (7 × 7) → (1 × 1) transition [86]. The high density of adatoms on the (111) surface was also confirmed with STM [87] and AFM [88] investigations of a silicon surface after quenching from 900 °C (i.e., after rapid cooling).

By using an *in-situ* UHV-REM technique, the reversible transition (2 × 1) ↔ (1 × 2) due to the adatom electromigration effect has been examined on Si(100) [60, 89]. It is well established that the spreading of regular steps does not occur during resistivity sample heating under AC, but that step distribution is changed drastically under DC heating [3]. In fact, DC heating at elevated temperatures causes neighboring steps to pair by producing a preferential type of reconstructed domain. The majority of the surface is covered by either (1 × 2) or (2 × 1) domains, depending on the direction of heating current (AC or DC). Generally, a step pairing on the silicon(001) surface may be initiated by two effects: surface tension induced by sample deformation [90]; and the electromigration of adatoms facilitated by DC heating of the sample [89]. The first of these effects should be ignored since stress sample deformation was minimized during construction of the sample holder.

The effect of step bunching and electromigration was monitored on both Si(001) [91–96] and Si(5512) [97] surfaces, with two scenarios of the initial stages of step bunching being shown to depend on the average step–step distance [92]. From successive REM images of bunching, the step number and the average distance between the coupled steps in a bunch were measured in relation to the annealing time and the coupled step number in the bunch, respectively. In spite of essential differences in the structural properties of the Si(001) and Si(111) surfaces, the step numbers in the bunch may be described by the same root-square function of the annealing time [98].

4.7
Epitaxial Growth

At the initial stages of epitaxy, competition exists between the attachment of adatoms into the steps and the formation of growth islands. The transition between growth modes becomes more complicated with the appearance of growth islands on the terraces, because competition then occurs between the capture of adatoms by steps or islands and the nucleation of new islands. Thus, both step-flow and nucleation modes of growth appear simultaneously. The transition between the two regimes of growth has been intensively analyzed experimentally and theoretically [99–103]. Typically, a description of the nucleation processes includes a so-called characteristic growth length or minimal step–step distance, where the transition between step-flow growth and two-dimensional (2-D) nucleation occurs. There exist several procedures to measure a characteristic growth length. First, the disappearance of RHEED oscillations due to a decrease of the atom flux or increasing the substrate temperature allows an estimation to be made of the characteristic length, assuming a uniform distribution of the steps [104]. Second, this length can be measured directly as the average distance between 2-D islands or as the size of so-called "denuded zones," which are areas free from growth islands near the steps [105, 106]. The third means of determining the growth length is to deposit atoms on the surface with various interstep distances, as shown in Refs [107, 108].

According to the exponential dependence of the adatom diffusion length on the temperature, the measurements of critical distance using the methods described above become complicated at high substrate temperatures. One crucial problem relates to the preliminary formation of a surface containing monatomic steps with large step–step distances (up to several microns). Thus, a new procedure was proposed for measurement of the minimal interstep distance for 2-D nucleation by using surface morphology with step bunches and anti-bunches [109]. During the deposition of atoms on the surface, a movement of bunch and anti-bunch in opposite directions was observed due to the migration and attachment of adatoms into monatomic steps at the step-flow growth mode. As a result, the width between the step bunch and anti-bunch was increased to the critical distance, d_{crit}. According to the BCF theory, the rate of step bunch motion depends on the number of monatomic steps in a bunch or anti-bunch. Consequently, the motion was measured experimentally mainly for anti-bunches, as the number of steps in the bunch was drastically larger than in anti-bunch. When the distance between bunch and anti-bunch is increased to more than d_{crit}, then the formation of 2-D islands should be expected. In fact, simultaneously with step displacement, an island nucleation was observed on those surface areas with the largest distances between bunch and anti-bunch.

Island nucleation on the terrace provides evidence that the migration length of adatoms is at least twofold less than the terrace width. Therefore, silicon deposition on the vicinal silicon surface with step bunch and anti-bunch morphology would allow the determination of the critical interstep distance for island nucleation, which is related to the adatom diffusion length. A comparative analysis of various measurement procedures has been given in Ref. [110]. In addition, the characteristic growth lengths of adatom migration on the Si(111) surface have been measured at various temperatures, and fluxes deposited for silicon and germanium [111]. These data, when drawn in Arrhenius plots, allow a deduction of the activation energy (1.3 ± 0.2 eV) for adatom diffusion. The latter point is an important parameter when describing mass transport on a silicon surface, as knowledge of this activation energy provides an ability to control the epitaxial mechanisms and to improve the quality of the growth film.

4.8
Thermal Oxygen Etching

A real crystal surface contains both adsorbed atoms and vacancies, which can easily migrate along a surface to provide a mass transfer. Ordinarily, the adatom behavior is

4.8 Thermal Oxygen Etching

under detailed consideration, whereas the surface vacancies may be neglected due to their small concentrations at low substrate temperatures. In order to clarify the role of vacancies during surface transformations, the interaction of molecular oxygen with the silicon surface was examined over a wide range of temperatures and oxygen pressures. The interaction between oxygen and silicon is known to depend very heavily on the substrate temperature and the oxygen pressure [112]. For example, at low temperatures and high oxygen flux, the silicon surface will be passivated by a thin film of dioxide, whereas the surface will be thermally etched at a high temperature and a low oxygen pressure. Observations using REM identified a step movement in the step-up direction at low oxygen pressure. A monatomic step motion during oxygen treatment occurs due to the nucleation, diffusion, and interaction of vacancies with monatomic steps [113]. The step rate was increased with an increasing flux of oxygen, and was linearly dependent on the width of the neighboring terraces [114]. According to the classical BCF theory, steps with wide terraces move faster than those with narrow terraces [46]. Consequently, a linear dependence of step motion on terrace width would indicate that the vacancy diffusion length is larger than the interstep distance. This, in turn, would suggest that the main diffusion species on the silicon(111) surface during thermal etching were surface vacancies.

At a high supersaturation of the vacancies on the surface, negative island nucleation and enlargement would occur during oxygen exposure (Figure 4.9). The existence of denuded zones free from islands along atomic steps may be noted, and all vacancies formed inside this zone would disappear through an interaction with

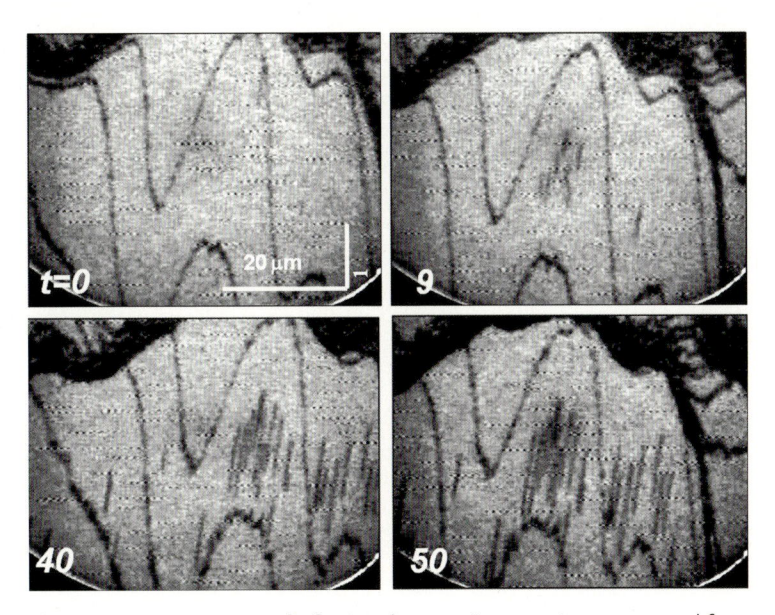

Figure 4.9 Consequent set of reflection electron microscopy images captured from a video of the same area of silicon(111) surface during exposure for 0, 9, 40 and 50 s in an oxygen atmosphere ($P > P_{crit}$).

monatomic steps. Moreover, negative island nucleation was not observed at the narrow terraces. Evidently, the occurrence of this 2-D mechanism of silicon thermal etching requires the terrace width to be more than twice the surface diffusion length of vacancies. Further exposure of the silicon surface to oxygen causes an enlargement of the negative islands, and a step motion in the step-up direction. The interaction of the moving steps with islands leads to drastic changes of step shape, and to the disappearance of islands. An activation energy for surface vacancy diffusion was estimated to be 1.35 ± 0.15 eV [115].

The nucleation of 2-D negative islands involves the removal of atoms from the top monolayer of the silicon surface. During the etching process, the intensity of the reflected beam of electrons will be changed, as the reflection coefficient of the surface is changed from maximum (clean surface) to minimum (when the islands comprise half of the surface area). Thus, oxygen exposure was shown to cause a periodic intensity change in the reflected electron beam from the silicon(111) surface (Figure 4.10). One period of the oscillations of the reflected beam should correspond to the removal of one monolayer of substrate atoms. The period of the oscillations does not depend on the substrate temperature over a range of 540 to 835 °C, but rather is proportional to the oxygen flux; this means that the oxygen flux may cause a kinetic limitation of the surface reactions. At this point, it should be noted that the period of RHEED intensity oscillations depends on the oxygen pressure, whereas the temperature range of the 2-D mechanism of thermal etching of the silicon surface was related both to the oxygen pressure and the interstep distance.

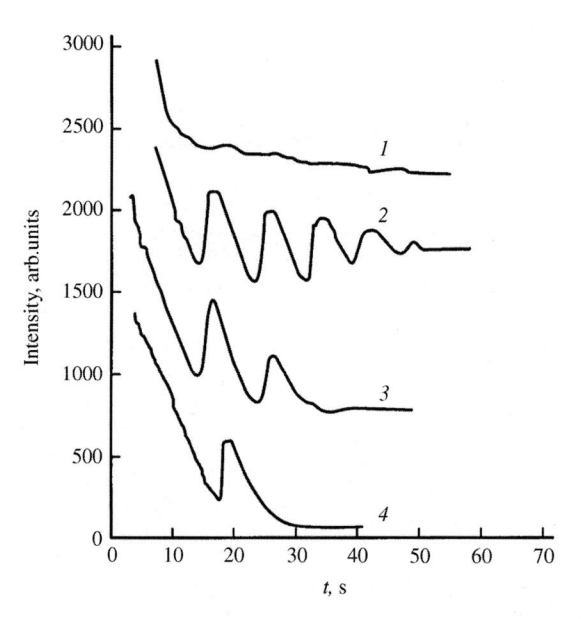

Figure 4.10 Temporal dependences of intensity of the specula-reflected electron beam measured during etching of the silicon surface at constant oxygen pressure and temperatures of 835 °C (1), 775 °C (2), 745 °C (3), and 730 °C (4).

4.9
Conclusions

In conclusion, the *in-situ* observation of monatomic step behavior on the silicon surfaces by using UHV-REM allows the evaluation of many important parameters for the surface processes of sublimation, growth, thermal annealing, adsorption, gas reaction, and phase transitions. UHV-REM allows experiments to be performed with a high spatial resolution, *in-situ*, on the various physical phenomena of a crystal surface, in order to optimize the methods of surface cleaning, to study the desorption kinetics of various coverings from a surface, to investigate the crystal structure and local stress of atomically clean surfaces on a monolayer level, to analyze the superstructural phase transitions on a surface, to examine atomic mechanisms of evaporation and epitaxial growth, to explore processes of reorganization of a surface, and to form epitaxial and dielectric films.

Acknowledgments

The author would like to thank his former coworkers, including A.L. Aseev, S.S. Kosolobov, E.E. Rodyakina, L.V. Litvin, and A.B. Krasilnikov (Institute of Semiconductor Physics, Novosibirsk), and K. Yagi, Y. Tanishiro, and H. Minoda (Tokyo Institute of Technology, Tokyo). These studies were conducted at the Collective Center for the Use of Nanostructures, and partly supported by the Russian Ministry of Education and Science.

References

1 Hsu, T. (guest ed.) (1992) Current research on reflection electron microscopy. *Microsc. Res. Tech.*, **20** (Special issue), 317–464.
2 Wiesendanger, R. and Guntherodt, H.J. (eds) (1992–1993) *Scanning Tunnelling Microscopy*, Springer-Verlag, Berlin, I-3.
3 Sheglov, D.V., Kosolobov, S.S., Rodyakina, E.E., and Latyshev, A.V. (2005) *Microscopy and Analysis*, **19**, 9–11.
4 Yagi, K. (1995) *J. Electron. Microsc. (Tokyo)*, **44**, 269–280.
5 Bethge, H. (1964) *Surf. Sci.*, **3**, 33–41.
6 Bethge, H., Hoche, H., Katzer, D., and Keller, K.W. (1980) *J. Crystal Growth*, **48**, 9–18.
7 Katzer, D. and Safran, G. (1984) *Ultramicroscopy*, **15**, 135–138.
8 Ishikawa, J., Ikeda, N., Kenmochi, M., and Ichinokawa, T. (1985) *Surf. Sci.*, **159**, 256–264.
9 Cherns, D. (1974) *Philos. Mag.*, **30**, 549–556.
10 Kambe, K. and Lehmpfuhl, G. (1975) *Optik*, **42**, 187–194.
11 Iijima, S. (1977) *Optik*, **47**, 437–452.
12 Smith, D.J., Bovin, J.O., Bursill, L.A., Petford-Long, A.K., and Ye, H.Q. (1987) *Surf. Interface Anal.*, **10**, 135–141.
13 Takayanagi, K., Tanishiro, Y., Kobayashi, K., Akiyama, K., and Yagi, K. (1987) *Jpn. J. Appl. Phys.*, **26**, L957–L960.
14 Smith, D.J. (1986) *Surf. Sci.*, **178**, 462–474.
15 Xu, P., Dunn, D., Zhang, J.P., and Marks, L.D. (1993) *Surf. Sci.*, **285**, L479–L485.

16 Takayanagi, K., Yagi, K., Kobayashi, K., and Honjo, G. (1978) *J. Phys. E. Sci. Instrum.*, **11**, 441–448.

17 Heinemann, K. and Poppa, H. (1986) *J. Vac. Sci. Technol. A*, **4**, 127–136.

18 Poppa, H. (1975) Examination of thin films, in *Epitaxial Growth* (ed. J.W. Mathews), Academic Press, New York, Part A, pp. 215–279.

19 Gibson, J.M. (1988) in *Surface and Interface Characterization by Electron Optical Method* (eds A. Howie and U. Valdre)) Plenum Press, New-York, pp. 55–76.

20 Uchida, Y. and Lehmpfuhl, G. (1987) *Surf. Sci.*, **188**, 364–377.

21 Kondo, Y., Ohi, K., Shibashi, Y.I., Hirano, H., Harada, Y., Takayanagi, K., Tanishiro, Y., Kobayashi, K., and Yagi, K. (1991) *Ultramicroscopy*, **35**, 111–118.

22 Metois, J.J., Nitsche, S., and Heylaud, J.C. (1989) *Ultramicroscopy*, **27**, 349–358.

23 Wilson, R.J. and Petroff, P.M. (1983) *Rev. Sci. Instrum.*, **54**, 1534–1537.

24 Valdre, U., Pashley, D.W., Robinson, E.A., Stowell, M.J., and Law, T.J. (1970) *J. Phys.*, **E3**, 501–506.

25 Smith, D.J., Podbrdsky, J., Swann, P.R., and Jones, J.S. (1989) *Mater. Res. Soc. Symp. Proc.*, **139**, 289–294.

26 Pasemann, H. and Pchelyakov, O. (1982) *J. Crystal Growth*, **58**, 288–289.

27 Yagi, K. (1993) *Surf. Sci. Rep.*, **17**, 305–362.

28 Latyshev, A.V. (1999) *Phys. Low-Dimens. Struct.*, **3–4**, 169–180.

29 Ruska, E. (1933) *Z. Phys.*, **83**, 492–497.

30 Cowley, J.M. and Hojlund Nielsen, P.E. (1975) *Ultramicroscopy*, **1**, 145–150.

31 Osakabe, N., Tanishiro, Y., Yagi, K., and Honjo, G. (1981) *Surf. Sci.*, **102**, 424–442.

32 Takayanagi, K., Yagi, K., Kobayashi, K., and Honjo, G. (1974) *Jpn. J. Appl. Phys. Suppl.*, **1**, 533–536.

33 McDonald, M.L., Gibson, J.M., and Unterwald, F.C. (1989) *Rev. Sci. Instrum.*, **60**, 700–707.

34 Ishiguro, Y., Naruse, M., and Yanagihara, T. (1993) *J. Electron Microsc.*, **42**, 64–71.

35 Kroshkov, A.A., Baranova, E.A., Yakushenko, O.A., Latyshev, A.V., Aseev, A.L., and Stenin, S.I. (1985) *Prib. Tech. Eksper*, **1**, 199–202 (in Russian).

36 Aseev, A.L., Latyshev, A.V., Krasilnikov, A.B., and Stenin, S.I. (1985) Reflection electron microscopy study of the structure of atomic clean silicon surface. Proceedings. VIIth International School on Defects in Crystals, May 22–27, Szczyrk, Poland, pp. 231–237.

37 Ino, S. (1977) *Jpn. J. Appl. Phys.*, **16**, 891–908.

38 Latyshev, A.V., Aseev, A.L., Krasilnikov, A.B., and Stenin, S.I. (1990) *Surf. Sci.*, **227**, 24–34.

39 Latyshev, A.V., Aseev, A.L., Gorokhov, E.B., and Stenin, S.I. (1984) *Poverhnost*, **2**, 89–92 (in Russian).

40 Kosolobov, S.S., Nasimov, D.A., Sheglov, D.V., Rodyakina, E.E., and Latyshev, A.V. (2002) *Phys. Low-Dimens. Struct.*, **5/6**, 231–239.

41 Aseev, A.L., Latyshev, A.V., Krasilnikov, A.B., and Stenin, S.I. (1987) in *Defects in Crystals* (ed. E. Mizera), World Scientific Publ. Co., Singapore, pp. 231–237.

42 Shuman, H. (1977) *Ultramicroscopy*, **2**, 361–369.

43 Osakabe, N., Tanishiro, Y., Yagi, K., and Honjo, G. (1980) *Surf. Sci.*, **97**, 393–408; (1981) **102**, 424–442.

44 Latyshev, A.V., Aseev, A.L., Krasilnikov, A.B., and Stenin, S.I. (1989) *Phys. Status Solidi A*, **113**, 421–430.

45 Sheglov, D.V., Prozorov, A.V., Nasimov, D.A., Latyshev, A.V., and Aseev, A.L. (2002) *Phys. Low-Dimens. Struct.*, **5/6**, 239–247.

46 Burton, W.K., Cabrera, N., and Frank, F.C. (1951) *Philos. Trans. R. Soc. London, Ser.*, **A243**, 299–358.

47 Latyshev, A.V., Aseev, A.L., Krasilnikov, A.B., and Stenin, S.I. (1989) *Surf. Sci.*, **213**, 157–169.

48 Alfonso, C., Heyraud, J.C., and Metois, J.J. (1993) *Surf. Sci. Lett.*, **291**, L745–L749.

49 Williams, E.D. (1994) *Surf. Sci.*, **299/300**, 502–510.

50 Bartelt, N.C., Goldberg, J.L., Einstein, T.L., Williams, E.D., Heyraud, J.H., and Metois, J.J. (1993) *Phys. Rev. B*, **48**, 15453–15456.

51. Latyshev, A.V., Minoda, H., Tanishiro, Y., and Yagi, K. (1996) *Phys. Rev. Lett.*, **76**, 94–97.
52. Latyshev, A.V., Minoda, H., Tanishiro, Y., and Yagi, K. (1996) *Surf. Sci.*, **357/358**, 550–554.
53. Latyshev, A.V., Minoda, H., Tanishiro, Y., and Yagi, K. (1995) *Jpn. J. Appl. Phys.*, **34**, 5768–5773.
54. Schwoebel, R.L. and Shipsey, E.J. (1969) *J. Appl. Phys.*, **37**, 3682–3686.
55. Uwaha, M. and Saito, Y. (1993) *J. Crystal Growth*, **128**, 87–92.
56. Saito, Y. and Uwaha, M. (1994) *Phys. Rev. B*, **49**, 10677–10692.
57. Pimpinelli, A., Elkinani, I., Karma, A., Misbah, C., and Villain, J. (1994) *J. Phys.: Condens. Matter*, **6**, 2661–2680.
58. Alfonso, C., Bermond, J.M., Heyraud, J.C., and Metois, J.J. (1992) *Surf. Sci.*, **262**, 371–381.
59. (a) Stoyanov, S. (1990) *Europhys. Lett.*, **11**, 361–366; (b) Stoyanov, S. (1990) *Jpn. J. Appl. Phys.*, **29**, L659–L662.
60. Yagi, K. (1993) *Surface Sci. Report*, **17**, 305–362.
61. Homma, Y., McClelland, R.J., and Hibino, H. (1990) *Jpn. J. Appl. Phys.*, **29**, L2254–L2256.
62. Krug, J. and Dobbs, H. (1994) *Phys. Rev. Lett.*, **73**, 1947–1050.
63. Ichikawa, M. and Doi, T. (1992) *Appl. Phys. Lett.*, **60**, 1082–1084.
64. Pierre-Louis, O. (2003) *Surf. Sci.*, **529**, 114–134.
65. Natori, A., Fujimura, H., and Fukuda, M. (1992) *Appl. Surf. Sci.*, **60/61**, 85–91.
66. Uwaha, M. (1992) *Phys. Rev. B*, **46**, 4364–4366.
67. Kandel, D. and Weeks, J.D. (1995) *Phys. Rev. Lett.*, **74**, 3632–3635.
68. Houchmandzadeh, B., Misbah, C., and Pimpinelli, A. (1994) *J. Phys. France*, **4**, 1843–1853.
69. Suga, N., Kimpara, J., Wu, N.-J., Yasunaga, H., and Natori, A. (2000) *Jpn. J. Appl. Phys.*, **39**, 4412–4416.
70. Sato, M. and Uwaha, M. (2001) *Surf. Sci.*, **493**, 494–498.
71. Latyshev, A.V., Krasilnikov, A.B., and Aseev, A.L. (1993) *Ultramicroscopy*, **48**, 377–380.
72. Latyshev, A.V., Krasilnikov, A.B., and Aseev, A.L. (1990) *Surf. Sci.*, **227**, 24–34.
73. Krasilnikov, A.B., Latyshev, A.V., Aseev, A.L., and Stenin, S.I. (1992) *J. Cryst. Growth*, **116**, 178–184.
74. Homma, Y., Suzuki, M., and Hibino, H. (1992) *Appl. Surf. Sci.*, **60/61**, 479–484.
75. Yagi, K. and Tanishiro, Y. (1995) *J. Phase Transitions*, **53**, 197–204.
76. Jung, T.M., Phaneuf, R.J., and Williams, E.D. (1994) *Surf. Sci.*, **301**, 129–135.
77. Krasilnikov, A.B., Latyshev, A.V., and Aseev, A.L. (1993) *Surf. Sci.*, **290**, 232–238.
78. Latyshev, A.V., Minoda, H., Tanishiro, Y., and Yagi, K. (1998) *Appl. Surf. Sci.*, **130–132**, 60–66.
79. Latyshev, A.V., Minoda, H., Tanishiro, Y., and Yagi, K. (1998) *Surf. Sci.*, **401**, 22–33.
80. Kosolobov, S.S., Song, S.A., Fedina, L.I., Gutakovskii, A.K., and Latyshev, A.V. (2005) *JETP Lett.*, **81**, 149–153.
81. Latyshev, A.V., Nasimov, D.A., Savenko, V.N., and Aseev, A.L. (2000) *Thin Solid Films*, **367**, 142–148.
82. Latyshev, A.V., Nasimov, D.A., Savenko, V.N., Kosolobov, S.S., and Aseev, A.L. (2002) *Inst. Phys. Conf. Ser.*, **169**, 153–162.
83. Latyshev, A.V., Aseev, A.L., and Stenin, S.I. (1988) *JETP Lett.*, **47**, 530–532.
84. Latyshev, A.V., Krasilnikov, A.B., Sokolov, L.V., Aseev, A.L., and Stenin, S.I. (1991) *Surf. Sci.*, **254**, 90–96.
85. Kohmoto, S. and Ichimiya, A. (1989) *Surf. Sci.*, **223**, 400–412.
86. Iwatsuki, M., Kitamura, S., Sato, T., and Sueyoshi, T. (1992) *Appl. Surf. Sci.*, **60/61**, 580–586.
87. Yang, Y.N. and Williams, E.D. (1994) *Phys. Rev. Lett.*, **72**, 1862–1865.
88. Nasimov, D.A., Sheglov, D.V., Rodyakina, E.E., Kosolobov, S.S., Fedina, L.I., Teys, S.A., and Latyshev, A.V. (2003) *Phys. Low-Dimens. Struct.*, **3/4**, c.157–c.166.
89. Latyshev, A.V., Krasil'nikov, A.B., Aseev, A.L., and Stenin, S.I. (1988) *JETP Lett.*, **48**, 526–529.
90. Webb, M.B., Men, F.K., Swartzentruber, B.S., Kariotis, R., and Lagally, M.G. (1991) *Surf. Sci.*, **242**, 23–26.
91. Kahata, H. and Yagi, K. (1989) *Jpn. J. Appl. Phys.*, **28**, L858–L861.

92 Latyshev, A.V., Litvin, L.V., and Aseev, A.L. (1998) *Appl. Surf. Sci.*, **130/132**, 139–145.

93 Enta, Y., Suzuki, S., and Kono, S. (1989) *Phys. Rev. B*, **39**, 5524–5526.

94 Litvin, A.V., Krasilnikov, A.B., and Latyshev, A.V. (1991) *Surf. Sci. Lett.*, **244**, L121–L124.

95 Nielsen, J.-F., Pettersen, M.S., and Pelz, J.P. (2001) *Surf. Sci.*, **480**, 84–96.

96 Doi, T., Ichikawa, M., and Hosoki, S. (2002) *Surf. Sci.*, **499**, 161–173.

97 Suzuki, T., Metois, J.J., and Yagi, K. (1995) *Surf. Sci.*, **339**, 105–113.

98 Rodyakina, E.E., Kosolobov, S.S., Sheglov, D.V., Nasimov, D.A., Se Ahn Song, and Latyshev, A.V. (2004) *Phys. Low-Dimens. Struct.*, **1/2**, 9–18.

99 Stoyanov, S. (1989) *J. Cryst. Growth*, **94**, 751–756.

100 Venables, J.A. (1994) *Surf. Sci.*, **299/300**, 798–812.

101 Mo, Y.-W., Kleiner, J., Webb, M.B., and Lagally, M.G. (1992) *Surf. Sci.*, **268**, 275–295.

102 Nakahara, H., Ichikawa, M., and Stoyanov, S. (1993) *Ultramicroscopy*, **48**, 417–424.

103 Iwanari, S. and Takayanagi, K. (1992) *J. Cryst. Growth*, **119**, 229–236.

104 Sakamoto, K., Miki, K., and Sakamoto, T. (1990) *J. Cryst. Growth*, **99**, 510–513.

105 Mo, Y.-W., Kleiner, J., Webb, M.B., and Lagally, M.G. (1992) *Surf. Sci.*, **268**, 275–295.

106 Voigtlander, B., Zinner, A., Weber, T., and Bonzel, H.P. (1995) *Phys. Rev. B*, **51**, 7583–7591.

107 Latyshev, A.V., Aseev, A.L., Krasilnikov, A.B., and Stenin, S.I. (1989) *Phys. Status Solidi A*, **113**, 421–430.

108 Ichikawa, M. and Doi, T. (1987) *Appl. Phys. Lett.*, **50**, 1141–1143.

109 Latyshev, A.V., Krasilnikov, A.B., and Aseev, A.L. (1996) *Phys. Rev. B*, **54**, 2586–2589.

110 Latyshev, A.V., Kosolobov, S.S., Nasimov, D.A., Savenko, V.N., and Aseev, A.L. (2002) in *Atomistic Aspects of Epitaxial Growth* (eds M. Kotrla *et al.*, Kluwer Academic Publishers, pp. 281–299.

111 Latyshev, A.V., Krasilnikov, A.B., and Aseev, A.L. (1996) *Thin Solid Films*, **281-282**, 20–23.

112 Lander, J.J. and Morrison, J. (1962) *J. Appl. Phys.*, **33**, 2089–2092.

113 Schlier, R.E. and Fransworth., H.E. (1959) *J. Chem. Phys.*, **30**, 917–923.

114 Kosolobov, S.S., Aseev, A.L., and Latyshev, A.V. (2001) *Semiconductors*, **35**, 1038–1044.

115 Latyshev, A.V., Nasimov, D.A., Savenko, V.N., Kosolobov, S.S., and Aseev, A.L. (2002) *Inst. Phys. Conf. Ser.*, **169**, 153–162.

Part II
Growth and Interactions

5
Electron and Ion Irradiation
Florian Banhart

5.1
Introduction

The irradiation of materials with energetic particles has been a subject of interest for many decades. The technology of fission or fusion reactors needed a thorough understanding of the response of materials to irradiation. The same is of importance for electronic components in satellites that are exposed to a high radiation level in space. Particle irradiation is also applied deliberately in technological processes such as ion implantation in semiconductor devices, and knowledge about the formation of radiation defects is of paramount importance for the functioning of electronic components. Besides technological applications, particle irradiation has also shown a large number of fundamentally interesting phenomena in materials, for example, the occurrence of self-organization effects. The basis of irradiation is always the formation of atomic defects, and irradiation experiments are therefore particularly suited to study defects in detail.

Irradiation with electrons is generally unavoidable in electron microscopy because an electron beam is necessary for the formation of an image of a specimen. It is therefore obvious that the same beam is used for *in-situ* irradiation and imaging. Many irradiation phenomena have been observed accidentally as alterations of the specimens when they were inspected under the electron beam. On the other hand, many dedicated irradiation studies with a focused electron beam of high intensity have been carried out over several decades. The energy of the electron beam in a typical transmission electron microscope ranges between 100 and 300 keV and is high enough to cause persistent structural alteration in a large number of materials. Dedicated high-voltage electron microscopes with electron energies up to 3 MeV have been used for irradiation studies of even radiation-hard materials.

Ion irradiation of specimens in the electron microscope needs an extensive setup to direct an ion beam from an accelerator onto the field of view on the specimen. Only a few instruments of this type have been built to date, but very useful insights into the behavior of materials under ion irradiation, for example, for ion implantation technology, have been obtained. Due to the limited number of *in-situ* ion irradiation experiments, the main focus of this chapter will be on electron irradiation.

In-situ Electron Microscopy: Applications in Physics, Chemistry and Materials Science, First Edition.
Edited by Gerhard Dehm, James M. Howe, and Josef Zweck.
© 2012 Wiley-VCH Verlag GmbH & Co. KGaA. Published 2012 by Wiley-VCH Verlag GmbH & Co. KGaA.

The first in-situ irradiation studies were carried out in the 1960s when high-voltage electron microscopes became available and radiation effects during microscopy were unavoidable. Atomic defects in solids such as vacancies or interstitials were generated, but the resolving power of the microscopes did not allow the detection of individual point defects. However, the agglomeration of atomic defects leads to structures such as voids or dislocation loops that can be seen at moderate resolution. The formation of defect agglomerates became a field of intense research for many years, and dedicated stages, allowing the heating of the specimens to high temperatures, have been used. The annealing of defects has been studied and quantitative information about the mobility of interstitials and vacancies has been obtained. Since the 1980s, electron microscopes have allowed the resolution of crystal lattices, and *in-situ* electron microscopy has been extended towards atomic resolution. Irradiation phenomena became visible on a smaller scale, and meanwhile the observation of even single vacancies or interstitials appears to be feasible.

Irradiation with a focused electron beam can be used to induce structural transformations of materials on a very small scale. Nowadays, modern electron microscopes are equipped with field emission guns, allowing the concentration of the beam onto a spot of diameter <1 nm. With aberration-corrected condenser systems, beam spots with a diameter of even <0.1 nm have been achieved. The application of focused electron beams can be used for the structuring of materials on a scale that has only been accessible to surface-sensitive techniques such as scanning tip microscopy. Hence structural transformations of materials can now be carried out and studied in the same instruments at atomic resolution.

Many studies of nanostructured materials have shown that irradiation does not just generate damage but may also have beneficial effects. Irradiation with focused particle beams can be used for the engineering of materials on the nanometer or even subnanometer scale. Self-assembly or self-organization effects have been demonstrated, for example, in graphitic nanoparticles, and offer routes towards new nanostructures. The atomic structure can now be modified by irradiation on a very small scale, leading to morphological, mechanical, electronic, or magnetic changes or to phase transformations.

This chapter treats the physical principles of irradiation and the formation of defects. A summary of the key experiments on the *in-situ* irradiation of materials is given with focus on nanomaterials that are of current interest in novel technological concepts. Due to their unique ability to reconstruct after electron or ion irradiation, graphitic systems such as carbon nanotubes are particularly suited for irradiation-induced structuring experiments and are discussed in detail.

5.2
The Physics of Irradiation

5.2.1
Scattering of Energetic Particles in Solids

The interaction between energetic particle beams and solids is governed on the atomic scale by scattering processes [1]. Electrons or ions from the beams are scattered at electrons or nuclei in the specimens. Scattering can be elastic, that is, without energy transfer, or inelastic, with a certain energy transfer to the electrons or nuclei of the specimen. Elastic scattering cannot lead to alterations of the specimen but is used for the formation of an image in the electron microscope. Inelastic scattering of a particle at the electron system of the specimen leads to ionization or electronic excitations, whereas inelastic scattering at the nuclei may cause displacements of atoms. Inelastic electron–electron scattering is used for analytical electron microscopy in electron energy loss spectroscopy (EELS) or energy-dispersive X-ray (EDX) spectroscopy to obtain information about the chemical composition of the specimens, bonding characteristics, or the density of electron states.

The energy transfer in an inelastic scattering event is governed by the rules of momentum conservation. Energy transfer is efficient when both particles involved in the scattering have similar masses. For example, electrons with an energy of several hundred keV can transfer only a very small fraction of their energy to nuclei, typically some tens of electronvolts. Energetic ions, on the other hand, transfer much more energy to nuclei and can lead to extended displacement cascades in the specimens. When atoms close to the surface of a specimen are displaced and their energy is high enough to leave the specimen, this leads to sputtering [2].

The quantity determining the probability of scattering is the respective cross-section σ (given in units of barns; 1 barn $= 10^{-28}$ m^2). The probability of scattering p is then given by the product $p = \sigma j$, where j is the beam current density. Typical electron beam current densities in transmission electron microscopy (TEM) range between 10 A cm^{-2} (imaging conditions) and 10^4 A cm^{-2} (irradiation by a fully focused beam). If we take graphitic structures as an example ($\sigma \approx 20$ barn), we obtain values between 10^{-3} and 1 displacements of each atom per second [3]. We always have to distinguish between scattering events leading to persistent radiation damage in the specimen (for example, atom displacements) and events leading to only a short and localized disturbance which is quenched immediately (for example, inter-band excitations). The latter are not of interest in radiation studies because they do not lead to structural alterations of the material.

An important question in irradiation experiments is the total deposition of energy in the specimen. The major fraction of the energy dissipates as heat and only a small fraction (if any) is stored as persistent defects. Heating of specimens under irradiation can cause phase transformations (melting, evaporation) or dissociation. Whereas ions have a short range and are easily stopped in thin specimens such as used in TEM studies (typical specimens have thicknesses of 10–100 nm), beam electrons in TEM have a range of the order of millimeters and therefore leave thin TEM specimens without much energy dissipation. Heating is therefore non-negligible in ion irradiation experiments whereas in electron irradiation heating only has to be considered in insulators such as polymers with low thermal conductivity and high sensitivity to heating. It is a counter-intuitive fact that heating of thin specimens decreases with increasing particle energy.

5.2.2
Scattering of Electrons

Several mechanisms govern the inelastic scattering of an electron beam at *electrons* in the specimen [3]. Individual scattering events lead to electronic excitation or ionization of atoms or to the breaking of bonds. Collective excitations lead to plasmons that dissipate into phonons within a short time. The cross-sections for plasmon excitation or ionization are generally much higher than for inter-band excitations [1]. The respective cross-sections depend on the energy of the electron beam and generally decrease with increasing beam energy. Thus, electronic excitations dominate at low beam energies. The presence of conduction electrons normally quenches electronic excitations within a short time so that damage of the specimen (for example, bond breaking) does not occur. Therefore, metals do not show any damage caused by electron–electron scattering, whereas insulators with localized electrons are sensitive to electronic damage.

The inelastic electron scattering of electrons at *nuclei*, on the other hand, dominates at high electron energies. Although the cross-section of single scattering events decreases towards higher energies, the occurrence of displacement cascades leads, in most systems, to an overall increasing rate of defect formation at higher electron energies. Atom displacements are the only source of radiation damage in metals.

An inelastic scattering event is shown schematically in Figure 5.1. The electron is scattered at a nucleus by a certain angle Θ. The transferred energy T is given by

$$T = T_{max} \cos^2 \Theta \tag{5.1}$$

where T_{max} is the maximum transferred energy in a central collision ($\Theta = 0$). From geometric considerations, it is clear that high-angle scattering events with low energy transfer dominate whereas central collisions with high energy transfer are unlikely. The maximum transferred energy for a relativistic particle (electron or ion with mass m and energy E) can be derived from the rules of momentum conservation:

$$T_{max} = \frac{2ME(E + 2mc^2)}{(m+M)^2 c^2 + 2ME} \tag{5.2}$$

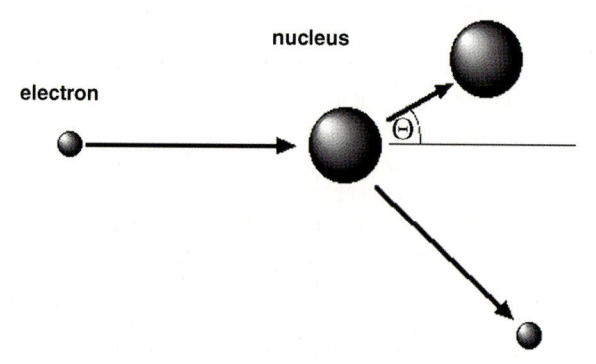

Figure 5.1 Scattering of an electron at a nucleus.

where M is the mass of the nucleus. For electron irradiation ($m_e \ll M$, $E \ll Mc^2$), Eq. (5.2) can be simplified:

$$T_{max} = \frac{2E(E+2m_e c^2)}{Mc^2} \qquad (5.3)$$

Whereas the scattering event happens within 10^{-21} s, the dissipation of energy until stable defects have formed occurs on a time scale of 10^{-11} s.

5.2.3
Scattering of Ions

Inelastic scattering of ions may occur at bound or free electrons in the specimen [4]. Principally, the same mechanisms of energy transfer as in electron–electron scattering occur, for example, ionization, generation of phonons or plasmons, and interband excitations. Ion–electron scattering dominates at high ion energies whereas ion–nucleus scattering dominates at low ion energies and large ion masses and leads to atom displacements. The energy transfer is determined by the screened Coulomb interaction and given by Eq. (5.2). The cross-over between the dominance of ion–electron and ion–nucleus scattering depends on the ion mass. For example, in a carbon target the cross-over for Ar ions is at 100 keV and for Xe ions is at 1 MeV. For protons, ion–electron scattering dominates at practically all ion energies.

5.3
Radiation Defects in Solids

5.3.1
The Formation of Defects

Radiation defects in solids are generated by ballistic atom displacements or by the breaking of bonds [5]. Since the latter has not been studied in much detail and did not show phenomena of comparable interest, we focus on defects generated by knock-on displacements of atoms. When an atom is displaced by an energetic electron or ion, a vacancy–interstitial pair (Frenkel pair) is generated. When the separation between vacancy and interstitial is low, the pair can recombine within a short time and this does not lead to any structural changes. At larger separations and/or when the mobility of at least one of the two species is low, persistent radiation damage has been generated and may lead to structural changes that are observable in the electron microscope.

A certain threshold energy has to be transferred to the atom to be separated at a minimum distance from its vacancy so that spontaneous recombination does not occur [6]. The displacement threshold is a characteristic energy for each material and also depends on the direction of incidence relative to the crystal orientation [7]. For reasons of momentum conservation, the threshold energy of beam electrons is orders of magnitude larger than that of ions. It is common to give the energy of the displaced atom and to calculate the respective energy of the projectile by Eq. (5.2) or (5.3). A compilation of values for different materials was given by Jung [6]. To give an example, the displacement threshold for atoms in graphenic materials such as carbon nanotubes is ~17 eV [8] whereas it is 30 eV in diamond crystals. Although the same type of atoms has to be displaced, the much denser packing of diamond requires a clearly higher energy of the atom to reach a position that is stable against recombination. The corresponding electron energies as calculated from Eq. (5.3) are 82 keV for graphenic structures and 200 keV for diamond.

5.3.2
The Migration of Defects

Approximately 10^{-11} s after the displacement of an atom, the migration behavior of the atom (now as an interstitial) and its vacancy determine the structural evolution of the specimen. The diffusivity D of an atom (or a vacancy) depends on the temperature T according to the Arrhenius law:

$$D = D_0 \exp\left(-\frac{E_a}{kT}\right) \tag{5.4}$$

where D_0 is a prefactor (given by the attempt frequency for jumps, the jump distance, and a geometric factor) and E_a is the activation energy. At higher temperature, the diffusivity of defects is fast enough to lead to either annihilation of defects of the same sign or to the agglomeration of defects of opposite sign. It is always a dynamic picture that appears during the formation of defects and their annealing and depends on the defect production rate (beam intensity) and temperature.

Systems under intense electron or ion irradiation are always far from thermal equilibrium. The dissipation of energy is high because only a small fraction of the energy that is transferred to the system is stored as persistent defects. The overwhelming fraction dissipates as heat, so that the energy flux through the system is considerable. Under these conditions, self-organization processes may appear [9] and have indeed been observed. The graphite–diamond transformation under electron irradiation [10, 11] can be considered as such a process, but other impressive phenomena such as the formation of void lattices [12] have also been observed.

It has to be kept in mind that defects can only be studied when they are visible in the electron microscope. Single vacancies or interstitial atoms have been hardly visible in previous studies, but the application of modern techniques of microscopy promise the imaging of single defects, as has already been demonstrated [13–15]. However, a precondition for defect imaging is that the defects do not migrate during the exposure time of the images (typically 0.05–1 s). Defect agglomerates, on the other hand, are much less mobile than single point defects and also much easier to observe in TEM images. Examples of visible agglomerates are voids or dislocation loops that have already been studied by conventional imaging techniques. The observation of the formation or decay of defect agglomerates gives us the possibility to study the diffusion of atoms or vacancies quantitatively. This was carried out in detail in the 1970s and 1980s.

5.4
Setup in the Electron Microscope

5.4.1
Electron Irradiation

Electron irradiation experiments can be done with almost every TEM system. When the electron energy is higher than the threshold energy for atom displacements, radiation effects occur. The ability of the microscope to concentrate the electron beam onto a very small and intense spot facilitates the irradiation of small specimen regions. This is achieved in microscopes with field emission guns and improved further when aberration correctors for the illumination system are available. Whereas the optimum conditions for imaging in the TEM are a parallel and coherent beam of moderate intensity, dedicated irradiation experiments often require a much brighter beam. This can be achieved when large condenser apertures are selected; the first condenser lens is weakly excited and, hence, the spot size is large. In a real *in-situ* experiment, the alterations of the specimen should be visible during irradiation, that is, the conditions for both imaging and irradiation should be fulfilled at the same time. Therefore, beam brightness, diameter, and coherence have to be adjusted so as to find an acceptable compromise. On the other hand, alternately irradiating and imaging the specimen give us the possibility of irradiating with a fully focused beam spot (then the illuminated area of the specimen is much too small for imaging) and imaging under optimum conditions (with a large beam diameter). The structuring of specimens with an electron beam of less than 1 nm in diameter has hardly been done to date and is an interesting challenge for future studies. The ability of modern TEM and scanning transmission electron microscopy (STEM) systems with a field emission gun and aberration-corrected condenser to focus the electron beam onto a spot 1 Å in diameter gives us the possibility to modify specimens on the atomic scale. Hence this could be a first step towards sub-nanometer engineering of materials.

To observe the defect evolution, it is often mandatory to use a heated specimen stage. These stages are commercially available and allow heating of the specimen up to 1000–1300 °C during imaging and irradiation. Tilting the specimen around two axes is also possible in some stages. Although the temperature of the stage is measured by a thermocouple, the actual temperature of the specimen detail under the beam is often difficult to estimate. The thermal conductivity of the specimen material and radiative heat losses have to be taken into account. A particular problem in high-temperature microscopy is the thermal expansion of the stage and the specimen material itself, leading to a drift of the image that can be severe in high-resolution microscopy. Water cooling of the rod of the specimen holder is usually applied to keep the temperature of the body low, but causes mechanical vibrations of the system. Electronic drift compensation by an adjustable piezo-driven specimen shift or an image shift by deflection coils can be used to overcome this problem.

The preparation of specimens has to be planned for the respective conditions of irradiation and heating. Cross-sectional specimens can be made by using special epoxies that are stable at temperatures up to 800 °C. Special grids should be used that do not melt, evaporate, weld to, or form alloys with the specimen material or the holder and do not react chemically with the specimen material. Molybdenum grids fulfill these conditions for most applications. Nanoparticles can be attached directly to the metal grids. Amorphous carbon films (holey/lacey carbon grids) should be avoided because they are often unstable under irradiation and at high temperature and overlap the specimen material in the projection, thus deteriorating the image.

5.4.2
Ion Irradiation

In-situ ion irradiation experiments in the electron microscope need an experimental setup where the beam line of an ion accelerator is attached to the specimen chamber of the electron microscope [16, 17]. In some setups, only small ion guns have been attached to the TEM column [18]. These are particularly suited to study ion implantation for applications in semiconductor technology. In other laboratories, large ion accelerators in separate rooms are connected to the TEM via a beam line. Such facilities have been realized, for example, at the Argonne National Laboratory (USA), the National Institute of Materials Science (NIMS) in Tsukuba (Japan), and the Center of Nuclear Spectrometry and Mass Spectrometry in Orsay (France). Several other facilities have been operated in Japan. The electron microscope itself does not need many modifications; an opening at the specimen chamber of the TEM can be made by the manufacturer of the TEM and does not significantly reduce the image resolution. The ion beam current can be measured by a Faraday cup on the specimen holder. Typical ion beam current densities are 10^{-7}–5×10^{-5} $A\,cm^{-2}$ and, therefore, 7–9 orders of magnitude lower than the electron beam current densities in the TEM.

As an example, the ion irradiation facility at the NIMS in Tsukuba is shown schematically in Figure 5.2. [19]. An ion accelerator with a 200 keV source is connected via a long beam line to a high-voltage electron microscope. The ion beam is directed at an angle of ~45° onto the specimen surface. Other facilities have different setups but the overall principle is the same. Ion accelerators with energies in the range from a few keV up to 2 MeV (Argonne) have been applied (even higher energies are planned at the facility in Orsay).

5.5
Experiments

In this section, a number of examples for *in-situ* irradiation experiments in electron microscopes are given. Due to limited space, only a selection of experiments can be presented, and the focus is on modern techniques such as high-resolution *in-situ* TEM of nanomaterials. Electron irradiation experiments from the 1960s to the 1980s to study, for example, the formation and annealing of vacancies in metals or the

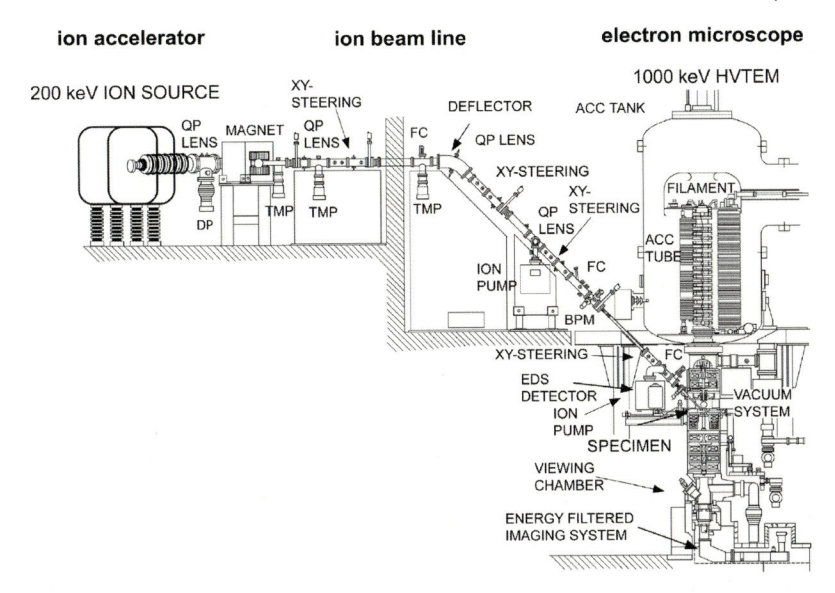

Figure 5.2 Setup of the ion irradiation facility at the NIMS in Tsukuba (Japan) [19]. An ion accelerator is connected by a beam line to the specimen chamber of a high-voltage electron microscope. Courtesy of K. Furuya.

growth of dislocation loops, can be found in the literature [5, 7]. The focus here is on electron rather than on ion irradiation, not only due to the author's expertise in this field but also because ion irradiation can only be carried out in a few facilities worldwide whereas an electron beam for irradiation experiments is available in every TEM.

5.5.1
Electron Irradiation

The scientific importance of electron irradiation studies was initially the understanding of defect formation, migration, and annealing. Nowadays, irradiation studies are aimed at generating or transforming structures with a precision that cannot be attained by conventional structuring techniques such as lithography. The engineering of nanosystems needs tools that are smaller than the systems themselves, and this can be achieved by using focused electron beams in electron microscopes. For efficient control of the process, the *in-situ* observation of the structuring is indispensable. However, real sub-nanometer structuring has only been demonstrated in a few examples hitherto, and the technique still needs to be developed further. The ultimate vision would be a downscaling of the established technology of focused ion beam (FIB) structuring towards the atomic scale. This could, in principle, already be achieved with the 1 Å electron beams in modern TEMs with corrected illumination systems, but the experiments are still in the early stages. Nevertheless, it has already been demonstrated that even electron beams with a much

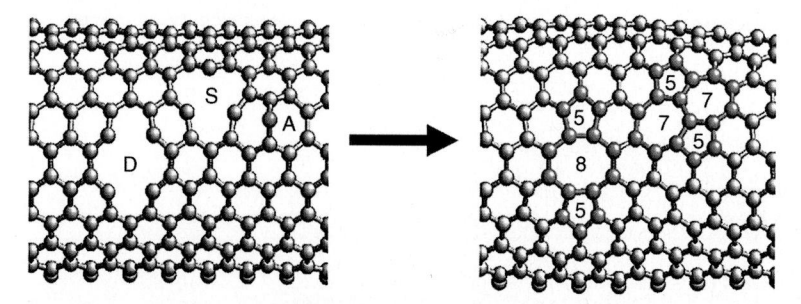

Figure 5.3 Reconstruction of a defective carbon nanotube as obtained from a computational study [4]. A single vacancy (S), a divacancy (D), and an adatom (A) reconstruct by forming pentagonal, heptagonal, and octagonal rings. Courtesy of A. Krasheninnikov.

larger diameter than the size of nanostructures can be beneficial in nanostructuring processes, as is shown in the following examples.

Nanoparticles based on graphitic carbon such as carbon nanotubes are the material system which has so far, shown the largest variety of structural transformations under electron irradiation. This is due to the unique ability of graphene layers to reconstruct after the formation of vacancies [4]. Perfect graphene (one monolayer of graphite) is built up by a network of hexagonal rings. Pentagonal, heptagonal, or octagonal rings are structurally stable in the graphene lattice but introduce curvature, so that closed graphenic cages (for example, the fullerenes) can be made. Whereas a radiation-induced monovacancy in graphene is stable, divacancies can close the lattice by reconstruction of the bonds so that no dangling bonds are left. The reconstruction of the lattice of a carbon nanotube is shown as an example in Figure 5.3 [4]. Upon removal of atoms, the tube changes its curvature locally and shrinks, but the shell remains coherent. Such a reconstruction is only possible above a certain temperature when vacancies are sufficiently mobile; a value of 200–300 °C has been found experimentally [3].

The shrinkage of graphitic cages under electron irradiation is responsible for a number of surprising phenomena. The first discovery was the formation of spherical carbon onions in 1992 [20]. These are concentric multi-shell fullerene clusters and form when graphitic nanoparticles are exposed to intense electron irradiation. A few years later, it was observed that carbon onions self-compress when irradiated at higher specimen temperatures and develop high pressure in their cores [21]. This can lead to the transformation of the graphitic core to a diamond crystal, as shown in Figure 5.4. Hence carbon onions can be used as nanoscopic compression cells when subjected to electron irradiation. Relatively simple techniques are available to place metal crystals inside carbon onions, for example, by co-evaporation of carbon and a metal in an arc discharge. In these core–shell arrangements, the compression of single nanometer-sized crystals can be carried out while the crystals are imaged at atomic resolution. By measuring the lattice spacings of metal crystals inside contracting carbon onions, pressures of more than 20 GPa have been obtained [22].

A contraction under the electron beam has also been observed in carbon nanotubes. Sustained electron irradiation of a nanotube leads to its collapse [23–25]. This is particularly interesting when the tube is filled with a metallic nanowire [26].

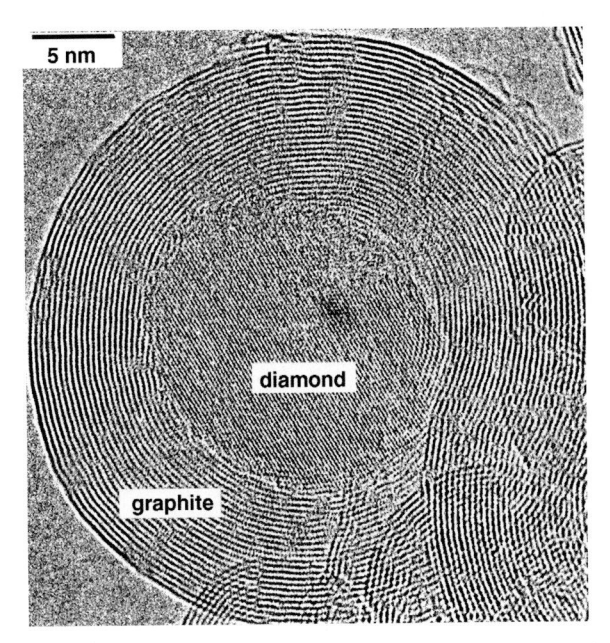

Figure 5.4 Nucleation of a diamond crystal inside a graphitic carbon onion under electron irradiation at 700 °C [21].

The collapse of the graphitic cylinders exerts pressure on the metal wire (also of the order 20 GPa) so that they are heavily deformed and squeezed through the shrinking channel, as shown in Figure 5.5 for the example of a cobalt crystal. This is a useful phenomenon because it allows us to study the deformation behavior of individual nanometer-sized crystals. Of course, the crystals must be robust against electron irradiation, that is, the displacement threshold of the crystal should be higher than the applied electron energy.

An experiment devoted to the deformation of nanocrystallites is shown in Figure 5.6. A gold crystal was encapsulated in a carbon onion and displaced through a small hole [27]. While the hole was "drilled" by a focused electron beam, the contraction of the carbon onion was induced by irradiating the whole particle with a larger beam. The absence of visible deformation defects at the temperature of this example (600 °C) and the appearance of defects at lower temperatures (not shown here) allows conclusions to be drawn on the role of defects during the deformation. Several other irradiation experiments on metal crystals encapsulated in graphitic shells have been undertaken. When the carbon shells are closed and the metal crystal is set under pressure, the diffusion of metal atoms outwards through the shells can be observed until the whole crystal has disappeared [28]. As a further example, the reaction between the graphitic shells and an encapsulated iron crystal has been induced by high temperature and compression and observed in detail [29].

The irradiation of carbon nanotubes with focused electron beams has been used to tailor the structure of the tubes. For example, multi-walled nanotubes can be peeled

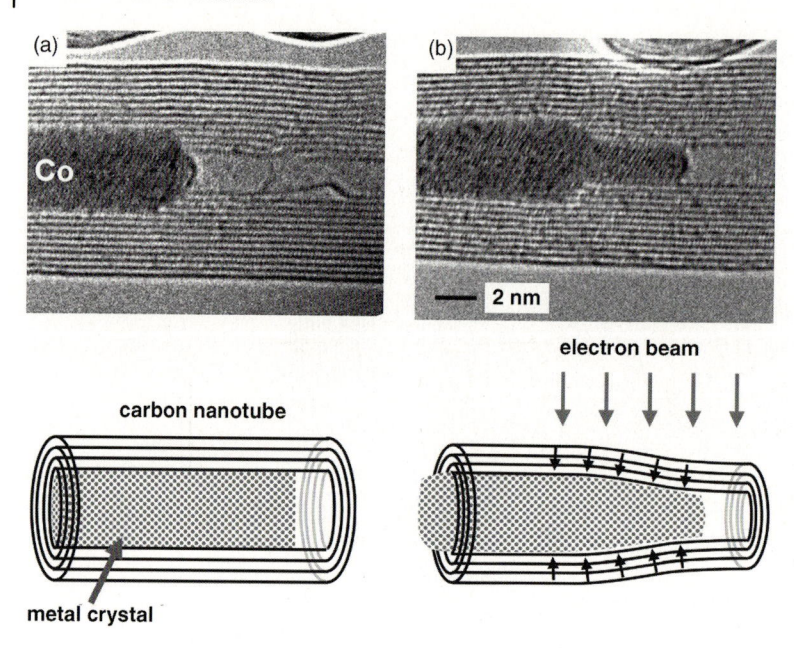

Figure 5.5 The electron irradiation of a multi-walled carbon nanotube filled with a crystalline Co wire leads to the collapse of the tube and the deformation and extrusion of the Co crystal (specimen temperature: 600 °C) [26].

or bent when the beam is focused on one side of the tube, as shown in Figure 5.7 [30]. Bundles of single-walled nanotubes can be cut by a focused electron beam and their open ends closed with caps. The particular behavior of the tubes during cutting allows us to draw conclusions about the mobility of carbon atoms inside the tubes [31].

Single-walled nanotubes can be welded by an electron beam so as to form a molecular junction, as shown in Figure 5.8 [32]. Although two perfect tubes would never join because the energy of such a junction is clearly higher than those of two individual tubes, the presence of vacancies can promote the coalescence. The coalescence of parallel single-walled tubes within a bundle has also been achieved by applying electron irradiation [33].

When atoms in composite structures are displaced by electron irradiation, this can lead to intermixing of the components. An example of a carbon nanotube filled with a metal crystal is shown in Figure 5.9 [34]. Here the carbon atoms are sputtered by the beam into the metal crystal. Since the solubility of carbon in transition metals is low and their diffusion is fast, the segregation of carbon on the surface of the metal occurs after a short time. In the particular geometry of this example, a new nanotube sprouts from the end of the metal crystal into the hollow channel of the host tube. This *in-situ* experiment allows us to observe the nucleation and growth of single- or multi-walled carbon nanotubes in real time and at high resolution.

The effects of electron irradiation on carbon materials have also been studied on an atomic scale by using lattice imaging with the highest resolution. Single vacancies in

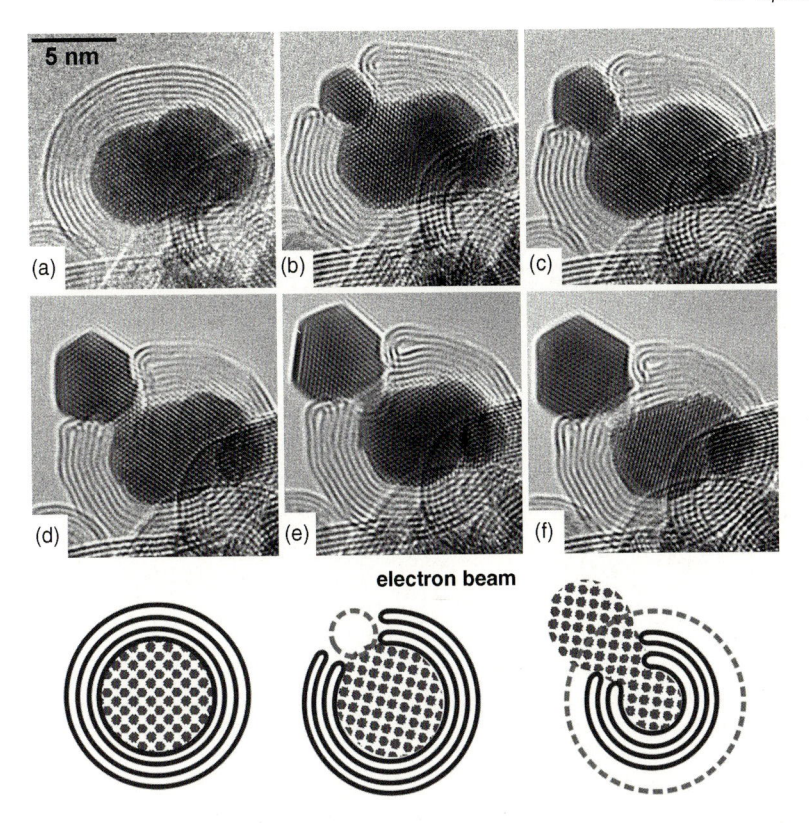

Figure 5.6 Plastic deformation of an Au crystal encapsulated by a spherical carbon onion at 600 °C. The graphitic shells were first punctured by a focused electron beam (b) and then irradiated with a uniform electron beam of large diameter. The shrinkage of the graphitic shells extrudes the Au crystal [27].

nanotubes or related graphitic nanoparticles have been generated deliberately and imaged [13]. In double-walled carbon nanotubes, inter-layer defects, presumably individual interstitial atoms, appeared under irradiation [14]. The generation and imaging of single point defects will certainly be a productive field for upcoming *in-situ* irradiation studies.

The displacement of atoms by an electron or ion beam can change the phase of certain materials. Crystalline structures can be disordered and finally amorphized when the mobility of atoms is low [35, 36]. Conversely, metastable amorphous solids can relax into the stable crystalline state under electron or ion irradiation when the activation energy is supplied by the energy transfer from the beam to the atoms. This has been shown in irradiation studies of Si [37]. However, even the transformation of a stable into a metastable crystalline phase can be achieved when the two phases coexist. This is shown in Figure 5.10 for the example of the transformation of graphite to diamond under electron irradiation. Although the free energy of the graphite phase is lower, the kinetics at the interface can favor the growth of diamond at the expense of

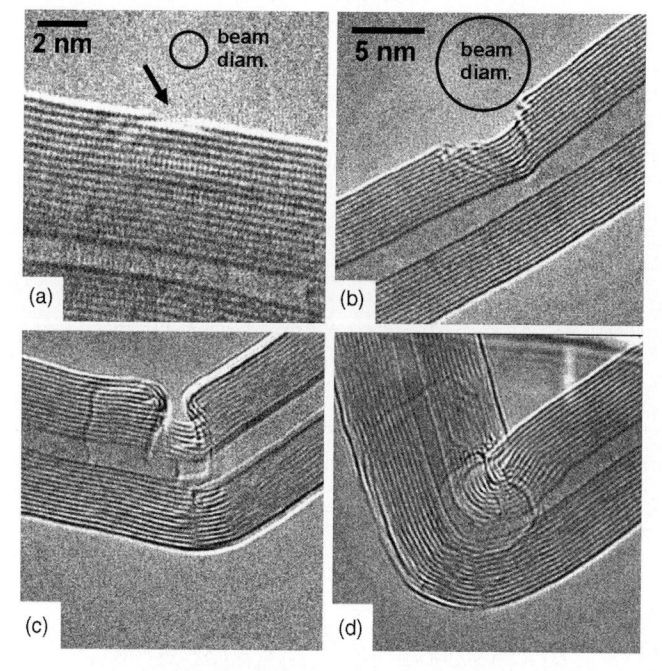

Figure 5.7 (a) By focusing an intense electron beam spot on to the periphery of a multi-walled carbon nanotube, the outermost shell can be removed locally. Irradiation with a larger electron beam (b–d) causes the bending of the tube (specimen temperature: 600 °C) [30].

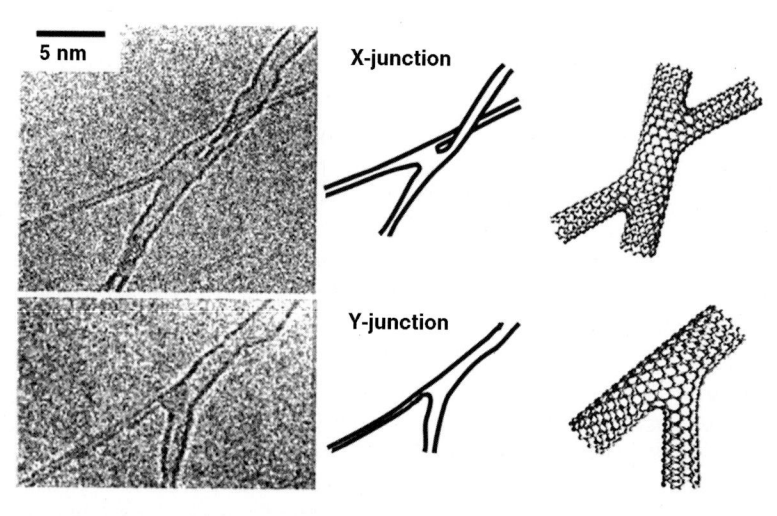

Figure 5.8 Electron irradiation of two crossing single-walled carbon nanotubes at 800 °C caused the welding of the tubes so as to form an X-junction (top images). By sputtering in the electron beam, one arm of the junction was removed so that a Y-junction was left (bottom images) [32].

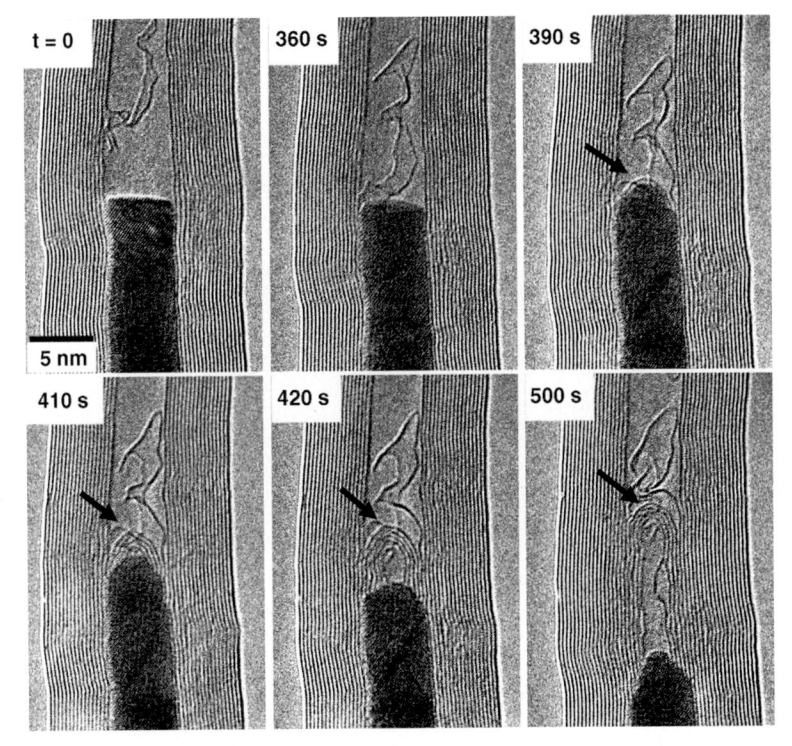

Figure 5.9 A multi-walled carbon nanotube filled with an FeCo crystal was exposed to electron irradiation at 600 °C. A new nanotube (arrowed) sprouts from the metal crystal as soon as the end of the metal crystal is hemispherical. The irradiation time is indicated [34].

graphite. This is achieved when the displacement rate in the metastable phase (diamond) is lower than that in the stable phase (graphite). Due to its dense packing, diamond has a higher radiation hardness, so that interfacial carbon atoms that aggregate at the diamond crystal survive longer before being displaced again than in the graphite crystal. In a certain temperature range, this leads to the growth of diamond under the beam [10, 11].

Electron irradiation has been shown to cause a number of interesting phenomena in nanometer-sized metal crystals. Melting or evaporation of nanocrystals can be promoted by electron irradiation [38]. Even when there is no transformation into the liquid phase, crystals can change their shape considerably under electron irradiation due to surface migration of atoms [39–41]. Electron irradiation of a dense arrangement of metal crystals on a substrate can cause Ostwald ripening, that is, the growth of larger at the expense of smaller crystals. This is due to the diffusion of metal atoms and the reduction of surface energy, and can also be observed when neighboring crystals coalesce under the beam [42]. This also happens in bulk materials composed of nanometer-sized crystallites (these are commonly denoted "nanocrystalline materials") and leads to grain growth or sintering [43]. Sputtering effects can be used for the structuring of metallic materials. This has been shown in studies where metal

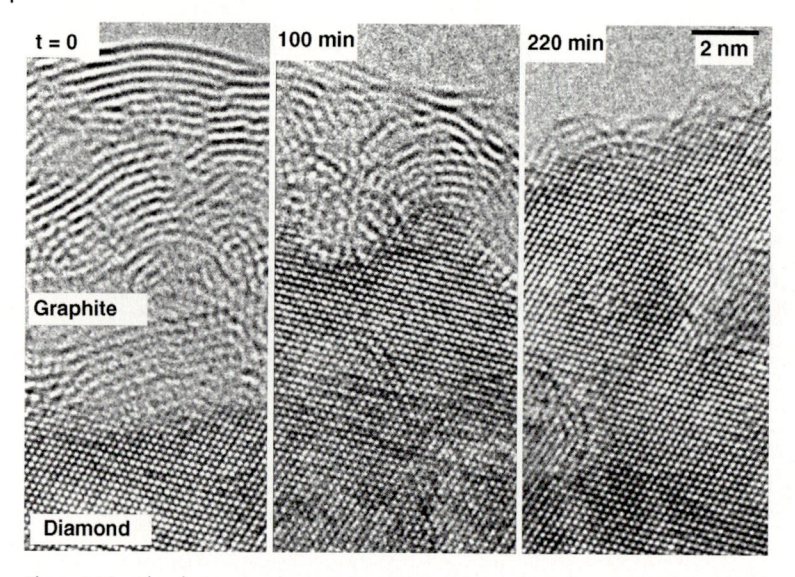

Figure 5.10 The electron irradiation of a graphite–diamond interface at 730 °C leads to the growth of diamond at the expense of graphite. The irradiation time is indicated. The metastable diamond phase grows although no pressure is applied [11].

wires were thinned by a focused electron beam [44–47] or holes were drilled into nanowires [48].

5.5.2
Ion Irradiation

Besides the differences in the technical setup, ion irradiation differs in many respects from electron irradiation. The high momentum transfer in an ion impact leads to a low range of the ions and, thus, to a much higher energy deposition, leading to considerable heating of even thin TEM specimens. In contrast to energetic electrons, most of the impinging ions are trapped in the sample and may form precipitates. Due to the high energy loss in a short distance, the density of radiation defects is much higher than under electron irradiation.

Ion irradiation experiments in electron microscopes have given detailed insight into the processes occurring during ion implantation and precipitation of the implanted species. An example is the implantation of Pb and Cd ions in an Al target. The formation of precipitates consisting of a PbCd alloy was observed *in-situ* [49]. The melting and solidification behavior of these alloy precipitates has been studied by varying the temperature of the specimen in the TEM.

Spectacular effects were observed when inert gases, for example, Ar, Kr, or Xe, were implanted in Al samples [50]. These gases are not soluble in Al but form precipitates. Solidification of Xe precipitates occurs due to the high pressure, of the order of several GPa, acting on the precipitate when displacing the metal matrix. This is shown in Figure 5.11 for Xe crystals in Al at room temperature. The lattice and the faceting of the Xe crystals can be clearly seen. Normally, these precipitates have the

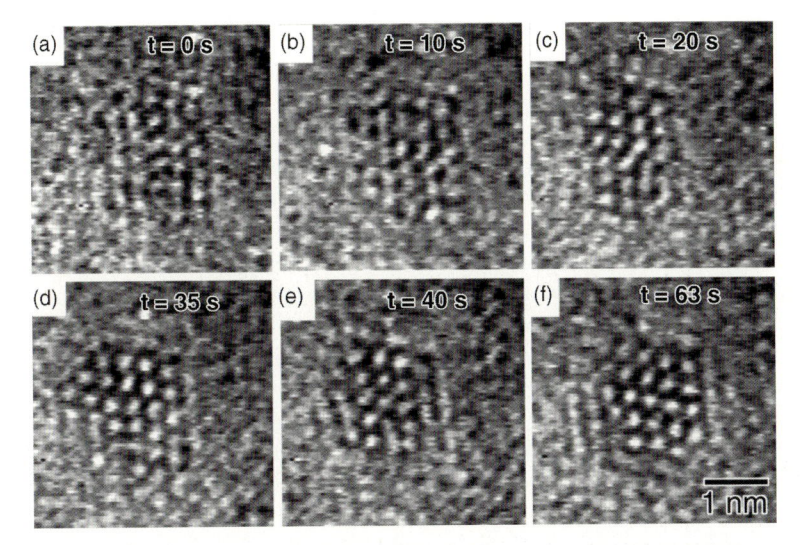

Figure 5.11 An Xe crystal nucleates and grows in an Al matrix during Xe implantation at room temperature. The ion irradiation time is indicated [50]. Courtesy of K. Furuya.

same phase (fcc) and the same crystal orientation as the metal matrix. Several *in-situ* experiments have been undertaken to study the properties and stability of Xe crystals. The formation of crystal defects such as stacking faults, amorphization, and defect annealing was observed, in addition to coalescence of the crystals. The influence of electron irradiation on the precipitates has also been studied.

Phase transformations are caused more efficiently by ion than by electron irradiation. Due to the rapid accumulation of radiation defects, amorphization of many materials can be achieved under high-dose irradiation. This has been carried out by irradiation with light and heavy inert gas ions of different energies in technically important alloys, for example, AlTi [51], at different temperatures. Besides amorphization, transformation to new crystalline phases has also been observed. Another example of a phase transformation is the surface transformation of steel from an fcc to a bcc phase under irradiation with P ions [52].

The formation of carbon onions, their self-compression, and the transformation of their cores to diamond (Section 5.5.1) has, after being carried out under an electron beam, also been successful by irradiation with Ne^+ ions (3 MeV) [53]. Although the ion irradiation was done *ex situ* in a separate ion accelerator, the similarity to the result of *in-situ* electron irradiation showed that the mechanisms of morphological and phase transformation are the same.

5.6
Outlook

The irradiation of materials with energetic electrons or ions has resulted in a large number of interesting phenomena on the nanoscale. However, not much would be known without the possibility of observing the processes *in-situ* in the electron microscope. The appeal of *in-situ* electron irradiation studies is the possibility of using the same electron beam for imaging and irradiation, and this can be done in almost every electron microscope. Hence it is not surprising that the unavoidable irradiation of materials with electrons during every inspection in electron microscopes has often led to accidental observations that have later emerged in dedicated irradiation studies.

Structural changes are nowadays studied over a wide range of specimen temperatures and with lattice resolution. However, the resolution of electron microscopes has been improved considerably in recent years by the implementation of aberration correctors, which will give the field of *in-situ* electron microscopy a new impetus. On the one hand, aberration-corrected objective lenses not only have a higher point resolution, but can also be made with larger gaps for *in-situ* experimentation. On the other hand, aberration correctors nowadays allow the focusing of an extremely bright electron beam on to a spot 0.1 nm in diameter, so that irradiation studies can be carried out with a selectivity on the real atomic scale. The sub-nanometer engineering of materials by focused electron beams promises to emerge as a new field of electron irradiation studies.

Acknowledgments

The examples shown in this chapter are from work in collaboration with A. Krasheninnikov, P.M. Ajayan, M. Terrones, J.X. Li, L. Sun, J.A. Rodriguez-Manzo, and Y. Lyutovich. K. Furuya contributed with information on ion irradiation and provided two figures.

References

1 Reimer, L. (1989) *Transmission Electron Microscopy*, Springer, Berlin.
2 Cherns, D., Minter, F.J., and Nelson, R.S. (1976) *Nucl. Instrum. Methods*, **132**, 369–376.
3 Banhart, F. (1999) *Rep. Prog. Phys.*, **62**, 1181–1221.
4 Krasheninnikov, A. and Banhart, F. (2007) *Nat. Mater.*, **6**, 723–733.
5 Urban, K. (1979) *Phys. Status Solidi A*, **56**, 157–168.
6 Jung, P. (1991) in *Landolt-Börnstein, New Series III-25* (ed. H. Ullmaier), Springer, Berlin, p. 1.
7 Vajda, P. (1977) *Rev. Mod. Phys.*, **49**, 481–521.
8 Smith, B.W. and Luzzi, D.W. (2001) *J. Appl. Phys.*, **90**, 3509 3515.
9 Seeger, A., Jin, N.Y., Phillipp, F., and Zaiser, M. (1991) *Ultramicroscopy*, **39**, 342–354.
10 Zaiser, M. and Banhart, F. (1997) *Phys. Rev. Lett.*, **79**, 3680–3683.
11 Lyutovich, Y. and Banhart, F. (1999) *Appl. Phys. Lett.*, **74**, 659–660.
12 Loomis, B.A., Gerber, S.B., and Taylor, A. (1977) *J. Nucl. Mater.*, **68**, 19–31.
13 Hashimoto, A., Suenaga, K., Gloter, A., Urita, K., and Iijima, S. (2004) *Nature*, **430**, 870–873.
14 Urita, K., Suenaga, K., Sugai, T., Shinohara, H., and Iijima, S. (2005) *Phys. Rev. Lett.*, **94**, 155502.
15 Zobelli, A., Ewels, C.P., Gloter, A., Seifert, G., Stephan, O., Csillag, S., and Colliex, C. (2006) *Nano Lett.*, **6**, 1955–1960.

16 Allen, C.W. (1994) *Ultramicroscopy*, **56**, 200–210.
17 Birtcher, R.C., Kirk, M.A., Furuya, K., Lumpkin, G.R., and Ruault, M.-O. (2005) *J. Mater. Res.*, **20**, 1654–1683.
18 Hojou, K., Furuno, S., Ohtsu, H., Izui, K., and Tsukamoto, T. (1988) *J. Nucl. Mater.*, **155–157**, 298–302.
19 Furuya, K., Mitsuishi, K., Song, M., and Saito, T. (1999) *J. Electron Microsc.*, **48**, 511–518.
20 Ugarte, D. (1992) *Nature*, **359**, 707–709.
21 Banhart, F. and Ajayan, P.M. (1996) *Nature*, **382**, 433–435.
22 Sun, L., Rodriguez-Manzo, J.A., and Banhart, F. (2006) *Appl. Phys. Lett.*, **89**, 263104.
23 Crespi, V.H., Chopra, N.G., Cohen, M.L., Zettl, A., and Louie, S.G. (1996) *Phys. Rev. B*, **54**, 5927–5931.
24 Ajayan, P.M., Ravikumar, V., and Charlier, J.-C. (1998) *Phys. Rev. Lett.*, **81**, 1437–1440.
25 Banhart, F., Li, J.X., and Krasheninnikov, A. (2005) *Phys. Rev. B*, **71**, 241408.
26 Sun, L., Banhart, F., Krasheninnikov, A., Rodriguez-Manzo, J.A., Terrones, M., and Ajayan, P.M. (2006) *Science*, **312**, 1199–1202.
27 Sun, L., Krasheninnikov, A., Ahlgren, T., Nordlund, K., and Banhart, F. (2008) *Phys. Rev. Lett.*, **101**, 156101.
28 Banhart, F., Füller., T., Redlich, Ph., and Ajayan, P.M. (1997) *Chem. Phys. Lett.*, **269**, 349–355.
29 Sun, L. and Banhart, F. (2006) *Appl. Phys. Lett.*, **88**, 193121.
30 Li, J.X. and Banhart, F. (2004) *Nano Lett.*, **4**, 1143–1146.
31 Gan, Y., Kotakoski, J., Krasheninnikov, A.V., Nordlund, K., and Banhart, F. (2008) *New J. Phys.*, **10**, 023022.
32 Terrones, M., Terrones, H., Banhart, F., Charlier, J.-C., and Ajayan, P.M. (2000) *Science*, **288**, 1226–1229.
33 Terrones, M., Grobert, N., Banhart, F., Charlier, J.-C., Terrones, H., and Ajayan, P.M. (2002) *Phys. Rev. Lett.*, **89**, 075505.
34 Rodriguez-Manzo, J.A., Terrones, M., Terrones, H., Sun, L., Banhart, F., and Kroto, H.W. (2007) *Nat. Nanotechnol.*, **2**, 307–311.
35 Carpenter, G.J.C. and Schulson, E.M. (1978) *J. Nucl. Mater.*, **23**, 180–189.
36 Mori, H. and Fujita, H. (1982) *Jpn. J. Appl. Phys.*, **21**, L494–L496.
37 Lulli, G., Merli, P.G., and Antisari, M.V. (1987) *Phys. Rev. B*, **36**, 8038–8042.
38 Lee, J.-G., Lee, J., Tanaka, T., and Mori, H. (2006) *Phys. Rev. Lett.*, **96**, 075504.
39 Smith, D.J., Petford-Long, A.K., Wallenberg, L.R., and Bovin, J.-O. (1986) *Science*, **233**, 872–875.
40 Ajayan, P.M. and Marks, L.D. (1998) *Phys. Rev. Lett.*, **63**, 279–282.
41 Narayanaswamy, D. and Marks, L.D. (1993) *Z. Phys. D*, **26**, S70–S72.
42 Flüeli, M., Buffat, P.A., and Borel, J.-P. (1988) *Surf. Sci.*, **202**, 343–353.
43 Chen, Y., Palmer, R., and Wilcoxon, J.P. (2006) *Langmuir*, **22**, 2851–2855.
44 Kondo, Y. and Takayanagi, K. (1997) *Phys. Rev. Lett.*, **79**, 3455–3458.
45 Takai, Y., Kawasaki, T., Kimura, Y., Ikuta, T., and Shimizu, R. (2001) *Phys. Rev. Lett.*, **87**, 106105.
46 Oshima, Y., Kondo, Y., and Takayanagi, K. (2003) *J. Electron Microsc.*, **52**, 49–55.
47 Zandbergen, H.W., van Duuren, R.J.H.A., Alkemade, P.F.A., Lientschnig, G., Vasquez, O., Dekker, C., and Tichelaar, F.D. (2005) *Nano Lett.*, **5**, 549–553.
48 Zhan, J., Bando, Y., Hu, J., and Golberg, D. (2006) *Appl. Phys. Lett.*, **89**, 243111.
49 Johnson, E., Touboltsev, V.S., Johansen, A., Dahmen, U., and Hagège, S. (1997) *Nucl. Instrum. Methods Phys. Res. B*, **127–128**, 727–733.
50 Song, M., Mitsuishi, K., and Furuya, K. (2001) *Mater. Sci. Eng. A*, **304–306**, 135–143.
51 Song, M., Mitsuishi, K., Takeguchi, M., Furuya, K., Tanabe, T., and Noda, T. (2000) *Philos. Mag. Lett.*, **80**, 661–668.
52 Johnson, E., Wohlenberg, T., and Grant, W.A. (1979) *Phase Transit.*, **1**, 23–33.
53 Wesolowski, P., Lyutovich, Y., Banhart, F., Carstanjen, H.D., and Kronmüller, H. (1997) *Appl. Phys. Lett.*, **71**, 1948–1950.

6
Observing Chemical Reactions Using Transmission Electron Microscopy
Renu Sharma

6.1
Introduction

Understanding the chemical reaction processes involved in materials synthesis and their functioning is central to our ability to control them. A number of technologies, such as thermogravimetric analysis, calorimetry, and temperature-programmed reduction, etc., are routinely used to determine the optimum reaction conditions such as temperature and pressure. Similarly, X-ray diffraction, Raman spectroscopy, infrared spectroscopy, scanning probe microscopy (SPM), scanning electron microscopy (SEM), transmission electron microscopy (TEM), and energy-dispersive X-ray spectroscopy (EDS), are examples of some of the techniques employed to characterize reactants and products before and after the reaction. However, it has been clear for some time that measurements performed on reactants and products are often not sufficient to determine subtle changes in the reaction mechanisms and kinetics. Therefore, *in-situ* observations of chemical reactions using the various techniques mentioned above are now commonly employed.

The need for nanoscale measurements stems from the rigorous control of synthesis conditions that is essential for nanofabrication. Continuing progress in nanotechnology requires that nanomaterials with a desired property must be produced in significant quantities. Moreover, the properties of nanomaterials are often controlled by their nanoscale structure and will be adversely affected by any structural modification occurring during their operation. Since TEM is one of the most powerful techniques for atomic-scale characterization, a number of modifications to the TEM sample holders and to the column have been made over the years that enabled chemical reaction kinetics and mechanisms to be followed at the nanoscale. The insights provided by these observations can be exploited to facilitate robust scaling of nanoscale synthesis processes to the manufacturing scale. The quest to observe chemical reactions at or near the atomic scale has been the driving force behind recent advances in both TEM specimen holders and TEM columns [1–13]. For example, solid-phase chemical reactions resulting from increase or decrease in temperature can be observed using suitable heating or cooling holders.

In-situ Electron Microscopy: Applications in Physics, Chemistry and Materials Science, First Edition.
Edited by Gerhard Dehm, James M. Howe, and Josef Zweck.
© 2012 Wiley-VCH Verlag GmbH & Co. KGaA. Published 2012 by Wiley-VCH Verlag GmbH & Co. KGaA.

More extensive modifications of the TEM column and/or the holders permit gas–solid reactions to be followed at the atomic level.

This chapter provides a brief description of currently available instrumentation, the types of chemical reactions that can be followed, experiment planning strategies, and examples of using TEM-related techniques to measure structural and chemical transformations and reaction kinetics. A discussion of the limitations and directions for future development is also given.

6.2
Instrumentation

Transmission electron microscopes require high vacuum (better than 1.34×10^{-4} Pa (10^{-6} Torr)) in order to avoid loss of image contrast/resolution arising from the multiple scattering of the incident electrons by the gas molecules. Recent developments in instrumentation have made it possible to introduce gas or liquid, confined to the sample area, to study various types of gas–solid or liquid–solid reactions at the nanoscale without losing image contrast or resolution. Broadly speaking there are two types of modifications that have been successfully employed for *in-situ* observations: specialized sample holders or differentially pumped TEM columns. Holders for heating and cooling are commercially available and have been extensively used to follow solid-state reactions such as phase transformations, decomposition, reduction, and dehydroxylation. Tan *et al.* have also incorporated the capability to measure electric field effects on the movement of domain boundaries in piezoelectric and ferroelectric single crystals at elevated temperatures [14, 15].

Several other groups have also constructed holders that confine fluids using electron-transparent windows. The basic principle of the windowed design is shown in Figure 6.1, where the sample is sandwiched between two electron-transparent windows and the flow of liquid or gas can be regulated using external pumps. Ross *et al.* have used such custom-built windowed holders to follow the etching process in Si by HF and the nucleation and growth of Cu particles during electrolysis of copper sulfate solution under static conditions (without flow) [16, 17]. Similar TEM holders have also been successfully employed to follow particle motion in liquids [18, 19]. Figure 6.2 shows a schematic of a commercially available windowed holder [2]. Windowed grids are sealed to the cap plates at the top and bottom using O-ring seals (Figure 6.2). The main drawback of the windowed design is that the window material must be thick enough, typically 50 nm, to withstand the pressure differential between the gas/liquid and microscope vacuum. The additional thickness of the confining windows tends to degrade the spatial resolution of the image, typically making high-resolution (lattice) imaging impossible. However, Creemer *et al.* [11] have recently demonstrated high-resolution images of catalyst particles in 1×10^5 Pa ($\sim$760 Torr) of H_2 at 500 °C using a windowed holder. This holder, which they call a nanoreactor, is based on microelectromechanical system (MEMS) technology, and features micrometer-scale gas flow channels and a heating device in a silicon chip, as shown in

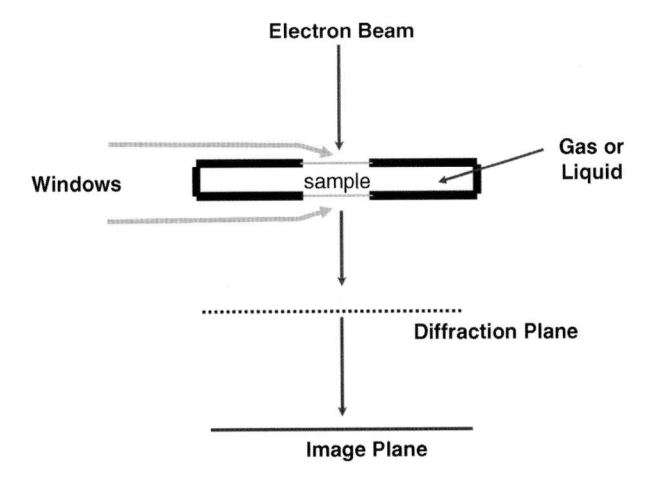

Figure 6.1 General principle of using electron-transparent windows to keep the gas or liquid confined to the area around the sample. The electron beam is transmitted through the windows, and interaction with the liquid/gas and solid samples forms a diffraction pattern at the diffraction plane and images on the image plane as in a regular TEM instrument.

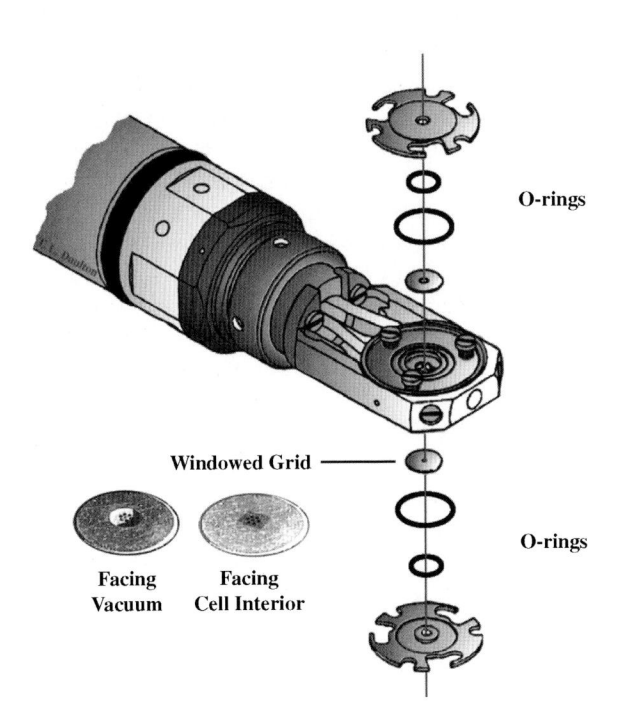

Figure 6.2 Schematic representation of a commercially available windowed holder consisting of two window grids (above and below the samples), O-rings for sealing, and top and bottom screw-fittings to keep the assembly in place. Figure reproduced with permission from Ref. [2]; © 2001, Springer-Verlag.

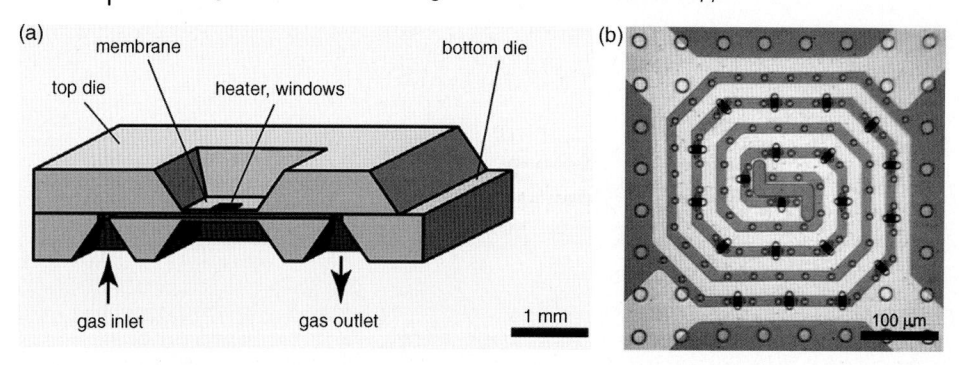

Figure 6.3 (a) Schematic cross-section of the nanoreactor showing micron-sized holes for gas inlet and outlet. (b) Optical close-up of the nanoreactor membrane. The bright spiral is the Pt heater. The small ellipses are the electron-transparent windows. The circles are the SiO_2 spacers that define the minimum height of the gas channel. Figure reproduced with permission from Ref. [11]; © 2008, Elsevier.

Figure 6.3. The high-resolution performance was achieved by thinning the electron-transparent windows locally (~10 nm diameter) to 10 nm thickness.

Alternatively, a gas-injection system can be incorporated into a heating holder such that a small amount of gas can be directly delivered in the vicinity of the samples, which are mounted on a wire-heating holder, as shown in Figure 6.4 [9]. This design has been successfully applied to follow catalytic reactions and the growth of $W_{18}O_{49}$ nanowires [10, 20, 21]. Such an approach relies on careful control of the gas flow to avoid the degradation of the column vacuum. Therefore, modifying the TEM column to follow gas–solid interactions is currently more popular than using modified holders as they allow high pressure to be achieved for an unobstructed view of the sample. Also, most of the modified holders for heating, cooling, indentation,

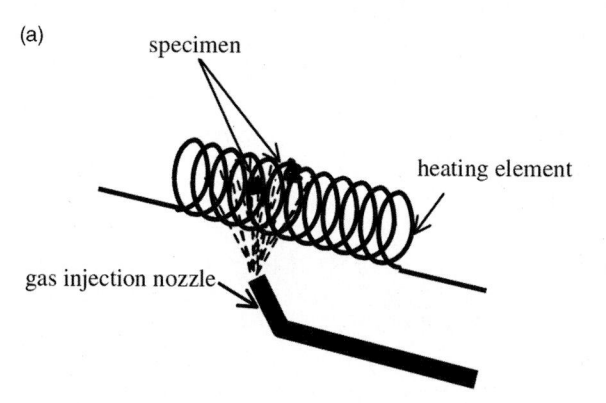

Figure 6.4 (a) Schematic diagram showing the location of the gas injector with respect to the heating coil on which a powder sample is loaded. (b) An optical image of the gas-injection/specimen heating holder. Figure reproduced with permission from Ref. [10]; © 2005, Oxford University Press.

(b)

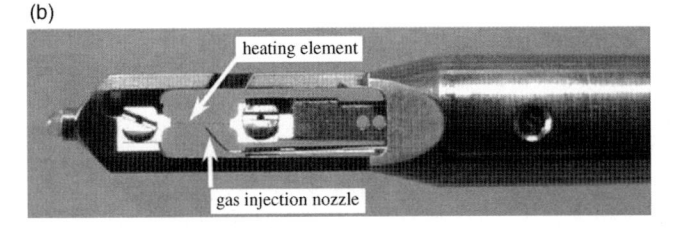

Figure 6.4 *(Continued)*

tomography, and so on can be used in these microscopes. The main drawback of a differentially pumped system is that the upper limit on achievable temperature and pressure are 900 °C and 2.5×10^3 Pa (approx. 1.9 Torr), respectively [22]. Therefore, we may not be able to achieve the same reaction conditions in the TEM column as used for most industrial applications.

Currently, two types of TEM instruments, with modified column, capable of *in-situ* observation of gas–solid interactions are commercially available. First are ultra-high vacuum (UHV) TEM systems, where a gas-injection system is incorporated in the TEM column to introduce low pressures, around 1.33×10^{-4} Pa (10^{-6} Torr), of gas and/or vapor. Such microscopes have been successfully used to follow the nucleation and growth of Ge islands on Si [23], growth of Si, Ge and InP nanowires [24–29], nitridation of Al_2O_3 [30], growth of carbon nanotubes [31], and so on. These microscopes are ideally suited for studying gas interactions with clean surfaces due to the low base pressures [1.33×10^{-8} Pa (10^{-10} Torr)], readily achieved in the column. Second are differentially pumped TEM systems that employ additional pumping in the objective pole-piece area to permit the introduction of gas into the sample region. These are usually referred to as environmental transmission electron microscopes (ETEM)s. This design was first proposed by Swann and Tighe [4] and has been improved constantly since then [1, 3, 7, 32–34]. The basic principle of a differential pumping system is shown in Figure 6.5. Gas is introduced into the sample area and then its leak rate into the rest of the column is reduced by using apertures above and below the sample area and pumping the gas leaking through these apertures using a turbomolecular pump (TMP) (Figure 6.5, 1st level of pumping). The gas leak rate is further reduced by introducing another TMP attached to the second outlet between the condenser aperture and top of the pole-piece for the upper half of the column, and between the selected area aperture and the differential pumping aperture, located above the viewing screen (Figure 6.5, 2nd level of pumping), for the lower half of the column. The pre-gun area is further evacuated using another pump, typically an ion pump (Figure 6.5, 3rd level of pumping). As a result, samples can be exposed to 1.33×10^3–2.66×10^3 Pa (10–20 Torr) gas pressure while the gun chamber is kept at 1.33×10^{-8} Pa (10^{-10} Torr). During the last 20 years, a number of groups have worked on this design and such microscopes are now commercially available. Details about the design and functioning of these microscopes have been published in a number of reviews [23, 35–39] and will not be covered in this chapter.

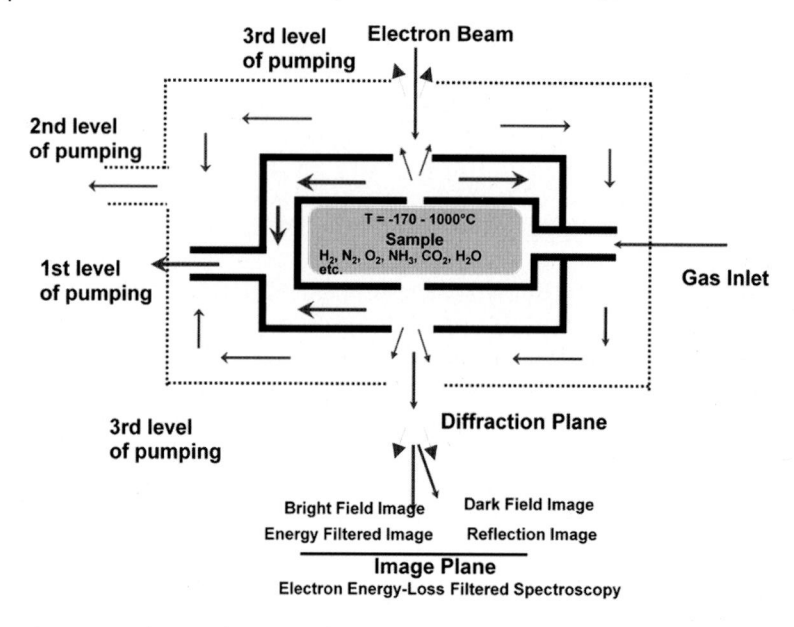

[Figure: Schematic flow-chart of a three-stage differential pumping system showing electron beam entry, sample chamber (T = -170 - 1000°C, with H₂, N₂, O₂, NH₃, CO₂, H₂O etc.), gas inlet, 1st/2nd/3rd levels of pumping, diffraction plane, and image plane with bright field image, dark field image, energy filtered image, reflection image, and Electron Energy-Loss Filtered Spectroscopy.]

Figure 6.5 Schematic flow-chart of a three-stage differential pumping system that is used to convert a TEM instrument to an ETEM. Gas is introduced in the sample area and the leak rate into the microscope column is restricted by a set of small apertures (~100 μm diameter), placed above and below the sample. Gas leaked through these apertures is pumped using a magnetically levitated turbomolecular pump (TMP; 1st level of pumping). The space between the condenser aperture and viewing chamber is pumped using another TMP (2nd level of pumping). The region between condenser aperture and gun chamber is pumped by an ion pump (3rd level of pumping).

6.3
Types of Chemical Reaction Suitable for TEM Observation

We can broadly categorize chemical reactions based on the physical state of the reactants and products, that is, gas, liquid or solid. TEM is often most suited to characterize solids; therefore, one of the reactants or products should be solid so that the structural and chemical changes occurring during the reaction can be followed using electron diffraction, imaging, and spectroscopic techniques. The following sections provide a short description of reaction types studied to date.

6.3.1
Oxidation and Reduction (Redox) Reactions

Oxidation is one of the most commonly occurring reactions, with everyday examples including the rusting of Fe and discoloration of Cu. Oxidation and reduction reactions are also fundamental to the function of a number of technologically important processes for energy generation and storage, such as fuel cells, electrochemical cells, and photocatalysts. Many of these processes involve oxidation/reduction cycles that are generally referred to as redox reactions. An important example of such a reaction is the functioning of the three-way catalyst system used in automobile exhausts. This catalyst system typically consists of Pt or Rh nanoparticles supported on ceria–zirconia mixed oxides. The function of a catalytic converter is to convert pollutant gases (NO_x, CO, and unreacted hydrocarbons) into relatively benign products (N_2, CO_2, and H_2O). During the process, ceria from the support is either reduced to give lattice oxygen to convert CO to CO_2 and unreacted hydrocarbons to CO_2 and H_2O, or oxidized by converting NO_x to N_2 [40, 41]. The redox process is accompanied by changes in the chemical composition and structure of the ceria support. *In-situ* observations of individual nanoparticles have revealed that the redox behavior is controlled by their chemical composition [42]. Details of the experimental protocol for this reaction are given in Section 6.5.2. The redox behavior of Au-Cu single crystals is described in Chapter 8.

6.3.2
Phase Transformations

The term "phase transformation" generally refers to transitions of states of matter, that is, from solid to liquid to gas or vice versa. However, when discussing solids, the phrase describes a change in the atomic positions such as occurs during the crystallization of amorphous material or martensitic atomic displacement during the quenching of a steel. Such transformations are often accompanied by changes in material properties; for example, the band gap in TiO_2 changes from 2.98 to 3.2 eV as a result of the structural transformation from rutile to anatase with temperature [43]. Similarly, the change in magnetization configuration in the ferromagnetic Ni_2MnGa shape memory alloy is accompanied by a cubic-to-tetragonal phase transformation at low temperatures [44]. Tsuchiya *et al.* [45] also observed the formation of intermediate structure in Ni–Mn–Ga alloys using *in-situ* electron diffraction. These studies generally combine Lorentz microscopy observations with electron diffraction to obtain images of magnetic domain structures during heating or cooling. The effect of electric fields on piezoelectric and ferroelectric transformation has also been observed using a TEM holder capable of heating samples when voltages of up to 600 V were applied. Tan *et al.* used this holder to determine the relationship between structural transformations and materials properties in Nb-doped lead zirconium titanate (PZT) [46].

6.3.3
Polymerization

During the last century, polymers found application in everyday products such as clothing fabric, ropes, and plastics. Polymers are long, covalently bonded chains consisting of multiple repetitions of relatively small organic molecules that act as a repeat unit. The formation of these large molecules is termed polymerization and is often assisted by a catalyst. Understanding the reaction process is fundamental to controlling the reaction conditions for maximum yield, control of molecular weight, polydispersity, and so on. Oleshko et al. used an ETEM to follow the mechanism for the synthesis of polypropylene from propylene using a Ziegler–Natta catalyst [47, 48].

6.3.4
Nitridation

Epitaxial growth of nitride films is often achieved by direct nitridation of III–V materials that have been deposited on a suitable substrate such as sapphire. The presence of impurities such as oxygen has been reported to degrade the functional properties of nitride films. Yeadon et al. used a UHV TEM instrument with a gas-injection system to follow the formation of AlN from Al_2O_3 films by heating the samples to 950 °C in NH_3 [30]. Based on their observations, they were able to propose a diffusion-limited reaction model involving transport of oxygen and nitrogen ions through the AlN epilayer growing between the free surface and the unreacted α-Al_2O_3.

The nitridation reaction is also a key process in the production of integrated circuits in the semiconductor industry. In order to stop the metal used for interconnects (on-chip wiring), such as Cu, Cr, or Ti, from diffusing into the Si substrate and forming unwanted silicides at high (operating) temperatures, Ti or Cr nitride is used as a barrier layer. The nitridation reactions of these metals have been studied by recording time- and temperature-resolved selected-area electron diffraction (SAED) as the samples were heated in $\sim$266 Pa (2 Torr) of NH_3 (a nitrogen source) in an ETEM [49]. Time- and temperature-resolved SAED patterns show that the nitridation temperature for Cu–Ti thin films decreases with increasing Cu content in the films. Measured reaction rates, obtained from low-magnification images, also agree with the reaction-controlled model for the nitridation reaction [50].

6.3.5
Hydroxylation and Dehydroxylation

Water plays an important role in many naturally occurring reactions. For example, clays or minerals, such as brucite and semactite, can take in water (hydroxylate) or give up water (dehydroxylate) as the environmental conditions change. The phenomenon results in structural transformation of materials as they are converted from oxide to hydroxide and back to oxide. Dehydroxylation generally occurs upon heating whereas hydroxylation can occur at room temperature. It has been shown that hydroxylation processes depend upon the structure and morphology of the starting materials [51, 52]. For example, we have shown that the hydroxylation rate for crystalline magnesium oxide (MgO) cubes (5–10 nm) is much lower than for amorphous MgO nanoparticles (1–5 nm) freshly prepared by dehydroxylation of magnesium hydroxide [$Mg(OH)_2$] [53].

Water molecules in the atmosphere are also responsible for the deliquescence and efflorescence of atmospheric nanoparticles. Freney et al. used the ETEM to understand the effects of humidity, particle size, and chemical nature of atmospheric particles on deliquescence and efflorescence phenomena relevant to aerosols [54, 55]. Their results provided new data for aqueous electrolytes that aid in our understanding of atmospheric science [56, 57].

6.3.6
Nucleation and Growth of Nanostructures

One approach to the effective use of one-dimensional nanostructures, such as nanotubes and nanowires in devices, is to synthesize them *in-situ* during the device fabrication process in such a way that they have the precise electronic properties required. Understanding the nucleation and growth mechanisms is an essential step for optimizing their synthesis conditions. During the last decade, *in-situ* TEM observations have been extensively used to understand the nucleation and growth of a variety of nanoparticles, nanowires, and nanotubes [26, 29, 58–65]. Growth mechanisms for oxide nanostructures such as $W_{18}O_{49}$ nanowires [21] and BiO whiskers [66] have been revealed using a gas-injection heating holder.

Figure 6.6a shows a series of images, extracted from a digital video, recorded as a low pressure [0.013 Pa (10^{-4} Torr)] of disilane (Si_2H_4) was introduced into the ETEM column over an Au/SiO_2 thin-film sample heated to 590 °C. It is clear from these images that the Si nanowires nucleate from supersaturated Au–Si eutectic liquid. Careful measurements show (Figure 6.6b) that the incubation period R is dependent on the Au particle size (cross-sectional area A_0) with wires nucleating from small particles earlier than from the large particles [26, 29]. Based on the data shown in Figure 6.6b, it can be concluded that the incubation period is approximately equal to the square root of the cross-sectional area or radius of the particles ($R \approx \sqrt{A}$) [29].

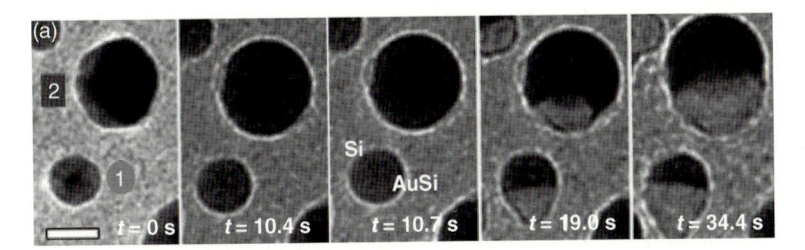

Figure 6.6 An ETEM image sequence of Au nanoparticles supported on SiO_x during exposure to ~0.13 Pa of Si_2H_6 at 590 °C (scale bar = 10 nm), extracted from a video. The time elapsed between images with $t=0$ s roughly equal to the onset of disilane exposure is marked on individual images. Note that the nucleation onset (incubation) times (R) for particles 1 and 2 are different. (b) Incubation time (R), given in seconds for Si precipitation versus A_0, the cross-sectional area of the initial Au crystal with data points for particles marked as 1 and 2, respectively. Figure reproduced with permission from Ref. [30]; © 2008, Nature Publishing Group.

(b)

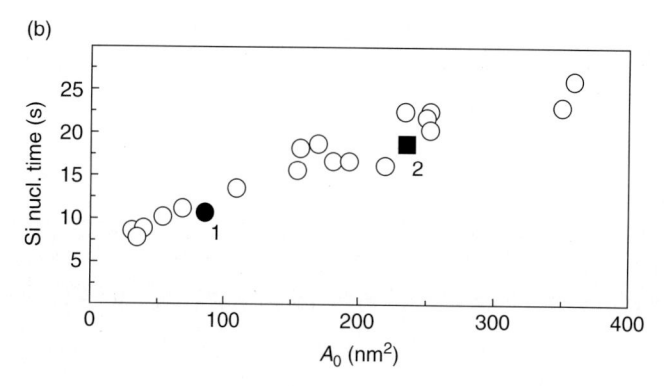

[Figure: scatter plot with y-axis "Si nucl. time (s)" ranging 0–25, x-axis "A_0 (nm^2)" ranging 0–400, showing two groups of data points labeled 1 and 2]

Figure 6.6 (Continued)

6.4
Experimental Setup

It is imperative to emphasize that for *in-situ* TEM observations of chemical reactions, the column acts as both reaction cell and characterization tool. In other words, we perform experiments in the TEM column and characterize reactants and products concurrently; therefore, we need to pay special attention to the choice of the TEM grid/support material, heating/cooling holders, and the possible interactions of ambient gas (liquid) with these components. Moreover, it is important to evaluate the effect of the electron beam on the reaction path. In the following sections, some important factors to consider before performing *in-situ* TEM experiments are given.

6.4.1
Reaction of Ambient Environment with Various TEM Components

It is important to know the nature of the materials used for various components of the microscope column and specimen holders, such as the body of the holder, body of the heating furnace, washers, and wiring material. Some of this information may be proprietary and therefore not readily available. In this case, the manufacturer should be consulted to make sure that the liquids or gases to be used or produced during reaction will not harm the instrumental components. For example, Ta heating holders will oxidize when exposed to oxygen or air at high temperature, and Pt heating wires will form silicides when heated above 600 °C in the presence of silane or disilane. Also, the relevant phase diagrams should be examined to make sure that the experiments can be safely performed using the available instrumentation.

6.4.2
Reaction of Grid/Support Materials with the Sample or with Each Other

Special attention must be paid to choice of the material for the specimen support (e.g., TEM grid). For example, Cu is one of the most commonly used TEM grid materials

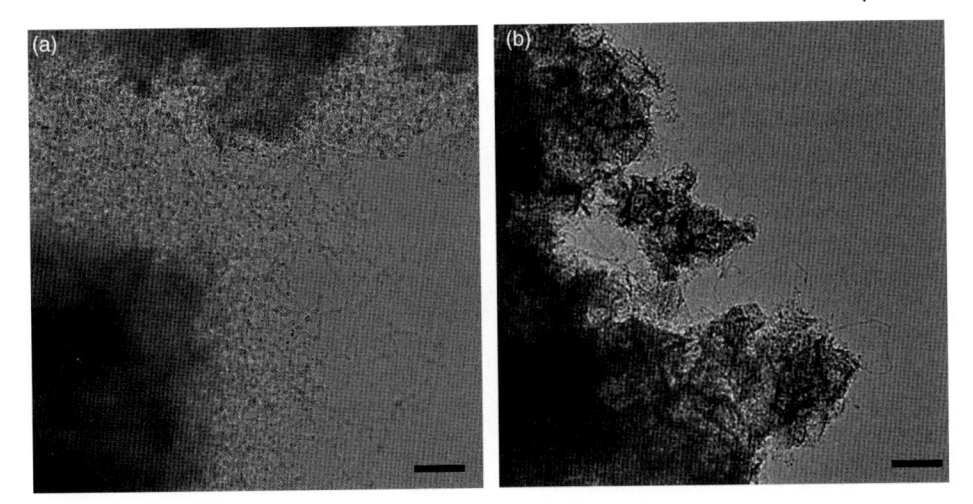

Figure 6.7 Low-magnification images recorded after CNT growth, using ESTEM, when an Ni/SiO$_2$ sample was dry loaded on to (a) an Au and (b) an Ni support grid under otherwise identical conditions. Note the increased yield of CNTs formed on the Au grid. Scale bar = 100 nm.

for loading samples. It is obvious that one should not use Cu grids for reactions above the melting point (1083 °C), but it has been shown that metal atoms can start to diffuse at temperatures as low as half the melting temperature (as measured in kelvin; the Taman temperature), which is ∼400 °C for Cu. The diffusion of metal atoms on to the samples near or above the Taman temperature can affect the results. For example, we have found that the yield of carbon nanotube formation increased noticeably at ∼500 °C when Ni/SiO$_2$ catalyst was loaded directly on to Au grids instead of Ni grids, as shown in Figure 6.7. We believe that since the reaction temperature was above the Taman temperature for Au (396 °C), some Au atoms diffused to the catalyst particles and changed their reactivity. Controlled experiments with varying (0.1–0.8 mole fraction) confirmed that Ni doped with less than 0.2 mole fraction of Au increases the CNT yield considerably. Another example is CO reacting with Ni grids above 800 °C to form nickel carbonyl, a volatile product.

6.4.3
Temperature and Pressure Considerations

The maximum achievable temperature depends first and foremost upon the modified heating holder and varies between 1000 and 1500 °C, depending on the source and design. For example, Kamino and Saka designed holders that are stable above 1500 °C [9]. Allard *et al.* recently reported a heating holder capable of cycling temperatures from ambient to above 1000 °C in 1 ms [13]. The steady-state temperature at the sample using this holder is stable enough to allow the capture of atomic resolution images in both TEM and scanning transmission electron microscopy (STEM) modes. Other heating holders are also commercially available with approximately the same temperature range.

However, for *in-situ* observations of gas–solid interactions, the temperature limit depends not only on the heating holder, but also on the components of the TEM column. For example, some of the internal components of the commercially available differentially pumped ETEMs or environmental scanning TEMs (ESTEMs) cannot withstand high temperatures. As gases transport the heat from the sample to other parts of the column, the upper limit for achievable temperature with gas flow is 900 °C in a modern commercial ESTEM, even if the heating holder is capable of achieving higher temperatures.

6.4.4
Selecting Appropriate Characterization Technique(s)

It is important to note that all of the TEM-related techniques such as imaging (bright-field, dark-field, low- and high-resolution), electron diffraction (SAED, convergent beam, and electron nanodiffraction), STEM (annular dark-field and high-angle-annular dark-field), electron energy-loss and energy-dispersive spectroscopy, tomography, and holography can be used for *in-situ* observations. Some of these techniques can be combined within the same experiment, depending on the temporal resolution of the desired technique relative to the reaction rate of the chemical process under observation. Ideally, we should combine more than one technique to identify unequivocally each step of the reaction process. For example, structural information can be obtained from either diffraction patterns or high-resolution images, but spectroscopic data need to be collected to determine the chemical changes. Examples given in this chapter (and elsewhere in this book) can be used as guides for selecting a technique or a set of techniques that should be used to obtain as much information as possible for a given chemical reaction. However, it should be kept in mind that the power of *in-situ* TEM lies in providing atomic-scale information on the reaction mechanism, the relationship between local composition and reactivity, the relationship between local structure and properties, and so on, and not the bulk behavior.

6.4.5
Recording Media

The ability to obtain useful information from *in-situ* TEM imaging or spectroscopic data is strongly dependent upon available temporal resolution of the recording media. Both digital and analog cameras, currently available for recording high-resolution images, are limited to a frame rate of $\sim 30\,\mathrm{s}^{-1}$ [frames per second (fps)]. Improving the time resolution for high-resolution electron microscopy imaging is not a trivial challenge as it depends upon both the detector efficiency and electron dose [67]. Typical beam currents are between 1 and 10 nA; assuming recording takes place at a video rate of 30 fps or 0.03 s per frame, the number of electrons per frame is between 6×10^9 and 6×10^{10}. Assuming that images have the standard National Television System Committee (NTSC) resolution of 440×480 pixels, there are on average 750 electrons per pixel. Assuming that one wants to detect a minimum of eight contrast levels, 64 electrons per pixel would be necessary on a perfect detector.

In order to reduce the electron dose, we need to improve the detection quantum efficiency (DQE) of the camera. Typical camera DQEs are between 0.07 and 0.7 across the spatial frequency and kV range of interest, thus requiring between 100 and 1000 electrons per pixel. This means that, at the resolution and sensitivity of current cameras, we are already at (or close to) the practical limit of the frame rate. Increasing image integration times is frequently not an option as the required dose may alter (damage) many samples of interest. Increases in either the frame rate or the resolution in terms of number of pixels per frame will require improvements in detector sensitivity (DQE) and possibly also new designs for the illumination systems of microscopes to maximize the beam current.

In 2005, a research group at Lawrence Livermore National Laboratory modified the TEM column to incorporate laser pulses to initiate a transient process to be investigated, for example, phase transformations or chemical reactions, and timed it with the electron probe, also stimulated by a laser pulse, precisely such that snapshots could be recorded with 15 ns temporal resolution [68]. This microscope has been successfully employed to understand thermal annealing processes in thin films and the laser ablation mechanism for the synthesis of nanowires [69], and chapter 3 of this book.

The temporal resolution that can be achieved for spectroscopy is lower than that for imaging: 2–10 s or more compared with 0.34 s for video imaging. Therefore, spectroscopic techniques are often used to analyze a sample before and then after a reaction, unless the process is sufficiently slow. However, the recent introduction of a new generation of the Gatan GIF Quantum series of imaging energy filters, which allow the efficient collection of spectra at rates of up to $1000\,s^{-1}$, is beginning to help overcome this limitation.

6.4.6
Independent Verification of the Results and the Effects of the Electron Beam

Limitations of *in-situ* TEM characterization include the effect of the electron beam and the low achievable signal statistics of the data. Moreover, as samples suitable for TEM are usually of nanometer-scale thickness, they may not represent the reaction mechanisms as they occur in bulk materials. Also, the reactions conditions, such as temperature and pressure that can be produced in the TEM system, may not represent the real-life situation. Therefore, it is important to verify the thermodynamic and kinetic parameters using other techniques such as XRD, thermogravimetric analysis, and Raman spectroscopy. Also, the structure and chemistry of reactants and products subjected to the same reaction conditions as used in the TEM system should be tested on bulk samples.

6.5
Available Information Under Reaction Conditions

In order to understand the reaction mechanisms, thermodynamics, and kinetics, it is necessary to make different sets of measurements depending upon the information needed. Most studies employ TEM to record data from one or a combination of imaging and spectroscopic techniques applied to a sample under reaction conditions. The following examples provide a guide to selecting the specific combination of techniques most suitable to follow a particular reaction process. Detailed information about individual reactions can be found in the references provided in this chapter and in other chapters of this book.

6.5.1
Structural Modification

Identifying structural changes during a reaction is the most frequently used technique for following reaction mechanisms and/or paths. These changes may occur during any of the reactions mentioned in Section 6.3. Time- and temperature-resolved high-resolution imaging and/or electron diffraction can be used to determine changes in local structure. The time resolution is dependent on the recording media but, as discussed above, is currently limited to video frame rates, that is, 30 or $0.034\,s^{-1}$, whereas the temperature resolution is dependent on the heating holder and TEM configuration used, as explained in Section 6.4.

6.5.1.1 Electron Diffraction

SAED patterns, recorded at various temperatures and time intervals (temperature- and time-resolved), are commonly used to identify the onset of reaction and the formation of intermediate phases. Schoen et al. performed controlled heating experiments to follow the formation of $CuInSe_2$ nanowires by solid-state diffusion of Cu in α-In_2Se_3 nanowires oriented along different directions with respect to the Cu source [70]. Although there was no appreciable change in structure and morphology, the diffraction patterns changed due to incorporation of Cu in the In_2Se_3 lattice and the transformation from the In_2Se_3 to the $CuInSe_2$ structure. The appearance of streaking in the diffraction patterns along 0001, indicates that the transformation to $CuInSe_2$ proceeds through the formation of a disordered intermediate structure at temperatures as low as 225 °C as Cu is incorporated within the In_2Se_3 phase (Figure 6.8a and b). An ordered $CuInSe_2$ structure (Figure 6.8c) that is stable after cooling (Figure 6.8d) was observed to form upon further heating to 350 °C. Conversely, In_2Se_3 nanowires oriented along the $\langle11\bar{2}0\rangle$ direction with respect to the Cu source transform directly to the crystalline $CuInSe_2$ phase above 470 °C, indicating that Cu diffusion in In_2Se_3 is dependent upon the crystallographic orientation. It also shows that the synthesis temperature of nanowires with anisotropic structures can be controlled by their crystallographic orientation.

Other examples of determining structural transformations using electron diffraction include the nitridation reaction of Cu–Cr and Cu–Ti thin films [49], the initial stages of oxidation of Cu–Au alloy [71], the structural phase transition in $Ca_2Fe_2O_5$ [72], the microstructural evolution of Ni–Al thin films with temperature [73], and the electric field-induced phase transition in Nb-doped $Pb(Zr_{0.95}Ti_{0.05})O_3$ [46].

6.5.1.2 High-Resolution Imaging

Time- and temperature-resolved high-resolution imaging is another method for following structural changes during chemical reaction. Sayagués and Hutchison

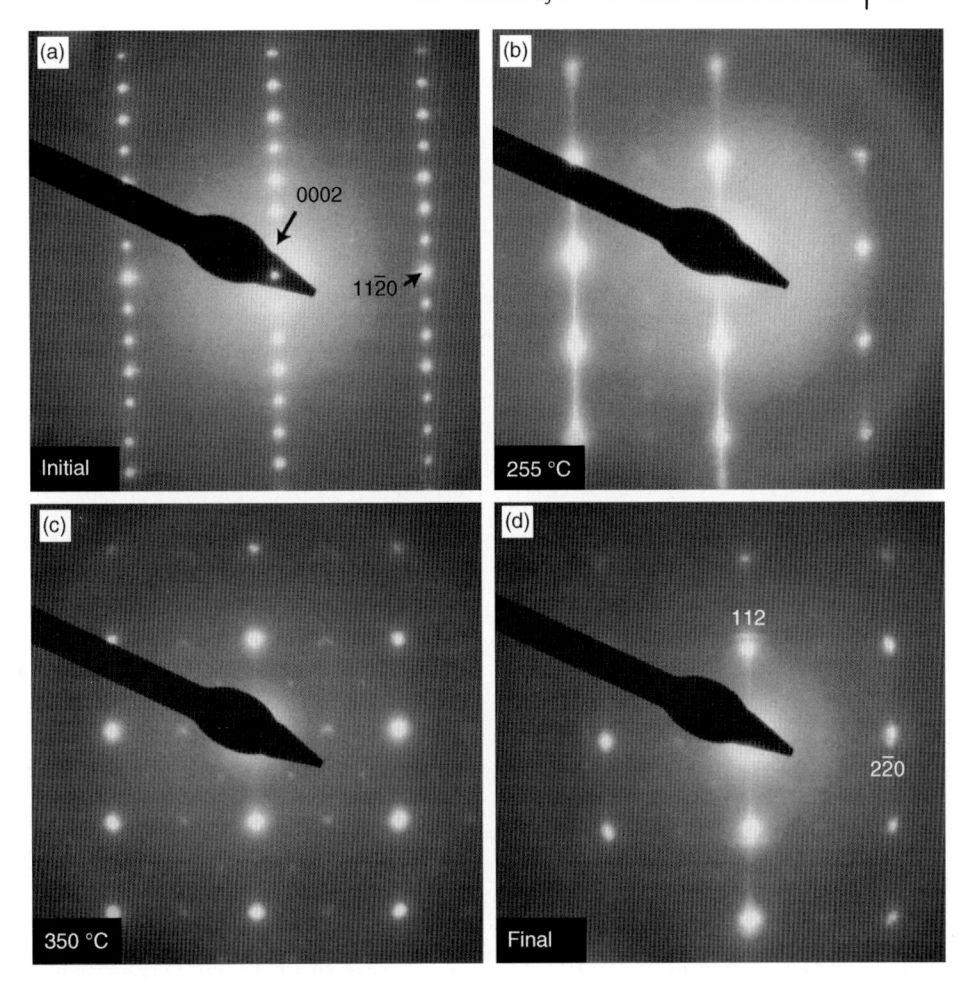

Figure 6.8 Selected-area electron diffraction patterns of an In_2Se_3 nanowire with Cu contacts, acquired during *in-situ* heating to 350 °C. (a) Prior to heating, the pattern indicates a hexagonal structure with a well-defined superstructure normal to the basal planes (0001). (b) The streaking along 0001 at 225 °C indicates the start of Cu diffusion in this plane. (c) A defect-free $CuInSe_2$ nanowire with a cubic structure formed upon further heating to 350 °C. (d) The cubic structure was stable upon cooling to room temperature. Figure reproduced with permission from Ref. [71]; © 2009, American Chemical Society.

at the University of Oxford successfully used an ETEM to determine the structural modifications occurring during oxidation reactions of $Nb_{12}O_{29}$ to $Nb_{22}O_{54}$ [74]. Moreover, they also synthesized a new compound using the chemical and structural information of the defect structure formed during the process [75]. Temperature-resolved high-resolution imaging has also revealed the appearance and disappearance of an ordered superlattice during temperature cycling from 600 to 700 °C for ceria (CeO_2) [76]. These structures were formed due to oxygen vacancy ordering in CeO_2 during reduction in H_2 at 730 °C and disappeared as the crystal reoxidized upon cooling to 600 °C. This was a direct observation of the redox behavior of CeO_2 and a measure of its oxygen storage capacity [76].

Nucleation-and-growth studies of a number of nanostructures such as nanowires and nanotubes have also been reported. High-resolution images have shown that these materials nucleate and grow via a vapor–liquid–solid (VLS) or vapor–solid–sold (VSS) mechanism depending upon the chemical system [28, 29, 61, 62]. For example, Si nanowires grow via a VLS mechanism when silane or disilane is introduced in the TEM column over an Au catalyst heated above the Au–Si eutectic temperature. On the other hand, they grow via a VSS mechanism below the eutectic temperature using Pd as catalyst. Whereas Au–Si liquid acts as the catalyst in the former case, solid $PdSi_2$ is the catalyst in the latter [29].

CNTs have also been reported to grow via a VSS mechanism [58, 65]. High-resolution images have recently been used to determine the physical and chemical state of iron-based catalyst particles during CNT growth [58, 77]. Sharma et al. have also shown the structural transformations occurring in the iron catalyst particles before CNT nucleation and growth [77]. They used the column of an ESTEM as a flow reactor for both Fe particle and CNT synthesis. First, arrays of equidistant Fe particles were deposited on a perforated SiO_2 thin film suspended on an Si wafer support by electron beam-induced decomposition (EBID) of nonacarbonyldiiron [$Fe_2(CO)_9$] vapor at room temperature in the column of the ETEM. In-situ electron energy-loss spectroscopy (EELS) data confirmed the formation of Fe-containing particles and a small amount of carbon. Next, these particles were heated to the reaction temperature (650 °C) in hydrogen to remove the co-deposited carbon. The hydrogen was then replaced by acetylene, leading to the formation of CNTs [38].

Structural transformations in the catalyst particles could be deduced from the diffractograms of individual high-resolution images extracted from the video sequence. Fe particles were found to oxidize in the low vacuum (0.001 Pa) of the TEM column to form magnetite (Fe_3O_4) upon heating (Figure 6.9a), but to reduce when 0.08 Pa of acetylene at 650 °C was introduced in the sample region. The face-centered cubic (fcc) magnetite structure transformed to a body-centered cubic (bcc) oxide structure before reducing to ferrite (bcc α-Fe; Figure 6.9b). In the next stage, the particles were carburized to iron carbide (cementite, Fe_3C; Figure 6.9c) before nucleating the CNTs (Figure 6.9d). Time-resolved high-resolution imaging confirmed that the particles remained crystalline throughout these phase transformations. High-resolution images were also used to obtain an atomic-scale understanding of the nucleation of graphene sheets on the Fe-terminated (001) surface of cementite. The following reaction sequence was deduced using high-resolution images recorded as the reactions proceeded:

$$Fe_2(CO)_9 \rightarrow Fe \text{ and } C \xrightarrow{H_2/H_2O/650°C} Fe_3O_4 \text{ (fcc)} \xrightarrow{C_2H_2/650°C} \text{Iron oxide (bcc)} \xrightarrow{C_2H_2/650°C}$$

$$\alpha\text{-Fe} + (CO_2 + H_2) \xrightarrow{C_2H_2/650°C} Fe_3C \xrightarrow{C_2H_2/650°C} Fe_3C + CNT$$

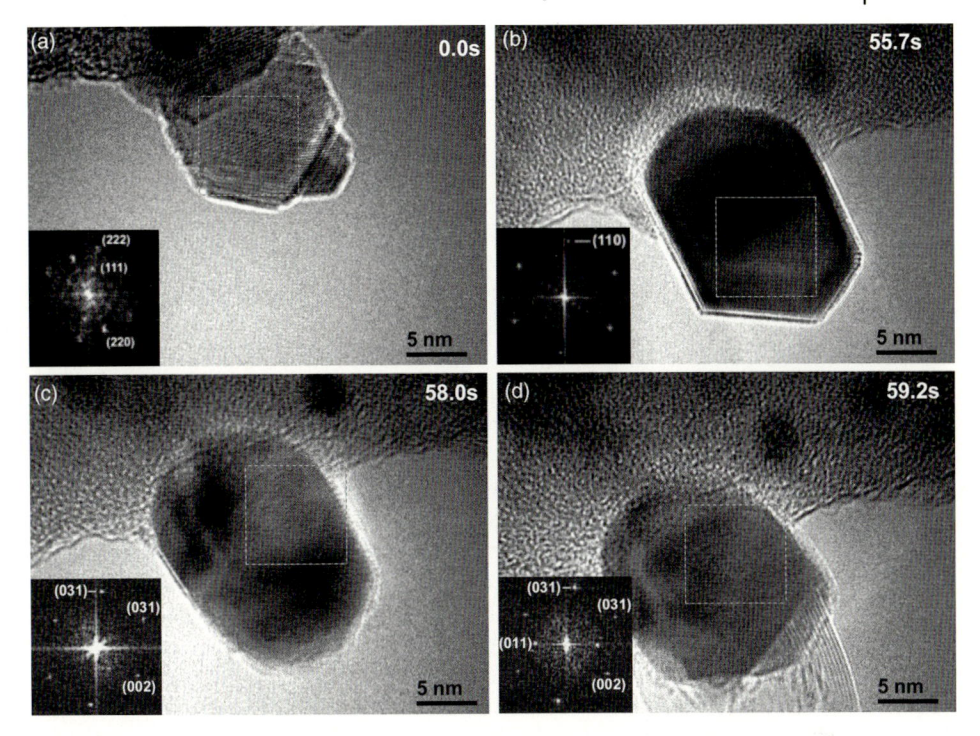

Figure 6.9 Time-resolved high-resolution images extracted from a digital video recorded at 650 °C in ~1.33 Pa (10^{-2} Torr) of flowing acetylene (C_2H_2). Fe-containing catalyst particles were deposited *in-situ* by electron beam-induced decomposition of $Fe_2(CO)_9$ vapor. Fast Fourier transforms (FFTs), or diffractograms, of the area of the particle indicated by the square selection in each frame are shown in the lower left-hand corner; the elapsed time is shown in the upper right-hand corner. Diffractograms could be indexed (marked on diffractograms) as (a) Fe_3O_4, (b) α-Fe, and (c, d) Fe_3C. The iron carbide phase persists after CNT formation. Figure reproduced with permission from Ref. [78]; © 2009, American Chemical Society.

The catalytic decomposition of hydrocarbons by Fe nanoparticles can result either in the formation of a graphitic layer encapsulating the catalyst, thereby deactivating it, or in the formation of CNTs. This technique is also relevant for understanding other catalytic reactions.

6.5.2
Chemical Changes

While both electron-diffraction patterns and high-resolution images provide information about structural changes, spectroscopic techniques are needed to determine the chemical composition and oxidation state of the reactant, product, or intermediate compound. Spectroscopic data can be obtained along with structural information or independently using either X-ray energy-dispersive spectroscopy (EDS) or

EELS. EDS is ill-suited for *in-situ* studies during the course of a chemical reaction because (a) infrared radiation from the heated specimen tends to swamp the detector at elevated temperatures, and (b) high gas pressures can damage the thin windows in front of the detector. EDS is therefore best suited to obtain chemical information on reactants and products before and after reaction. On the other hand, EELS is suitable for following the chemical changes during gas–solid interactions, such as reduction, oxidation, and nitridation, since the EELS detector is spatially remote from the specimen area and most of the EELS signal can be collected within a few milliradians of the incident beam direction, and is therefore not significantly attenuated by the differential pumping apertures of the ETEM.

Spectroscopic techniques can be used to obtain qualitative or quantitative information. For example, EELS was used to determine the presence of N to confirm that the N was incorporated within the nanowires formed after Au/Ga droplets had been exposed to NH_3 at 800 °C [78]. Moreover, the near-edge fine structure for selected elements, such as the $L_{2,3}$ edges (measuring transitions from 2p to 3d bound states) of third-row transition metals and the $M_{4,5}$ edges (3d to 4f transition) of the lanthanide family of elements is sensitive to their oxidation state and can be used to follow the changes in oxidation state during redox reactions [79]. A shift in the position and/or in the relative intensity of L_2 and L_3 (M_4 and M_5 for lanthanides) of the characteristic edge (also referred to as white lines) is also indicative of a change in oxidation state. The extent of reduction in ceria and doped ceria has been determined quantitatively as described below.

Intrinsic and doped ceria (CeO_2) are often used as a catalyst or catalyst support due to their ability to release lattice oxygen (reduction) or to take up environmental oxygen (oxidation) depending on the oxygen potential in the ambient environment. In simple terms, Ce can coexist and switch between the $+3$ and $+4$ oxidation states, and therefore exhibits a property called oxygen storage capacity (OSC) [80]. Figure 6.10a shows the change in the Ce M_4 and M_5 white-line intensities with temperature when heated in ~260 Pa of flowing hydrogen [81]. It is also known that the oxygen-to-cerium ratio should change as the material loses oxygen due to reduction that corresponds to CeO_2 (Ce^{4+}) converting to $CeO_{1.5}$ (Ce^{3+}). However, these intensity ratios cannot be used directly to quantify the Ce oxidation state as the measurements of the Ce $M_{4,5}$ cross-sections in the literature differ by a factor of almost two [82, 83]. Moreover, for ceria nanoparticles, surface adsorbents such as water may also contribute to the observed oxygen signal. The change in the Ce M_5/M_4 ratio was plotted against the oxidation states, determined from the Ce/O signal, to obtain the quantitative change in oxidation state with temperature (Figure 6.10b) [81]. These results show that Ce M_5/M_4 ratio can be used to determine the oxidation state and this approach has been successfully employed to follow the redox behavior of individual nanoparticles in mixed cerium–zirconium oxides [84].

Crozier and Chenna [85] have recently reported a procedure for quantitative measurement of the composition of gas mixtures using EELS. They have shown that valence-loss (or low-loss) region of EELS data can provide the composition of gas mixture used for understanding the catalytic processes in the ESTEM.

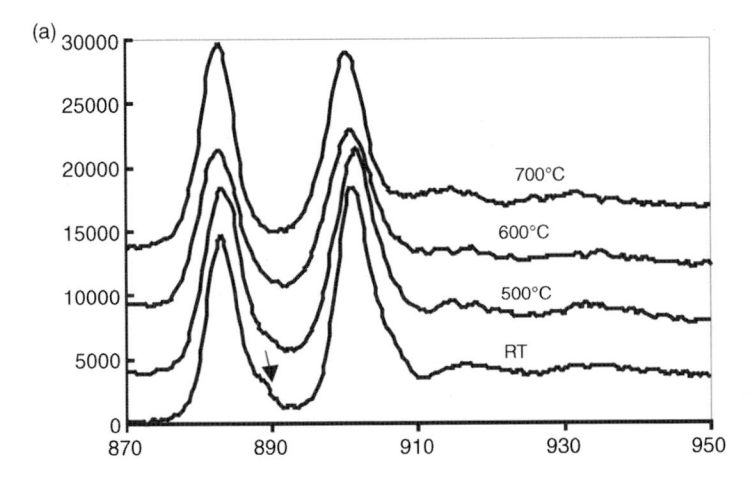

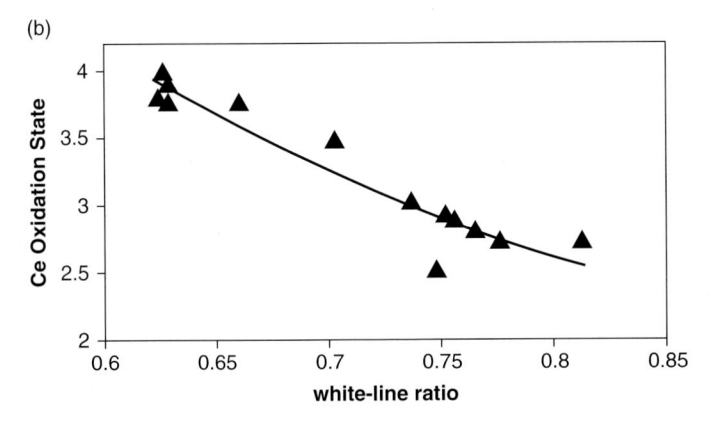

Figure 6.10 (a) Background-subtracted Ce $M_{4,5}$ peaks, extracted from electron energy-loss spectroscopic data recorded during heating of CeO_2 nanoparticles. Note the change in relative white-line intensity with temperature and disappearance of the small shoulder, indicated by the arrow, with reduction. (b) Plot showing the relationship between Ce oxidation state and white-line intensity ratio with temperature.

Energy-filtered transmission electron microscopy (EFTEM) imaging, using an in-column or post-column filter, can also be employed to follow the progression of a reaction front. Recently, EFTEM has been used to show that a 2 nm amorphous layer of Ni–Si is present in as-deposited samples due to interdiffusion of Ni and Si at room temperature [86]. *In-situ* imaging acquired at elevated temperature showed an increase in the thickness of this reaction layer as Ni reacts with Si to form nickel silicides, as confirmed by diffraction and imaging.

6.5.3
Reaction Rates (Kinetics)

Reaction kinetics at the nanoscale can be completely different from those in the bulk due to differences in surface area, structure of the bonding planes, and so on. For example, theoretical calculations have shown that the carbon diffusion rate is different for various crystallographic planes of bcc and fcc iron [87]. *In-situ* measurements can provide direct evidence of such subtle differences. Reaction rates can be measured by following changes in length, diameter, area, volume, or chemical composition with time and temperature. The measured rates at various temperatures are then used to calculate the activation energy for the particular reaction using the Arrhenius equation. Baker *et al.* have done pioneering work on calculating activation energies for CNT formation for different transition metal catalysts [88]. Sinclair's group has used *in-situ* measurements extensively for solid-state reactions in microelectronic materials [89–91]. Various examples of measuring reaction kinetics can be found in a recent review paper and the references therein [39].

6.6
Limitations and Future Developments

It is not possible to follow each and every chemical reaction using *in-situ* TEM-related techniques. Some examples that are beyond the reach of current technology are the following:

1) **Corrosive Gases**: Reactions involving gases such as NO_x, SO_x, H_2S, and halogens (F, Cl, Br) as reactants or products will corrode the materials currently used in TEM columns and holders.
2) **Temperature**: The design of heating holders and the microscope configuration limit achievable temperatures. For example, although samples can be heated up to 1500 °C using filament heating holders, current differentially pumped TEM instruments limit the highest attainable temperature to 900 °C when gases are introduced into the sample chamber. Other issues with currently available commercial heating holders pertain to a relative lack of control over heating and cooling rates and thermal drift that make it difficult to collect good quality data. For example, samples continue to drift due to expansion of various components for 15–20 min after reaching the desired observation temperature. Therefore, high-resolution images or nanoscale spectra cannot be recorded during this period, making it difficult to obtain temperature-resolved data. Some new commercial designs to overcome some of these difficulties have been reported recently [13].
3) **Temporal Resolution**: Observable reaction rates are dependent on the recording method used. Therefore, any reaction that occurs faster than the recording rate cannot be captured. Currently, this limit for video imaging is ~1/30 s and for EDS it is ~20 s. However, there are some recent developments that may help in overcoming this limitation. For example, Kim *et al.* recently reported the ability to

capture data for faster reaction rate processes by combining pulsed heating with pulsed electron beam imaging or diffraction [70]. Post-column EELS detectors capable of recording 1000 energy-loss spectra in 1 s have recently been introduced, but the feasibility of obtaining data with good signal-to-noise ratios for nanoparticles is not yet certain.

It is worth noting that *in-situ* measurements are not only limited in their ability to follow fast reactions, but are also not suited for following reaction processes that take weeks or months, such as the effect of thermal cycles on materials used in solid oxide fuel cells or the processes leading to the deactivation of catalysts after multiple cycles.

Acknowledgements

Stimulating discussions and scientific contributions from Professors Peter Crozier, Maria Gajdarziska, Michael Treacy, and Peter Rez, Drs. Ruigang Wang and See Wee Chee, and Michael McKelvy are gratefully acknowledged. Mr. Karl Wise has been an integral part of ASU's ETEM group. Funding from the NSF and DOE for the purchase of the ETEM, modifications, and various research projects is also gratefully acknowledged. The author is also grateful to Andrew Berglund, J. Alexander Liddle, Mark Stiles and Ian M. Anderson of NIST for their review and comments for improving this chapter.

References

1 Sharma, R. and Weiss, K. (1998) Development of a TEM to study in situ structural and chemical changes at an atomic level during gas–solid interactions at elevated temperatures. *Microsc. Res. Tech.*, **42** (4), 270–280.

2 Daulton, T.L., Little, B.J., Lowe, K., and Jones-Meehan, J. (2001) In situ environmental cell-transmission electron microscopy study of microbial reduction of chromium(VI) using electron energy loss spectroscopy. *Microsc. Microanal.*, **7**, 470.

3 Doole, R.C., Parkinson, G.M., and Stead, J.M. (1991) High resolution gas reaction cell for the JEM 4000. *Inst. Phys. Conf. Ser.*, **119**, 157–160.

4 Swann, P.R. and Tighe, N.J. (1972) Performance of differentially pumped environmental cell in the AE1 EM7, in Proceedings of 5th European Regional Congress on Electron Microscopy.

5 Parkinson, G.M. (1989) High resolution, in situ controlled atmosphere transmission electron microscopy (CTEM) of heterogeneous catalysts. *Catal. Lett.*, **2**, 303–307.

6 Parkinson, G.M. (1991) Controlled environment transmission electron microscopy (CTEM) of catalysis. *Inst. Phys. Conf. Ser.*, **119**, 151–156.

7 Robertson, I.M. and Teter, D. (1998) Controlled environment transmission electron microscopy. *Microsc. Res. Tech.*, **42** (4), 260–269.

8 Gai, P.L. (2002) Development of wet environmental TEM (Wet-ETEM) for in situ studies of liquid–catalyst reactions on the nanoscale. *Microsc. Microanal.*, **8**, 21.

9 Kamino, T. and Saka, H. (1993) Newly developed high resolution hot stage and its applications to materials science. *Microsc. Microanal.*, **4**, 127–135.

10 Kamino, T., Yaguchi, T., Konno, M., Watabe, A., Marukawa, T., Mima, T.,

Kuroda, K., Saka, H., Arai, S., Makino, H., Suzuki, Y., and Kishita, K. (2005) Development of a gas injection/specimen heating holder for use with transmission electron microscope. *J. Electron Microsc.*, **54** (6), 497–503.

11 Creemer, J.F., Helveg, S., Hoveling, G.H., Ullmann, S., Molenbroek, A.M., Sarro, P.M., and Zandbergen, H.W. (2008) Atomic-scale electron microscopy at ambient pressure. *Ultramicroscopy*, **108** (9), 993–998.

12 Takeo, K., Toshie, Y., Mitsuru, K., Akira, W., and Yasuhira, N. (2006) Development of a specimen heating holder with an evaporator and gas injector and its application for catalyst. *J. Electron Microsc.*, **55** (5), 245–252.

13 Allard, L.F., Bigelow, W.C., Jose-Yacaman, M., Nackashi, D.P., Damiano, J., and Mick, S.E. (2009) A new MEMS-based system for ultra-high-resolution imaging at elevated temperatures. *Microsc. Res. Tech.*, **72** (3), 208–215.

14 Tan, X., Xu, Z., Shang, JK., and Han, P. (2000) Direct observations of electric field-induced domain boundary cracking in <001> oriented piezo-electric $Pb(Mg_{1/3}Nb_{2/3})O_3$–$PbTiO_3$ single crystal. *Appl. Phys. Lett.*, **77**, 1529–1531.

15 Qu, W., Zhao, X., and Tan, X. (2007) Evolution of nanodomains during the electric field-induced relaxor to normal phase transition in an Sc-doped $Pb(Mag_{1/3}Nb_{2/3})O_3$ ceramic. *J. Appl. Phys.*, **102** (1–8), 084101.

16 Williamson, M.J., Tromp, R.M., Vereecken, P.M., Hull, R., and Ross, F.M. (2003) Dynamic microscopy of nanoscale cluster growth at the solid–liquid interface. *Nat. Mater.*, **2**, 532–536.

17 Ross, F.M. and Searson, P.C. (1995) *Dynamic Observation of Electrochemical Etching in Silicon*, IOP Publishing, Bristol.

18 Zheng, H.M. *et al.* (2009) Observation of single colloidal platinum nanocrystal growth trajectories. *Science*, **324** (5932), 1309–1312.

19 Zheng, H.M. *et al.* (2009) Nanocrystal diffusion in a liquid thin film observed by in situ transmission electron microscopy. *Nano Lett.*, **9** (6), 2460–2465.

20 Sakai, N., Xiong, Y.P., Yamaji, K., Kishimoto, H., Horita, T., Brito, M.E., and Yokokawa, H. (2006) Transport properties of ceria–zirconia–yttria solid solutions $\{(CeO_2)_x(ZrO_2)_{1-x}\}_{1-y}(YO_{1.5})_y$ ($x = 0$–1, $y = 0.2, 0.35$). *J. Alloys Compd.*, **408–412**, 503–506.

21 Chen, C.L. and Mori, H. (2009) In situ TEM observation of the growth and decomposition of monoclinic $W_{18}O_{49}$ nanowires. *Nanotechnology*, **20** (28), 6.

22 Sharma, R. (2005) An environmental transmission electron microscope for in situ synthesis and characterization of nanomaterials. *J. Mater. Res.*, **20** (7), 1695–1707.

23 Kammler, M. *et al.* (2003) Lateral control of self-assembled island nucleation by focused-ion-beam micropatterning. *Appl. Phys. Lett.*, **82** (7), 1093–1095.

24 Lang, C., Kodambaka, S., Ross, F.M., and Cockayne, D.J.H. (2006) Real time observation of GeSi/Si(001) island shrinkage due to surface alloying during Si capping. *Phys. Rev. Lett.*, **97** (22), 4.

25 Portavoce, A., Kammler, M., Hull, R., Reuter, M.C., and Ross, F.M. (2006) Mechanism of the nanoscale localization of Ge quantum dot nucleation on focused ion beam templated Si(001) surfaces. *Nanotechnology*, **17** (17), 4451–4455.

26 Kim, B.J., Tersoff, J., Wen, C.Y., Reuter, M.C., Stach, E.A., and Ross, F.M. Determination of size effects during the phase transition of a nanoscale Au–Si eutectic. *Phys. Rev. Lett.*, **103** (15), 4.

27 Dick, K.A., Deppert, K., Samuelson, L., Wallenberg, L.R., and Ross, F.M. (2008) Control of GaP and GaAs nanowire morphology through particle and substrate chemical modification. *Nano Lett.*, **8** (11), 4087–4091.

28 Kim, B.J., Tersoff, J., Kodambaka, S., Reuter, M.C., Stach, E.A., and Ross, F.M. (2008) Kinetics of individual nucleation events observed in nanoscale vapor–liquid–solid growth. *Science*, **322** (5904), 1070–1073.

29 Hofmann, S., Sharma, R., Wirth, C.T., Cervantes-Sodi, F., Ducati, C., Kasama, T., Dunin-Borkowski, R.E., Drucker, J., Bennet, P., and Robertson, J. (2008) Ledge-flow controlled catalyst interface

dynamics during Si nanowire growth. *Nat. Mater.*, **7** (5), 372–375.

30 Yeadon, M., Marshall, M.T., Hamdani, F., Pekin, S., Morkoc, H., and Gibson, J.M. (1998) In situ transmission electron microscopy of AlN growth by nitridation of (0001) α-Al_2O_3. *J. Appl. Phys.*, **83** (5), 2847–2850.

31 Lin, M., Tan, J.P.Y., Boothroyd, C., Loh, K.P., Tok, E.S., and Foo, Y.-L. (2007) Dynamical observation of bamboo-like carbon nanotube growth. *Nano Lett.*, **7** (8), 2234–2238.

32 Birnbaum, H.K. and Sofronis, P. (1993) Hydrogen enhanced local plasticity – a mechanism for hydrogen-related fracture. *Mater. Sci. Eng.*, **A176**, 191.

33 Gai, P.L. (1999) Environmental high resolution electron microscopy of gas–catalyst reactions. *Top. Catal.*, **8**, 97–113.

34 Hansen, T.W., Wagner, J.B., Hansen, P.L., Dahl, S., Topsoe, H., and Jacobsen, J.H. (2001) Atomic-resolution in situ transmission electron microscopy of a promoter of a heterogeneous catalyst. *Science*, **294**, 1508–1510.

35 Sharma, R. and Crozier, P.A. (2005) Environmental transmission electron microscopy in nanotechnology, in *Transmission Electron Microscopy for Nanotechnology* (eds. N. Yao and Z.L. Wang), Springer, Berlin and Tsinghua University Press, Beijing, pp. 531–565.

36 Gai, P.L., Sharma, R., and Ross, F.M. (2008) Environmental (S)TEM studies of gas–liquid–solid interactions under reaction conditions. *MRS Bull.*, **33** (2), 107–114.

37 Sharma, R. (2008) Observation of dynamic processes using environmental transmission and scanning transmission electron microscope, in *In Situ Electron Microscopy at High Resolution* (ed. F. Banhart), World Scientific, Singapore.

38 Sharma, R. (2009) Kinetic measurements from in situ TEM observations. *Microsc. Res. Tech.*, **72** (3), 144–152.

39 Stach, E.A. (2008) Real-time observations with electron microscopy. *Mater. Today*, **11**, 50–58.

40 Fornasiero, P., Di Monte, R., Ranga Rao, G., Kaspar, J., Meriani, S., Trovarelli, A., and Graziani, M. (1995) Rh-loaded CeO_2–ZrO_2 solid solutions as highly efficient oxygen exchangers: dependence of the reduction behavior and the oxygen storage capacity on the structural properties. *J. Catal.*, **151**, 168–177.

41 Fornasiero, P., Balducci, G., Kaspar, J., Meriani, S., Di Monte, R., and Graziani, M. (1996) Metal-loaded CeO_2–ZrO_2 solid solutions as innovative catalysts for automotive catalytic converters. *Catal. Today*, **29**, 47–52.

42 Wang, R., Crozier, P.A., Sharma, R., and Adams, J.B. (2006) Nanoscale heterogeneity in ceria zirconia with low temperature redox properties. *J. Phys. Chem. B* **110**, 18278–18285.

43 Naik, V.M., Haddad, D., Naik, R., Benci, J., and Auner, G.W. (2002) Solid-state chemistry of inorganic materials IV, *Proc. Mater. Res. Soc. Symp.*, **755**, 413–417.

44 de Graaf, M., Willard, M.A., McHenry, M.E., and Zhu, Y.M. (2001) In situ Lorentz TEM cooling study of magnetic domain confuguration in Ni_2MnGa. *IEEE Trans. Magnet.*, **37** (4), 2663–2665.

45 Tsuchiya, K., Yamamoto, K., Hirayama, T., Nakayama, H., Todaka, Y., and Umemoto, M. (2003) TEM observation of phase transformation and magnetic structure in ferromagnetic shape memory alloys, presented at Electron Microscopy: Its Role in Materials Science – The Mike Meshii Symposium, March 2–6, 2003, San Diego, CA.

46 Qu, W.G., Tan, X.L., and Yang, P. (2009) In situ transmission electron microscopy study on Nb-doped $Pb(Zr_{0.95}Ti_{0.05})O_{-3}$ ceramics. *Microsc. Res. Tech.*, **72** (3), 216–222.

47 Oleshko, V.P., Crozier, P.A., Cantrell, R.D., and Westwood, A.D. (2001) In situ and ex situ study of gas phase propylene polymerization over a high activity $TiCl_4$–$MgCl_2$ heterogeneous Ziegler–Natta catalyst. *Macromol. Rapid Commun.*, **22**, 34–40.

48 Oleshko, V.P., Crozier, P.A., Cantrell, R.D., and Westwood, A.D. (2002) In situ real-time environmental TEM of gas phase Ziegler–Natta catalytic polymerization of propylene. *J. Electron. Microsc.*, **51** (Suppl.), S27–S29.

49. Atzmon, Z., Sharma, R., Mayer, J.W., and Hong, S.Q. (1993) An in situ transmission electron microscopy study during NH$_3$ ambient annealing of Cu–Cr thin films. *Proc. Mater. Res. Soc. Symp.*, **317**, 245–250.
50. Atzmon, Z., Sharma, R., Russell, S.W., and Mayer, J.W. (1994) Kinetics of copper grain growth during nitridation of Cu–Cr and Cu–Ti thin films by in situ TEM. *Proc. Mater. Res. Soc. Symp.*, **337**, 619–624.
51. Sharma, R., McKelvy, M.J., Béarat, H., Chizmeshya, A.V.G., and Carpenter, R.W. (2004) In situ nanoscale observations of the Mg(OH)$_2$ dehydroxylation and rehydroxylation mechanisms. *Philos. Mag.*, **84**, 2711–2729.
52. McKelvy, M.J., Sharma, R., Chizmeshya, A.V.G., Carpenter, R.W., and Streib, K. (2001) Magnesium hydroxide dehydroxylation: in situ nanoscale observations of lamellar nucleation and growth. *Chem. Mater.*, **13** (3), 921–926.
53. Gajdardziska-Josifovska, M. and Sharma, R. (2005) Interaction of oxide surfaces with water: environmental transmission electron microscopy of MgO hydroxylation. *Microsc. Microanal.*, **11** (6), 524–533.
54. Freney, E.J., Martin, S.T., and Buseck, P.R. (2009) Deliquescence and efflorescence of potassium salts relevant to biomass-burning aerosol particles. *Aerosol. Sci. Tech.*, **43** (8), 799–807.
55. Wise, M.E., Martin, S.T., Russell, L.M., and Buseck, P.R. (2008) Water uptake by NaCl particles prior to deliquescence and the phase rule. *Aerosol. Sci. Tech.*, **42** (4), 281–294.
56. Wise, M.E., Biskos, G., Martin, S.T., Russell, L.M., and Buseck, P.R. (2005) Phase transitions of single salt particles studied using a transmission electron microsocpe with an environmental cell. *Aerosol. Sci. Technol.*, **39**, 849–856.
57. Semeniuk, T.A., Wise, M.E., Martin, S.T., Russell, L.M., and Buseck, P.R. (2007) Hygroscopic behavior of aerosol particles from biomass fires using environmental transmission electron microscopy. *J. Atmos. Chem.*, **56** (3), 259–273.
58. Yoshida, H., Takeda, S., Uchiyama, T., Kohno, H., and Homma, Y. (2008) Atomic-scale in situ observation of carbon nanotube growth from solid state iron carbide nanoparticles. *Nano Lett.*, **9** (11), 3810–3815.
59. Terrones, M., Hsu, W.K., Kroto, H.W., and Walton, D.R.M. (1999) Nanotubes: a revolution in materials science and electronics, *Top. Curr. Chem.*, **190**, 189–234.
60. Silvis-Cividjian, N., Hagen, C.W., Kruit, P., van der Stam, M.A.J., and Groen, H.B. (2003) Direct fabrication of nanowires in an electron microscope. *Appl. Phys. Lett.*, **82** (20), 3514–3416.
61. Kodambaka, S., Hannon, J.B., Tromp, R.M., and Ross, F.R. (2006) Control of Si nanowire growth by oxygen. *Nano Lett.*, **6** (6), 1292–1296.
62. Kodambaka, S., Tersoff, J., Reuter, M.C., and Ross, F.M. (2007) Germanium nanowire growth below the eutectic temperature. *Science*, **316** (5825), 729–732.
63. Sharma, R. and Zafar, I. (2004) In situ observations of carbon nanotube formation using environmental electron microscopy (ETEM). *Appl. Phys. Lett.*, **84**, 990–992.
64. Hofmann, S., Sharma, R., Ducati, C., Du, G., Mattevi, C., Cepek, C., Mirco, C., Pisana, S., Parvez, A., Cervantes-Sodi, F., Ferrari, A.C., Dunin-Borkowski, R., Lizzit, S., Petaccia, L., Goldoni, A., and Robertson, J. (2007) In situ observations of catalyst dynamics during surface-bound carbon nanotube nucleation. *Nano Lett.*, **7** (3), 602–608.
65. Helveg, S., Lopez-Cartes, C., Sehested, J., Hansen, P.L., Clausen, B.S., Rostrup-Nielsen, J.R., Abild-Pedersen, F., and Norskov, J. (2004) Atomic-scale imaging of carbon nanofibre growth. *Nature*, **427**, 426.
66. Mima, T., Takeuchi, Y., Arai, S., Kishita, K., Kuroda, K., and Saka, H. (2008) In situ transmission electron microscopy observation of the growth of bismuth oxide whiskers. *Microsc. Microanal.*, **14** (3), 267–273.
67. Alani, R. and Pan, M. (2001) *In Situ Transmission Electron Microscopy Studies and Real-Time Digital Imaging*, Blackwell Science, Oxford.

68 King, W.E., Campbell, G.H., Frank, A., Reed, B., Schmerge, J.F., Siwick, B.J., Stuart, B.C., and Weber, P.M. (2005) Ultrafast electron microscopy in materials science, biology, and chemistry. *J. Appl. Phys.*, **97**, 111101.

69 Kim, J.S., LaGrange, T., Reed, B.W., Taheri, M.L., Armstrong, M.R., King, W.E., Browning, N.D., and Campbell, G.H. (2008) Imaging of transient structures using nanosecond in situ TEM. *Science*, **321** (5895), 1472–1475.

70 Schoen, D.T., Peng, H.L., and Cui, Y. (2009) Anisotropy of chemical transformation from In_2Se_3 to $CuInSe_2$ nanowires through solid state reaction. *J. Am. Chem. Soc.*, **131** (23), 7973–7975.

71 Wang, L., Zhou, G.W., Eastman, J.A., and Yang, J.C. (2006) Initial oxidation kinetics and energetics of $Cu_{0.5}Au_{0.5}$ (001) film invstigated by in situ ultrahigh vacuum transmission electron microscopy. *Surf. Sci.*, **600** (11), 2372–2378.

72 Kruger, H. *et al.* (2009) High-temperature structural phase transition in $Ca_2Fe_2O_5$ studied by in situ X-ray diffraction and transmission electron microscopy. *J. Solid State Chem.*, **182** (6), 1515–1523.

73 Li, P. Y., Lu, H. M., Tang, S. C., and Meng, X. K. (2009) An in situ TEM investigation on microstructure evolution of Ni–25 at.% Al thin films. *J. Alloys Compd.*, **478** (1–2), 240–245.

74 Sayagués, M.J. and Hutchison, J.L. (1999) From $Nb_{12}O_{29}$ to $Nb_{22}O_{54}$ in a controlled environment high resolution microscope. *J. Solid State Chem.*, **146**, 202.

75 Sayagués, M.J. and Hutchison, J.L. (1999) A new niobium tungsten oxide as a result of an in situ reaction in a gas reaction cell microscope. *J. Solid State Chem.*, **143**, 33.

76 Crozier, P.A., Wang, R.G., and Sharma, R. (2008) In situ environmental TEM studies of dynamic changes in cerium-based oxides nanoparticles during redox processes. *Ultramicroscopy*, **108** (11), 1432–1440.

77 Sharma, R., Moore, E.S., Rez, P., and Treacy, M.M.J. (2009) Site-specific fabrication of Fe particles for carbon nanotube growth. *Nano Lett.*, **9** (2), 689–694.

78 Diaz, R.E., Sharma, R., Jarvis, K., Zhang, Q.-L., and Mahajan, S. (2012) Direct observation of nucleation and early stages of growth of GaN nanowires. *J. Cryst. Growth.*, **341** (1), 1–6.

79 Chueh, Y.L., Lai, M.W., Liang, J.Q., Chou, L.J., and Wang, Z.L. (2006) Systematic study of the growth of aligned arrays of $\alpha\text{-}Fe_2O_3$ and Fe_3O_4 nanowires by a vapor–solid process. *Adv. Funct. Mater.*, **16** (17), 2243–2251.

80 Logan, A.D. and Shelf, M. (1994) Oxygen availability in mixed cerium/praseodymium oxides and the effect of noble metals. *J. Mater. Res.*, **9** (2), 468–475.

81 Sharma, R., Crozier, P.A., Kang, Z.C., and Eyring, L. (2004.) Observation of dynamic nanostructural and nanochemical changes in ceria-based catalysts during in situ reduction. *Philos. Mag.*, **84**, 2731–2747.

82 Hofer, F., Golob, P., and Brunegger, A. (1988) EELS quantification of the elements Sr to W by means of $M_{4,5}$ edges. *Ultramicroscopy*, **25** (1), 81–84.

83 Manoubi, T. and Colliex, C. (1990) Quantitative electron energy loss spectroscopy on M_{45} edges in rare earth oxides. *J. Electron Spectrosc.*, **50** (1), 1–18.

84 Wang, R., Crozier, P.A., Sharma, R., and Adams, J.B. (2006) Nanoscale heterogeneity in ceria zirconia with low temperature redox properties. *J. Phys. Chem.B*, **110** (37), 18278–18285.

85 Crozier, P.A. and Chenna, S. (2011) In situ analysis of gas composition by electron energy-loss spectroscopy for environmental transmission electron microscopy. *Ultramicroscopy*, **111**, 177–185.

86 Alberti, A., Bongiorno, C., Mocuta, C., Metzger, T., Spinella, C., and Rimini, E. (2009) Low temperature formation and evolution of a 10 nm amorphous Ni–Si layer on [001] silicon studied by in situ transmission electron microscopy. *J. Appl. Phys.*, **105** (9), 093506.

87 Jiang, D.E. and Carter, E.A. (2003) Carbon dissolution and diffusion in ferrite and austenite from first principles. *Phys. Rev. B*, **67** (21) 214103.

88 Baker, R.T.K., Harris, P.S., Thomas, R.B., and Waite, R.J. (1973) Formation of

filamentous carbon from iron, cobalt and chromium catalyzed decomposition of acetylene. *J. Catal.*, **30**, 86–95.

89 Sinclair, R., Parker, M.A., and Kim, K.B. (1987) In situ high-resolution electron-microscopy reactions in semiconductors. *Ultramicroscopy*, **23** (3–4), 383–395.

90 Sinclair, R., Min, KH., and Kwon, U. (2005) Applications of in situ HREM to study crystallization in materials. *Curr. Res. Adv. Mater. Proc.*, **494**, 7–11.

91 Min, K.-H., Sinclair, R., Park, I.-S., Kim, S.-T., and Chung, U.-I. (2005) Crystallization behavior of ALD-Ta_2O_5 thin films: the application of in situ TEM. *Philos. Mag.*, **85** (18), 2049–2063.

7
In-Situ TEM Studies of Vapor- and Liquid-Phase Crystal Growth

Frances M. Ross

7.1
Introduction

Electron microscopy is a powerful tool when used for the *ex-situ* analysis of the results of crystal growth experiments. Yet *in-situ* crystal growth experiments can provide an even deeper insight into these growth processes. Such experiments are based on time-resolved observations of the changes that occur in a specimen as material is deposited onto it. *In-situ* transmission electron microscopy (TEM) can be used to make quantitative measurements that lead to information concerning the physical mechanisms, diffusion pathways, rate-limiting steps, and transient structures that play a role during crystal growth. The ability to record structural and compositional changes as a function of time, combined in many cases with quantitative information on the environment to which the specimen is exposed, provides a unique window through which crystal growth processes can be examined.

Crystal growth has both academic and industrial importance. An understanding of the physics of crystal growth has implications for the design of nanostructured materials, the modification of surfaces, the properties of composite materials, and the development of materials with specific electronic properties, to name just a few applications. And given today's intense interest in nanotechnology, *in-situ* TEM provides one of the few ways in which we can examine the process of nanoscale self assembly – the use of spontaneously occurring physical processes to form nanoscale structures with a desired shape or composition.

In this chapter, we will show some examples in which *in-situ* TEM has provided quantitative information on the thermodynamics and kinetics of crystal growth processes, with emphasis placed on growth from both liquid and vapor phases. The examples are selected based on our own research interests, so this is in no sense a complete review; rather, it emphasizes the rapid development that is taking place worldwide in this area. More complete reviews of *in-situ* crystal growth experiments are provided in Refs [1–3]; here, the objective is to describe the experimental requirements for setting up a crystal growth experiment, and to illustrate through examples how *in-situ* TEM enables the quantitative study of

---

In-situ Electron Microscopy: Applications in Physics, Chemistry and Materials Science, First Edition.
Edited by Gerhard Dehm, James M. Howe, and Josef Zweck.
© 2012 Wiley-VCH Verlag GmbH & Co. KGaA. Published 2012 by Wiley-VCH Verlag GmbH & Co. KGaA.

crystal growth from both vapor and liquid phases. We will emphasize in particular the issues associated with obtaining quantitative information, and discuss the future developments that promise to make this area one of the most exciting for *in-situ* TEM.

7.2
Experimental Considerations

7.2.1
What Crystal Growth Experiments are Possible?

Some crystal growth processes can be studied simply by heating a sample in the microscope. By heating and cooling solid specimens, it is possible to observe the nucleation and growth of crystalline phases from amorphous structures, as well as the formation of new phases via solid-state reactions. The growth of crystalline Si from an amorphous specimen, for example, was one of the earliest reactions to be quantified *in-situ* in the TEM [4, 5]. More recent studies have examined silicide formation and metal-induced crystallization (for a review, see Ref. [3]).

But for many other interesting crystal growth processes, including chemical vapor deposition (CVD) and molecular beam epitaxy (MBE), the growth material needs to be supplied to the sample during the experiment. Several growth techniques are compatible with the geometry of the TEM, enabling a wide range of processes to be examined.

One experimental approach is to supply the growth material by evaporation. For this, a small evaporator (or perhaps two) can be mounted onto a sample holder, so that the deposition occurs directly onto the nearby substrate. The evaporator is transferred in and out of the TEM with the specimen. This has advantages and disadvantages: it is easy to change source materials, but not possible to bake the source without contaminating the specimen, and it can also be difficult to measure the incident flux. Alternatively, an evaporator can be placed above the pole-piece or on a port on the side of the microscope. This is feasible if a line of sight exists. In this case, the deposition rate will typically be low, as the evaporator will be physically further from the sample, but the flux can be calibrated more easily. *In-situ* evaporation has been used to examine silicide and germanide growth and grain growth in polycrystalline films, and also to deposit metals to serve as catalysts for subsequent CVD processes [6–11].

Suppling the growth material from a reactive gas source is experimentally much simpler. This requires a gas cylinder connected to tubing that directs the gas either towards the specimen region, or to a small volume around the specimen enclosed by windows or apertures. A more detailed discussion of such equipment can be found in Chapter 3. A TEM equipped with this type of modification can be used to study CVD processes, which are of great industrial importance. Such an approach has enabled the growth kinetics of carbon [12, 13] and semiconductor [14, 15] nanostructures to be examined *in-situ*. We will illustrate this technique by discussing the growth of Ge quantum dots and Si nanowires in Section 7.3.

Finally, it is possible to supply the growth material from a liquid phase. This approach is essential when studying processes such as liquid-phase epitaxy, electrochemical deposition and polymer growth from liquid precursors. As the liquid must be kept in place and prevented from evaporating, a variety of techniques including window cells, liquid injection, and the use of low-vapor-pressure liquids have been used. These techniques will be discussed in Section 7.4.

7.2.2
How Can These Experiments be Made Quantitative?

Quantitative studies of growth processes require both a well-characterized specimen and the ability to impose a growth environment that is well calibrated. The experimental effort required for calibration of the experiment is worthwhile, since the payoff is the ability to distinguish between competing crystal growth models, and to make measurements of key parameters such as activation energies.

The issues associated with providing a well-characterized starting substrate naturally depend on the material under study. Both the morphology of the sample and its surface chemistry must be controlled. The overall morphology of a sample is especially important for interpreting *in-situ* TEM results. In a thin TEM sample the most obvious effects may be related to the strain fields present. Since many growth processes are affected by strain, the relaxation allowed by the thin geometry could easily affect growth. Calculations can help to understand thin sample strain effects, so that the *in-situ* experiment can be related to *ex situ* results [16]. But other thin sample effects may also be important. For example, diffusion pathways along and through the sample may compete with each other, and the nearby surface may serve as a sink for species during growth [17].

The morphology of the sample surface is also important: surface step distribution, orientation, and grain size can all play a role in growth kinetics. A flat starting surface is used for most crystal growth experiments but a pattern may be imposed on the sample to determine its effect on the growth process.

To control the substrate surface chemistry of reactive materials such as Si, it is necessary to remove surface oxide and contamination after the sample is introduced into the microscope. Often, this can be achieved by heating the sample to very high temperatures (ca. 1250 °C in the case of Si). Etching with a reactive gas is another possibility. Carrying out the growth experiment in a microscope with an ultra-high vacuum (UHV) base pressure will avoid contamination by oxygen or organic materials once the surface has been cleaned. For less reactive surfaces some cleaning may be necessary, but the microscope environment may not have to be UHV for the growth experiment to be successful.

Calibration of the growth environment is essential for obtaining quantitative results. As noted above, the use of a UHV base pressure will avoid effects such as oxidation of reactive surfaces during growth. Even in a non-UHV system, it is important to know which gases are present. Trace amounts of different materials can have dramatic effects on growth morphology and kinetics; hence, the background gases present in the column during the growth experiment should be measured and controlled as far as practical. A mass spectrometer attached to the column is a useful tool in any growth experiment.

The sample temperature must also be well known. In the TEM, heat may be supplied to the sample in different ways, such as via a conventional furnace heater, by passing heating current directly through the sample, or with a local heater built using microelectromechanical system (MEMS) technology. Temperature calibration is a problem in any type of *in-situ* experiment, but it is especially difficult with conventionally prepared (polished or milled) samples, in which the thickness can vary locally. It is sometimes possible to carry out a series of growth experiments at the same location on such samples, thus minimizing errors in temperature. Careful calibration is certainly possible, but is time consuming. Thus, the increasing use of MEMS specimens with a defined geometry [19, 20] suggests exciting future possibilities for growth at well-controlled local hotspots, perhaps also using local temperature probes built onto the specimen.

When growth is carried out by supplying material from the gas phase it is, of course, important to measure the pressure at the sample as accurately as possible. Unfortunately, the pressure gauge may be some distance from the sample and so provide less reliable readings. Closed cells, in which the gas is confined between membranes [e.g. Ref. 19], may be more reproducible but an accurate pressure measurement is still difficult to achieve. Perhaps the only way to address this problem would be to calibrate the flow rate by using a materials growth system with well-known kinetics. In experiments where the material is supplied by evaporation rather than as a reactive gas, the problem is instead to measure the deposition rate at the sample. This can be achieved by using a crystal monitor at the sample position, or by making *ex-situ* measurements on test samples. Calibration issues for liquid-phase growth are discussed below.

Beam effects will always require some forethought by the microscopist for any *in-situ* experiment, and especially when the aim is to acquire quantitative information. By examining regions of the specimen that were not exposed to the beam, it may be possible to show that the beam is not significant to the particular growth process under study. If this is not the case, then beam effects can be evaluated by comparing results at different intensities and accelerating voltages. Such effects may even be useful, for example for growing carbon structures and for beam-induced deposition, especially if the dose can be quantified.

In summary, calibration of the growth environment is difficult and occasionally tedious. However, if it is performed accurately, so that measurements made from the TEM images can be combined with a thorough knowledge of the environment to which the sample is exposed, then the results will be especially useful. In the vapor phase, for example, the growth rate can be evaluated as a function of pressure and temperature, to determine activation energies and growth mechanisms. The beam dependence and the interactions that lead to growth in different materials systems can also be measured. In the liquid phase, growth kinetics of electrodeposited crystals can be determined as a function of applied potential to test electrochemical growth models. Clearly, the key to quantitative results is a thorough understanding of the whole experiment.

7.2.3
How Relevant Can These Experiments Be?

It is certainly worth considering how closely *in-situ* TEM growth experiments can mimic the real world, because there are important differences between the growth conditions used in conventional reactors and those available *in-situ*. Most significantly, gas-phase processes (such as CVD) often take place at a pressure that is too high to achieve in the TEM. This "pressure gap" is a serious limitation of *in-situ* TEM. Recent advances in microscope design and micromachined cells have allowed ever higher pressures to be sustained around the specimen (e.g., Ref. [19]; see Chapter 2). Clearly, such developments will provide an important route to achieving realistic growth conditions, for both vapor- and liquid-phase growth.

Apart from the pressure gap, some temperatures of interest are difficult to access, such as the high temperatures required when sintering ceramics. Beam effects may be impossible to eliminate completely, as the beam can interact with both the gas and the substrate. Moreover, limitations to specimen design may reduce the range of substrates that are available, or add unwanted effects such as strain relaxation.

It is a challenge to make TEM growth experiments relevant to the real world, but it is necessary to design our experiments critically in the context of achieving meaningful results. We can measure parameters such as activation energy and compare with *ex situ* results, we can extrapolate growth rates to higher pressures, and we can design specimens and TEM cells which are capable of an extended range of pressure and temperature. By relating our results to real world growth conditions, we will maximise their importance for technological applications while also advancing basic understanding of the fascinating physics of crystal growth.

7.3
Vapor-Phase Growth Processes

Deposition from the gas phase represents the simplest means of replicating crystal growth processes *in-situ*. Typically, a heated sample is exposed to a reactive precursor, such as silane or disilane for Si deposition, or methane for C deposition. When the gas cracks on the surface, the growth species is released and diffuses until it binds at a favorable location. This CVD process is of both industrial and academic interest, and *in-situ* TEM has provided information on epitaxial thin films (e.g. in Si and Ge, [21, 22]), polycrystalline layers (e.g., Al [23]) and self-assembled nanostructures (C, Ge, and Si; [12–15, 21, 24–29]). To illustrate the quantitative nature possible for such growth experiments, we review three examples: the growth of Ge nanostructures (quantum dots) on Si, and the nucleation and growth of Si nanowires.

7.3.1
Quantum Dot Growth Kinetics

Self-assembled island growth is a fascinating phenomenon that has been extensively studied *ex situ* in semiconductor systems. During the epitaxial growth of a semiconductor on a lattice-mismatched substrate, such as Ge on Si(001), the first few layers may grow as a flat film. However, as the strain energy builds up it becomes energetically favorable for small islands to form because they can partially relieve the strain. These islands, or quantum dots (QDs), which typically have dimensions of tens of nanometers, show modified electronic properties compared to the bulk material. Quantum dot formation is the preferred way to relieve strain over a range of growth rates, mismatch and temperature; under other conditions, a rippled surface or a dislocated flat film may be formed instead.

Figure 7.1 shows the nucleation kinetics of Ge islands on Si(001) [30, 31], obtained by exposing a cleaned, heated Si specimen to a flux of digermane (Ge_2H_6), and recording the growth of the islands via strain contrast. On a uniform Si substrate [30], the islands nucleate at random locations (Figure 7.1a). As the islands grow a coarsening process becomes visible, as seen from the diameters of the individual islands (Figure 7.2a). In order to model this somewhat unexpected behavior (Figure 7.2b), it proved necessary to understand both the growth kinetics of the individual islands and the development of the size distribution of the island ensemble. Growth kinetics turned out to be closely linked to island shape. Since shapes are not well resolved in the strain contrast images, complementary growth experiments in the low-energy electron microscope were used to show that each island changes from a square-based pyramid, through a series of transition shapes, to a multifaceted domelike shape [32].

It is clear from this example that a continuous observation of the size and shape of individual structures, combined with measurements on the ensemble of structures, is a key feature of *in-situ* studies of self assembly. This type of information makes such experiments very useful in elucidating growth mechanisms. By understanding growth kinetics, growth conditions can be optimized, for example to produce islands with the narrowest possible size distribution.

As well as growth kinetics, *in-situ* studies can address nucleation sites. This is of key importance since many applications of self-assembled quantum dots require placement at specific locations. To go beyond the random arrangement in figure 7.1a, we can pattern the surface (Figures 7.1b and 1c). In this case, the patterning was done by using a focused ion beam (FIB) to write spots containing a low dose of Ga^+ ions, then annealing the sample *in-situ* [33]. This results in a series of small pits on the

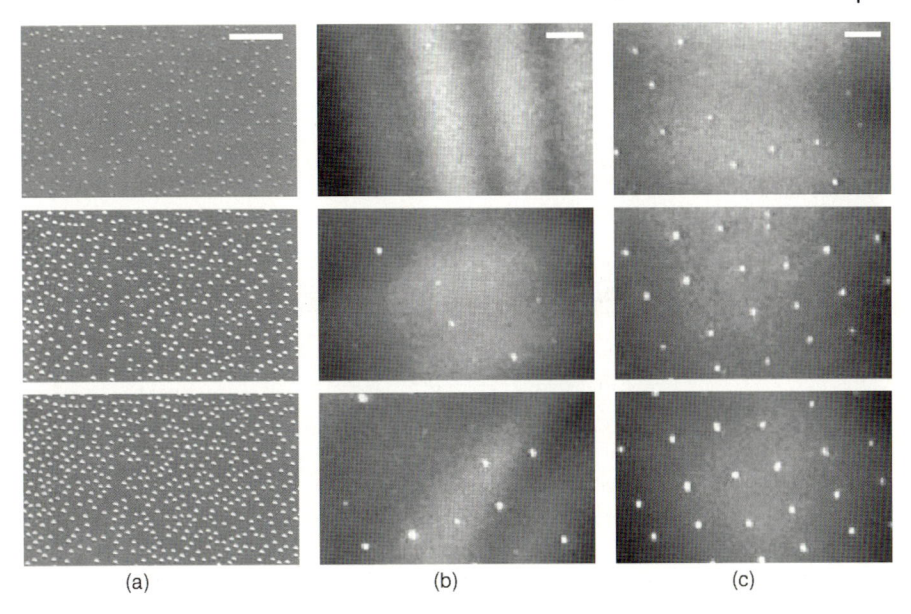

Figure 7.1 (a) Images obtained during Ge deposition at 2.3×10^{-7} Torr digermane and a substrate temperature of 640 °C in a UHV-TEM system. The images were obtained 21 s, 51 s, and 98 s after "nucleation" (i.e., the time at which distinct strain contrast is first seen). Nucleation occurred after a dose of approximately 50 L of digermane. The flux remained on during this sequence. Scale bar = 500 nm. Adapted from Ref. [30]; (b, c) Nucleation of Ge on patterned Si(001). A focused ion beam (in a connected UHV chamber) was used to create arrays of sites at low and high doses, 0.1 and 0.2 ms per spot, respectively. Annealing *in-situ* at 550 °C for 1 min was followed by Ge deposition at 3×10^{-8} Torr digermane and a substrate temperature of 500 °C. Images were obtained 223 s, 269 s, and 304 s after the digermane flux was turned on. Slower island nucleation can be seen on (b) compared to (c). Scale bars = 150 nm. Adapted from Ref. [31].

surface which serve as preferential sites for subsequent growth [34]. *In-situ* experiments allow the progression of pattern filling to be followed as growth proceeds: by comparing results for different FIB parameters, the optimum patterning for full occupancy can be determined [29, 31]. The island shape and size, which differ from those on the unprocessed surface, can be studied *in-situ*, while the lack of nucleation off site can also be explained [34]. This understanding of nucleation may help to develop patterned arrays of QDs for a variety of exciting applications, including quantum cellular automata and photonic structures.

7.3.2
Vapor–Liquid–Solid Growth of Nanowires

Nanostructures with an elongated, wirelike shape are also of interest for applications ranging from logic elements, interconnects and sensors to battery anodes.

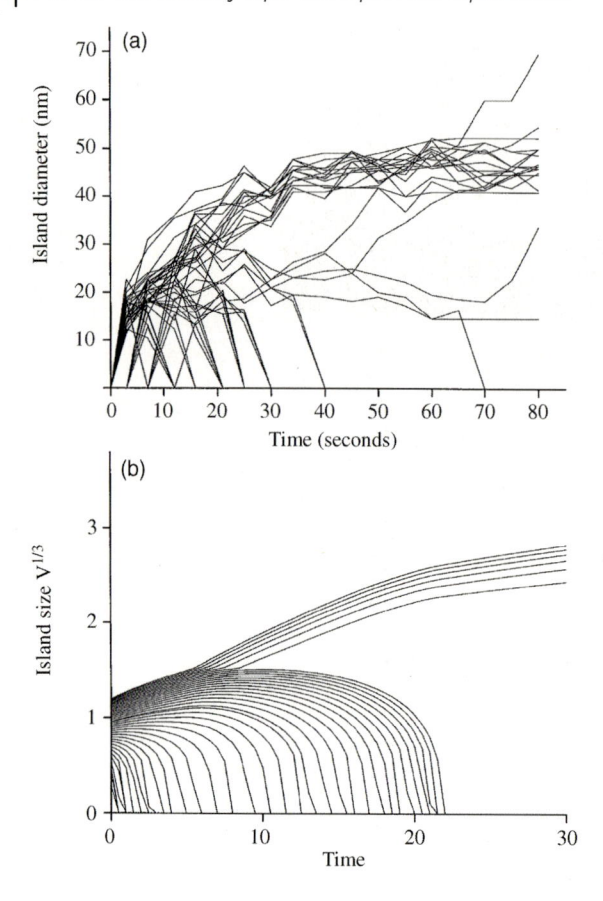

Figure 7.2 (a) The evolution of every Ge island within an area of 0.25 μm² on a Si(001) substrate. The growth conditions were 5×10^{-7} Torr digermane and 650 °C. The island showing rapid growth near the end of the sequence has probably formed a dislocation. Note that measurements were only obtained every few seconds from the growth movie, which accounts for the jumpiness in the lines; (b) Simulation showing the fate of islands with different initial sizes. The model does not include dislocated islands, so at large times it predicts a higher growth rate than is observed experimentally.

These nanowires can be created by using a catalyzed CVD process, in which the growth rate is accelerated at certain locations by small catalytic particles. As with self-assembled QDs, structural control is helped by an understanding of the growth mechanism.

Nanowire formation in several materials has been examined using *in-situ* TEM [15, 35–37]. Here we focus on Si wires grown using Au catalysts, as an illustration of the unexpected results and interesting possibilities available from *in-situ* microscopy [14]. Disilane (Si_2H_6) is used as the precursor gas, while a reflection mode sample geometry allows the wire growth to be observed directly [38]. An example of nanowire imaging and measurement of growth kinetics is shown in Figure 7.3.

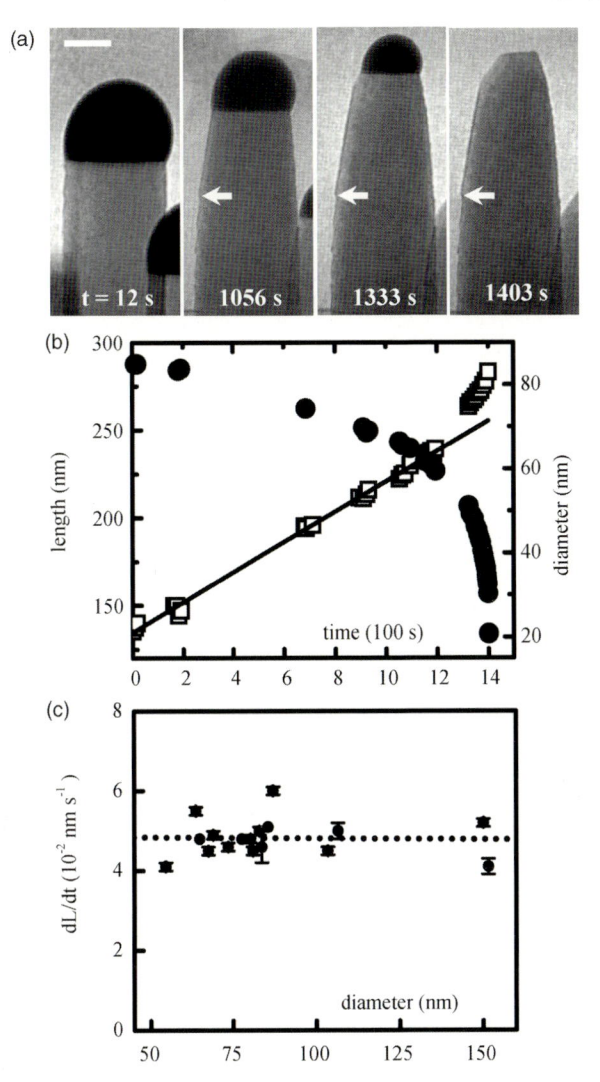

Figure 7.3 (a) Representative bright-field TEM images of a Si wire acquired at four successive times t (as indicated on the images) during deposition at 635 °C and 1×10^{-6} Torr disilane. t is the time since measurements on this wire began, although growth had been taking place for 4 h 17 min at $t = 0$. For clarity, the white arrows highlight a reference point on the wire sidewall. Scale bar = 40 nm; (b) Length (open squares) and diameter (solid circles) of the same wire as a function of t. The straight line is a least-squares fit to the first 1200 s. The wire growth rate is constant during this time; (c) dL/dt versus diameter for an ensemble of Si wires grown at 1×10^{-6} Torr disilane and 655 °C. The solid line is a least-squares fit. The error bars represent measurement uncertainties in growth rates. Adapted from Ref. [39].

Images recorded during growth provide structural details that cannot be captured *ex situ*, such as the growth interface and the wire surface structure shown in Figure 7.3a. The catalyst state can be obtained directly (here, it is a liquid Au–Si eutectic), and the catalyst stability can be measured. In Figure 7.3a, an unanticipated catalyst instability is visible: the surface migration of Au driven by coarsening [40] changes the droplet volume and hence the nanowire diameter. Observations made during growth, before cooling and exposure to the atmosphere, showed that the state of the catalyst particle can change when the growth flux is switched off [41], and the nanowire surface structure can change as the sample cools and oxidizes [14].

Kinetic measurements allow the growth mechanism to be examined in more detail. The dependence of nanowire growth rate on diameter, temperature, and pressure serves as a sensitive probe of the reaction pathway ([39]; Figure 7.3b and c). Growth kinetics can, of course, also be measured *ex situ* for ensembles of nanowires, but the ability to follow a single nanostructure as it grows provides a powerful means of determining growth physics.

A final benefit of *in-situ* TEM is the ability to examine the results of small modifications to the environment. For example, effects due to background gases can be investigated, provided that the microscope base pressure is well controlled, as in a UHV instrument. Low levels of oxygen added during growth prevent the catalyst migration shown in Figure 7.3a, and lead to wires of a constant diameter [41]. In Figure 7.4 we show another effect of oxygen, which is to change the nanowire growth direction. A reduction in the overall pressure can lead to similar nanowire direction changes [15]. Thus, *in-situ* experiments may suggest ways in which nanowire structures can be controlled, as well as providing an understanding of the processes that are important during crystal growth.

7.3.3
Nucleation Kinetics in Nanostructures

In nanotechnology, it is essential to control the place and time of nucleation if individual self-assembled nanostructures are to be used as the functional elements of a circuit. *In-situ* TEM is one of the few tools available to examine individual nucleation events. It therefore provides the exciting opportunity of testing nucleation theories that were developed based on studies of statistical ensembles of events [42, 43]. Observations of the nucleation of phases on cooling and heating can be used to probe parameters such as critical nucleus size, growth kinetics at different interfaces, and size effects [3]. However, crystal growth processes normally occur at a constant temperature, with a gradually increasing supersaturation triggering the nucleation event. Consequently, experiments carried out at constant temperature, but with increasing concentration, may provide direct information on this important phenomenon.

The reaction between Au and Si at constant temperature is shown in Figure 7.5. Here, as the Si supersaturation increases, the sequence of transformations begins with formation of the AuSi liquid eutectic (Figure 7.5a and b), followed by the

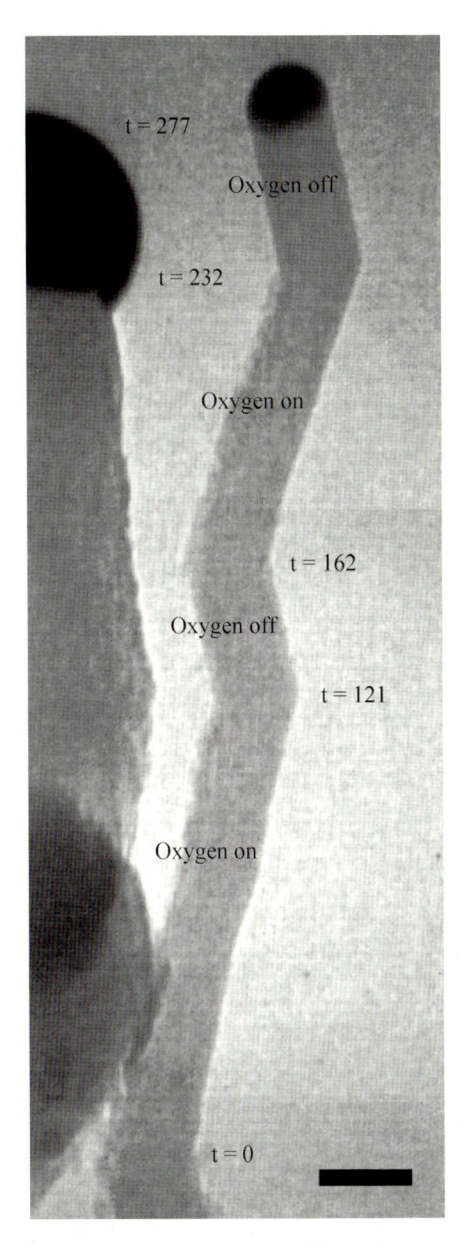

Figure 7.4 Experimental reversible switching using O_2. A 30 nm-diameter Si nanowire is imaged during growth at 700 °C and 2×10^{-6} Torr disilane. 2×10^{-7} Torr oxygen was added at the times indicated (in minutes; arbitrary zero). The growth changes direction by ~33°, consistent with switching between $\langle 110 \rangle$ and $\langle 111 \rangle$ growth directions. The small features at the base of the wire are polycrystalline Si particles that grow on the oxidized surfaces of the wire. Scale bar = 40 nm. Adapted from Ref. [59].

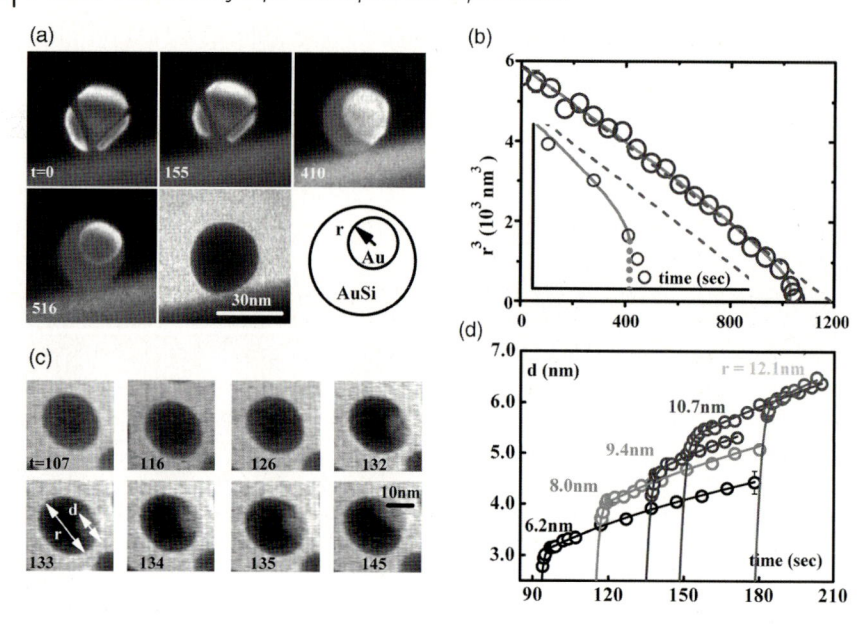

Figure 7.5 (a) The reaction of an aerosol Au particle with disilane, observed in dark-field during the solid to liquid transformation at 525 °C and 1.5×10^{-6} Torr disilane, with time (t) in seconds. The bright-field image shows the droplet after the transformation is complete. Inset is an idealized schematic of the geometry.; (b) Size of Au versus time t for data obtained at 500 °C and 1.5×10^{-6} Torr disilane. The straight dashed line is a least-squares fit excluding the last four datum points; the curved line is a model fit. From Ref. [44]; (c) Bright-field images extracted from a video obtained during nucleation of Si from a 20 nm-diameter AuSi droplet. The time is shown in seconds since disilane was introduced. The growth conditions were 585 °C and 8×10^{-7} Torr; (d) Linear dimension, d, of several Si nuclei versus time t for droplets of different initial radii R (as indicated) at 585 °C and 1.5×10^{-6} Torr. The growth rate is initially high but then decreases dramatically. Solid lines are fits from a growth model. The error bar indicates measurement errors of ~5%. From Ref. [45].

nucleation and growth of Si from AuSi (Figure 7.5c and d). The experimental approach is similar to that described in Section 3.2 for nanowire growth studies, but here a plan view SiN membrane is used. The Au may be either an agglomerated film [44] or size-selected particles [45]. Exposure to disilane then drives the phase transformations.

Quantitative measurements can be obtained from such experiments, allowing transformation parameters to be extracted. In Figure 7.5b, for example, the increase in reaction rate at small Au volumes is driven by the energetic cost of the AuSi/Au interface; hence, a value for this energy can be determined [45]. In Figure 7.5d, for the subsequent nucleation of Si, the nucleation times and growth rates agree well with a model [44] in which Si is absorbed through the droplet surface, until nucleation occurs at a surprisingly high supersaturation, followed by rapid deposition of excess Si. This type of detailed analysis is difficult to achieve without observations of individual nucleation events. It is clear that *in-situ* TEM can open exciting windows into nucleation in nanoscale systems, and many other materials systems await study.

7.4
Liquid-Phase Growth Processes

We have seen that crystal growth from the vapor phase, using a reactive gas or an atomic flux, can be examined *in-situ* by using customized sample geometries or microscopes modified for gas delivery. Crystal growth from the liquid phase is an equally important and interesting phenomenon, with processes such as particle agglomeration, liquid-phase epitaxy and electrochemical deposition useful in a variety of applications. Working with liquids in the transmission electron microscope is more difficult than working with gas-phase sources, however, and consequently liquid-phase crystal growth experiments have been relatively uncommon. When designing experiments to probe liquid-phase crystal growth, the sample and holder must be customized to encapsulate and control the liquid, while the imaging mode must be optimized for relatively thick samples. Some of the key issues involved with observations of liquid-phase crystal growth are summarized below, illustrated with an example of liquid-phase growth under electrochemical control.

7.4.1
Observing Liquid Samples Using TEM

Some liquids are encapsulated within solids, while others have a low vapor pressure. Either type can be prepared and examined just as if they were solid specimens. The full range of analytical and structural probes is available, with the only constraints being that the liquid can not be too mobile during observation or too thick for imaging. Heating and cooling experiments involving encapsulated or low-vapor-pressure liquids have provided insights into solidification and melting processes, allowing phase transformations to be examined directly and the effect of size on phase diagrams to be quantified (as in Section 7.2.2). Even larger volumes of liquids that are too thick for imaging can be used *in-situ*, an example being the ionic liquid droplets used to observe lithiation reactions [46]. In terms of crystal growth itself, the liquid-phase epitaxy experiments described in Sections 7.3.2 and 7.3.3, involving droplets of liquid AuSi and similar materials, did not require any special precautions, because of the low (ca. 10^{-10} Torr) vapor pressure at typical growth temperatures.

In many crystal growth experiments, however, it is more difficult to prepare a sample to observe the liquid phase reaction of interest. High vapour pressure liquids may cause problems for the microscope vacuum, or evaporate too rapidly for observations; some reactants may be air-sensitive; while others may react immediately on contact with a substrate or catalyst, so must be kept separate until the imaging conditions are set up. For such experiments different strategies have been developed. High vapor pressure liquids, most importantly aqueous solutions, are enclosed in a "liquid cell" to prevent evaporation. Crystal growth reactions in such cells will be discussed below. Air-sensitive samples, such as ionic liquids, require a glove box and vacuum transfer capabilities into the microscope [46]. And liquids that react on contact can be injected onto the growth substrate *in-situ* using a modified sample holder. Catalytic reactions involving liquid precursors, for example the stages of nylon formation, have been examined in this way [47]. These experiments have provided quantitative information, allowing *in-situ* TEM to have a significant impact on the development of commercial polymerization processes [48].

7.4.2
Electrochemical Nucleation and Growth in the TEM System

The challenge when studying electrochemical deposition, and other reactions that take place in an aqueous environment, is that quantitative analysis requires a stable liquid environment of known geometry. With a well-defined liquid geometry, the considerable prior effort that has been applied to modeling growth processes in bulk liquids can be applied to the *in-situ* experiments. Aqueous films evaporate rapidly in the TEM environment, and the strategy of injecting the liquid onto the substrate does not provide a sufficiently stable geometry. A different approach is therefore required, in which the liquid layer is encapsulated in a cell between two electron-transparent membranes or windows. Electrodes built into the cell and connected externally to a potentiostat allow the same degree of control as in a conventional electrochemical experiment, enabling quantitative crystal growth experiments to be conducted.

Electrochemical liquid cells are conveniently fabricated using silicon nitride as the window material [49, 50]. The electrolyte, 1–2 μm in thickness, completely fills the volume between the windows, creating a well defined liquid geometry. The working electrode, on which crystal growth takes place, extends over the lower window, so that growth can be observed in plan view. The other electrodes can be out of the field of view. A TEM resolution in the range 10 to 20 nm is possible for imaging at 300 kV although, remarkably, imaging using the STEM mode, with appropriate collector angles, has been shown to resolve 5 nm Au particles through several micrometers of water [51].

An example of electrochemical crystal growth is shown in Figure 7.6. Here, the cell is filled with an acidified copper sulphate solution. A constant voltage drives copper to deposit as clusters on the polycrystalline Au working electrode. The nucleation and growth of individual clusters is visible in the bright-field images, and the applied voltage and resulting current can be recorded at the same time. Between experiments, an opposite polarity is applied to strip the copper from the working electrode.

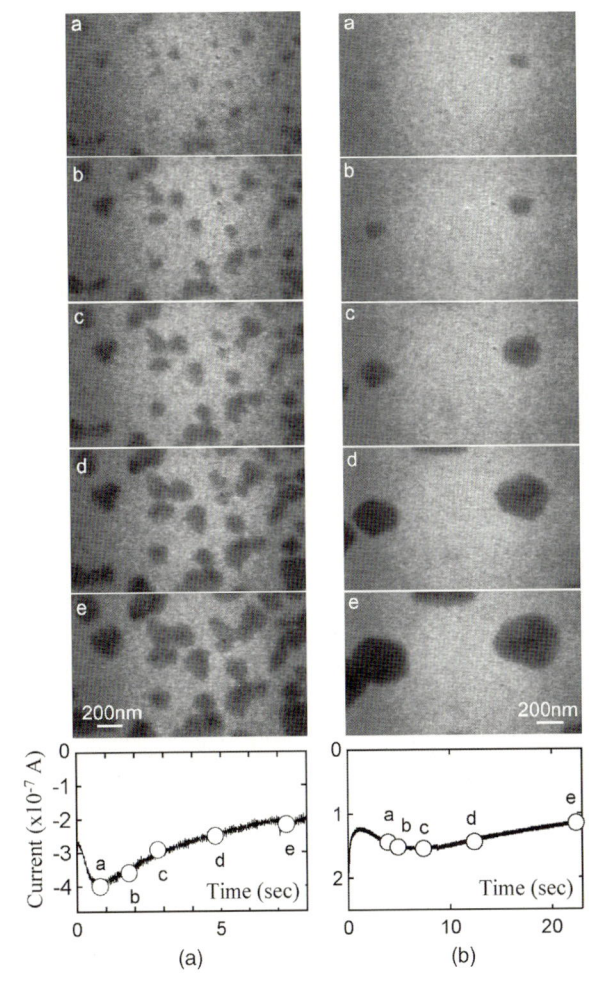

Figure 7.6 Image sequences recorded during the electrochemical deposition of Cu onto a polycrystalline Au electrode from acidified copper sulfate solution at applied potentials of (a) −0.08 V and (b) −0.06 V. The current transient curve below each sequence shows the time at which each image was recorded. From Ref. [52].

Such experiments provide a rich supply of data for quantitative analysis. Electrochemical data (here, for example, current versus time for different potentials) can be fitted using cluster growth models [53], while the extent to which the growth of individual clusters actually follows the predictions of these models is visible. Nucleation times [52] and preferential sites [49] can be measured directly from the video frames, while growth kinetics can be analyzed for individual clusters (Figure 7.7) [54]. This type of quantitative analysis has allowed the development and modification of electrochemical growth models [50].

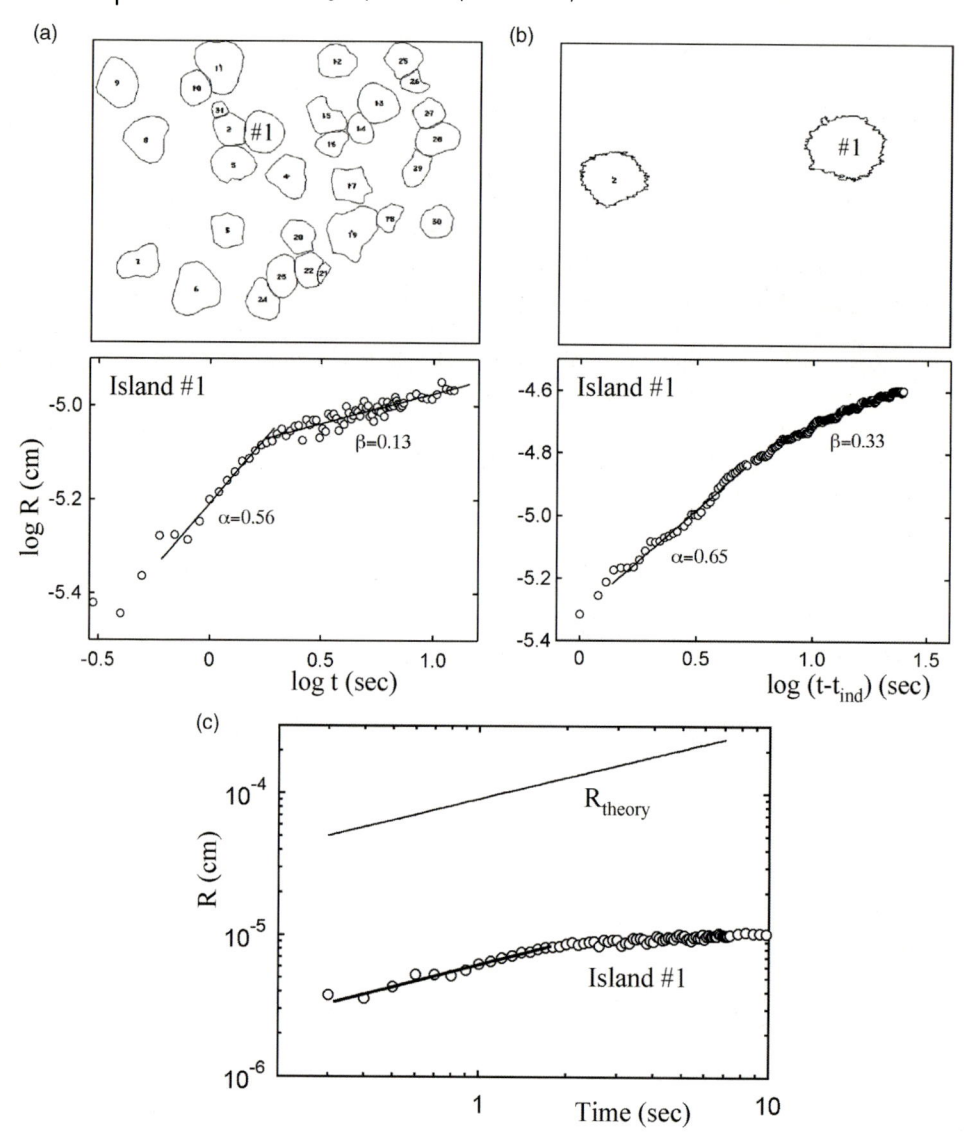

Figure 7.7 (a) Contour plot of copper islands after deposition for 4.8 s at −0.08 V (the image is 1.85 × 1.42 μm), extracted from the video sequence shown in Figure 7.6a, and log(R) versus log(t) plot for the growth of the island indicated. The island growth shows two regimes with different exponents, as predicted qualitatively from a diffusion-limited growth law; (b) Contour plot of copper islands after deposition for 22 s at −0.06 V with log(R) versus log(t) plot for the growth of the island indicated. Two regimes are visible, although the difference in exponents is reduced, as expected from the growth conditions; (c) Comparison of the log(R) versus log(t) plot for growth of the island in panel (a) with the prediction from a diffusion-limited growth law. The straight line fit through the data at early times is the diffusion-limited growth exponent. The growth rate of each cluster is tenfold slower than expected, suggesting a more complex diffusion pathway for the copper ions. From Ref. [54].

Liquid-phase crystal growth processes that do not require electrochemical connections can be examined in similar (though more simple) liquid cells. In this case, the crystal growth process is not stimulated by an applied voltage, but rather by heating or by the electron beam. Observations of beam-induced growth of Pt nanoparticles [55] from a Pt^+-containing solution enable two coexisting growth mechanisms to be distinguished, helping to provide an understanding of the size distribution of the nanoparticles. For individual nanoparticles in solution, Brownian motion is visible [51, 56, 57], and Brownian motion theories can be tested by imaging during slow evaporation [58]. Beam-induced assembly into aggregates has also been observed [56]. This suggests exciting possibilities for the study of metacrystals formed by the self-assembly of nanoparticles into flat or three-dimensional arrangements.

Liquid cell experiments are complex to set up, and it is important to consider beam effects and the effects that the small liquid volume may have on the results. However, the effort is worthwhile since no other technique combines TEM's spatial and temporal resolution for liquid phase growth processes. Liquid cells are available commercially and new designs are under development that incorporate microscopic heaters, spacers to control the membrane separation, and pumps to move the liquid through the cell or mix liquids from different reservoirs (see Refs [19, 56]). This area is likely to see a rapid expansion in the near future, with exciting consequences for our understanding of liquid-phase crystal growth.

7.5 Summary

Both vapor-phase and liquid-phase crystal growth can be studied successfully using *in-situ* TEM. The imaging and analytical opportunities provided by TEM enable quantitative results to be obtained that can expand the present understanding of crystal growth. It is worth noting that *in-situ* TEM growth experiments are best not performed in isolation, since results obtained with other techniques, such as scanning tunneling microscopy, atomic force microscopy, low-energy electron microscopy and scanning electron microscopy, as well as averaged measurements such as electrochemical parameters, provide complementary information and allow us to evaluate the TEM results.

The results of TEM crystal growth studies have been limited to a relatively narrow range of materials and growth conditions. Even though the results to date are of both industrial and academic importance, our challenge is to make *in-situ* TEM studies of crystal growth even more relevant to the real world. Recent developments in aberration correction are rapidly driving microscope performance, and this enhanced performance will be important for crystal growth studies in complex materials. But other areas for development will include improvements in high pressure and liquid cells for the TEM, the design of microscopes with larger pole-piece gaps that may include pressure, temperature and other sensors, better detectors that enable faster image acquisition, and the exciting possibility of including microelectromechanical technologies to increase the functionality of the specimen itself.

Acknowledgments

The author would like to acknowledge M.C. Reuter, R.M. Tromp, and A.W. Ellis for developing the experimental tools required for these studies. These studies were partially funded by the National Science Foundation and the Defense Advanced Research Projects Agency.

References

1. Butler, E.P. and Hale, K.F. (1981) *Dynamic Experiments in the Electron Microscope, Practical Methods of Electron Microscopy*, vol. **9**, North-Holland, Amsterdam.
2. Poppa, H. (2004) *J. Vac. Sci. Technol. A*, **22**, 1931.
3. Ross, F.M. (2006) *Chapter 6 in Science of Microscopy* (eds P.W. Hawkes and J.C. Spence), Springer.
4. Parker, M.A. and Sinclair, R. (1986) *Nature*, **322**, 531.
5. Sinclair, R., Parker, M.A., and Kim, K.B. (1987) *Ultramicroscopy*, **23**, 383.
6. Gibson, J.M., Batstone, J.L., and Tung, R.T. (1987) *Appl. Phys. Lett.*, **51**, 45.
7. Kleinschmit, M.W., Yeadon, M., and Gibson, J.M. (1999) *Appl. Phys. Lett.*, **75**, 3288.
8. Ross, F.M., Bennett, P.A., Tromp, R.M., Tersoff, J., and Reuter, M. (1999) *Micron*, **30**, 21.
9. Chong, R.K.K., Yeadon, M., Choi, W.K., Stach, E.A., and Boothroyd, C.B. (2003) *Appl. Phys. Lett.*, **82**, 1833.
10. Tanaka, M., Han, M., Takeguchi, M., and Furuya, K. (2004) *Philos. Mag.*, **84**, 2699.
11. Sun, H.P., Chen, Y.B., Pan, X.Q., Chi, D.Z., Nath, R., and Foo, Y.L. (2005) *Appl. Phys. Lett.*, **86**, 71904.
12. Sharma, R. and Iqbal, Z. (2004) *Appl. Phys. Lett.*, **84**, 990.
13. Helveg, S., Lopez-Cartes, C., Sehested, J., Hansen, P.L., Clausen, B.S., Rostrup-Nielsen, J.R., Abild-Pedersen, F., and Norskov, J.K. (2004) *Nature*, **427**, 426.
14. Ross, F.M. (2010) *Rep. Prog. Phys.*, **73**, 114501.
15. Madras, P., Dailey, E., and Drucker, J. (2009) *Nano Lett.*, **9**, 3826.
16. Hull, R. and Bean, J. (1994) *Mater. Res. Soc. Bull.*, **19** (6), 32.
17. Sinclair, R. (1994) *Mater. Res. Soc. Bull.*, **19** (6), 26.
18. Stach, E.A., Hull, R., Bean, J.C., Jones, K.S., and Nejim, A. (1998) *Microsc. Microanal.*, **4**, 294.
19. Creemer, J.F., Helveg, S., Kooyman, P.J., Molenbroek, A.M., Zandbergen, H.W., and Sarro, P.M. (2010) *J. Microelectromech. Syst.*, **19**, 254.
20. Kallesøe, C., Wen, C.-Y., Mølhave, K., Bøggild, P., and Ross, F.M. (2010) *Small*, **6**, 2058.
21. LeGoues, F.K., Tersoff, J., Reuter, M.C., Hammar, M., and Tromp, R. (1995) *Appl. Phys. Lett.*, **67**, 2317.
22. Stach, E.A., Hull, R., Tromp, R.M., Reuter, M.C., Copel, M., LeGoues, F.K., and Bean, J.C. (1998) *J. Appl. Phys.*, **83**, 1931.
23. Drucker, J., Sharma, R., Weiss, K., and Kouvetakis, J. (1995) *J. Appl. Phys.*, **77**, 2846.
24. LeGoues, F.K., Reuter, M.C., Tersoff, J., Hammar, M., and Tromp, R.M. (1994) *Phys. Rev. Lett.*, **73**, 300.
25. LeGoues, F.K., Hammar, M., Reuter, M.C., and Tromp, R.M. (1996) *Surf. Sci.*, **349**, 249.
26. Hammar, M., LeGoues, F.K., Tersoff, J., Reuter, M.C., and Tromp, R.M. (1996) *Surf. Sci.*, **349**, 129.
27. Sharma, R., Rez, P., Treacy, M.M.J., and Stuart, J. (2005) *J. Electron Microsc.*, **54**, 231.
28. Hofmann, S., Sharma, R., Wirth, C.T., Cervantes-Sodi, F., Ducati, C., Kasama, T., Dunin-Borkowski, R.E., Drucker, J., Bennett, P., and Robertson, J. (2008.) *Nat. Mater.*, **7**, 372.

29 Gherasimova, M., Hull, R., Reuter, M.C., and Ross, F.M. (2008) *Appl. Phys. Lett.*, **93**, 023106.

30 Ross, F.M., Tersoff, J., and Tromp, R.M. (1998) *Phys. Rev. Lett.*, **80**, 984.

31 Portavoce, A., Hull, R., Reuter, M.C., and Ross, F.M. (2007) *Phys. Rev. B*, **76**, 235301.

32 Ross, F.M., Tromp, R.M., and Reuter, M.C. (1999) *Science*, **286**, 1931.

33 Kammler, M., Hull, R., Reuter, M.C., and Ross, F.M. (2003) *Appl. Phys. Lett.*, **82**, 1093.

34 Portavoce, A., Kammler, M., Hull, R., Reuter, M.C., and Ross, F.M. (2006) *Nanotechnology*, **17**, 4451.

35 Wu, Y. and Yang, P. (2001) *J. Am. Chem. Soc.*, **123**, 3165.

36 Zhou, G., Yang, J.C., Xu, F., Barnard, J.A., and Zhang, Z. (2002) *Mater. Res. Soc. Symp. Proc.*, **737**, F6.3.

37 Stach, E.A., Pauzauskie, P.J., Kuykendall, T., Goldberger, J., He, R., and Yang, P. (2003) *Nano Lett.*, **3**, 867.

38 Ross, F.M., Tersoff, J., and Reuter, M.C. (2005) *Phys. Rev. Lett.*, **95**, 146104.

39 Kodambaka, S., Tersoff, J., Reuter, M.C., and Ross, F.M. (2006) *Phys. Rev. Lett.*, **96**, 096105.

40 Hannon, J.B., Kodambaka, S., Ross, F.M., and Tromp, R.M. (2006) *Nature*, **440**, 69.

41 Kodambaka, S., Hannon, J.B., Tromp, R.M., and Ross, F.M. (2006) *Nano Lett.*, **6**, 1292.

42 Turnbull, D. and Fisher, J.C. (1949) *J. Chem. Phys.*, **17**, 71.

43 Turnbull, D. and Cech, R.E. (1950) *J. Appl. Phys.*, **21**, 804.

44 Kim, B.J., Tersoff, J., Reuter, M.C., Kodambaka, S., Stach, E.A., and Ross, F.M. (2008) *Science*, **322**, 1070.

45 Kim, B.J., Tersoff, J., Wen, C.-Y., Reuter, M.C., Stach, E.A., and Ross, F.M. (2009) *Phys. Rev. Lett.*, **103**, 155701.

46 Huang, J.Y., Zhong, L., Wang, C.M., Sullivan, J.P., Xu, W., Zhang, L.Q., Mao, S.X., Hudak, N.S., Liu, X.H., Subramanian, A., Fan, H., Qi, L., Kushima, A., and Li, J. (2010.) *Science*, **330**, 1515.

47 Gai, P.L. (2002) *Microsc. Microanal.*, **8**, 21.

48 Gai, P.L. and Calvino, J.J. (2005) *Annu. Rev. Mater. Res.*, **35**, 465.

49 Williamson, M.J., Tromp, R.M., Vereecken, P.M., Hull, R., and Ross, F.M. (2003) *Nat. Mater.*, **2**, 532.

50 Radisic, A., Vereecken, P.M., Hannon, J.B., Searson, P.C., and Ross, F.M. (2006) *Nano Lett.*, **6**, 238.

51 de Jonge, N., Poirier-Demers, N., Demers, H., Peckys, D.B., and Drouin, D. (2010) *Ultramicroscopy*, **110**, 1114.

52 Radisic, A., Vereecken, P.M., Searson, P.C., and Ross, F.M. (2006) *Surf. Sci.*, **600**, 1817.

53 Scharifker, B. and Hills, G. (1983) *Electrochim. Acta*, **28**, 879.

54 Radisic, A., Ross, F.M., and Searson, P.C. (2006) *J. Phys. Chem. B*, **110**, 7862.

55 Zheng, H., Smith, R.K., Jun, Y.-W., Kisielowski, C., Dahmen, U., and Alivisatos, A.P. (2009) *Science*, **324**, 1309.

56 Grogan, J.M. and Bau, H.H. (2010) *J. Microelectromech. Syst.*, **19**, 885.

57 Klein, K.L., Anderson, I.M., and de Jonge, N. (2011) *J. Microsc.*, **242**, 117.

58 Zheng, H., Claridge, S.A., Minor, A.M., Alivisatos, A.P., and Dahmen, U. (2009) *Nano Lett.*, **9**, 2460.

59 Schwarz, K.W., Tersoff, J., Kodambaka, S., Chou, Y.-C., and Ross, F.M. (2011) *Phys. Rev. Lett.*, **107**, 265502.

8
In-Situ TEM Studies of Oxidation
Guangwen Zhou and Judith C. Yang

8.1
Introduction

An extensive interest has existed for several decades in an understanding of the mechanism of surface oxidation, mainly because of its critical role in areas such as environmental stability, high-temperature corrosion, electrochemistry, catalytic reactions, gate oxides, and thin-film growth. Yet, surprisingly little is known about the mesoscopic processes that take place between the surface chemisorption of oxygen on a clean surface, and the formation of a continuous oxide layer. Classic oxidation theories such as those of Cabrera and Mott [1] and Wagner [2] are commonly cited, but these suffer serious deficiencies because of the lack of structure considerations. For example, Yang *et al.* recently noted that [3–6], while these classic theories assume uniform layer-by-layer growth of an oxide phase, starting with a continuous monolayer, several reports have indicated that the initial oxidation of many surfaces actually occurs via the nucleation, growth, and coalescence of oxide islands [7–12].

A key stumbling block that had held back progress in understanding the atomic-level and mesoscopic processes that occur during surface oxidation has been the scarcity of structural and chemical information obtained *in situ* during oxidation. This has been due, in large part, to an inability of traditional analytical techniques to be used for *in-situ* measurements of structures, chemistry, and kinetics at the nanoscale, as the reactions progress under well-controlled environments. More recently, *in-situ* ultrahigh-vacuum (UHV) environmental transmission electron microscopy (TEM) has been identified as an ideal tool to obtain important structural and chemical information, based on which a fundamental understanding can be acquired of the oxidation mechanism on a nanometer scale or below, through time- and temperature-resolved high-resolution images, electron diffraction, and electron energy-loss spectroscopy (EELS) during the reaction with controlled surface conditions and environments. Indeed, by using *in-situ* UHV-TEM techniques the nucleation, growth, coalescence and morphological evolution of oxide films can be visualized directly during the oxidation of atomically clean surfaces under gas environments. As a consequence, essential insights can be provided into active

In-situ Electron Microscopy: Applications in Physics, Chemistry and Materials Science, First Edition.
Edited by Gerhard Dehm, James M. Howe, and Josef Zweck.
© 2012 Wiley-VCH Verlag GmbH & Co. KGaA. Published 2012 by Wiley-VCH Verlag GmbH & Co. KGaA.

sites, transient states, reaction mechanisms, kinetics and thermodynamics [5, 13], which are difficult to obtain via other experimental methods yet are essential to understand oxidation kinetics. The overall reaction kinetics of early-stage oxidation are especially complex, due to a convolution of various kinetic parameters, including both the nucleation and growth of individual oxide islands. Whilst most techniques used to measure reaction kinetics cannot distinguish between these two processes, *in-situ* environmental TEM can be used to study them simultaneously by following the nucleation events and the growth rate of individual oxide islands.

8.2
Experimental Approach

The majority of TEM studies of oxidation have been carried out with oxide films stripped from the substrate [14]. However, the disadvantages of this approach are that the relative position of the oxide film on the substrate is changed, and an accurate determination of orientation relationships between the oxide and the underlying substrate is prohibited. Furthermore, the initial growth stages of the oxide film cannot be effectively studied, since any discontinuous layers cannot be stripped off. The importance of *in-situ* oxidation has been recognized however and, indeed, some TEM investigations have involved a side chamber for oxidation processing, after which the oxidized samples are transferred to the microscope column for observation, without being exposed to the air [15]. This semi-*in-situ* oxidation approach has the disadvantages of interrupting oxidation and not following the structural changes of the oxide film in real time. Consequently, important information such as the behavior of oxide nucleation, growth and coalescence may be missed, which would be critical for an atomic-scale understanding of oxidation mechanisms.

8.2.1
Environmental Cells

The *in-situ* study of surface oxidation is inherently difficult with TEM, as the operation of an electron source requires a vacuum better than 10^{-6} Torr. In order to conduct oxidation experiments *in situ* in the transmission electron microscope, the oxygen gas must be confined to the sample region in order to allow the column and gun vacuum to be maintained at operable conditions. In designing the gas confinement system, two types of environmental cell have been developed: (i) the thin window; and (ii) differential pumping.

The first of these designs involves the creation of a sealed sample holder to isolate the reaction cell from the column vacuum, where the oxygen gas is introduced to oxidize a sample placed between two electron-transparent windows. Besides being electron-transparent, the window material must be strong enough to sustain the cell in the surrounding vacuum. In principle, pressures of up to one atmosphere can be attained without affecting the electron gun, although the instrument's performance is partly sacrificed by having to use thick window materials that will tolerate such high pressures. As the use of such a window material also superimposes additional structural information on the image and diffraction of the sample, materials with a weak electron diffraction, such as amorphous C and SiN thin films, are normally preferred for windows.

The second approach used for gas confinement is to place small apertures above and below the samples, with a differential pumping system being applied through the small apertures inside the objective pole-piece. In this way, oxygen gas can be introduced directly into the sample region, where the pressure close to the gun is much lower than that in the vicinity of the sample region. For such a differential pumping system, the oxygen gas pressure to be maintained in the sample region is determined by the leak rate through the apertures; this can be calculated by using the kinetic theory of gases and the pumping speed in their vicinity.

Each of these techniques has both advantages and disadvantages. In general, window-type environmental cells cannot withstand heating, which makes them unsuitable for high-temperature oxidation experiments. In addition, the combined thickness of the windows and the sample may reduce the contrast, image resolution and analytical capabilities, including energy-dispersive X-ray spectroscopy (EDS). As the oxygen gas is confined within the sample holder, *in-situ* oxidation experiments using window environmental cells can be easily conducted using either commercial or home-built sample holders, without needing to modify the microscope, while the gas pressure applied may reach one atmosphere.

In contrast, the differential pumping approach requires the electron microscope to be reconfigured and is, therefore, much more expensive. The gas pressure in the sample region is usually limited to 20 Torr, without causing any deleterious effects to the electron gun and the microscope. The main advantages of the differential pumping environmental cell TEM compared to the window-type include a better resolution and a greater range of sample tilting. The latter advantage is due to the windows-type environmental cell requiring not only a larger sample holder but also allowing a much greater versatility of potential environmental experiments, using specialized sample holders (e.g., mechanical testing, magnetic, high-temperature holders, etc.). The design of the differential pumping environmental cell also permits an incorporation of the UHV capabilities to provide the necessary controlled surface conditions when performing fundamental and quantitative oxidation studies at nanoscale, and below.

8.2.2
Surface and Environmental Conditions

These improved experimental techniques, especially when performed in a UHV environment, permit the investigation of surfaces that are atomically clean at the start of the experiments. This is an extremely important point if the oxidation kinetics that exist especially at the initial stages of surface oxidation are to be understood on a quantitative basis [16, 17]. The UHV environment is necessary to provide an atomically clean surface, because impurities play a dramatic role in oxidation kinetics [18] and any surface impurities must be controlled in order to acquire both quantitative and fundamental insights. Typically, bakeable components can be added to a differential pumping environmental cell microscope in order to achieve UHV base pressures that will allow the surface oxidation to be examined where the surface contamination is minimized [19]. On most surfaces, due to air exposure, there will be a thin layer of native oxide, the presence of which can complicate the kinetic studies of surface oxidation. The native oxide may be removed *in situ* inside the microscope, either by high-temperature annealing under vacuum or by introducing reactive gases into the microscope chamber; these gases react with the oxide layer to form gaseous species that can be pumped out from the microscope. An example of this approach was described by Yang *et al.* [20, 21], who produced clean Cu surfaces in an *in-situ* UHV-TEM study of the initial oxidation kinetics of copper surfaces by introducing methanol gas *in situ* that would react with the native copper oxide to form gaseous formaldehyde and water vapor (Figure 8.1). The situation is more problematic for some other materials, such as Al, when the native oxides are thermodynamically stable and not easily removed, either by high-temperature annealing or by using reactive gases. To overcome this problem, a sputtering clean system can be built into the environmental TEM that will permit materials systems to be studied under clean surface conditions. Recently, just such a differential pumping environmental TEM was developed by FEI™ which not only had a built-in plasma cleaner but was also linked to a mass spectrometer.

8.2.3
Gas-Handling System

The control of both gas pressure and gas purity is important when performing *in-situ* UHV-TEM studies of surface oxidation. In order to establish any quantitative

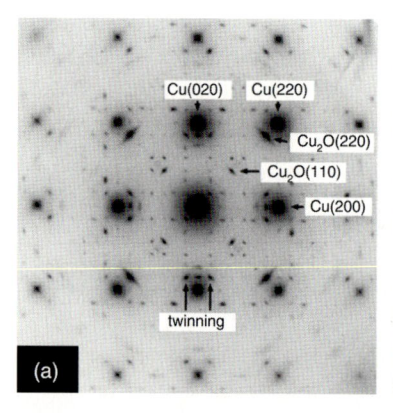

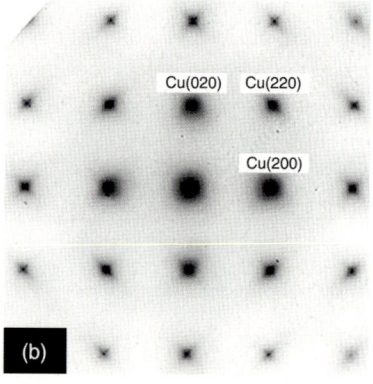

Figure 8.1 *In-situ* cleaning of a Cu surface with native Cu oxide by methanol vapor at 5×10^{-5} Torr column pressure and 350 °C. (a) The electron diffraction pattern from the Cu film before cleaning, where the small diffraction spots are from the native Cu oxide formed due to air exposure; (b) The electron diffraction from the Cu film after annealing in methanol vapor; disappearance of the small spots suggests that the Cu_2O is removed by the methanol reaction [20, 21].

correlation between the oxidation kinetics and the oxygen pressure, and to determine accurately the oxidation thermodynamics – such as the phase boundary between oxidation and reduction and the critical oxygen partial pressure for oxide nucleation – a gas-handling system is required that can provide an accurate control of the oxygen partial pressures and the oxidizing environments (e.g., H_2/H_2O, CO/CO_2). The installation of a residual gas analyzer (RGA) detector close to the sample region would also be needed to provide an accurate monitoring of the gas composition and partial pressures during the surface oxidation process.

Oxidation studies can be carried out in dry oxygen gas as well as under wet conditions that involve the exposure of samples to mixtures of water vapor and other gases. In this case, a water bubbler can be used to produce water vapor, while the partial pressure and gas flow rate of the gas mixture (O_2, H_2O, H_2, CO_2, CO, etc.) can be controlled via a series of mass-flow controllers.

8.2.4
Limitations

Although *in-situ* environmental UHV-TEM represents a powerful technique to provide an understanding of oxidation mechanisms based on high-resolution structural and chemical information obtained *in situ*, several other factors must also be considered when correlating the information obtained from TEM samples with effects in bulk materials:

- As the TEM samples are typically 100 nm thick (or less), this raises questions as to the surface effects on oxidation behavior and differences from bulk oxidation. However, the geometry of thin-film TEM samples may not be an issue in the initial oxidation stages, as these are controlled mostly by surface processes that include surface absorption, diffusion, and oxide nucleation.
- A second point is related to the pressure gap, which is one of the challenges that arise during *in-situ* environmental TEM studies of gas reactions. As noted above, the gas pressure in the sample region for the differential pumping environmental TEM permits the *in-situ* study of surface oxidation at low pressures of up to about 20 Torr, whereas the industrially important reactions usually take place at high pressures, such as one atmosphere. It is generally unclear whether it is feasible to extrapolate over this pressure gap, from results obtained under low-oxygen-pressure conditions to real life.
- The third issue related to *in-situ* oxidation is the effect of electron beam irradiation, which may not only cause sample local heating but also alter the rate of oxygen surface absorption and dissociation and displace surface atoms from their lattice sites [22]. The effects of electron beam irradiation on the oxidation kinetics can be inferred by comparing the oxidation behavior of surfaces that have been oxidized with and without electron illumination.

Despite these potential difficulties, considerable advances have been made in the present understanding of surface oxidation by performing *in-situ* oxidation experiments in real time and at high spatial resolution, using UHV-TEM.

8.3
Oxidation Phenomena

8.3.1
Surface Reconstruction

The general sequence of surface oxidation is: (i) oxygen chemisorption; (ii) nucleation and growth of the oxide; and (iii) bulk oxide growth. Usually, the oxygen surface chemisorption leads to surface reconstruction, the role of which is important to determine in oxide nucleation, since it has been speculated that an oxygen surface saturated layer must be formed before the onset of oxide formation. By using *in-situ* environmental UHV-TEM, Yang *et al.* observed a pronounced strain contrast induced by the anisotropic Cu $(\sqrt{2} \times 2\sqrt{2})R45°$ surface reconstruction during the oxidation of Cu(100) surface at 450 °C, as shown in Figure 8.2 [13]. Oxide nucleation occurs initially on the reconstructed Cu–O surface, after which the surface conditions – including the surface topology and roughness, impurities, and any orientation that might influence the surface reconstruction – should cause significant alterations to the initial stages of oxidation.

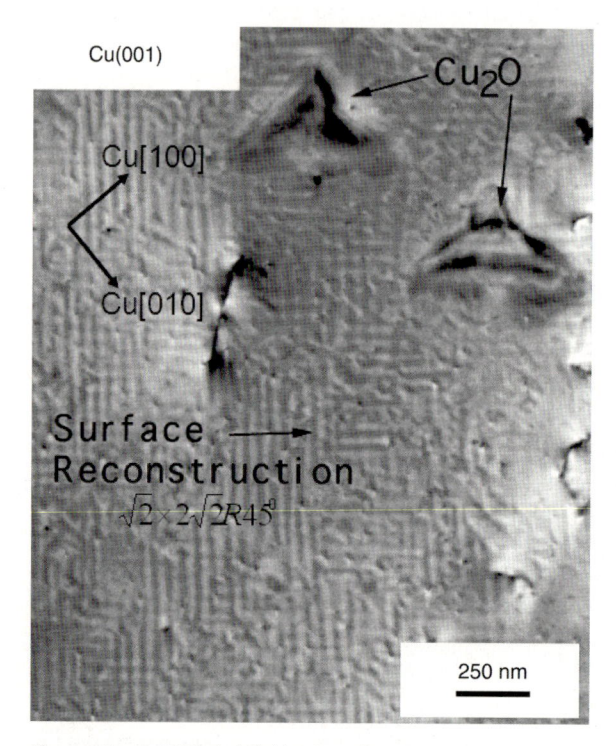

Figure 8.2 Cu(200) dark-field image after oxygen is introduced into the microscope at 5×10^{-4} Torr oxygen and 450 °C. A pronounced strain contrast due to surface reconstruction is clearly visible [13].

8.3.2
Nucleation and Initial Oxide Growth

Oxygen surface chemisorption has been investigated extensively using surface science methods, which mostly examine the adsorption of up to about one monolayer of oxygen. In contrast, the growth of a continuous oxide layer during the later stages of oxidation has been investigated by bulk oxidation method, such as thermogravimetric analysis (TGA). Consequently, the least well-understood regime in the oxidation process is the nucleation and initial growth of oxides. *In-situ* TEM has the unique ability to provide a direct visualization of the oxide nucleation and initial growth, which is inaccessible to both surface science methods and traditional bulk oxidation studies, but is essential for a fundamental understanding of atomistic oxidation kinetics.

Previously, Yang *et al.* have examined the initial nucleation stages of Cu(100) surfaces over a wide range of oxygen pressure and temperature conditions. A sequence of dark-field (DF) TEM images, taken from $Cu_2O(110)$ reflection and showing the nucleation events of Cu_2O islands in the oxidation of Cu(100) surfaces at 350 °C and an oxygen pressure of 5×10^{-4} Torr, is shown in Figure 8.3 [5]. Following the introduction of oxygen gas, no oxide islands appeared within the first couple of minutes, but they were then observed to nucleate rapidly, followed by island growth. After about 22 min of the oxidation, the islands reached the saturation number density of the nuclei. The number density of the oxide islands with respect to the oxidation time is shown in Figure 8.4, where the solid line corresponds to the fitting by oxygen surface diffusion-limited nucleation processes. As oxidation theories had assumed the formation of a uniform and continuous oxide film during metal oxidation, this result represents a critical departure from classic oxidation theories, and is important in understanding corrosion processes.

The evolution of the island area during island growth is recorded *in situ* [13]. The *in-situ* observations of the growth of individual islands during the oxidation of Cu(100) at an oxygen pressure of 5×10^{-4} Torr at 350 °C is shown in Figure 8.5, while the quantitatively measured island areas as a function of oxidation time are shown in

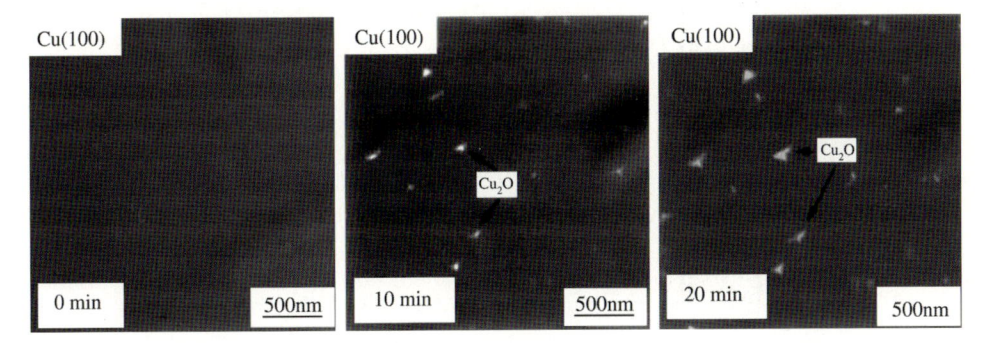

Figure 8.3 Cu_2O dark-field images taken at 0, 10, and 20 min after O_2 is introduced into the microscope in the oxidation of Cu(100) thin films at a pressure of 5×10^{-4} Torr at 350 °C [5].

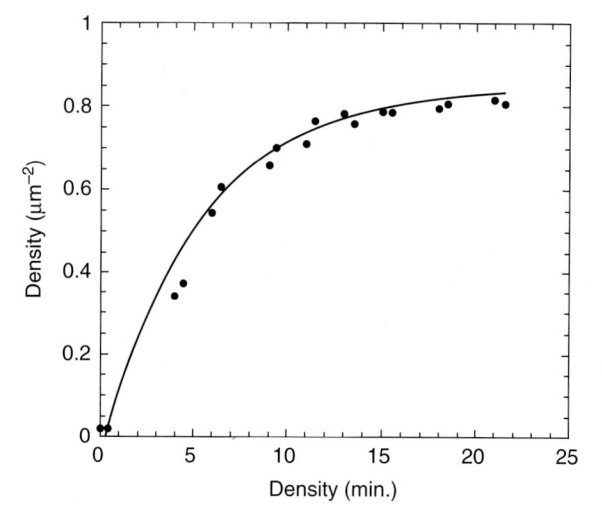

Figure 8.4 Cu$_2$O island density as a function of oxidation time at 350 °C and oxygen pressure of 5 × 10^{-4} Torr. The solid line corresponds to the theoretical fit based on the nucleation processes limited by oxygen surface diffusion [5].

Figure 8.6. The best power-law fit to the island growth was $t^{1.3}$, which was slightly higher than t – the predicted power-law dependence for three-dimensional (3-D) growth by oxygen surface diffusion [4]. To account for the slight deviation from the linear growth, Yang et al. have considered the direct impingement or bulk diffusion of oxygen that contributes to the island growth, and obtained a good fit to the experimental data [4]. This is shown in Figure 8.6, where the solid line corresponds to the fitting by oxygen surface diffusion plus oxygen direct impingement.

8.3.3
Role of Surface Defects on Surface Oxidation

It is generally believed that a number of important surface properties – for example, chemical reactivity – depend critically on surface defects. Hence, a complete theory of important processes such as oxidation and corrosion must take into account the role of such defects.

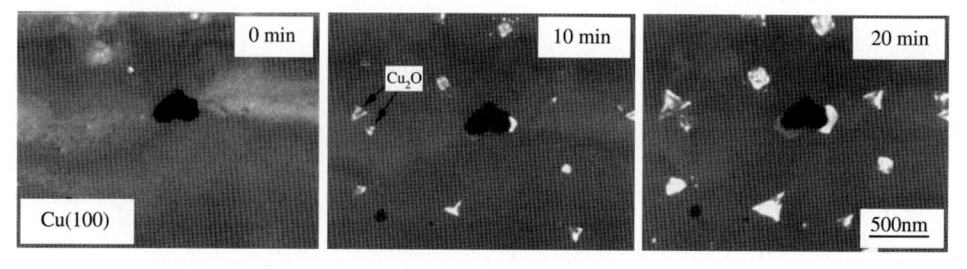

Figure 8.5 Dark-field images taken from the Cu$_2$O(110) reflection in the oxidation of Cu(100) thin film at 350 °C and oxygen pressure of 5 × 10^{-4} Torr [4].

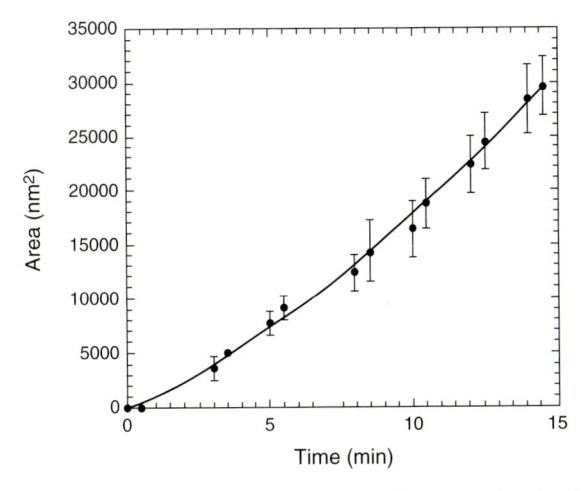

[Graph showing oxide island area vs time, with data points and error bars from 0 to ~15 min, area rising from 0 to ~30000 nm²]

Figure 8.6 Comparison of the experimentally measured oxide island area with the theoretical fitting of the oxygen surface diffusion and direct impingement/bulk diffusion for the 3-D growth of Cu_2O islands in the oxidation of Cu(100) at 350 °C and oxygen pressure of 5×10^{-4} Torr [4].

It is feasible to directly image surface steps with weak beam DF images by TEM [23], and therefore to provide the opportunity of clarifying the role of surface steps in oxide formation. By using weak beam techniques with *in situ* UHV-TEM, Ross and Gibson examined the role of the surface steps and terraces in silicon oxidation [24]. The results obtained (see Figure 8.7) revealed clearly that a flat oxide layer would form, but that the surface steps would not move as the oxide grew during the oxidation of Si(111). This result was unexpected, since it had been supposed that oxidation would occur preferentially at the step edges, but is important for controlling interface roughness, for example when growing gate oxides [19].

Milne and Howie have proposed that surface steps play a significant role in the nucleation of copper oxides during the oxidation of copper [15]. Milne has observed large faceted islands along surface steps by scanning reflection electron microscopy, but was unable to observe any nucleation at these steps [15]. By using *in-situ* UHV environmental TEM techniques, Yang *et al.* [25] were able to observe the very early-stage oxidation of Cu surfaces, and their results revealed that surface steps are not the preferential sites for oxide nucleation (see Figure 8.8). Similarly, Zhou and Yang also failed to observe dislocations as being the preferential sites for oxide nucleation (Figure 8.9) [26].

8.3.4
Shape Transition During Oxide Growth in Alloy Oxidation

By using *in-situ* environmental TEM techniques, Zhou *et al.* have examined the effect of alloying on the oxide structure continuity in Cu–Au(100) oxidation [27]. The *in-situ* TEM observation of the growth of one oxide island during the oxidation of Cu–5%Au (100) at 600 °C is shown in Figure 8.10. In this case, the oxide island was seen to

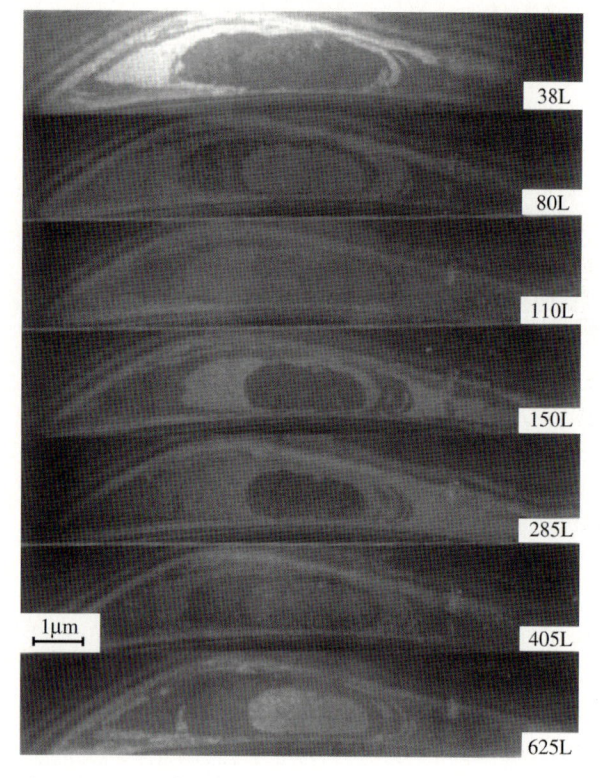

Figure 8.7 Part of a series of forbidden reflection images showing terrace contrast changes upon Si(111) exposure at room temperature to 2.5×10^{-7} Torr H_2O. Doses are given in Langmuirs (L; 1 L = 10^{-6} Torr s^{-1}). The image resolution is limited by the objective aperture to about 2 μm [24].

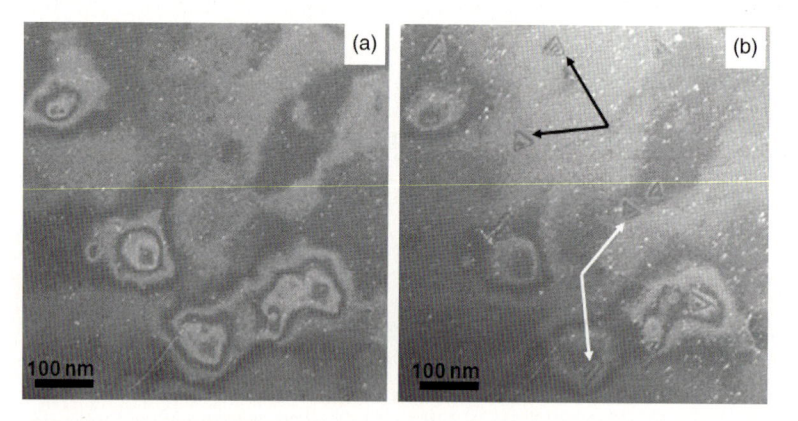

Figure 8.8 Cu(200) dark-field images taken at (a) 0 min and (b) 20 min after oxidation at 290 °C. In (b), the black arrows point to the Cu_2O Islands that are not nucleated at steps, while the white arrows point to islands that are nucleated at a step position [25].

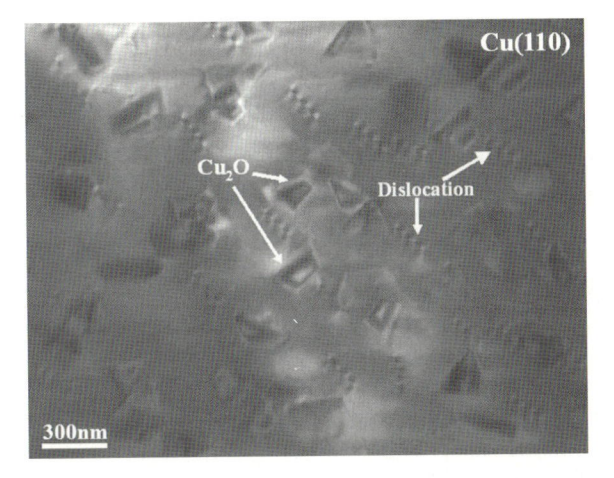

Figure 8.9 The bright-field image of a Cu(110) film oxidized at 450 °C and oxygen pressure of 0.01 Torr, where many threading dislocations are present on the surface. The oxide islands have a random distribution, which is not uncorrelated with the location of the dislocations. This suggests that the threading dislocations are not the preferred sites for oxide nucleation [26].

nucleate with a square shape and to retain this shape during the initial growth. The island then exhibited a gradual transition to a dendritic shape as the growth slowed along the normal to the island edges (i.e., along Cu$_2$O <110> directions), while maintaining a faster growth rate along the directions of the island corners (i.e., along Cu$_2$O<100> directions). In contrast, the oxide islands formed during the oxidation of pure Cu(001), under the same conditions, showed an initially square shape that transformed to a rectangular morphology as growth proceeded [28, 29]. As seen in Figure 8.10, a nonuniform dark contrast developed around the island with the island growth, while EDS analysis revealed that the Au molar fraction was highly enhanced in the region with dark contrast [30]. Typically, Au does not form a stable oxide under most conditions, and will not pass into solution in the oxide phase due to its stable

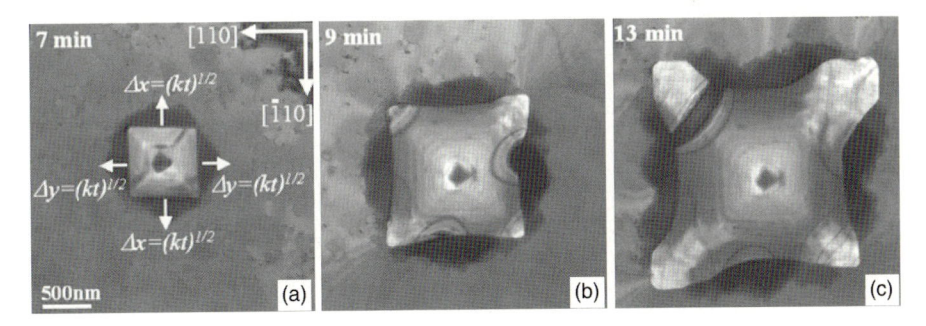

Figure 8.10 *In-situ* TEM observation of the growth of a Cu$_2$O island on a (001)Cu–5%at Au sample at 600 °C at pO$_2$ = 5 × 10^{-4} Torr. A crossover from an initially compact structure (a) to dendritic growth (b, c) is observed as time progresses. The total oxidation time is noted on each image [27].

electronic configuration. As a result, the oxidation of Cu–Au alloys will result in a selective oxidation of Cu and an ejection of Au from the metal–oxide interface, leading in turn to an Au enrichment in the remaining alloy. Consequently, the formation of a dark contrast around the oxide island is related to the accumulation of Au atoms segregated from the metal–oxide interface during the oxide growth. The contrast features in Figure 8.10 revealed that the alloy film regions adjacent to the island edges would become Au-rich, but that there was almost no excess Au accumulation in the alloy film adjacent to the island corners. The contrast features around the oxide island revealed there to be a nonuniform partition of Au atoms in the alloy around the island, while alloy regions adjacent to the island edges had a greater Au accumulation than that adjacent to the island corners, as confirmed by the EDS analysis [30]. This result proved to be important to understand the effect of alloying on the formation of a protective oxide layer. For example, dendritic transition causes a discontinuous morphology of the oxide, which is undesirable in many technological applications. Hence, in order to grow continuous, protective oxide films, it will be necessary to nucleate a large number density of oxide islands, which in turn will lead to a full coalescence of the oxide islands prior to their dendritic transition.

8.3.5
Effect of Oxygen Pressure on the Orientations of Oxide Nuclei

It has been shown previously that the oxidation of copper and many other metals proceeds via the nucleation of oxide islands [8, 10, 17, 28, 31–33], which is assumed to have thermodynamically controlled orientations [34]. By using *in-situ* TEM techniques, however, this has been shown as the case only if the metal surface is oxidized under a relatively low oxygen pressure (pO_2), since an increasing oxygen pressure would lead to the nucleation of randomly oriented islands [35].

Bright-field TEM images of Cu(100) surfaces oxidized at 350 °C and different levels of pO_2 for 10 min, are shown in Figure 8.11. In this case, oxide islands were observed to form on the surface, while the density of the oxide nuclei increased with increasing pO_2. Selected-area electron diffraction (SAED) patterns from the oxidized surfaces revealed that Cu_2O islands that nucleated under the lower pO_2 (i.e., <5 Torr) had a cube-on-cube epitaxy with the Cu(100) substrate (i.e., (011)Cu_2O//(011)Cu and Cu_2O//Cu). In contrast, oxidation at $pO_2 = 150$ Torr and above (for the oxidation under the oxygen pressure over $\sim 8 \times 10^{-4}$ Torr, the electron gun must be isolated from the microscope column by a vacuum valve) resulted in the nucleation of nonepitaxial Cu_2O islands, as revealed by the presence of a Cu_2O diffraction ring pattern (Figure 8.11c). The intensity distribution over the diffraction rings was rather uniform, which suggested that the oxide islands were oriented at random. The appearance of additional diffraction spots or rings surrounding Cu reflections in the electron diffraction patterns shown in Figure 8.11 was caused by a double diffraction of Cu and Cu_2O islands. These *in-situ* TEM observations indicated that the epitaxial nucleation of oxide islands cannot be maintained within the whole range of oxygen pressures. The change in nucleation orientation under different oxygen pressures

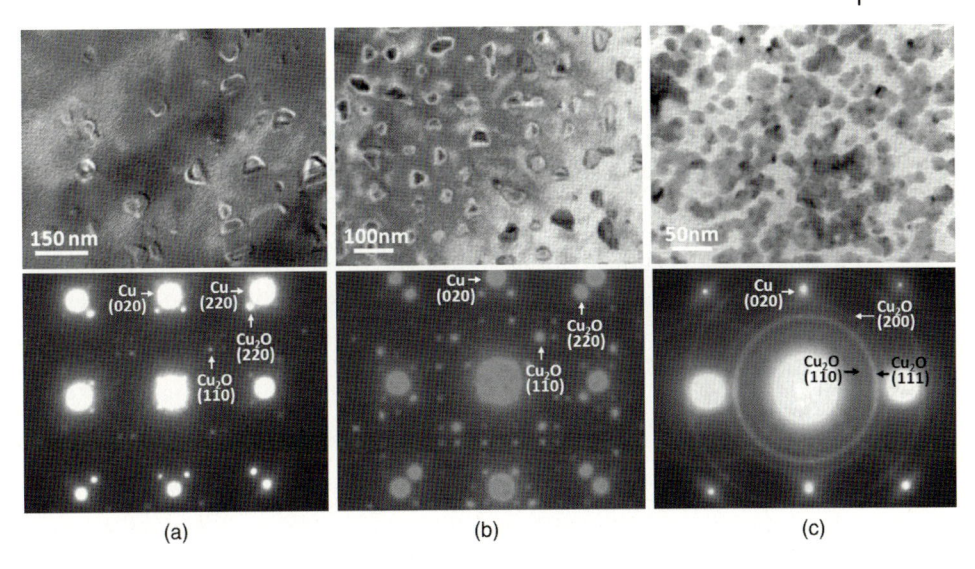

Figure 8.11 Upper panels: Bright-field TEM images of Cu_2O islands formed from Cu(100) oxidized at 350 °C and different oxygen pressures for 10 min. (a) $pO_2 = 5 \times 10^{-4}$ Torr, (b) $pO_2 = 0.5$ Torr, (c) $pO_2 = 150$ Torr. Lower panels: SAED patterns from the corresponding oxidized Cu(100) surfaces, where the additional reflections are due to double diffraction of electron beams by Cu and Cu_2O. A transition from nucleating epitaxial oxide islands to randomly oriented Cu_2O islands occurs upon increasing the oxygen pressure [35].

can be physically understood as follows [35]. At low oxygen pressure, the nucleation barrier is very high, and the top priority for accelerating the nucleation kinetics is to lower the nucleation barrier. Therefore, an epitaxial nucleation of oxide islands would be kinetically favored. Conversely, at high oxygen pressures the nucleation barrier is reduced and the issue of effective collisions of oxygen atoms becomes important. The nucleation of oxide islands with weak interaction and poor structural match (i.e., nonepitaxial) with the metal substrate is enhanced. Kinetically speaking, in order to obtain the epitaxial oxide film on the metal substrate by oxidation, the oxygen pressure should be relatively low. If the oxygen pressure is too high, however, the kinetics leads to a deviation of nucleating oxide islands from the orientation of the metal substrate and the epitaxial relation will be lost.

8.3.6
Oxidation Pathways Revealed by High-Resolution TEM Studies of Oxidation

The *in-situ* visualization of oxidation processes at near-atomic resolution is very important for understanding the structural aspects of oxidation mechanisms. The information limit of the transmission electron microscope may be compromised in the differential pumping environmental cell microscope, because the small apertures below the samples may block any high-angle diffraction. Moreover, the resolution of the microscope may deteriorate due to an inelastic scattering of

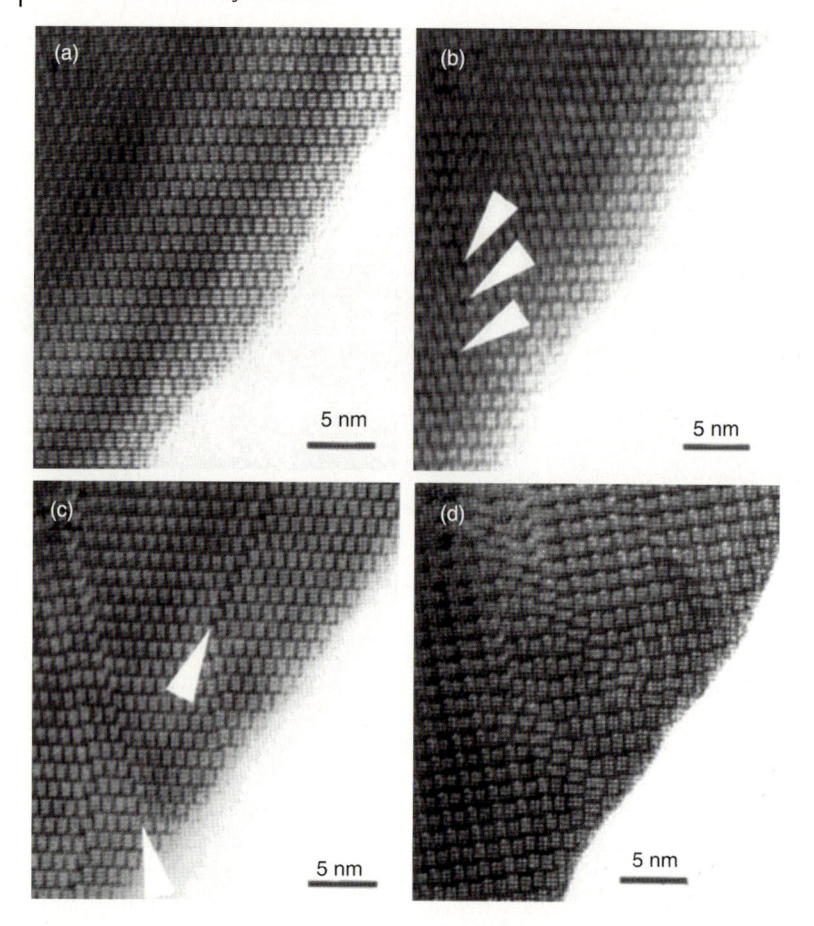

Figure 8.12 Oxidation of $Nb_{12}O_{29}$. (a) High-resolution lattice image under vacuum; (b) The lattice image in 10^3 Pa of air, where the initial oxidation sites are indicated by arrowheads; (c) The formation of planar defects (marked by arrowheads) due to the formation of $NbO_{2.5}$ phase on the continued oxidation; (d) The lattice image showing various oxide structures after final oxidation [22, 36].

the electron beam by the oxygen gas in the sample region, and also by the window materials if window cells are used. Despite such restrictions, the *in-situ* high-resolution TEM of oxidation processes has been achieved in the same experimental systems.

Some examples of *in-situ* high-resolution TEM observations of the oxidation of $Nb_{12}O_{29}$ by Hutchison *et al.* are shown in Figure 8.12 [36]. Here, the oxidation-induced structural changes are indicated by the presence of black dots (marked by white arrowheads) in the 4×3 structure of $Nb_{12}O_{29}$. Further oxidation results in the formation of a metastable phase of $NbO_{2.5}$, as evidenced by the planar defects of 3×3 blocks, while continued oxidation leads to the formation of higher oxide domains. Such *in-situ* high-resolution TEM observations provide direct experimental data relating to oxidation pathways that are not accessible by examining post-oxidation samples.

8.4
Future Developments

Although, during recent years, significant progress has been made in the fundamental understanding of surface oxidation by using *in-situ* environmental TEM techniques, many fundamental questions remain unresolved:

- How is the oxide phase formed from the oxygen chemisorbed surface?
- How are dislocations generated and propagated at the metal–oxide interface during metal oxidation?
- What is the atomic mechanism involved in the conversion of substrate atoms to the oxide phase at the interface between the oxide phase and its underlying substrate?

To address these questions will require improvements to be made in the spatial and temporal resolution of the transmission electron microscope, as well as an improved environment control. The incorporation of recent developments such as lens aberration correction into the *in-situ* environmental TEM cell will extend the spatial resolution to sub-Ångstrom levels, which should in turn have a significant impact on the fundamental aspects of oxidation studies, including oxide nucleation mechanisms, interface dynamics, and oxidation pathways. At present, the temporal resolution of available environmental cell TEM technologies is limited by the video-frame rate resolution (typically 30 frames per second). Yet, many processes – including atom migration, dislocation generation and propagation, as well as some phase transitions in oxidation – occur at much faster rates than presently available detectors can cope with, due to their limited acquisition speeds. Clearly, the development of image detectors with much faster acquisition rates, perhaps up to 3000 frames per second, would be very welcome.

To date, most *in-situ* TEM studies of surface reactions have involved images of the sample in two-dimensional (2-D) projection. Consequently, the development of *in-situ* 3-D imaging techniques (e.g., electron tomography) will provide critical insights into oxide growth mechanisms, including the shape dynamics of 3-D oxide islands during early-stage oxidation and the temporal evolution of the Burgers vectors of 3-D dislocation arrays generated at the interface between the oxide layer and its underlying substrate during oxidation.

The development of more modular approaches to *in-situ* TEM, such as thin-film processing under UHV conditions and simultaneous surface science characterization, will create unprecedented scientific opportunities to study oxidations. For example, both Auger electron spectroscopy (AES) and X-ray photoelectron spectroscopy (XPS) are inherently surface-sensitive techniques that are capable of providing complementary information to that obtained with TEM. Moreover, the incorporation of AES or XPS with *in-situ* thin-film processing (such as thin-film synthesis and doping within the *in-situ* TEM chamber) should provide unique opportunities to investigate how chemical modifications may affect the oxidation mechanism. These integrated *in-situ* capabilities should also permit the simultaneous monitoring of various chemical species and oxidation states on the surface by AES or XPS. In addition, determination of the reaction kinetics and structural evolution (via TEM imaging and electron diffraction) of the oxidation progresses should provide significant improvements to the present understanding of fundamental oxidation phenomena.

8.5
Summary

In this chapter, the powerful role of *in-situ* environmental UHV-TEM in elucidating oxidation mechanism has been highlighted, based on time- and temperature-resolved images, diffraction, and spectroscopy at high spatial resolution. Previously, *in-situ* environmental UHV-TEM has been applied successfully to investigations of oxidation phenomena, including oxygen chemisorption-induced surface reconstructions, the initial oxidation stages of oxide nucleation and initial growth, the effect of surface defects and oxygen pressures on oxide formation, the shape transition of oxide islands during alloy oxidation, and oxidation pathways.

Unfortunately, *in-situ* environmental UHV-TEM studies of oxidation have encountered certain limitations, the most problematic of which have included the effects of sample geometry, electron beam effects, and difficulties in accurately calibrating the experimental parameters, including oxidation temperatures and oxygen pressures. Nonetheless, recent advances in microscope design, including the development of aberration correction and electron detectors, have allowed significant improvements to be made in the spatial and temporal resolution of *in-situ* TEM which, together, hold the promise of major enhancements in the capabilities of this technique.

References

1 Cabrera, N. and Mott, N.F. (1949) Theory of the oxidation of metals. *Rep. Prog. Phys.*, **12**, 163–184.
2 Wagner, C. (1933) Beitrag zur theorie des anlaufvorgangs. *Z. Phys. Chem.*, **B21**, 25–41.
3 Yang, J.C., Kolasa, B., Gibson, J.M., and Yeadon, M. (1998) Self-limiting oxidation of copper. *Appl. Phys. Lett.*, **73**, 2841–2843.
4 Yang, J.C., Yeadon, M., Kolasa, B., and Gibson, J.M. (1997) Oxygen surface diffusion in three-dimensional Cu_2O growth on Cu(001) thin films. *Appl. Phys. Lett.*, **70**, 3522–3524.
5 Yang, J.C., Yeadon, M., Kolasa, B., and Gibson, J.M. (1998) The homogeneous nucleation mechanism of Cu_2O on Cu(001). *Scripta Mater.*, **38**, 1237–1241.
6 Yang, J.C., Evan, D., and Tropia, L. (2002) From nucleation to coalescence of Cu_2O islands during in situ oxidation of Cu(001). *Appl. Phys. Lett.*, **81**, 241–243.
7 Thurmer, K., Williams, E., and Reutt-Robey, J. (2002) Autocatalytic oxidation of lead crystallite surfaces. *Science*, **297**, 2033–2035.
8 Hajcsar, E.E., Underhill, P.R., and Smeltzer, W.W. (1995) Initial stages of oxidation on Co-Ni alloys: Island

nucleation and growth. *Langmuir*, **11**, 4862–4872.

9 Holloway, P.H. and Hudson, J.B. (1974) Kinetics of the reaction of oxygen with clean nickel single crystal surfaces. *Surf. Sci.*, **43**, 123–140.

10 Holloway, P.H. and Hudson, J.B. (1974) Kinetics of the reaction of oxygen with clean nickel single crystal surfaces. *Surf. Sci.*, **43**, 141–149.

11 Marikar, P., Brodsky, M.B., Sowers, C.H., and Zaluzec, N.J. (1989) In situ HVTEM studies of the early stages of oxidation of nickel and nickel chromium-alloys. *Ultramicroscopy*, **29**, 247–256.

12 Shinde, S.R., Ogale, A.S., Ogale, S.B., Aggarwal, S., Novikov, V.A., Williams, E.D., and Ramesh, R. (2001) Self-organized pattern formation in the oxidation of supported iron thin films. I. An experimental study. *Phys. Rev. B.*, **64**, 035408.

13 Yang, J.C., Bharadwaj, M.D., Zhou, G.W., and Tropia, L. (2001) Surface kinetics of copper oxidation investigated by in-situ UHV transmission electron microscopy. *Microsc. Microanal.*, **7**, 486–493.

14 Birks, N., Meier, G.H., and Pettit, F.S. (2006) *Introduction to High Temperature Oxidation of Metals*, 2nd edn, Cambridge University Press, Cambridge.

15 Milne, R.H. and Howie, A. (1984) Electron microscopy of copper oxidation. *Philos. Mag. A*, **49**, 665–682.

16 Heinemann, K., Rao, D.B., and Douglass, D.L. (1975) Oxide nucleation on thin films of copper during in situ oxidation in an electron microscope. *Oxid. Met.*, **9** (4), 379–400.

17 Heinemann, K., Rao, D.B., and Douglas, D.L. (1975) In situ oxidation studies on (001)Cu-Ni alloy thin films. *Oxid. Met.*, **11** (6), 321–334.

18 Rakowski, J.M., Meier, G.H., and Pettit, F.S. (1996) The effect of surface preparation on the oxidation behavior of gamma TiAl-base intermetallic alloys. *Scripta Mater.*, **35** (12), 1417–1422.

19 Ross, F. (2006) A unique tool for imaging crystal growth. *Mater. Today*, **9** (4), 54–55.

20 Yang, J.C., Yeadon, M., Olynick, D., and Gibson, J.M. (1997) *Microsc. Microanal.*, **3**, 121–125.

21 Francis, S.M., Leibsle, F.M., Haq, S., Xiang, N., and Bowker, M. (1994) Methanol oxidation on Cu(110). *Surf. Sci.*, **315**, 284–292.

22 Robertson, I.M. and Teter, D. (1998) Controlled environment transmission electron microscopy. *Microsc. Res. Tech.*, **42**, 260–269.

23 Cherns, D. (1974) Direct resolution of surface atomic steps by transmission electron microscopy. *Philos. Mag.*, **30** (3), 549–556.

24 Ross, F.M. and Gibson, J.M. (1992) *Phys. Rev. Lett.*, **68**, 1782–1785.

25 Yang, J.C., Yeadon, M., Kolasa, B., and Gibson, J.M. (1999) *J. Electrochem. Soc.*, **146**, 2103.

26 Zhou, G.W., Wang, L., and Yang, J.C. (2005) Effect of surface topology on the formation of oxide islands on Cu surfaces. *J. Appl. Phys.*, **97**, 063509.

27 Zhou, G.W., Wang, L., Birtcher, R.C., Baldo, P.M., Pearson, J.E., Yang, J.C., and Eastman, J.A. (2006) Cu_2O island shape transition during Cu-Au oxidation. *Phys. Rev. Lett.*, **96**, 226108.

28 Zhou, G.W. and Yang, J.C. (2002) Formation of Quasi-one-dimensional Cu_2O structures by in situ oxidation of Cu(100). *Phys. Rev. Lett.*, **89**, 106101.

29 Zhou, G.W. and Yang, J.C. (2004) Temperature effects on the growth of oxide islands on Cu(110). *Appl. Surf. Sci.*, **222**, 357–364.

30 Zhou, G.W., Eastman, J.A., Birtcher, R.C., Pearson, J.E., Baldo, P.E., and Thompson, L.J. (2007) Composition effects on the early-stage oxidation kinetics of (001)Cu-Au alloys. *J. Appl. Phys.*, **101**, 033521.

31 Zhou, G.W. and Yang, J.C. (2003) Initial oxidation kinetics of copper (110) film investigated by in situ UHV-TEM. *Surf. Sci.*, **531**, 359–367.

32 Zhou, G.W., Fong, D.D., Wang, L., Fuoss, P.H., Baldo, P.M., Thompson, L.J., and Eastman, J.A. (2009) Nanoscale duplex oxide growth during early stages of oxidation of Cu-Ni(100). *Phys. Rev. B*, **80**, 134106.

33 Zhou, G.W. (2009) Nucleation thermodynamics of oxide during metal oxidation. *Appl. Phys. Lett.*, **94**, 201905.

34 Lawless, K.R. (1974) The oxidation of metals. *Rep. Prog. Phys.*, **37**, 231–316.

35 Luo, L.L., Kang, Y.H., Liu, Z.Y., Yang, J.C., and Zhou, G.W. (2011) Dependence of degree of orientation of copper oxide nuclei on oxygen pressure during initial stages of copper oxidation. *Phys. Rev. B*, **83**, 155418.

36 Hutchison, J.L., Holton, D., and Doole, R.C. (1993) In situ oxidation and reduction of complex oxides, in *Electron Microscopy and Analysis* (ed. A.J. Craven), Institute of Physics Conference Series, Bristol, pp. 465–468.

Part III
Mechanical Properties

In-situ Electron Microscopy: Applications in Physics, Chemistry and Materials Science, First Edition.
Edited by Gerhard Dehm, James M. Howe, and Josef Zweck.
© 2012 Wiley-VCH Verlag GmbH & Co. KGaA. Published 2012 by Wiley-VCH Verlag GmbH & Co. KGaA.

9
Mechanical Testing with the Scanning Electron Microscope
Christian Motz

9.1
Introduction

Modern materials generally exhibit a more or less complex microstructure that is designed to fulfill certain demands. For example, an understanding of the materials' mechanical performance requires a knowledge of the local processes occurring inside the microstructure during mechanical loading. There is, therefore, a need to investigate these local processes and to correlate them with the overall properties of the material. Such a "bottom-up" design process – whereby the basic mechanisms are understood and improved microstructures are produced on that basis – is essential for the development of new advanced materials.

A few decades ago, "*in-situ*" mechanical tests were usually performed using optical microscopy, or with the use of other optical instruments. For example, Staal and Elen [1] studied the fatigue crack growth behavior in a stainless steel by using optical methods, while Tungatt and Humphreys [2] investigated deformation twinning in sodium nitrate under an optical microscope, and Hay and Evans [3] investigated grain boundary migration. The main disadvantages of optical microscopy are the limited resolution and the small depth of focus, which come most into play when studying fine- and ultrafine-grained materials, or where a pronounced surface roughness is evolved during deformation. As a consequence, there is today an increasing tendency to perform mechanical tests *in-situ* with the scanning electron microscope which, in addition to having a better resolution and depth of focus, has certain other advantages – and disadvantages! A brief comparison of the characteristics of optical and scanning electron microscopies is provided in Table 9.1.

As can be seen from the data in Table 9.1, each of these techniques has both advantages and disadvantages. For example, biological samples or polymers are generally not very well suited to investigations with scanning electron microscopy (SEM) and are better examined using optical microscopy. Newly developed techniques, such as confocal laser microscopy [4] can however be used to overcome many of the shortcomings of conventional optical microscopy. In contrast, metallic materials are best suited for investigations with SEM, and this had led to increasing numbers of mechanical tests being performed nowadays *in-situ*, using this

In-situ Electron Microscopy: Applications in Physics, Chemistry and Materials Science, First Edition.
Edited by Gerhard Dehm, James M. Howe, and Josef Zweck.
© 2012 Wiley-VCH Verlag GmbH & Co. KGaA. Published 2012 by Wiley-VCH Verlag GmbH & Co. KGaA.

Table 9.1 Comparative characteristics of optical and scanning electron microscopies.

Optical microscopy (OM)	Scanning electron microscopy (SEM)
No special sample pre-requisition.	Sample must be vacuum proof and should be electrical conductive.
Usually no sample damage.	Electron beam may damage sample (e.g., polymers, biological sample).
Sample testing in many environments.	Only vacuum and some special environments possible (environmental SEM).
Limited resolution (ca. 0.5 µm) and depth of focus (in the range of the resolution).	Nanometer-scale resolution and large depth of focus (from several tens of µm to several 100 µm at low magnifications).
Usually enough space for loading stage (in principle, unlimited).	Space for the loading stage is limited by the vacuum chamber (typically several hundreds of mm for conventional SEMs).
Real-time imaging possible.	Image acquisition takes several seconds.

approach. Indeed, a combination of *in-situ* testing with additional measurement methods, which perhaps might include image correlation methods for strain mapping [5, 6] or electron back-scattering diffraction (EBSD) for local crystal orientation measurements [7] and, in the near future also for strain (stress) measurements [8], can provide a detailed view of the deformation processes to the point where they are today considered "state-of-the-art" techniques (see Chapter 1).

9.2
Technical Requirements and Specimen Preparation

In general, the use of SEM for *in-situ* testing has no special requirements, although certain properties of the microscope itself may be of interest. Notably, the specimen chamber should be large enough to include the loading device and to ensure correct stage operation. The additional weight of the loading device must also be taken into account to avoid any overloading of the specimen stage, and the pumping system should be sufficient so as to avoid excessive pumping times (the fresh installation of a massive loading device may involve several hours of pumping time). It should also be borne in mind that, due to the extension of the loading device, the working distance range may be limited. Depending on the size of the specimen chamber and the loading device, very small or very large working distances may not be achievable, and this will impact on the availability of SEM detectors and also on the image resolution. An example of a loading device placed in the microscope chamber is shown in Figure 9.1.

The preparation of a specimen depends on the specimen type and the information to be obtained. Nonconductive specimens, such as polymers, ceramics, and biological samples, might be covered with a thin conductive layer of gold or carbon. In this case, it must be ensured that the layer has a good adhesion to the underlying material, and shows no signs of cracking or peel-off during straining. Samples that are very

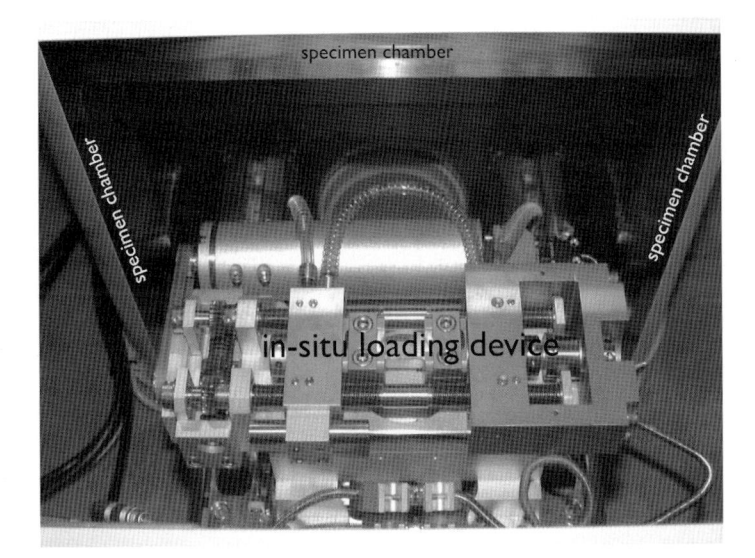

Figure 9.1 Example of a universal loading device placed in a scanning electron microscope chamber.

sensitive to the electron beam exposure must be prepared especially carefully. Possible beam damage may also be reduced by decreasing the accelerating voltage and the specimen current. With this type of material, a compromise must always be struck between specimen damage and image resolution. For metallic materials, special precautions are usually not required, although special surface preparation techniques may be used to increase the surface topology so as to obtain a better image contrast. For example, grain boundary etching will increase the contrast of the grain boundaries, which is useful when characterizing plastic deformation in polycrystalline metals. An example of grain boundary etching on an aluminum sample, before and after plastic straining, is shown in Figure 9.2. Usually, similar polishing and etching techniques can be applied with SEM as with optical microscopy, although

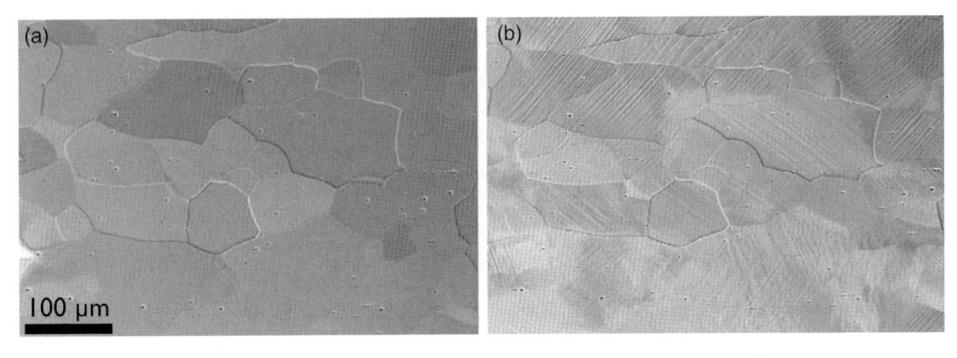

Figure 9.2 Scanning electron microscopy images of a polycrystalline aluminum sample.
(a) With a grain boundary etching in the undeformed state; (b) *In-situ* strained to 5% total strain.

other surface treatments may be required for specialized applications, such as image correlation techniques for local strain measurements or EBSD scans for determining local crystal orientations [9]. In addition, markers may be placed on the surface, perhaps as either microindentations or nanoindentations, by using focused ion beam (FIB) milling, to indicate regions of interest that may need to be relocated during further examinations with SEM.

9.3
In-Situ Loading of Macroscopic Samples

9.3.1
Static Loading in Tension, Compression, and Bending

Today, a wide range of commercially available loading devices exists for the *in-situ* testing of macroscopic samples. The principal layout of these devices is very similar, and consists of one or two cross-heads driven by sealed direct current (DC) or stepping motors via two screw spindles. Loading devices with two cross-heads that move in opposing directions have the advantage that the specimen remains in position at the center during loading, and almost no positioning corrections are necessary. A typical *in-situ* loading device (Kammrath & Weiss) is shown in Figure 9.3. In this case, the specimens are usually mounted directly between the cross-heads for tensile tests, whereas for compression tests they are fixed between two hard plates. The contact between the plates and the specimen can be lubricated, for example with

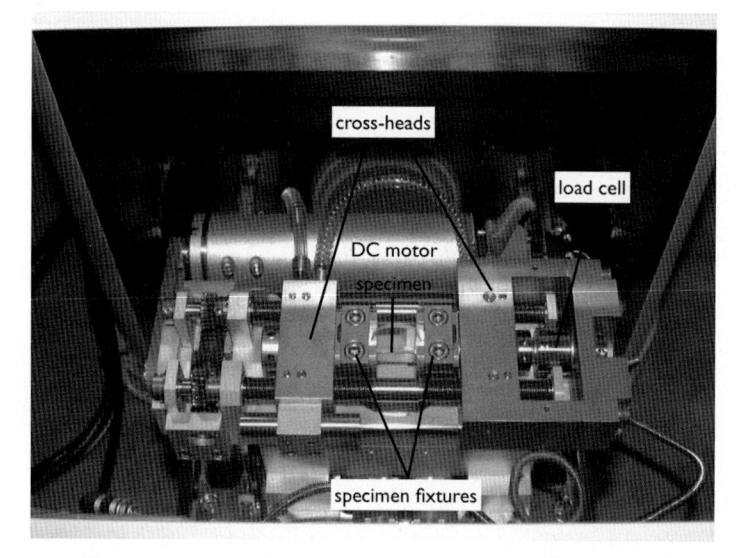

Figure 9.3 An *in-situ* loading device with two cross-heads with opposed moving directions. This can be used for tension, compression, and bending tests.

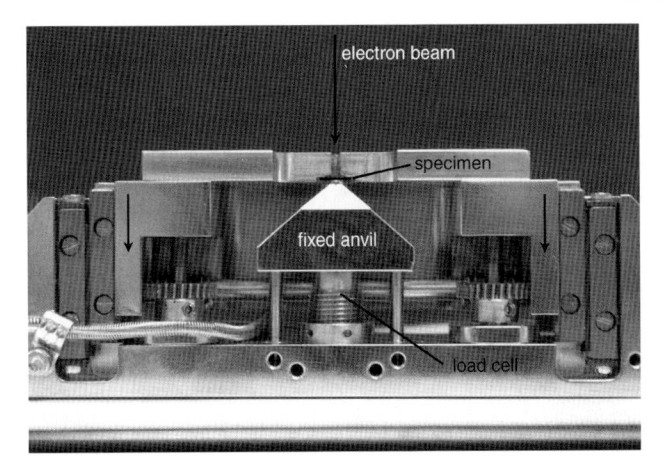

Figure 9.4 An *in-situ* bending device with vertically aligned loading axes to examine the top surface of the bending specimen during loading.

graphite, although it must be ensured that the lubricant used is suitable for a high-vacuum environment. Although, when investigating sample bending, a special fixture is needed to perform either three- or four-point bending tests, this normally permits only a side view of the specimen. In order to examine the top surface of the specimen during bending (where the highest stresses and strains are present) a special bending device is required, such that the loading axis is aligned vertically. An example of this type of bending device (Kammrath & Weiss) is shown in Figure 9.4.

The loading device, including the specimen, is usually mounted on the specimen stage of the scanning electron microscope. Due to the relatively high weight of the loading devices (they often weigh several kilograms), it must be ensured that the specimen stage of the microscope is not overloaded; for this reason, the specimen stage may occasionally need to be modified before the loading device can be mounted. At minimum, a lateral positioning of the specimen should be available, while a height adjustment of the specimen (using different working distances in SEM) may be important for special imaging conditions (e.g., for image correlation techniques). Other degrees of freedom, such as rotating or tilting the specimen, are generally not necessary.

The typical maximum load of these devices is in the range of 1 to 10 kN, which enables the testing of specimens with cross-sections of several square millimeters. The load is measured with conventional load cells, delivering a minimum load resolution on the order of 0.01 N, while the maximum displacement is often in the range of several millimeters. Usually, the cross-head displacement is measured using displacement gauges that are accurate in the submicrometer region. The average strain in the specimen can be calculated from the cross-head displacement, in good approximation, although for a more accurate strain determination it is better to use small clip-gauges, mounted directly on the specimen. Strain measurement is also possible by using image correlation methods (see Chapter 10), which allows the local measurement of surface strains with a relatively high accuracy.

9.3.2
Dynamic Loading in Tension, Compression, and Bending

In principle, the above-described "static" loading devices can also be used for dynamic loading, although the achievable frequencies are far below 1 Hz. In order to cycle fatigue specimen at higher frequencies, however, special loading devices are required. Unfortunately, servo-hydraulic devices cannot be used due to the high-vacuum environment in the microscope chamber, while electro-magnetic fatigue devices may interfere with the electron beam, so that the ability to image while the specimen is being cycled is limited. The most promising method is to use a piezo system (e.g., a piezo-stack) although, due to a reduced displacement of piezo systems, a mechanical multiplication is generally necessary to increase the displacement available for specimen loading. This is accompanied, however, by a reduction in the maximum load. A typical piezo-driven fatigue loading device is shown in Figure 9.5, for which the typical load range is ± 1000 N and the displacement range is $\pm 100\,\mu m$ at frequencies up to 10 Hz. Until now, *in-situ* fatigue experiments have been performed relatively infrequently, due mainly to the long testing period required inside the microscope, which is cost-intensive. Consequently, *in-situ* fatigue loading devices are rarely used and normally restricted to self-made devices.

9.3.3
Applications of *In-Situ* Testing

The *in-situ* testing of macroscopic samples allows the study of both local deformation and fracture/fatigue processes. Typically, the new materials are built up from complex microstructures that contain several phases, and in order to understand the mechanical performance of such materials, an extensive knowledge of the local

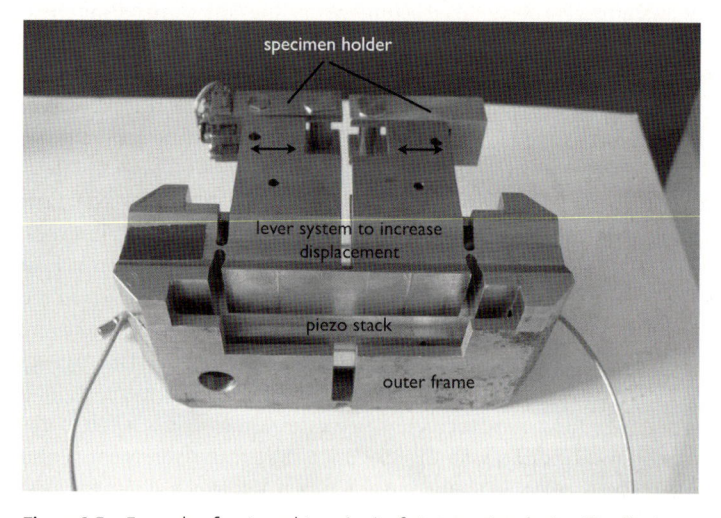

Figure 9.5 Example of a piezo-driven *in-situ* fatigue testing device. The displacement of the piezo stack is increased by a mechanical lever system.

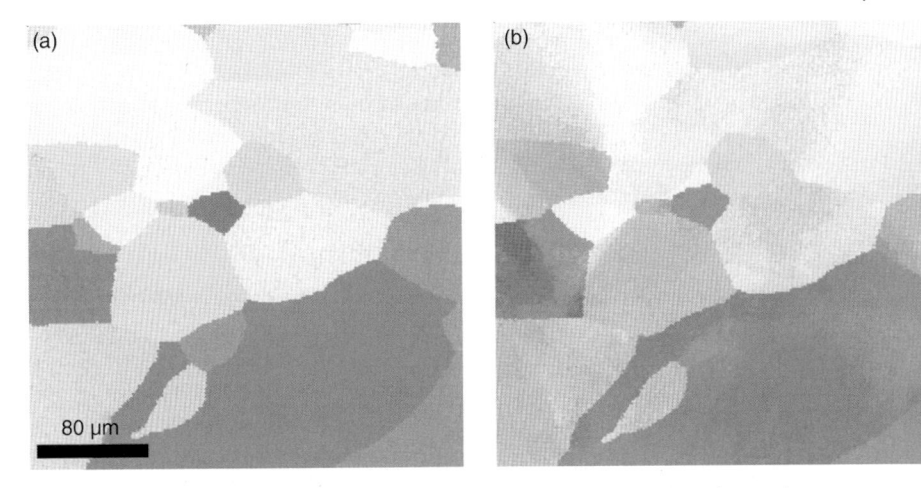

Figure 9.6 Inverse pole figure maps showing the change in crystal orientation during plastic deformation in a polycrystalline aluminum sample. (a) Undeformed sample; (b) 5% deformed sample; compare with Figure 9.2.

processes that occur during deformation or fracture is imperative. In particular, a combination of *in-situ* testing and modern analytical techniques, such as image correlation methods for local strain evaluation or EBSD mapping to determine crystal orientation changes during plastic deformation, have led to *in-situ* test methods becoming a powerful tool. As an example of this, the crystal orientation changes in the so-called inverse pole figure (IPF) maps of a polycrystalline aluminum sample are shown in Figure 9.6.

Previously, Tatschl and Kolednik [7, 9] investigated the deformation behavior of polycrystalline pure copper with EBSD mapping and local strain measurements, and also analyzed the slip behavior in individual grains with the data obtained via these two techniques. Musienko *et al.* [10] compared these results with a three-dimensional (3-D) finite-element simulation using crystal plasticity, while similar investigations were carried out by You *et al.* [11], in which the deformation behavior of a thin stainless steel wire with a bamboo grain structure was studied using *in-situ* testing and finite-element simulations. A more complex microstructure – a metal matrix compound (MMC) – was analyzed by Kolednik and coworkers [12], who examined local deformation behavior in the matrix and also the influence of hard particles. Likewise, *in-situ* fatigue tests were performed by both Vehoff *et al.* [13] and Bichler *et al.* [14] to monitor fatigue crack growth behavior, including crack closure and the influence of overloads.

9.4
In-Situ Loading of Micron-Sized Samples

The ongoing miniaturization in many fields of technology requires a detailed knowledge of the mechanical properties in small dimensions of the materials to be used. It has long been known that to reduce the internal length scale of a material, or the sample size, may impact on their mechanical performance. Thus, the present-day testing of micron-sized samples and investigations into size effects has become commonplace [15–17]. The determination of correct loading conditions, and investigation of the deformation processes that occur inside small samples, leads to the requirement for *in-situ* testing, using SEM.

9.4.1
Static Loading of Micron-Sized Samples in Tension, Compression, and Bending

The major problem encountered when testing micron-sized specimens relates to their handling. "Conventional" specimen preparation, transfer to the loading device and fixing the specimens in the device is usually not applicable; rather, the specimen is generally prepared directly in the bulk material and then transferred to the loading device with the bulk part. Although only the free part of the specimen is loaded and the remainder stays fixed in the bulk, this method avoids any complicated handling of small specimens. As an example, a micro-bending beam loaded with an *in-situ* nanoindenter is shown in Figure 9.7; it should be noted that one end of the beam is fixed in the bulk material, so that no specialized sample handling is necessary.

To date, because of the special specimen geometries resulting from the above-mentioned reasons, no standardized layout for loading devices has been developed; rather, individual research groups tend to use their own designs of loading device. Nonetheless, the most common approach is to modify an existing microindentation or nanoindentation device to be used inside the vacuum chamber of a scanning electron microscope (see Ref. [18]). An example of a nanoindenter mounted in a microscope chamber for *in-situ* testing is shown in Figure 9.8; this allows *in-situ*

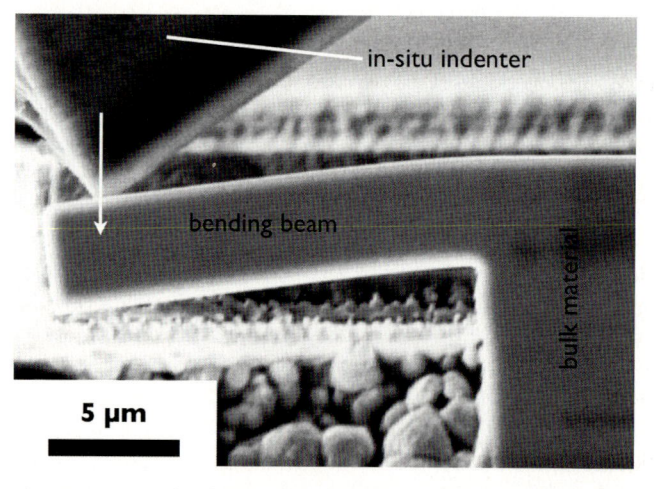

Figure 9.7 Example of a micro-bending beam "mounted" with one end in the bulk material to avoid complicated specimen transfers, which is *in-situ*-loaded with a nanoindenter.

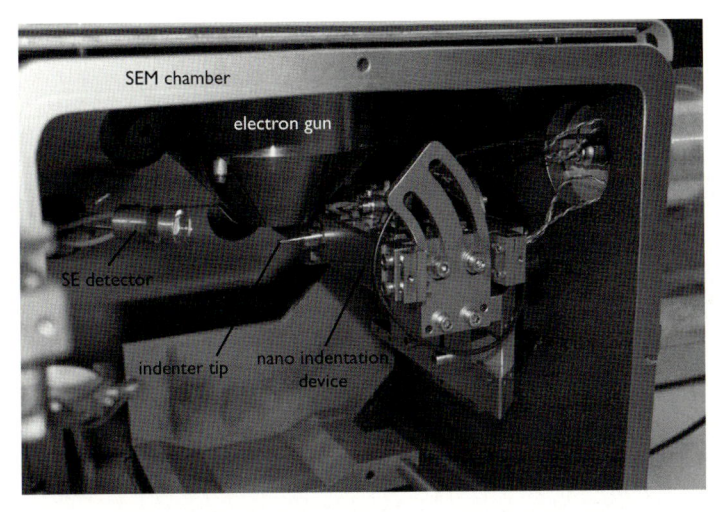

Figure 9.8 Example of an *in-situ* nanoindenter mounted in a scanning electron microscope chamber for testing micron-sized samples (specimen not shown here).

compression or tension tests to be carried out when the system is equipped with a flat-punch indenter or with special grippers [19]. In addition to the adapted indenters, special micro-electromechanical structure (MEMS) devices may also be used for the *in-situ* loading of micron-sized or nanosized specimens (see Ref. [20]). Due to an increasing need for *in-situ* tests on micron- or nanosized samples in order to characterize and understand their mechanical behaviors (size effects), new "universal micro-mechanical" loading devices are currently under development. A typical example, which can be used for tension, compression, and fatigue testing, is shown in Figure 9.9 [21].

Due to the adaption of microindenters and nanoindenters to be used inside the microscope chamber, the typical load range and resolution of these *in-situ* devices are similar to those of the indenters. Typically, the maximum load is in the range of 1 mN to several hundred mN, with a resolution of approximately 1 µN or less. For the load cell, different designs are used; the standard method is to measure the displacement of a spring and to calculate the corresponding load. This approach has the disadvantage that, to achieve a good accuracy, a relatively soft spring must be used to obtain sufficiently high displacements under small loads, and this results in the machine having a low degree of stiffness. Alternative methods for load measurement can be used to overcome this problem, however. For example, the load can be measured using an oscillating thin (e.g., tungsten) wire; measurement of the natural frequency of the wire, which changes as the loading is changed, can provide highly accurate load values because the time (frequency) measurements can be made with great precision. The typical displacement range of the loading devices is on the order of several tens of micrometers, with a typical resolution of <1 nm. The displacement measurement is usually made by measuring the tip (cross-head) displacement, as known from the microindenters and nanoindenters. A direct displacement measurement is restricted to optical methods (image correlation), due to the small sample sizes.

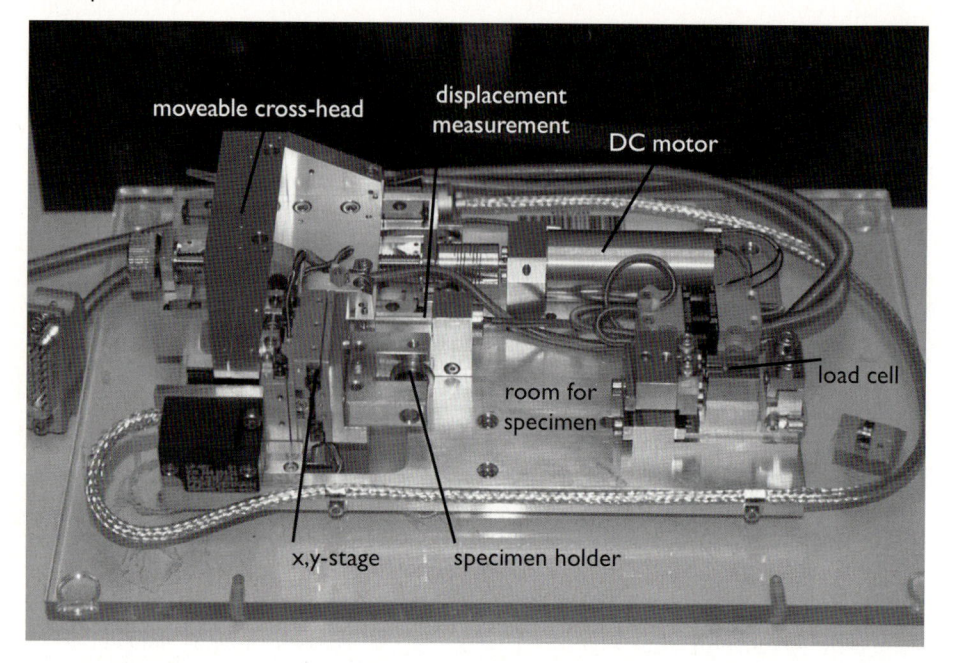

Figure 9.9 "Universal" loading device for tension, compression, or fatigue tests on small samples.

In this case, the loading can be achieved, for example, by using a standard stepping motor with a gearbox, by a piezo-stack, or by special plate capacitors. Typical displacement accuracies are several nanometers for the mechanical drive (stepping motor with gearbox), down to the subnanometer region for piezo drives.

Explicit fatigue tests conducted on small-scale samples are relatively rare. Usually, fatigue experiments are performed on thin films attached to substrates, in which case no special *in-situ* fatigue devices are required for "small" samples. Consequently, such *in-situ* fatigue devices for micron-sized fatigue samples are rarely available. In principle, the dynamic loading could be achieved using a piezo drive. The major problem encountered in small-scale fatigue testing is an induction of the dynamic load into the specimen.

9.4.2
Applications of *In-Situ* Testing of Small-Scale Samples

One advantage of the *in-situ* loading of small specimens is the reliability of correct loading; that is, any misalignment can be easily detected and readjustments made as needed – which may, in turn, significantly increase the yield on valid tests. Individual deformation events – such as strain bursts – may be associated with processes in the microstructure [18, 19, 21]. An example of an *in-situ* tensile test of a 25 μm-thick polycrystalline copper wire is shown in Figure 9.10. In this case, the falls in stress in the stress versus strain curve can be clearly associated with the appearance of new slip

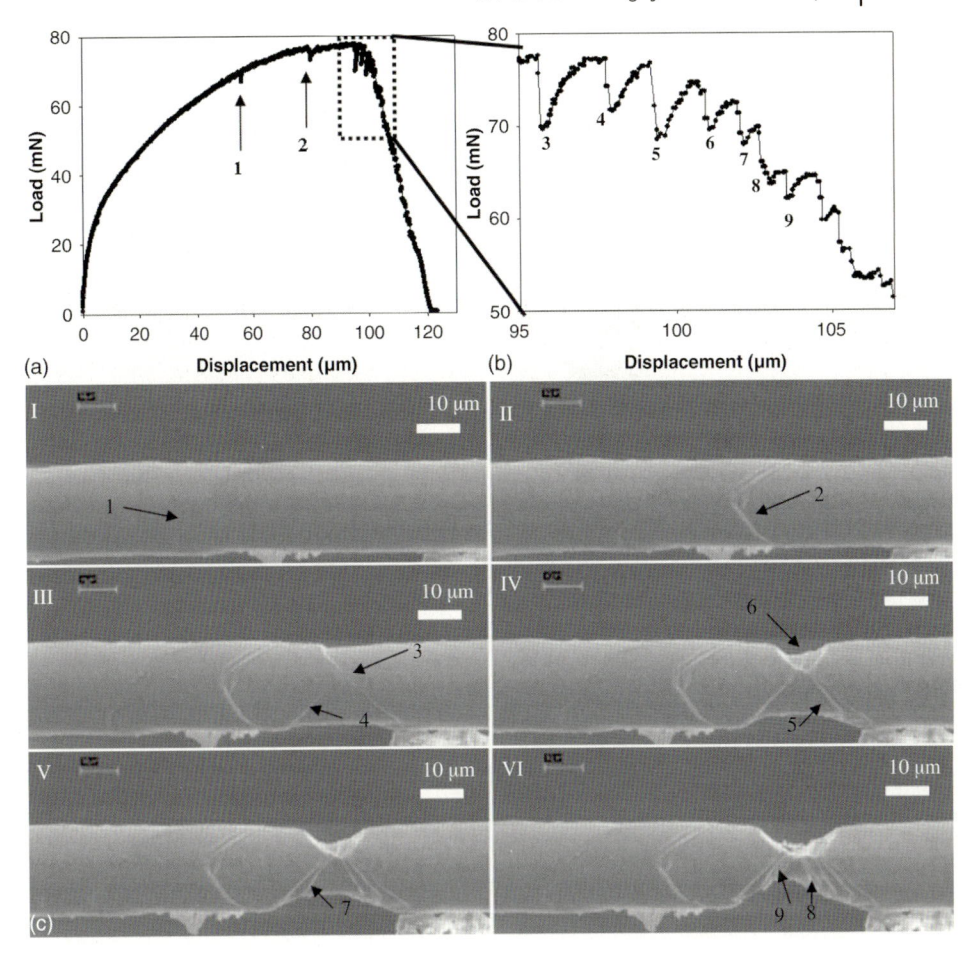

Figure 9.10 (a, b) Stress versus strain curves and (c) corresponding SEM images showing slip events (1 to 9) that are visible as stress drops in the stress versus strain curve. Figure reproduced with permission from Ref. [21]; © 2008, Carl Hanser Verlag.

bands on the sample surface. This may help to understand the principle deformation processes at the small scale, and may contribute to an understanding of size effects in mechanical properties.

Another example of *in-situ* testing relates to investigations of the fracture mechanical properties of micron-sized components. The *in-situ* loading of a notched single crystalline tungsten microbeam, and the corresponding load versus displacement curve, are shown in Figure 9.11. Here, besides a direct observation of the fracture processes, *in-situ* testing enables the measurement of any crack extensions that occur during the test. For samples in the micrometer region, the measurement of crack extension by using other methods is difficult; consequently, *in-situ* tests with optical crack length measurement have the potential to improve fracture mechanical tests on microsamples.

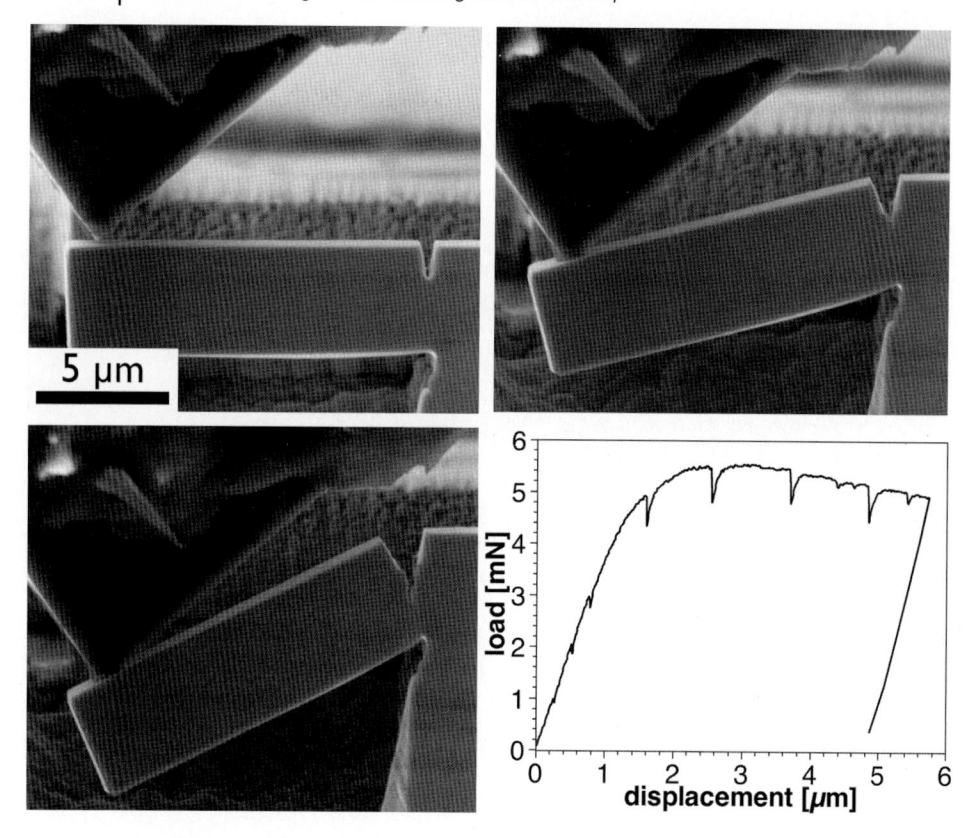

Figure 9.11 A series of micrographs of a notched, single crystalline tungsten microbeam loaded *in-situ* for the determination of fracture properties (load sequence from top left, to top right, to bottom left) and the according load versus displacement curve (bottom right). During the time of image acquisition, a relaxation of the specimen–indenter system occurs that is manifested in distinct load drops in the load versus displacement curve.

Nowadays, *in-situ* testing has become very popular in many fields of research, including the study of mechanical size effects in metals and alloys, the determination of the mechanical properties of biological structures [22], and validation of the correct operation of MEMSs.

9.4.3
In-Situ Microindentation and Nanoindentation

Both, microindentation and nanoindentation are common tools used to characterize the mechanical properties (i.e., hardness or Young's modulus) of small volumes. In the past, these experiments were generally conducted *ex-situ*, due mainly to the poor imaging conditions (typically, the contact area between the material and the indenter was not directly visible, such that only the surface in the vicinity of the indenter tip could be examined) and the need for high-resolution

SEM. During the past years, however, with the trend towards testing micron-sized samples, instrumented microindentation and nanoindentation devices have become available for operation inside the scanning electron microscope, and this has led to a renewed focus on *in-situ* microindentation and nanoindentation; for example, Michler *et al.* [23] have conducted *in-situ* scratch tests with a nanoindentation device. With the ongoing trend of performing micro-mechanical tests *in-situ* in the scanning electron microscope, the borders between typical indentation experiments and other investigations (such as the above-mentioned micro-compression tests) are merging into one another. Today, many of the newly developed test methods have notable advantages over indentation methods, in particular a simpler stress and strain state in the sample. In contrast, the main advantage of indentation tests is the simple and cheap specimen preparation, especially when compared to sample production using the focused ion beam technique.

9.5
Summary and Outlook

Today, *in-situ* testing in the scanning electron microscope spans a wide range of test methods, ranging from macroscopic mechanical testing to nanoindentation. With the increasing availability of scanning electron microscopes (today, they are almost "standard" equipment in materials science laboratories), *in-situ* testing is becoming increasingly attractive. Yet, *in-situ* testing has certain limitations compared to *ex-situ* mechanical testing. For example, due to the limited space inside the vacuum chamber, the load cells and strain gauges must be optimized with respect to their size, and this is usually accompanied by a loss in accuracy that must be taken into account. Especially, several issues must be considered in micro-mechanical testing, including a limited resolution in load and displacement measurements, the alignment of the specimen and loading device, "imperfect" specimen geometry, and machine stiffness (i.e., the stiffness of the entire set-up). Each of these issues may lead to major errors if not addressed.

Today, an increasing number of scanning electron microscope manufacturers offer *in-situ* loading devices while, with the improving resolution of modern next-generation microscopes, the testing of much smaller samples (down to the nanometer region) will also become possible. In addition, techniques such as high-resolution EBSD [8] to measure local elastic strains and stresses will open new fields for *in-situ* testing in the future.

References

1 Staal, H.U. and Elen, J.D. (1979) Crack closure and influence of cycle ratio R on fatigue crack growth in type 304 stainless steel at room temperature. *Eng. Fract. Mech.*, **11** (2), 275.

2 Tungatt, P.D. and Humphreys, F.J. (1981) An in-situ optical investigation of the deformation behaviour of sodium nitrate – an analogue for calcite. *Tectonophysics*, **78** (1-4), 661.

3 Hay, R.S. and Evans, B. (1992) The coherency strain driving force for CIGM in non-cubic crystals: Comparison with in situ observations in calcite. *Acta Metall. Mater.*, **40** (10), 2581.

4 Freyland, J.M., Eckert, R., and Heinzelmann, H. (2000) High resolution and high sensitivity near-field optical microscope. *Microelectron. Eng.*, **53** (1-4), 653.

5 Quinta da Fonseca, J., Mummery, P.M., and Withers, P.J. (2005) Full-field strain mapping by optical correlation of micrographs acquired during deformation. *J. Microsc.*, **218** (1), 9.

6 Nshanian, T., Dove, R., and Rajan, K. (1996) In-Situ strain analysis with high spatial resolution: A new failure inspection tool for integrated circuit applications. *Eng. Fail. Anal.*, **3** (2), 109.

7 Tatschl, A. and Kolednik, O. (2003) On the experimental characterization of crystal plasticity in polycrystals. *Mater. Sci. Eng. A*, **342** (1-2), 152.

8 Wilkinson, A.J., Meaden, G., and Dingley, D.J. (2006) High resolution mapping of strains and rotations using electron backscatter diffraction. *Mater. Sci. Technol.*, **22** (11), 1271.

9 Tatschl, A. and Kolednik, O. (2003) A new tool for the experimental characterization of micro-plasticity. *Mater. Sci. Eng. A*, **339** (1-2), 265.

10 Musienko, A., Tatschl, A., Schmidegg, K., Kolednik, O., Pippan, R., and Cailletaud, G. (2007) Three-dimensional finite element simulation of a polycrystalline copper specimen. *Acta Mater.*, **55** (12), 4121.

11 You, X., Connolley, T., McHugh, P.E., Cuddy, H., and Motz, C. (2006) A combined experimental and computational study of deformation in grains of biomedical grade 316LVM stainless steel. *Acta Mater.*, **54** (18), 4825.

12 Kolednik, O. and Unterweger, K. (2008) The ductility of metal matrix composites – Relation to local deformation behavior and damage evolution. *Eng. Fract. Mech.*, **75**, 3663.

13 Vehoff, H. and Neumann, P. (1979) In situ SEM experiments concerning the mechanism of ductile crack growth. *Acta Metall.*, **27** (5), 915.

14 Bichler, C. and Pippan, R. (1999) Direct observation of the residual plastic deformation caused by a single tensile overload, in *Advances in Fatigue Crack Closure Measurement and Analysis (ASTM STP 1343)* Vol. 2 (eds R.C. McClung, J.C. Newman Jr), ASTM International, pp. 191–206.

15 Uchic, M.D., Dimiduk, D.M., Florando, J.N., and Nix, W.D. (2004) Sample dimensions influence strength and crystal plasticity. *Science*, **305**, 986.

16 Greer, J.R., Oliver, W.C., and Nix, W.D. (2005) Size dependence of mechanical properties of gold at the micron scale in the absence of strain gradients. *Acta Mater.*, **53**, 1821.

17 Motz, C., Schöberl, T., and Pippan, R. (2005) Mechanical properties of micro-sized copper bending beams machined by the focused ion beam technique. *Acta Mater.*, **53**, 4269.

18 Kiener, D., Motz, C., Schöberl, T., Jenko, M., and Dehm, G. (2006) Determination of mechanical properties of copper at the micron scale. *Adv. Eng. Mater.*, **8**, 1119.

19 Kiener, D., Grosinger, W., Dehm, G., and Pippan, R. (2008) A further step towards an understanding of size-dependent crystal plasticity: In situ tension experiments of miniaturized single-crystal copper samples. *Acta Mater.*, **56**, 580.

20 Haque, M.A. and Saif, M.T.A. (2002) Mechanical behavior of 30–50 nm thick aluminum films under uniaxial tension. *Scr. Mater.*, **47**, 863.

21 Yang, B., Motz, C., Grosinger, W., Kammrath, W., and Dehm, G. (2008) Tensile behaviour of micro-sized copper wires studied by a novel fibre tensile module. *Int. J. Mater. Res.*, **99**, 716.

22 Huber, G., Mantz, H., Spolenak, R., Mecke, K., Jacobs, K., Gorb, SN., and

Arzt, E. (2005) *Proc. Natl Acad. Sci. USA*, **102** (45), 16293.

23 Michler, J., Rabe, R., Bucaille, J.-L., Moser, B., Schwaller, P., and Breguet, J.-M. (2005) Investigation of wear mechanisms through in situ observation during microscratching inside the scanning electron microscope. *Wear*, **259** (1-6), 18.

10
In-Situ TEM Straining Experiments: Recent Progress in Stages and Small-Scale Mechanics

Gerhard Dehm, Marc Legros, and Daniel Kiener

10.1
Introduction

Transmission electron microsocopy (TEM) studies to unravel dislocation mechanisms in materials have been employed on a vast scale since the 1960s, with *in-situ* deformation experiments having emerged during the same time period [1], and expanding to high-voltage TEM, mainly during the 1970s [2–7]. Initially, research was focused on dislocation glide and creep, dislocation reactions, and dislocation substructure formation in metals, alloys, and intermetallics (for a recent textbook/review, see Refs [8, 9]). During the 1980s dislocation mechanisms in semiconductor materials [10, 11] began to play a central role in advancing the performance of microelectronic devices for both computing and opto-electronic applications, as dislocations are detrimental to the performance of such systems. Some of the key achievements of *in-situ* TEM studies included a basic understanding of the accommodation of lattice mismatch in strained semiconductor films [12, 13], and the kinetics of misfit dislocations [14, 15].

During the past decade, *in-situ* deformation studies in TEM have begun to reattract immense research interest, due to unexpected mechanical size-effects in metals with (sub)micron dimensions or volumes [16–21]. This includeds miniaturized metallic objects, metal films and lines that are employed in flexible electronics, microelectronic devices, micro-electromechanical systems (MEMS), nano-electromechanical systems (NEMS), and medical devices, where constraints of geometric dimensions, surfaces and interfaces may influence the deformation mechanisms when these materials are exposed to combined thermo-mechanical strain [22] or pure mechanical strain [23].

In addition to *qualitatively* imaging the underlying dislocation mechanisms with television rate–time resolution, due to an availability of fast charge-coupled device (CCD) cameras with high dynamic contrast, today's *in-situ* deformation experiments also provide *quantitative* mechanical stress–strain data by employing novel stages and/or MEMS/NEMS devices [24–26]. Indeed, the latter have opened up a whole world of the "lab on the chip", mounted inside the transmission electron microscope.

In-situ Electron Microscopy: Applications in Physics, Chemistry and Materials Science, First Edition.
Edited by Gerhard Dehm, James M. Howe, and Josef Zweck.
© 2012 Wiley-VCH Verlag GmbH & Co. KGaA. Published 2012 by Wiley-VCH Verlag GmbH & Co. KGaA.

In this chapter, *in-situ* straining methods are first described, starting with thermo-mechanical straining in a conventional heating holder, pure mechanical loading in classical straining holders, moving on to novel developments regarding quantitative deformation stages, and finally presenting MEMS/NEMS devices which also allow the recording of quantitative load–displacement data. As these methods place different requirements to the loading of the specimen, the preparation of the required TEM samples is outlined in several cases, before recent results are reported in relation to the deformation mechanisms in miniaturized metallic single crystals, in thin metallic films, and in nanocrystalline metals.

10.2
Available Straining Techniques

10.2.1
Thermal Straining

The simplest approach to *in-situ* straining experiments exploits the differences in thermal expansion coefficients between different materials. This method has been widely used to study plasticity in metallic films [16, 27–30] and semiconductor films [14] on substrates, and requires simply a conventional heating stage and a suitable TEM sample design. The applied thermal strain, ε_{th}, can be calculated from the temperature increment, ΔT, and the mismatch in thermal expansion coefficients, α, between film and substrate:

$$\varepsilon_{th} = (\alpha_{film} - \alpha_{substrate})\Delta T. \tag{10.1}$$

The tests can be performed on plane-view samples, where the substrate is locally thinned to electron transparency by dimpling and etching (Figure 10.1a). The film is stretched biaxially in tension, much like the membrane on a drum. If the support from the substrate is insufficient, then buckling of the film may occur during

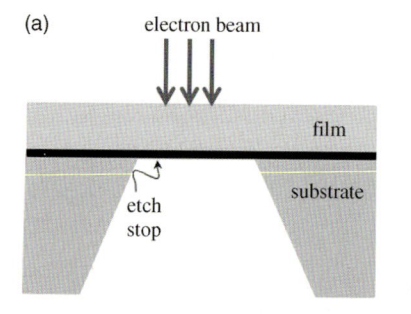

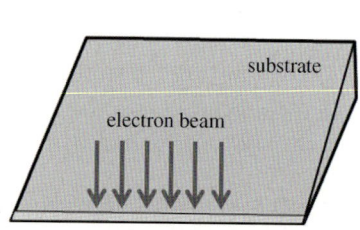

Figure 10.1 (a) Schematic of a plane-view TEM sample etched from the substrate side (e.g., Si) until an etch stop (e.g., SiN$_x$) is reached. For small film thickness, electron transparency is achieved in the center region of the sample (e.g., Al film thickness below ~0.6 μm on a ~50 nm- thick SiN$_x$ membrane, accelerating voltage 200 kV). (b) Wedge-shaped cross-sectional sample of a film on a substrate used for *in-situ* thermal straining experiments. The sample is made by tripod polishing. Figure redrawn from Refs [16, 18].

compression. However, this will be noticed immediately for crystalline samples, by the rapid and continuous motion of numerous bend contours through the region of interest. Thermal straining experiments can be also performed on cross-sectional TEM samples of thin films on substrates. In that case, either classical fabrication techniques involving gluing, grinding, and ion milling can be used [31], or tripod polishing of the cross-sectional sample into a wedge shape (Figure 10.1b) [18], or focused ion beam (FIB) structuring of the sample.

The main disadvantages of the thermal straining approach, besides the inherent coupling of thermal and mechanical stimuli, are the typically low strains imposed by thermal straining. As the thermal expansion coefficients are on the order of 10^{-6} to $10^{-5}\,K^{-1}$, and the maximum temperature difference is several hundreds of °C, the attainable thermal strain is less than about 1% for metals on Si. Moreover, any unwanted film–substrate interdiffusion or chemical reactions, unknown magnitude of stress in the sample, and activation of different deformation processes at different temperatures, will complicate an interpretation of the results. However, as this exposure may occur in many thin-film devices during application (e.g., sensors and switches in automotive applications), it is a relevant experimental approach to unravel the deformation and damage mechanisms of thin films and small structures.

Finally, it is worth noting that, instead of considering a real yield stress of the metallic film on its substrate, its "strength" is commonly taken as the stress value at the temperature at the end of a thermal cycle. This approximation [32] results from the facts that two different values of the yield stresses can be reached in tension and compression, and that these values could depend on the history of the film [33].

Recently, bimetallic actuators based on the same principles have been introduced to strain ligaments in tension inside the transmission electron microscope [34, 35]. The amount of strain developed by these actuators can reach a few percentage points, and the 3 mm TEM grids can be mounted inside a regular heating/cooling holder. While this provides an uniaxial stress state, the strain and temperature parameters will still act simultaneously.

10.2.2
Mechanical Straining

In order to decouple temperature and strain, isothermal straining experiments must be designed. The simplest approach is based on conventional TEM straining stages, for which the main design demand is the limited available space between the pole-pieces of the objective lens of the microscope (see Chapter 2); thus, most straining stages can operate only in single-tilt mode. Today only a few double-tilt stages exist, these are mainly self-made [2, 36] and may even operate at low and/or elevated temperatures. Recent developments regarding aberration-corrected microscopes have allowed for high-resolution imaging while maintaining a reasonable pole-piece width, and will hopefully stimulate new designs.

The transmission electron microscope sample is either machined as a single part that fits the holder, or is glued to a support and then screwed into the straining stage. Usually, only one crosshead of the straining rig is moveable (Figure 10.2),

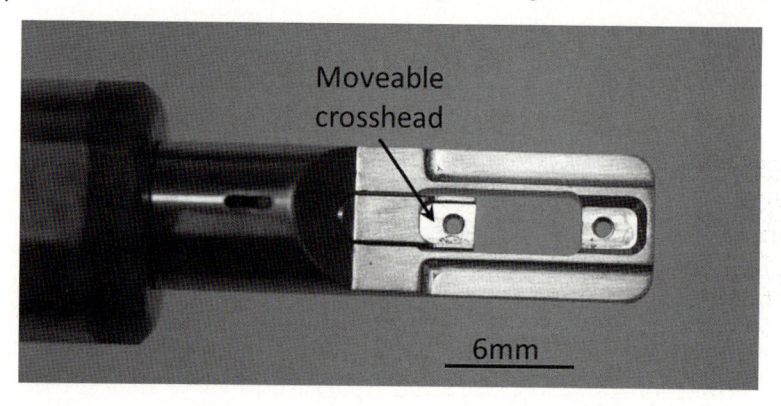

Figure 10.2 Optical image of a commercial, single-tilt TEM straining stage. The sample or sample support is fixed into the stage with screws.

which means that the area of interest will move during the experiment and must be followed by adjusting the stage. Straining is performed by a motor-driven mechanical system that consists of precision gears and spindles, and which provides the required precise crosshead displacement down to typically about 1 µm per displacement increment. As the mounting points of the sample are separated by several millimeters (typically 3–9 mm), only a small strain (ca. 10^{-4}) is exerted on the specimen when using the smallest displacement increment. The specimen is loaded stepwise with effective velocities ranging from several tens of nanometers per second up to several micrometers per second. The total elongation is measured by the displacement of the moveable crosshead, while the displacement of the region of interest of the sample can be only accessed by deducing the elongation from subsequent images, for example by applying image-correlation techniques. The load acting on the sample cannot be measured accurately due to frictional forces of the mechanically driven crosshead, as well as friction between the mounted specimen and the stage. Furthermore, in most cases the cross-sectional area of the sample remains unknown. Whilst these aspects prevent a correct measure of the stress, classical straining experiments nevertheless provide valuable information on the deformation mechanisms, and even allow insights into local stresses by using the imaged dislocation curvature as a local stress sensor (e.g.; [37]). An example is provided in Figure 10.3, where the radius of curvature, r, of the dislocations is used to calculate the resolved shear stress $\tau \approx Gb/r$ acting on the dislocation with Burgers vector b, where G denotes the shear modulus of the material. To apply this method, it is necessary to know the dislocation type as well as the angle of inclination of the glide plane with respect to the two-dimensional (2-D) image plane to correct for the image projection [37].

10.2.3
Instrumented Stages and MEMS/NEMS Devices

During recent years, nano-indentation stages and atomic force microscopy (AFM) stages have entered the *in-situ* TEM scene (e.g.; [38–41]). In TEM, the samples are

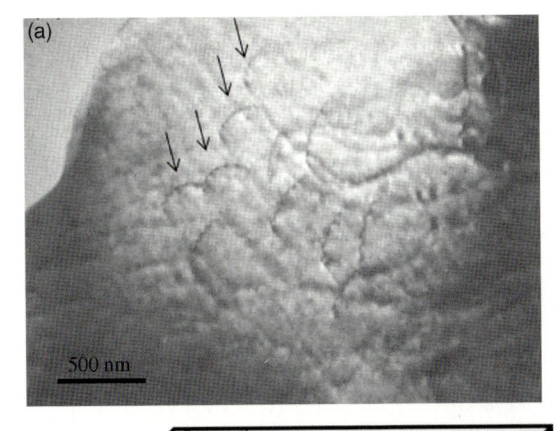

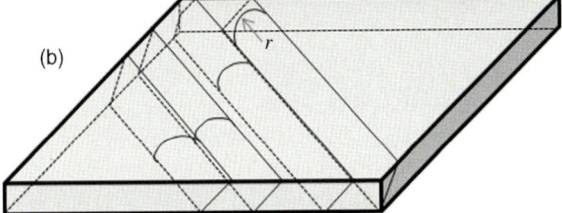

Figure 10.3 (a) TEM image of dislocations on an inclined (111) glide plane in an Al film on a Si substrate; (b) Schematic image of the dislocations highlighted by arrows in panel (a). In order to determine the dislocation radius r on the image, the angle of inclination must be known to correct for projection effects. Panel (a) reproduced with permission from Ref. [57]; © 2003, Elsevier.

mounted on a piezo-driven support within the stage, which allows their alignment with respect to the nano-indenter or the atomic force microscope tip (Figure 10.4a; see also Chapter 11). Here, either the tip or the sample is mounted on a load sensor (e. g., capacitive sensor, spring system) that permits quantitative load measurements to be made. Different load sensors with resolutions on the order of 10^{-4} are available for different regimes of interest. The indentation stages offer a maximum load of several mN, with a typical noise in the μN region, or with even better load resolutions in the nN region for the AFM sensors, at the expense of rather small applicable loads. The displacement is either again measured capacitively, or is deduced from the applied piezo-motion by assuming a linear piezo behavior.

The MEMS/NEMS devices are based on Si technology, using film deposition, lithography and etching techniques to design actuators and sensors on a chip [42]. One of the first successful MEMS/NEMS devices for *in-situ* TEM deformation studies of metallic lines was developed by Saif and colleagues [43, 44]. In this MEMS/NEMS device (see Figure 10.4b-d), which requires external actuation, loading of the specimen is achieved by pulling on the complete device inside a conventional TEM straining stage (see Figure 10.4b). The sample (a metal line) is co-fabricated with the MEMS/NEMS structure and suspended as a line on the device. Both, load and displacement sensing are performed by measuring the distance between certain Si

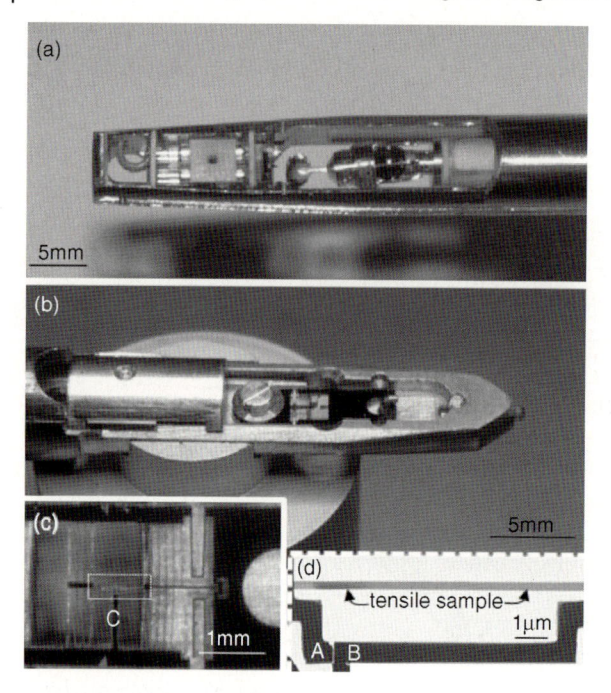

Figure 10.4 (a) Commercial nano-indentation stage with load and displacement sensor on the left-hand side and the sample mounted on the right-hand side, on a piezo-driven support; (b) MEMS/NEMS device from M.T.A. Saif and coworkers (e.g., Refs [26, 43, 44]) fixed into a conventional straining stage; (c) Enlarged photograph of the U-beams and springs which ensure accurate tensile loading; (d) Low-magnification TEM image of the boxed region of panel (c), with the actual tensile sample and the nearby load and displacement sensor beams A, B, in (d) and beam C in (c).

beams attached to the frame of the MEMS/NEMS device. During loading of the sample, the gap distance between the sensing beams is changed in proportion to their stiffness; the gap distances are recorded on TEM images, while the stiffness of the sensing beams must be measured *ex situ* prior to or after the experiment by nano-indentation, where the stiffness can be deduced from the load–displacement data for the known geometry of the beams.

While this device works well during *in-situ* scanning electron microsocopy (SEM) and TEM experiments, it possesses two major drawbacks [24]:

- First, the tensile sample is not located in the same field of view as the strain- and load-sensing beams. This requires moving between two different positions – the region of interest on the sample and the sensor elements – which in turn poses severe limitations to the achievable time resolution.
- Second, there is a limitation to thin-film materials being deposited on Si substrates followed by subsequent Si lithography technology, due to the co-fabricated process of sample and sensors as a MEMS/NEMS structure. Hence,

this approach was employed to study nanocrystalline materials [43, 45, 46]. In order to fix other samples on this device, a different design is needed.

Recent MEMS/NEMS devices have overcome both limitations. For example, the specimen can be transferred onto small support plates, where it is fixed by local material deposition via a precursor injected by a gas-injection system which reacts under the ion or electron beam in a FIB/SEM. This approach was taken by Pant et al. [47]. A very recent design by Saif et al. [48, 49] offeres the possibility to place dog-bone-shaped samples into an according counterpart in the loading apparatus, thereby completely removing the necessity of material deposition for sample fixation. This approach is especially desirable, as any unwanted contamination of the sample, as well as possible slip in the deposited grips, can thereby be excluded. In this new generation of MEMS/NEMS devices, the loading and sensing is performed by thermal actuators and capacitive sensors, respectively. Such MEMS/NEMS devices, as have also been developed by Espinosa et al. [50] and Zhang et al. [51], require special TEM holders with an electronic interface between the outside control electronics and the MEMS/NEMS device [26]. The load and displacement data are recorded simultaneously while imaging the actual sample, thus permitting a true *in-situ* experiment to be conducted. The main drawbacks of these new MEMS/NEMS devices reside in the tedious procedures needed to fabricate, align, and mount a sample via micromanipulators, in the limited strains available through the thermal expansion of Si beams, and the required adaptation of the stiffness/sensitivity of the MEMS machine to a given range of mechanical properties.

10.3
Dislocation Mechanisms in Thermally Strained Metallic Films

10.3.1
Basic Concepts

Thermal stresses often limit the lifetime of semiconductor devices, due either to thermo-mechanical fatigue of the metallization layers or by ratcheting effects of the metallization, which causes the cracking of adjacent passivation layers. In order to overcome such limitations and to establish quantitative material laws, a basic understanding of these underlying deformation phenomena is required. Thermal stresses in face-centered cubic (fcc) metal films (e.g., Al, Cu, Ni, Ag, and Au) were investigated for different film thicknesses that ranged typically from about 50 nm to about 2000 nm, mainly in the temperature range between ambient and 600 °C [52–59]. All studies revealed a tendency towards higher room temperature stresses ("film strength"; see Section 10.2.1) for smaller film thickness values. Following the observations of semiconductor films on semiconductor substrates, where threading dislocations channel through the film and deposit interfacial dislocations [12, 13] (see Figure 10.5a and c), Nix and Freund [52, 60] developed a quantitative model by balancing the energy of a threading dislocation gliding through the film and the work

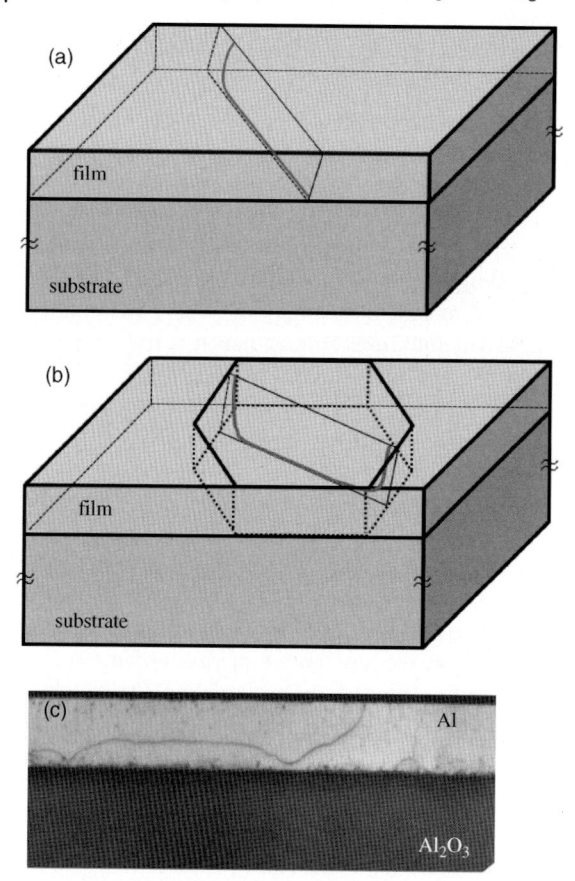

Figure 10.5 (a) Threading dislocation depositing an interfacial dislocation segment at the metal film/substrate interface, as suggested in the model of Nix and Freund [52, 60]; (b) For polycrystalline films on substrates Thompson assumed that interfacial dislocation segments are also deposited at the grain boundaries in addition to the film/substrate interface [61]; (c) TEM image of a threading dislocation in a thermally strained epitaxial Al film on α-Al$_2$O$_3$ substrate which drags behind a dislocation segment at some distance from the Al/α–Al$_2$O$_3$ interface. Figure reproduced with permission from Ref. [63]; © 2002, Journal of Materials Science and Technology/Allerton Press, New York.

done by the stress, predicting an increase in flow stress, σ, with decreasing film thickness, h:

$$\sigma_{Nix} = \frac{\sin\phi}{hm} \frac{G_{\text{eff}} b}{4\pi(1-\nu)} \ln\left(\frac{\beta_s}{b} h\right) \tag{10.2}$$

where ϕ is the angle between the film normal and the glide plane in a film of thickness h. Furthermore, m is the Schmid factor, b the magnitude of the Burgers vector, ν the Poisson's ratio of the film, G_{eff} the effective shear modulus of the film/substrate combination experienced by the interfacial dislocation segment, and $\beta_s = 1$–2 a numerical constant defining the cut-off radius of the stress field of the interfacial dislocation.

Later, Thompson [61] extended this model to polycrystalline films by considering the typical bamboo grain structure in annealed metallic films with grain diameters that were typically twice the film thickness, and assuming that the grain boundaries acted as additional obstacles for the threading dislocations. In Thompson's assumption, the dislocations are forced to deposit an interfacial dislocation segment at the grain boundaries as well as at the interface (Figure 10.5b), leading to an additional stress contribution, σ_{GB}, to the total stress calculated by Equation 10.2 with

$$\sigma_{GB} = \frac{2}{dm} \frac{G_{eff} b}{4\pi(1-\nu)} \ln\left(\frac{d}{b}\right), \tag{10.3}$$

where d corresponds to the distance between grain boundaries (i.e., grain size). A quantitative comparison of the two models with room temperature stress values of polycrystalline metal films on substrates revealed major discrepancies (e.g., [57, 62]), setting the ground for *in-situ* TEM experiments of thermally strained single crystalline and polycrystalline metallic films on substrates.

10.3.2
Dislocation Motion in Single Crystalline Films and Near Interfaces

In-situ TEM studies of thermally strained single crystalline metallic films on substrates confirmed the basic assumptions of Freund and Nix [52, 60]. Several cross-sectional thermal straining experiments of (111)-oriented Al films on (0001) α-Al$_2$O$_3$ substrates revealed the motion of threading dislocations on inclined {111} planes, laying down an interfacial dislocation segment at the epitaxial Al/α–Al$_2$O$_3$ interface (Figure 10.5c) [31, 63]. Performing similar experiments on polycrystalline Al and Cu films on Si substrates with an amorphous oxide or nitride diffusion barrier between the metal film and the substrate gave a completely different scenario. Dislocations were observed to advance through the grains without forming stable interfacial dislocation segments [28, 29, 64]. This behavior was unexpected, as dislocations cannot enter the amorphous diffusion barrier between film and substrate, and thus were thought to form stable interfacial dislocation segments. However, the *in-situ* TEM studies revealed that either no visible strain contrast from interfacial segments formed, or that the strain field of the interfacial dislocations vanished within seconds. This can only occur when the interfacial dislocations are able to "escape" at the film/interlayer interface, similar to the processes that occur at a free surface [28, 65]. However, at a free surface, a slip step corresponding to the edge component of the Burgers vector would form. It appears that the interface structure and bond strength of the metal film to the amorphous interlayer permits (diffusion driven) atomic rearrangements, allowing the dislocation core to spread, i.e. promoting the dislocation escape at the interface by local atomic rearrangements, without the necessity to form an abrupt slip step on the surface. This scenario is supported by the observations that both, different film/interlayer systems as well as the temperature, influence the escape time of the dislocations, i.e. the time interval until the dislocation contrast of the interfacial segment has faded during *in-situ* TEM observations. As this mechanism was not considered in the model of Thompson [61] for polycrystalline metal films (Equation 10.3, Figure 10.5b), it is not generally applicable to thermally strained metallic films.

10.3.3
Dislocation Nucleation and Multiplication in Thin Films

TEM studies have revealed that, for polycrystalline films, dislocation densities decrease with decreasing film thickness (and grain size). Thermal stress measurements of polycrystalline thin films on substrates revealed an increase in elastic strain and a decrease in plastic strain with decreasing film thickness (see e.g., [66] for data on Cu). As fewer and fewer dislocations are observed with decreasing film thickness and grain size, dislocation interactions and Taylor hardening cannot be the course of the high-flow stresses. In Al films with thicknesses below 200 nm, biaxial tensile stresses reach values of about 780 MPa, without any significant dislocation activity [58], which indicates that dislocation multiplication is hampered in small dimensions. This observation also holds true for polycrystalline films deposited on amorphous interlayer coated substrates. In contrast, epitaxial films on crystalline substrates [57] appear to possess sufficient dislocation sources, as the film stresses follow the prediction of Nix and Freund [52, 60] (see Figure 10.6). In this case, the

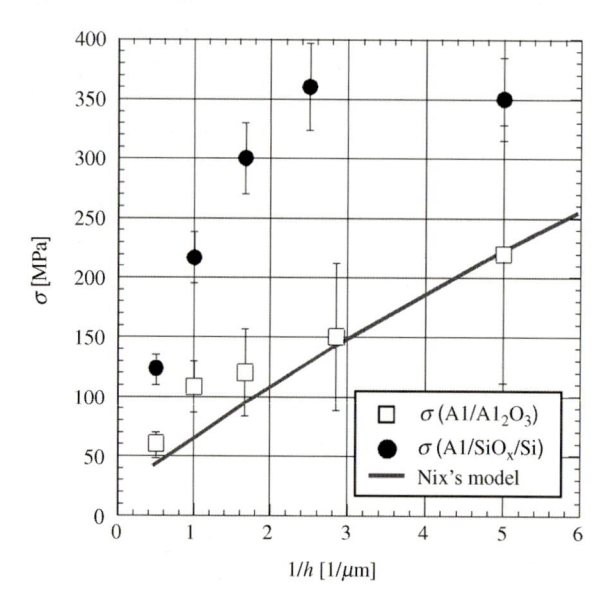

Figure 10.6 Stress values of 200 nm to 2000 nm-thick polycrystalline Al films (Al/SiO$_x$/Si) and (nearly) single crystalline Al films (Al/Al$_2$O$_3$) after cooling the samples from 400 °C to room temperature. The thick line indicates the calculated stress using the Nix–Freund equation (Equation 10.2; shear modulus of Al 26 GPa and 185 GPa for α-Al$_2$O$_3$). Data taken from Ref. [63].

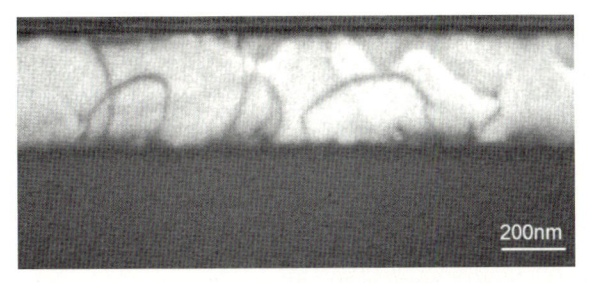

Figure 10.7 *In-situ* TEM image from dislocation half-loops emitted from the (111) Al/(0001) α-Al$_2$O$_3$ interface during thermal straining. Figure reproduced with permission from Ref. [63]; © 2002, Journal of Materials Science and Technology/Allerton Press, New York.

dislocations multiply from dislocation sources at the metal/substrate interface where, in addition to the thermal dislocation network, a misfit dislocation network also exists to relax the lattice mismatch strain between the metal lattice and the substrate lattice [31, 57, 64]. An example of dislocations emitted as half-loops from the interfacial dislocation network is shown in Figure 10.7. These types of interfacial source are not observed for polycrystalline films grown on amorphous templates, as no stable dislocation network exists at such interfaces. Dislocation nucleation is in that case observed to occur from the grain boundaries (e.g.; [67, 68]). As the grain size decreases with decreasing film thickness, the dislocations travel shorter distances for smaller grain sizes, which in turn poses a challenge for experimental observations. In addition, while a single dislocation crossing a grain compensates a rather large strain b/d within the grain, this is only a minor contribution to the global plastic strain [66], as nucleation events in many grains are needed for global yield. This explains the increase in elastic strains with decreasing dimensions (film thickness and grain size) for a given thermal strain.

The simplification used today in modeling the room temperature flow stresses in polycrystalline films is based on considering the optimum dimensions of a double-ended Frank–Read dislocation source situated inside a grain. The flow stress, σ, required to activate such a source scales with the smallest dimension, which can be the film thickness h or the grain size d:

$$\sigma \approx \frac{\beta G b}{h} \text{ or } \frac{\beta G b}{d}, \tag{10.4}$$

imposing the critical constraint (the smaller quantity) on the dislocation source [69, 70]. The factor $\beta = 3\ldots4$ considers that the dislocation segments initially bowing out between the anchor points need space to subsequently turn over at the pinning points in the opposite direction to form a full loop. If dislocations pile-up at impenetrable boundaries fully "encapsulating" the grain, then the evolving back-stresses will reduce the optimum source size to about $h/4$ or $d/4$, while a grain with a free surface will favor a source size with $h/3$ or $d/3$, according to the considerations of von Blanckenhagen *et al.* [70]. For free surfaces, single-ended sources (a Frank–Read source where only one segment is anchored at a pinning point) have also been observed [18, 71, 89], an example of which is shown in Figure 10.8. These

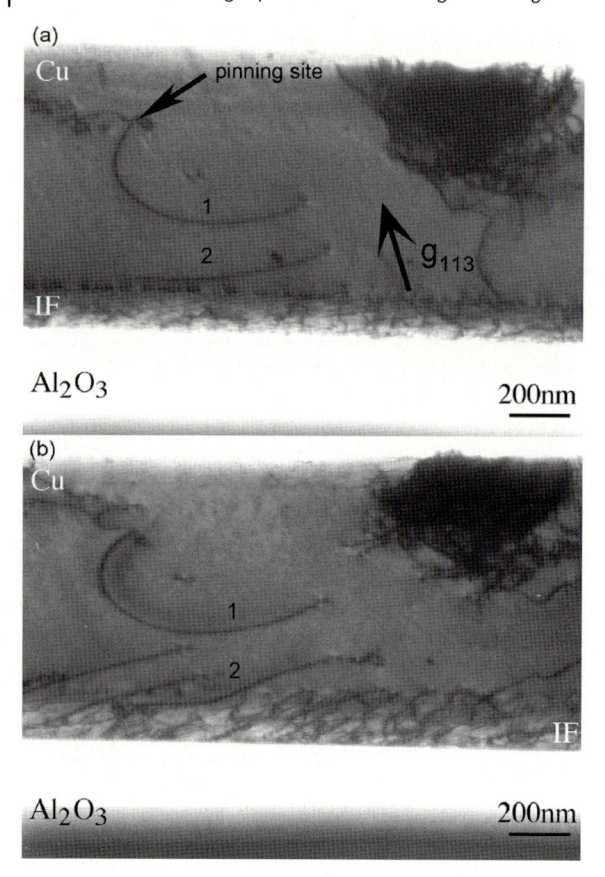

Figure 10.8 Dynamic observation of a dislocation spiral source (single-armed source, labeled as dislocation 1) operating in a (111) Cu film grown on (0001) α-Al_2O_3 substrate during thermal straining. The interface is imaged under inclination revealing the dislocation network located at the Cu/α-Al_2O_3 interface (IF). (a) The dislocation spiral source (1) and a previously emitted dislocation (2); (b) Dislocation 2 moves towards the interface and subsequently becomes incorporated into the interfacial dislocation network. This reduces the back-stress on the dislocation source, allowing it to revolve once again, emitting a new dislocation (not shown here). Figure reproduced with permission from Ref. [131]; © 2003, Materials Research Society.

single-ended sources may operate at approximately half the stress of a double-pinned source of the same size [72]. As such sources in small-grained materials are scarce, and the high-flow stresses lead to high dislocation velocities, it is difficult to observe multiplication phenomena within the time resolution of 25 to 30 images per second for (quantitative) *in-situ* TEM experiments. One of the few examples where a dislocation source operated inside the metal film is shown in Figure 10.8. Here, a single-ended spiral source was observed in a cross-sectional *in-situ* TEM study of an epitaxial Cu film on a (0001) α-Al_2O_3 substrate [18]. The spiral source stops revolving when feeling the back-stress of previously emitted dislocations piling-up at the interface (or when its length is minimum). If the stress becomes sufficiently high, the source overcomes the pile-up barrier (or the line length limitation) and revolves one turn, thereby emitting a new dislocation before being stopped again by the backstress from the pile-up. The pile-up at the Cu crystal/α-Al_2O_3 crystal interface does not increase in dislocation density, but remains constant with about two to three dislocations. The stress exerted on the leading dislocation closest to the interface is sufficiently high that it becomes incorporated into the pre-existing interfacial dislocation network, thereby reducing the pile-up by one dislocation and permitting the spiral source to emit the next dislocation (Figure 10.8).

10.3.4
Diffusion-Induced Dislocation Plasticity in Polycrystalline Cu Films

A peculiarity was observed for polycrystalline Cu films on substrates where, in addition to dislocation plasticity, grain boundary diffusional creep becomes an important stress compensation mechanism [73, 74]. Gao *et al.* [74] proposed that the stress field at grain boundaries due to grain boundary diffusion and the presence of a rigid film/substrate interface would cause shear stresses parallel to the interface, bearing similarities to the stress field ahead of a crack extending through the film towards the film/substrate interface. Balk *et al.* [66] observed in plan-view TEM studies that, during the cooling of the film from elevated temperatures, dislocations are emitted from some grain boundaries and glide on a {111} plane parallel to the interface (Figure 10.9). On an inclined {111} plane, these dislocations could not travel such long distances, as they would reach the top or bottom surface. Upon heating, however, the stress changed from tension to compression, and the previously emitted "parallel glide" dislocations [66] moved backwards and were re-absorbed by the grain boundary. This behavior proved to be reversible, with dislocation emission upon the next cooling cycle and dislocation absorption upon subsequent heating. It is believed that, upon cooling (tensile film stress), the surface atoms diffuse into the grain boundaries. Thereby, the tensile stress becomes locally compensated, while the atoms migrating into the grain boundaries are redistributed and partially "emitted" as parallel glide dislocations which advance by the shear stress next to the grain boundary/substrate intersection at glide planes parallel to the interface. Interestingly, this phenomenon vanishes if the surface of the polycrystalline Cu film is passivated by a continuous oxide layer, indicating that this dislocation mechanism is linked to the coupled surface- and grain-boundary diffusion. This interpretation is further supported by wafer curvature stress measurements of unpassivated and passivated Cu films of identical nominal film thickness and grain size, revealing higher stresses for passivated films where, again, a coupled surface-grain boundary diffusion is suppressed [75].

10.4
Size-Dependent Dislocation Plasticity in Metals

Plasticity in metals becomes size-dependent when the geometric or microstructural dimensions enter the size regime of dislocation glide distances and/or dislocation

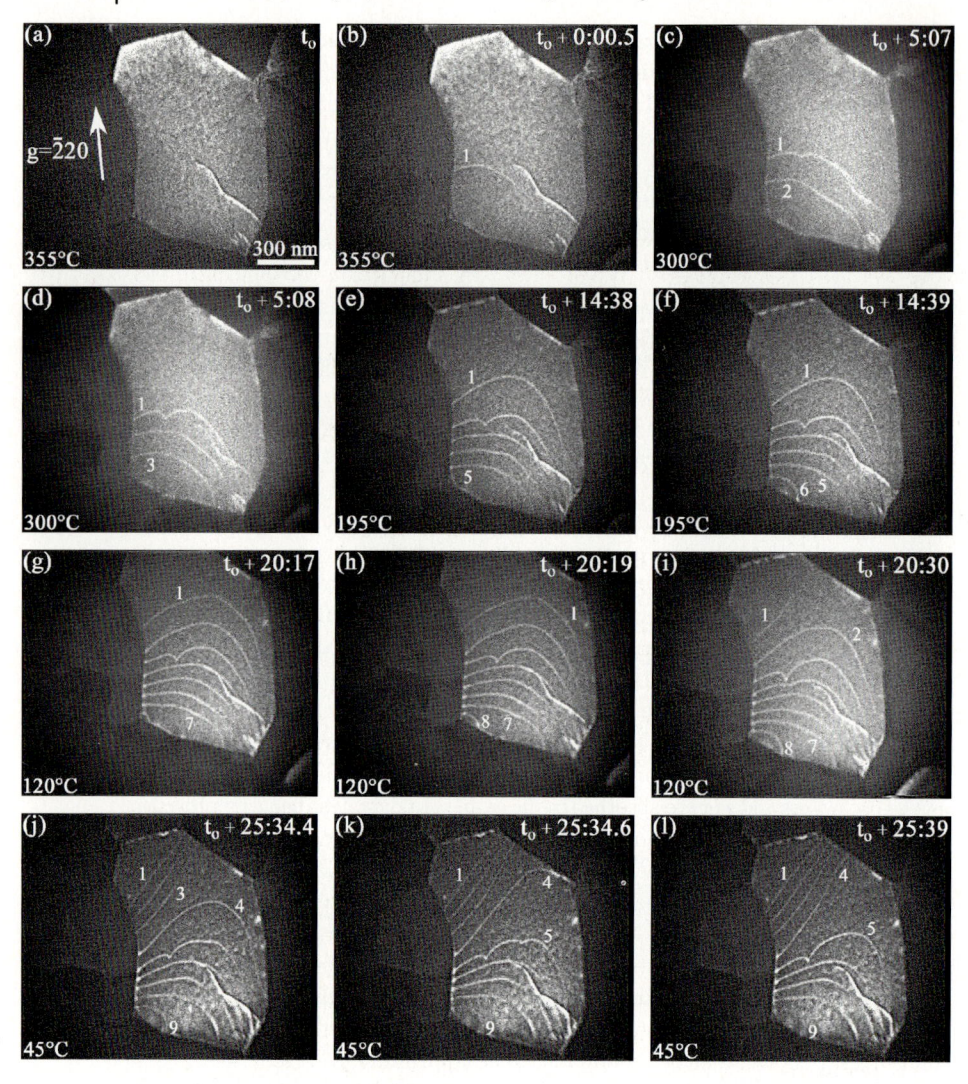

Figure 10.9 Emission of dislocations in a 200 nm-thick Cu film on a SiN$_x$-coated Si substrate during cooling from 355 to 45 °C, under a global biaxial tensile stress. The time frame is indicated with 25 min and 39 s between weak-beam image (a) and weak-beam image (l). The subsequently emitted dislocations (see numbers) advance on a (111) Cu plane parallel to the interface due to shear stresses created by grain boundary diffusion. See text for details. Figure reproduced with permission from Ref. [66]; © 2003, Elsevier.

source dimensions, as this may change the probability for dislocation–dislocation interactions [76]. Thus, dislocation mechanisms controlling plasticity in bulk metals such as Taylor hardening, formation of immobile dislocations, and dislocation multiplication [77] are hampered and other phenomena may play a central role. These aspects are at the core of recent *in-situ* TEM straining experiments aimed at shedding light on the deformation mechanisms of nanocrystalline and/or "nano-sized" metallic objects, as outlined below.

10.4.1
Plasticity in Geometrically Confined Single Crystal fcc Metals

For fcc single crystals, both micro-compression [78, 79] and micro-tensile testing [80] have revealed a strong size effect in flow stress, with smaller samples showing higher flow stresses along with stochastic variations in flow stresses for nominal identical samples and rather intermitted appearance of the flow curves (Figure 10.10). It was speculated that the crystals possess very few sources and lose all dislocations during early deformation, requiring dislocation nucleation from the sample surface for sustained plastic deformation ("dislocation starvation" [81]). Alternatively, Frank–Read sources could become truncated to single-armed spiral sources by the limited sample volume, with a smaller source size distribution for smaller crystals leading to higher stresses in smaller samples and dislocation escape rates at the surface occurring with similar rates as dislocation nucleation or multiplication [82–85]. The first in-situ TEM study to address this question was conducted on small (111)-oriented Ni pillars prepared by FIB structuring and loading with a flat punch using a commercial TEM nano-indentation stage [86] (see Chapter 11). The experiments were conducted inside the transmission electron microscope, simultaneously recording the load–displacement data while imaging the compression experiment [86]. The study revealed that the high density of dislocations initially present after FIB

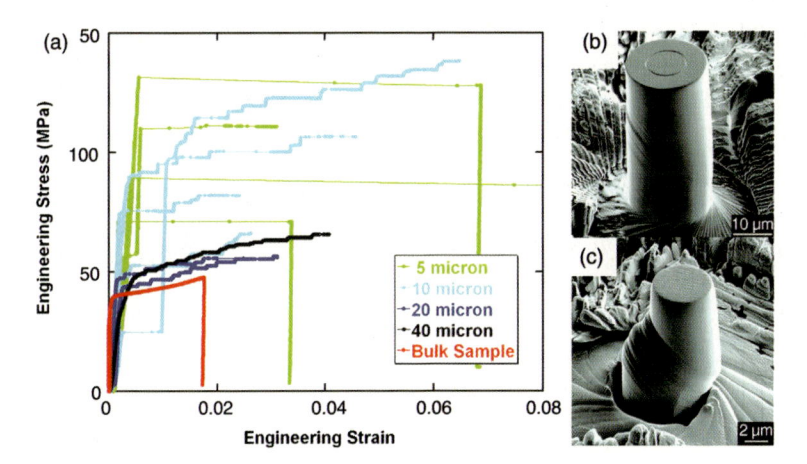

Figure 10.10 (a) Stress–strain curves of (134)-oriented Ni micro-compression pillars with diameters ranging from 40 to 5 μm. The micro-compression samples were cut into a single crystal using FIB cutting. Smaller sample diameters tend to show larger flow stresses. Note the intermitted stress strain response of the miniaturized samples in comparison to the bulk material; (b, c) Deformed micro-compression samples with 20 and 5 μm diameter, revealing single slip. Figure reproduced with permission from Ref. [78]; © 2004, The American Association for the Advancement of Science.

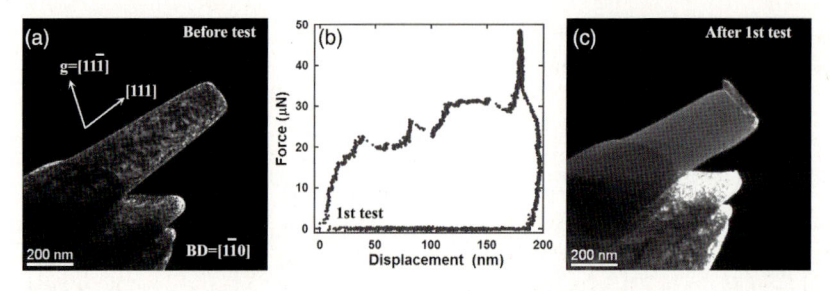

Figure 10.11 *In-situ* TEM compression experiment of Ni pillars. (a) The FIB-machined pillar contains a high density of surface near dislocations, (c) which disappear upon loading. This behavior was termed "mechanical annealing"; (b) Note the serrated flow in the load–displacement curves of the compression test. Figure reproduced with permission from Ref. [86]; © 2008, Nature Publishing Group.

cutting [87, 88] vanished upon loading the sample. This was interpreted as "mechanical annealing" (see Figure 10.11) [86], and as an experimental evidence of the starvation model described above. Upon further loading, no dislocation nucleation processes were observed; however, the load increased in a serrated manner indicating an atypical "hardening" [86], which implies a progressive activation and exhaustion of dislocation sources, although this remained unresolved in this *in-situ* experiment. In contrast to the study by Shan *et al.* [86], Oh *et al.* [71] performed a tensile test on a submicron Al single crystal which was FIB structured and strained in a conventional TEM straining stage (Figure 10.12). The FIB cutting was started from the polyimide back-side to thin the polyimide to a remaining thickness of about 1 μm, preventing Ga implantation into the Al film while already achieving electron transparency. Subsequently, lines were FIB-cut inside the electron-transparent film/polyimide area (Figure 10.12c), after which the sample was loaded inside the conventional TEM straining stage. The thin polyimide layer under the line ruptured at the boxed position (Figure 10.12c), leaving a free-standing Al

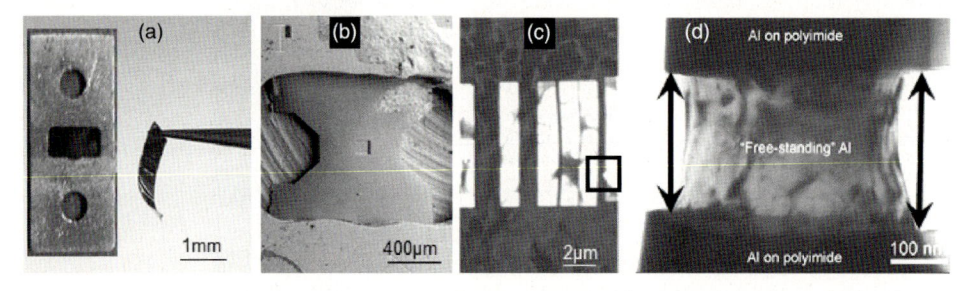

Figure 10.12 (a) A single-crystal Al film on polyimide is attached to a Cu support which fits into a conventional straining stage; (b) Using FIB cutting, the polyimide is locally thinned down to achieve electron transparency. Side grooves are FIB-cut to increase the stress in the electron-transparent region; (c) Lines are subsequently cut into the structure with a notch in the ~500 nm-wide line causing fracture of the polyimide and leaving (d) a free-standing Al section of the wire. Figure reproduced with permission from Ref. [71]; © 2009, Nature Publishing Group.

section with no visible FIB damage for the straining experiment (Figure 10.12d). The results of this *in-situ* experiment revealed that, throughout straining to failure at about 160% elongation, dislocations were present in the sample. The dislocation density remained rather constant at a strain rate of $10^{-4}\,\text{s}^{-1}$, as the dislocation escape rates at the crystal surface counterbalanced the dislocation nucleation rate [71]. Nucleation occurred from single-ended spiral sources which had minimal dimensions between about 50 and 150 nm [71]. This variation in source size is believed to explain the stochastic response and serrated flow [71, 89] of small crystal volumes to mechanical load (see Figure 10.10), as was also observed in discrete dislocation dynamic simulations [82, 83, 90]. Additionally, Oh et al. revealed that increasing the strain rate from $10^{-4}\,\text{s}^{-1}$ to $10^{-3}\,\text{s}^{-1}$ caused the activation of additional sources, which led to a noticeable increase in dislocation density as the nucleation rate exceeded the escape rate [71]. This observation also indicated that the deformation of crystals with small volume becomes strain-rate sensitive [71], as predicted from simulations [91]. This effect may be enhanced in Al as the native oxide slows down the escape of dislocations at free surfaces.

10.4.2
Size-Dependent Transitions in Dislocation Plasticity

A further reduction in dimensions, from about 100 nm to below about 50 nm, revealed a transition from perfect dislocations to partial dislocation plasticity and twinning [92, 93]. This transition was observed by Chen et al. [92] for nanocrystalline Al with grain size d of a few tens of nanometers. The phenomenon was explained by the lower stress required to nucleate a partial dislocation with Burgers vector b_p and an accompanying stacking fault with energy γ:

$$\tau_p = \frac{2\alpha G b_p}{d} + \frac{\gamma}{b_p} \qquad (10.5)$$

compared to nucleation of a full dislocation with Burgers vector b:

$$\tau_f = \frac{2\alpha G b}{d}. \qquad (10.6)$$

α amounts to 1.5 for screw and 0.5 for edge dislocations. The critical transition grain size d_c is obtained by setting $\tau_p = \tau_f$, resulting in [92]:

$$d_c = \frac{2\alpha G (b - b_p) b_p}{\gamma}. \qquad (10.7)$$

Interestingly, the same transition occurs in single crystalline (100)-oriented Au films when the film thickness is reduced from 160 nm to 40 nm, as observed during *in-situ* plane view straining experiments (see Figure 10.13) [94]. A reduction in film thickness caused a decrease in the height of the glide plane, which was equivalent to a reduction in grain size. This only differed from the previous study in that free surfaces were present instead of impenetrable grain boundaries and that, due to the

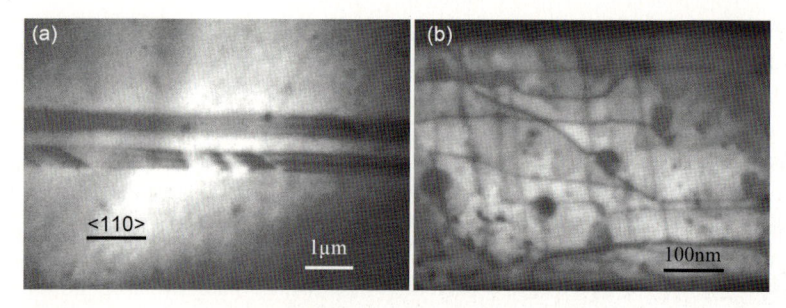

Figure 10.13 (a) The 80 nm-thick Au film deforms mainly by partial dislocations moving along <110> directions. Note the stacking fault contrast; (b) The 160 nm-thick Au film deforms by perfect dislocations. Figure reproduced with permission from Ref. [94]; © 2007, Elsevier.

absence of such obstacles, the perfect or partial dislocations could glide over long distances, facilitating *in-situ* observations. Dislocation nucleation was accomplished in the Au films by heterogeneous dislocation nucleation at the edge of pre-existing pores acting as stress concentrators [94]. Dislocations were emitted on the glide system with the highest Schmid factor experiencing the highest resolved shear stress [94]. Similar observations were reported recently for even smaller Au wires of a few nanometer diameter by Zheng *et al.* [95], where the nucleation of partial dislocations also occurred from pre-existing surface steps. These findings point to the increasing importance of surface steps and grain boundary ledges facilitating dislocation nucleation in the submicron dimensions regime, where high stresses and few dislocation sources commonly occur. Although this aspect was investigated via computational studies [96, 97], it has not yet received much attention from the microscopy community.

10.4.3
Plasticity by Motion of Grain Boundaries

Several research groups have proposed that, for nanocrystalline metals, not only partial dislocations but also grain boundary sliding, grain rotation, and grain boundary migration can occur [98–102]. These mechanisms were observed in molecular dynamics simulations of nanocrystalline metals, with some evidence obtained by *post mortem* TEM studies [103]. Grain boundary motion was directly observed via *in-situ* TEM straining experiments [104, 105]. The detailed *in-situ* TEM straining studies conducted by Legros *et al.* on nanocrystalline Al [104] and ultrafine-grained Al [106, 107] discarded thermally assisted grain boundary migration mechanisms (i.e., diffusional processes) as an explanation for the boundary motion. Rather, shear stress-induced migration was assumed to be the origin of the observed grain boundary motion (Figure 10.14), with measured velocities of up to 200 nm s^{-1} in nanocrystalline Al during tensile testing at room temperature [104]. The conclusion of these studies was based on the observation that stress, and not strain, drives the grain boundary motion (see Figure 10.14), as well as the high grain boundary velocities far exceeding the diffusion-induced grain boundary

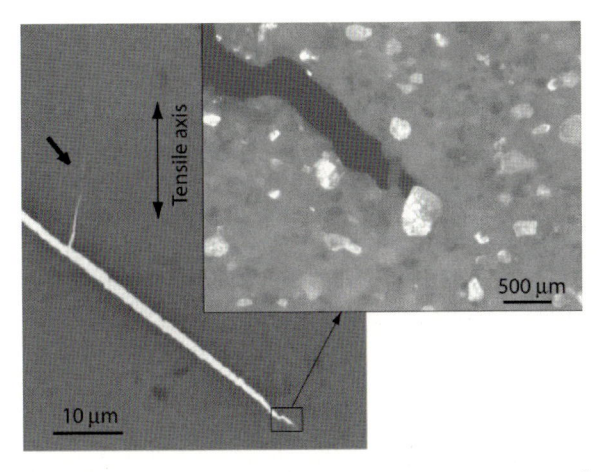

Figure 10.14 Stress-driven grain boundary motion in nanocrystalline Al causing grain growth ahead of the crack-tip. The boxed region is enlarged. See text for details. Figure reproduced with permission from Ref. [104]; © 2008, Elsevier.

motion [104]. This suggestion was recently confirmed in *ex-situ* TEM studies on nanocrystalline Ni films containing asymmetrical stress or strain concentrators [108]. In addition, dislocation plasticity was observed to occur down to grain sizes of about 40 nm, implying that several deformation mechanisms may be at play simultaneously [104].

10.4.4
Influence of Grain Size Heterogeneities

The MEMS/NEMS-based tensile testing of nanocrystalline Al and Au [46, 109, 110] revealed that the grain size distribution is also relevant to the deformation behavior. For wide or bimodal grain size distributions, different stress compensation mechanisms may operate, as particular compensation mechanisms may experience different activation thresholds in various grains. For example, larger dislocation sources in larger grains will require a lower stress to operate than smaller sources in smaller grains (Equation 10.4), assuming a scaling of sources with grain size (as indicated for fcc single-crystals in Section 10.4.1). As a consequence, smaller grains sustain higher elastic strains and higher yield stresses than larger grains. Deformation to strains where only the large grain fraction has undergone irreversible deformation by dislocation glide, while small grains experienced solely elastic deformation, can cause stress heterogeneities between small and large grains. During loading, large grains will yield earlier and subsequently deform plastically, while small grains may still be loaded elastically [46]. This leads to an extended transition region from elastic to fully plastic deformation, indicating that the number of grains reaching their yield stress continuously increases with strain [46, 111]. Upon unloading, a short region of elasticity would be followed by a Bauschinger-type behavior, with a deviation from elasticity indicating a reverse plasticity by compressive yielding of large grains, although the sample is still macroscopically in a tensile stress state. Such a behavior for a 215 nm-thick Al film with a mean grain size of 140 nm and a grain size distribution ranging from about 60 to 350 nm, is shown in Figure 10.15. *In-situ* TEM

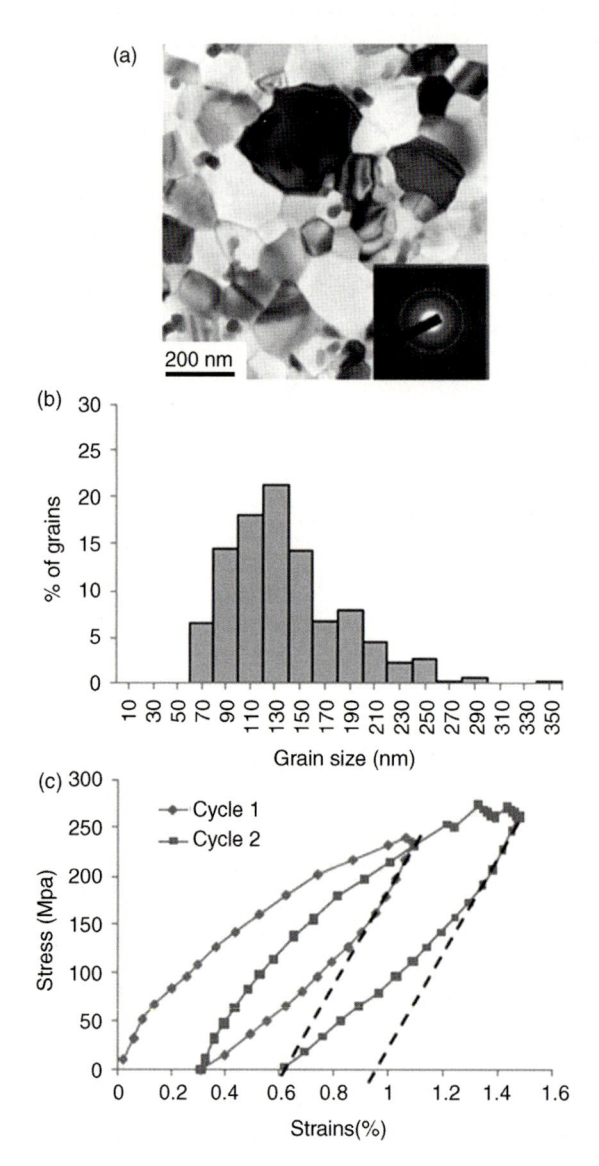

Figure 10.15 (a) The 215 nm thick Al film possesses no preferred texture (see inset for diffraction pattern) but reveals (b) a heterogeneous grain size distribution ranging from ~60 to ~350 nm; (c) Two subsequent stress–strain measurements of the Al sample. Note the extended microplastic region upon loading. Upon unloading, the sample behaves first elastically (see dashed line) followed by pronounced strain recovery due to a Bauschinger effect. Figure reproduced with permission from Ref. [46]; © 2010, Elsevier.

straining experiments conducted using the above-described MEMS/NEMS technology to monitor the global stress and strain (see Section 10.2.3) have confirmed the speculation that larger grains yield at lower stress and strain upon loading than do smaller grains. During unloading, very little dislocation activity is observed initially, whereas with further unloading an increasing amount of jerky dislocation motion is seen predominantly in the larger grains, while little or no dislocation activity for the small grains was observed in this TEM study [46]. The jerky dislocation motion indicated an unpinning from defects such as dislocation junctions, pinning sites at the surface, or point defects inside the grains and local variations in stresses – even within the larger grains. These results demonstrated that relating plasticity to average grain size may be oversimplified in certain cases, and that the full grain size distribution should be taken into account, especially in size regimes where different grains experience distinctly different yield stresses [112].

10.5
Conclusions and Future Directions

The mechanical properties of nanocrystalline metals and/or metals with miniaturized geometric dimensions are of immense importance for advanced devices and future applications of novel "nanomaterials". *In-situ* TEM studies provide the possibilities for direct observations of the key deformation mechanisms at play, their operation regime with respect to microstructure, geometric boundary conditions and applied mechanical stress, and thus provide the basis for establishing material models to improve reliability of devices and to identify design criteria for superior devices. *In-situ* TEM, complemented with other experimental techniques, especially *in-situ* X-ray and synchrotron experiments [113–120], as well as computational studies [72, 83, 84, 90, 121–124], have set the ground for the emerging understanding of plasticity in small dimensions. Whereas several aspects, such as the influence of external dimensions on dislocation nucleation and multiplication have, to some extent, been resolved, there remain open questions regarding the dynamics of dislocation interaction in small volumes, the effects of strain rate, and those of size or surface geometry on heterogeneous dislocation nucleation. Moreover, the present understanding of the mechanics of interfaces (including grain boundaries) remains incomplete, though these are of prime concern to applications of thin film/substrate structures and nanocrystalline materials. Unsolved questions include, for example:

- Under what conditions does grain boundary migration become the active deformation mechanism?
- What roles do grain boundary structure and chemistry play in grain boundary migration?
- When do interfaces behave brittle (interface fracture), and when ductile (emission and absorption of dislocations from interfaces, interface migration, etc.)?

- How do the mechanisms change with temperature, stress, strain rate, and stress state?
- Are magnetic and electronic properties influenced by the deformation mechanisms at play, and what is the coupling between these?

In addition, new fields are starting to emerge for *in-situ* mechanical testing inside the TEM. Tribology [125–127], fatigue [128, 129], and fracture [130] studies are now possible. This will shed new light on lifetime limiting aspects for materials at small dimensions by providing information on structural changes for well-defined loading conditions.

Acknowledgments

G.D. is grateful to J.T. Balk, B.J. Inkson, H.P. Karnthaler, S.H. Oh, J. Rajagopalan, C. Rentenberger, and M.T.A. Saif for fruitful collaborations on various *in-situ* TEM projects; M.L. thanks D. Caillard, F. Mompiou and K.J. Hemker, with whom much of the information reported here has been shared, and D.K. thanks G. Dehm, C. Scheu, A.M. Minor, and Z. Zhang for introducing him to various TEM techniques.

References

1 Hirsch, P., Howie, A., Nicholson, R., Pashley, D.W., and Whelan, M.J. (1967) *Electron Microscopy of Thin Crystals*, Krieger Publishing Company, Malabar (Florida) (reprinted).

2 Messerschmidt, U. and Appel, F. (1976) Quantitative tensile-tilting stages for the high voltage electron microscope. *Ultramicroscopy*, **1** (3), 223–230.

3 Martin, J.L. and Kubin, L.P. (1979) Optimum conditions for straining experiments in the HVEM. *Phys. Status Solidi A*, **56** (2), 487–494.

4 Martin, J.L. and Kubin, L.P. (1978) Discussion on the limitations of "in situ" deformation experiments in a high voltage electron microscope. *Ultramicroscopy*, **3**, 215–226.

5 Imura, T. and Saka, H. (1976) In-situ dynamic observations of dislocation behaviour in metals and alloys by high voltage electron microscopy. *Memoirs of the Faculty of Engineering, Nagoya University*, **28** (1), 54–112.

6 Fujita, H. (1986) Ultra-high voltage electron microscopy: Past, present, and future. *J. Electron Microsc. Tech.*, **3** (3), 243–304.

7 Saka, H. and Teshima, H. (1991) Recent progress in in-situ deformation experiments in HVEM. *Ultramicroscopy*, **39** (1-4), 81–85.

8 Messerschmidt, U. (2010) *Dislocation Dynamics During Plastic Deformation*, Springer Verlag, Berlin.

9 Caillard, D., Couret, A., Buschow, K.H., Jr, Robert, W.C., Merton, C.F., Bernard, I. et al. (2006) In situ deformation experiments: Technical aspects, in *Encyclopedia of Materials: Science and Technology*, Elsevier, Oxford, pp. 1–11.

10 Hagen, W. and Strunk, H. (1978) A new type of source generating misfit dislocations. *Appl. Phys.*, **17** (1), 85–87.

11 Haasen, P., Messerschmidt, U., and Skrotzki, W. (1986) Low energy dislocation structures in ionic crystals and semiconductors. *Mater. Sci. Eng.*, **81**, 493–507.

12 Matthews, J.W. and Blakeslee, A.E. (1974) Defects in epitaxial multilayers: I.

Misfit dislocations. *J. Cryst. Growth*, **27**, 118–125.

13 Matthews, J.W. and Blakeslee, A.E. (1975) Defects in epitaxial multilayers: II. Dislocation pile-ups, threading dislocations, slip lines and cracks. *J. Cryst. Growth*, **29** (3), 273–280.

14 Hull, R., Stach, E.A., Tromp, R., Ross, F., and Reuter, M. (1999) Interactions of moving dislocations in semiconductors with point, line and planar defects. *Phys. Status Solidi A*, **171** (1), 133–146.

15 Stach, E.A., Hull, R., Tromp, R.M., Ross, F.M., Reuter, M.C., and Bean, J.C. (2000) In-situ transmission electron microscopy studies of the interaction between dislocations in strained SiGe/Si (001) heterostructures. *Philos. Mag. A*, **80** (9), 2159–2200.

16 Dehm, G. (2009) Miniaturized single-crystalline fcc metals deformed in tension: New insights in size-dependent plasticity. *Prog. Mater. Sci.*, **54** (6), 664–688.

17 Uchic, M.D., Shade, P.A., and Dimiduk, D.M. (2009) Plasticity of micrometer-scale single crystals in compression. *Annu. Rev. Mater. Res.*, **39**, 361–386.

18 Legros, M., Cabie, M., and Gianola, D.S. (2009) In situ deformation of thin films on substrates. *Microsc. Res. Tech.*, **72** (3), 270–283.

19 Kraft, O., Gruber, A.P., Mönig, R., and Weygand, D. (2010) Plasticity in confined dimensions. *Annu. Rev. Mater. Res.*, **40**, 293–317.

20 Arzt, E., Dehm, G., Gumbsch, P., Kraft, O., and Weiss, D. (2001) Interface controlled plasticity in metals: dispersion hardening and thin film deformation. *Prog. Mater. Sci.*, **46**, 3–4.

21 Legros, M., Elliott, B.R., Rittner, M.N., Weertman, J.R., and Hemker, K.J. (2000) Microsample tensile testing of nanocrystalline metals. *Philos. Mag. A*, **80** (4), 1017–1026.

22 Baker, S.P. (2001) Plastic deformation and strength of materials in small dimensions. *Mater. Sci. Eng. A*, **321**, 16–23.

23 Arzt, E. (1998) Size effects in materials due to microstructural and dimensional constraints: a comparative review. *Acta Mater.*, **46** (16), 5611–5626.

24 Robertson, I.M., Ferreira, P.J., Dehm, G., Hull, R., and Stach, E.A. (2008) Visualizing the behavior of dislocations – Seeing is believing. *MRS Bull.*, **33** (2), 122–131.

25 Legros, M., Gianola, D.S., and Motz, C. (2010) Quantitative in situ mechanical testing in electron microscopes. *MRS Bull.*, **35** (5), 354–360.

26 Haque, M.A., Espinosa, H.D., and Lee, H.J. (2010) MEMS for in situ testing-handling, actuation, loading, and displacement measurements. *MRS Bull.*, **35** (5), 375–381.

27 Gerth, D., Katzer, D., and Krohn, M. (1992) Study of the thermal behaviour of thin aluminium alloy films. *Thin Solid Films*, **208** (1), 67–75.

28 Dehm, G. and Arzt, E. (2000) In situ transmission electron microscopy study of dislocations in a polycrystalline Cu thin film constrained by a substrate. *Appl. Phys. Lett.*, **77** (8), 1126–1128.

29 Legros, M., Dehm, G., Keller-Flaig, R.M., Arzt, E., Hemker, K.J., and Suresh, S. (2001) Dynamic observation of Al thin films plastically strained in a TEM. *Mater. Sci. Eng. A*, **15**, 463–467.

30 Stach, E.A., Dahmen, U., and Nix, W.D. (2000) Real time observations of dislocation-mediated plasticity in the epitaxial Al (011)/Si (100) thin film system, in *Proceedings, Materials Research Society meeting, April 23–26, 2000, San Francisco, CA, USA* (eds M. Yeadon, S. Chiang, R.F.C. Farrow, J.W. Evans, and O. Auciello), Materials Research Society, vol. **619**, p. 27.

31 Inkson, B.J., Dehm, G., and Wagner, T. (2002) In situ TEM observation of dislocation motion in thermally strained Al nanowires. *Acta Mater.*, **50** (20), 5033–5047.

32 Saada, G., Verdier, M., and Dirras, G. (2007) Elasto plastic behaviour of thin films. *Philos. Mag.*, **87**, 4878–4892.

33 Legros, M. (2006) Relaxation plastique des couches métalliques par dislocations et défauts étendus, in *Contraintes Mécaniques en Micro, nano et optoélectronique (Traité EGEM, série*

Electronique et Micro-Électronique) (ed. M. Mouis), Hermes Science Publications, Paris.

34 Zhang, Y., Han, X., Zheng, K., Zhang, Z., Zhang, X., Fu, J. *et al.* (2007) Direct observation of super-plasticity of Beta-SiC nanowires at low temperature. *Adv. Funct. Mater.*, **17** (17), 3435–3440.

35 Liu, P., Mao, S.C., Wang, L.H., Han, X.D., and Zhang, Z. (2011) Direct dynamic atomic mechanisms of strain-induced grain rotation in nanocrystalline, textured, columnar-structured thin gold films. *Scr. Mater.*, **64** (4), 343–346.

36 Pélissier, J. and Debrenne, P. (1993) In situ experiments in the new transmission electron microscopes. *Microsc. Microanal.*, **4** (2-3), 111–117.

37 Appel, F. and Messerschmidt, U. (1976) The estimation of the effective stress in aluminium foils deformed in the high-voltage electron microscope (HVEM). *Phys. Status Solidi A*, **34** (1), 175–181.

38 Nafari, A., Danilov, A., Rodjegard, H., Enoksson, P., and Olin, H. (2005) A micromachined nanoindentation force sensor. *Sens. Actuators, A*, **123–24**, 44–49.

39 Bobji, M.S., Ramanujan, C.S., Pethica, J.B., and Inkson, B.J. (2006) A miniaturized TEM nanoindenter for studying material deformation in situ. *Meas. Sci. Technol.*, **17** (6), 1324–1329.

40 Warren, O.L., Shan, Z., Asif, S.A.S., Stach, E.A., Morris, J.J.W., and Minor, A.M. (2007) In situ nanoindentation in the TEM. *Mater. Today*, **10** (4), 59–60.

41 Svensson, K., Jompol, Y., Olin, H., and Olsson, E. (2003) Compact design of a transmission electron microscope-scanning tunneling microscope holder with three-dimensional coarse motion. *Rev. Sci. Instrum.*, **74** (11), 4945–4947.

42 Hemker, K.J. and Sharpe, W.N. Jr (2007) Microscale characterization of mechanical properties. *Annu. Rev. Mater. Res.*, **37**, 93–126.

43 Haque, M.A. and Saif, M.T.A. (2002) In-situ tensile testing of nano-scale specimens in SEM and TEM. *Exp. Mech.*, **42** (1), 123–128.

44 Haque, M.A. and Saif, M.T.A. (2003) A review of MEMS-based miroscale and nanoscale tensile and bending testing. *Exp. Mech.*, **43** (3), 248–255.

45 Hattar, K., Han, J., Saif, M.T.A., and Robertson, I.M. (2005) In situ transmission electron microscopy observations of toughening mechanisms in ultra-fine grained columnar aluminum thin films. *J. Mater. Res.*, **20** (7), 1869–1877.

46 Rajagopalan, J., Rentenberger, C., Karnthaler, H.P., Dehm, G., and Saif, M.T.A. (2010) In situ TEM study of microplasticity and Bauschinger effect in nanocrystalline metals. *Acta Mater.*, **58** (14), 4772–4782.

47 Pant, B., Allen, B.L., Zhu, T., Gall, K., and Pierron, O.N. (2011) A versatile microelectromechanical system for nanomechanical testing. *Appl. Phys. Lett.*, **98**, 053506.

48 Kang, W. and Saif, M.T.A. (2010) A novel method for in situ uniaxial tests at the micro/nano scale – Part I: Theory. *J. Microelectromech. Syst.*, **19** (6), 1309–1321.

49 Kang, W., Han, J.H., and Saif, M.T.A. (2010) A novel method for in situ uniaxial tests at the micro/nanoscale – Part II: Experiment. *J. Microelectromech. Syst.*, **19** (6), 1322–1330.

50 Agrawal, R., Peng, B., and Espinosa, H.D. (2009) Experimental-computational investigation of ZnO nanowires strength and fracture. *Nano Lett.*, **9** (12), 4177–4183.

51 Zhang, D.F., Drissen, W., Breguet, J.M., Clavel, R., and Michler, J. (2009) A high-sensitivity and quasi-linear capacitive sensor for nanomechanical testing applications. *J. Micromech. Microeng.*, **19** (7), article 075003.

52 Nix, W.D. (1989) Mechanical properties of thin films. *Metall. Trans. A*, **20** (11), 2217–2245.

53 Venkatraman, R. and Bravman, J.C. (1992) Separation of film thickness and grain-boundary strengthening effects in Al thin-films on Si. *J. Mater. Res.*, **7** (8), 2040–2048.

54 Keller, R., Baker, S.P., and Arzt, E. (1998) Quantitative analysis of strengthening

mechanisms in thin Cu films: effects of film thickness, grain size, and passivation. *J. Mater. Res.*, **13** (5), 1307–1317.

55 Kobrinsky, M.J. and Thompson, C.V. (1998) The thickness dependence of the flow stress of capped and uncapped polycrystalline Ag thin films. *Appl. Phys. Lett.*, **73** (17), 2429–2431.

56 Leung, O.S., Munkholm, A., Brennan, S., and Nix, W.D. (2000) A search for strain gradients in gold thin films on substrates using X-ray diffraction. *J. Appl. Phys.*, **88** (3), 1389–1396.

57 Dehm, G., Balk, T.J., Edongué, H., and Arzt, E. (2003) Small-scale plasticity in thin Cu and Al films. *Microelectron. Eng.*, **70** (2-4), 412–424.

58 Eiper, E., Keckes, J., Martinschitz, K.J., Zizak, I., Cabie, M., and Dehm, G. (2007) Size-independent stresses in Al thin films thermally strained down to -100 degrees C. *Acta Mater.*, **55** (6), 1941–1946.

59 Taylor, A.A., Oh, S.H., and Dehm, G. (2010) Microplasticity phenomena in thermomechanically strained nickel thin films. *J. Mater. Sci.*, **45** (14), 3874–3881.

60 Freund, L.B. (1990) The driving force for glide of a threading dislocation in a strained epitaxial layer on a substrate. *J. Mech. Phys. Solids*, **38** (5), 657–679.

61 Thompson, C.V. (1993) The yield stress of polycrystalline thin films. *J. Mater. Res.*, **8** (2), 237–238.

62 von Blanckenhagen, B., Gumbsch, P., and Arzt, E. (2003) Dislocation sources and the flow stress of polycrystalline thin metal films. *Philos. Mag. Lett.*, **83** (1)

63 Dehm, G., Wagner, T., Balk, T.J., Arzt, E., and Inkson, B.J. (2002) Plasticity and interfacial dislocation mechanisms in epitaxial and polycrystalline Al films constrained by substrates. *J. Mater. Sci. Technol.*, **18** (2), 113–117.

64 Dehm, G., Inkson, B.J., Balk, T.J., Wagner, T., and Arzt, E. (2001) Influence of film/substrate interface structure on plasticity in metal thin films. *MRS, Mater. Res. Soc. Proc.*, **673**, 1–12.

65 McCarty, E.D. and Hackney, S.A. (1995) Local interface response to dislocation strain fields at the Al/SiO$_2$ and Al/C boundary region. *Mater. Sci. Eng. A*, **196** (1–2), 119–128.

66 Balk, T.J., Dehm, G., and Arzt, E. (2003) Parallel glide: unexpected dislocation motion parallel to the substrate in ultrathin copper films. *Acta Mater.*, **51** (15), 4471–4485.

67 Legros, M., Dehm, G., and Balk, T.J. (2005) In-situ TEM study of plastic stress relaxation mechanisms and interface effects in metallic films. Proceedings, Materials Research Society Symposium, March 28–April 1, 2005, San Francisco, CA, USA, vol. 875, O9.1.1-11.

68 Dehm, G., Motz, C., Scheu, C., Clemens, H., Mayrhofer, P.H., and Mitterer, C. (2006) Mechanical size-effects in miniaturized and bulk materials. *Adv. Eng. Mater.*, **8** (11), 1033–1045.

69 von Blanckenhagen, B., Gumbsch, P., and Arzt, E. (2001) Dislocation sources in discrete dislocation simulations of thin-film plasticity and the Hall–Petch relation. *Modell. Simul. Mater. Sci. Eng.*, **9** (3), 157–169.

70 von Blanckenhagen, B., Arzt, E., and Gumbsch, P. (2004) Discrete dislocation simulation of plastic deformation in metal thin films. *Acta Mater.*, **52** (3), 773–784.

71 Oh, S.H., Legros, M., Kiener, D., and Dehm, G. (2009) *In situ* observation of dislocation nucleation and escape in a submicrometre aluminium single crystal. *Nat. Mater.*, **8** (2), 95–100.

72 Rao, S.I., Dimiduk, D.M., Tang, M., Parthasarathy, T.A., Uchic, M.D., and Woodward, C. (2007) Estimating the strength of single-ended dislocation sources in micron-sized single crystals. *Philos. Mag.*, **87** (30), 4777–4794.

73 Thouless, M.D. (1993) Effect of surface diffusion on the creep of thin films and sintered arrays of particles. *Acta Metall. Mater.*, **41** (4), 1057–1064.

74 Gao, H., Zhang, L., Nix, W.D., Thompson, C.V., and Arzt, E. (1999) Crack-like grain-boundary diffusion wedges in thin metal films. *Acta Mater.*, **47** (10), 2865–2878.

75 Wiederhirn, G. (2007) The strength limits of ultra-thin copper films. PhD thesis, University of Stuttgart.

76 Benzerga, A.A. (2009) Micro-pillar plasticity: 2.5D mesoscopic simulations. *J. Mech. Phys. Solids*, **57** (9), 1459–1469.

77 Argon, A.S. (2008) *Strengthening Mechanisms in Crystal Plasticity*, Oxford University Press, New York.

78 Uchic, M.D., Dimiduk, D.M., Florando, J.N., and Nix, W.D. (2004) Sample dimensions influence strength and crystal plasticity. *Science*, **305** (5686), 986–989.

79 Volkert, C.A. and Lilleodden, E.T. (2006) Size effects in the deformation of sub-micron Au columns. *Philos. Mag.*, **86** (33-35), 5567–5579.

80 Kiener, D., Grosinger, W., Dehm, G., and Pippan, R. (2008) A further step towards an understanding of size-dependent crystal plasticity: In situ tension experiments of miniaturized single-crystal copper samples. *Acta Mater.*, **56** (3), 580–592.

81 Greer, J.R. and Nix, W.D. (2006) Nanoscale gold pillars strengthened through dislocation starvation. *Phys. Rev. B*, **73** (24), 245410.

82 Rao, S.I., Dimiduk, D.M., Parthasarathy, T.A., Uchic, M.D., Tang, M., and Woodward, C. (2008) Athermal mechanisms of size-dependent crystal flow gleaned from three-dimensional discrete dislocation simulations. *Acta Mater.*, **56** (13), 3245–3259.

83 Senger, J., Weygand, D., Gumbsch, P., and Kraft, O. (2008) Discrete dislocation simulations of the plasticity of micro-pillars under uniaxial loading. *Scr. Mater.*, **58** (7), 587–590.

84 Parthasarathy, T.A., Rao, S.I., Dimiduk, D.M., Uchic, M.D., and Trinkle, D.R. (2007) Contribution to size effect of yield strength from the stochastics of dislocation source lengths in finite samples. *Scr. Mater.*, **56** (4), 313–316.

85 Deshpande, V.S., Needleman, A., and Van der Giessen, E. (2005) Plasticity size effects in tension and compression of single crystals. *J. Mech. Phys. Solids*, **53** (12), 2661–2691.

86 Shan, Z.W., Mishra, R.K., Asif, S.A.S., Warren, O.L., and Minor, A.M. (2008) Mechanical annealing and source-limited deformation in submicrometre-diameter Ni crystals. *Nat. Mater.*, **73** (7), 115–119.

87 Kiener, D., Motz, C., Rester, M., and Dehm, G. (2007) FIB damage of Cu and possible consequences for miniaturized mechanical tests. *Mater. Sci. Eng. A*, **459** (1-2), 262–272.

88 Volkert, C.A. and Minor, A.M. (2007) Focused ion beam microscopy and micromachining. *MRS Bull.*, **32** (5), 389–395.

89 Kiener, D. and Minor, A.M. (2011) Source-controlled yield and hardening of Cu(100) studied by in situ transmission electron microscopy. *Acta Mater.*, **59** (4), 1328–1337.

90 Motz, C., Weygand, D., Senger, J., and Gumbsch, P. (2009) Initial dislocation structures in 3-D discrete dislocation dynamics and their influence on microscale plasticity. *Acta Mater.*, **57** (6), 1744–1754.

91 Zhu, T., Li, J., Samanta, A., Leach, A., and Gall, K. (2008) Temperature and strain-rate dependence of surface dislocation nucleation. *Phys. Rev. Lett.*, **100** (2)

92 Chen, M., Ma, E., Hemker, K.J., Sheng, H., Wang, Y., and Cheng, X. (2003) Deformation twinning in nanocrystalline aluminum. *Science*, **300**, 1275.

93 Rosner, H., Boucharat, N., Markmann, J., Padmanabhan, K.A., and Wilde, G. (2009) In situ transmission electron microscopic observations of deformation and fracture processes in nanocrystalline palladium and Pd90Au10. *Mater. Sci. Eng. A*, **525** (1-2), 102–106.

94 Oh, S.H., Legros, M., Kiener, D., Gruber, P., and Dehm, G. (2007) In situ TEM straining of single crystal Au films on polyimide: Change of deformation mechanisms at the nanoscale. *Acta Mater.*, **55** (16), 5558–5571.

95 Zheng, H., Cao, A.J., Weinberger, C.R., Huang, J.Y., Du, K., Wang, J.B. et al. (2010) Discrete plasticity in sub-10-nm-sized gold crystals. *Nat. Commun.*, **1**.

96 Rabkin, E. and Srolovitz, D.J. (2007) Onset of plasticity in gold nanopillar compression. *Nano Lett.*, **7** (1), 101–107.

97 Brochard, S., Hirel, P., Pizzagalli, L., and Godet, J. (2010) Elastic limit for surface step dislocation nucleation in face-centered cubic metals: Temperature and step height dependence. *Acta Mater.*, **58** (12), 4182–4190.

98 Weissmuller, J. and Markmann, J. (2005) Deforming nanocrystalline metals: New insights, new puzzles. *Adv. Eng. Mater.*, **7** (4), 202–207.

99 Wolf, D., Yamakov, V., Phillpot, S.R., Mukherjee, A., and Gleiter, H. (2005) Deformation of nanocrystalline materials by molecular-dynamics simulation: relationship to experiments? *Acta Mater.*, **53** (1), 1–40.

100 Schiotz, J. and Jacobsen, K.W. (2003) A maximum in the strength of nanocrystalline copper. *Science*, **301** (5638), 1357–1359.

101 Van Swygenhoven, H. and Derlet, P.A. (2001) Grain-boundary sliding in nanocrystalline fcc metals. *Phys. Rev. B*, **64** (22)

102 De Hosson, J.T.M., Soer, W.A., Minor, A.M., Shan, Z.W., Stach, E.A., Asif, S.A.S. et al. (2006) In situ TEM nanoindentation and dislocation-grain boundary interactions: A tribute to David Brandon. *J. Mater. Sci.*, **41** (23), 7704–7719.

103 Rosner, H., Markmann, J., and Weissmuller, J. (2004) Deformation twinning in nanocrystalline Pd. *Philos. Mag. Lett.*, **84** (5), 321–334.

104 Legros, M., Gianola, D.S., and Hemker, K.J. (2008) In situ TEM observations of fast grain-boundary motion in stressed nanocrystalline aluminum films. *Acta Mater.*, **56** (14), 3380–3393.

105 Minor, A.M., Lilleodden, E.T., Stach, E.A., and Morris, J.W. (2004) Direct observations of incipient plasticity during nanoindentation of Al. *J. Mater. Res.*, **19** (1), 176–182.

106 Mompiou, F., Caillard, D., and Legros, M. (2009) Grain boundary shear-migration coupling-I. In situ TEM straining experiments in Al polycrystals. *Acta Mater.*, **57** (7), 2198–2209.

107 Caillard, D., Mompiou, F., and Legros, M. (2009) Grain-boundary shear-migration coupling. II. Geometrical model for general boundaries. *Acta Mater.*, **57** (8), 2390–2402.

108 Rupert, T.J., Gianola, D.S., Gan, Y., and Hemker, K.J. (2009) Experimental observations of stress-driven grain boundary migration. *Science*, **326** (5960), 1686–1690.

109 Rajagopalan, J., Han, J.H., and Saif, M.T.A. (2007) Plastic deformation recovery in freestanding nanocrystalline aluminum and gold thin films. *Science*, **315** (5820), 1831–1834.

110 Rajagopalan, J., Han, J.H., and Saif, M.T.A. (2008) On plastic strain recovery in freestanding nanocrystalline metal thin films. *Scr. Mater.*, **59** (9), 921–926.

111 Gruber, P.A., Solenthaler, C., Arzt, E., and Spolenak, R. (2008) Strong single-crystalline Au films tested by a new synchrotron technique. *Acta Mater.*, **56** (8), 1876–1889.

112 Saada, G. and Kruml, T. (2011) Deformation mechanisms of nanograined metallic polycrystals. *Acta Mater.*, **59**, 2565–2574.

113 Tamura, N., Celestre, R.S., MacDowell, A.A., Padmore, H.A., Spolenak, R., Valek, B.C. et al. (2002) Submicron X-ray diffraction and its applications to problems in materials and environmental science. *Rev. Sci. Instrum.*, **73** (3), 1369–1372.

114 Pantleon, W., Poulsen, H.F., Almer, J., and Lienert, U. (2004) In situ X-ray peak shape analysis of embedded individual grains during plastic deformation of metals. *Mater. Sci. Eng. A*, **387-389** (1-2 Special issue), 339–342.

115 Gianola, D.S., Van Petegem, S., Legros, M., Brandstetter, S., Van Swygenhoven, H., and Hemker, K.J. (2006) Stress-assisted discontinuous grain growth and its effect on the deformation behavior of nanocrystalline aluminum thin films. *Acta Mater.*, **54** (8), 2253–2263.

116 Maaß, R., Van Petegem, S., Van Swygenhoven, H., Derlet, P.M., Volkert, C.A., and Grolimund, D. (2007) Time-resolved Laue diffraction of deforming micropillars. *Phys. Rev. Lett.*, **99** (14), 1–4.

117 Olliges, S., Gruber, P.A., Auzelyte, V., Ekinci, Y., Solak, H.H., and Spolenak, R. (2007) Tensile strength of gold nanointerconnects without the influence of strain gradients. *Acta Mater.*, **55** (15), 5201–5210.

118 Maaß, R., Van Petegem, S., Grolimund, D., Van Swygenhoven, H., Kiener, D., and Dehm, G. (2008) Crystal rotation in Cu single crystal micropillars: In situ Laue and electron backscatter diffraction. *Appl. Phys. Lett.*, **92** (7), 071905.

119 Kirchlechner, C., Kiener, D., Motz, C., Labat, S., Vaxelaire, N., Perroud, O. et al. (2011) Dislocation storage in single slip-oriented Cu micro-tensile samples: new insights via X-ray microdiffraction. *Philos. Mag.*, **91** (7-9), 1256–1264.

120 Kirchlechner, C., Keckes, J., Motz, C., Grosinger, W., Kapp, M.W., Micha, J.S. et al. (2011) Impact of instrumental constraints and imperfections on the dislocation structure in micron-sized Cu compression pillars. *Acta Mater.*, **59** (14), 5618–5626.

121 Derlet, P.M. and Van Swygenhoven, H. (2001) The role played by two parallel free surfaces in the deformation mechanism of nanocrystalline metals: a molecular dynamics simulation, in Proceedings, Materials Research Society, Structure and Mechanical Properties of Nanophase Materials Theory and Computer Simulation vs. Experiment. Symposium, November 28–30, 2000, Boston, Massachusetts, USA, vol. 634 (4), 1–6.

122 Yamakov, V., Wolf, D., Phillpot, S.R., and Gleiter, H. (2002) Deformation twinning in nanocrystalline Al by molecular-dynamics simulation. *Acta Mater.*, **50** (20), 5005–5020.

123 Van Swygenhoven, H., Derlet, P.M., and Froseth, A.G. (2006) Nucleation and propagation of dislocations in nanocrystalline fcc metals. *Acta Mater.*, **54** (7), 1975–1983.

124 Espinosa, H.D., Panico, M., Berbenni, S., and Schwarz, K.W. (2006) Discrete dislocation dynamics simulations to interpret plasticity size and surface effects in freestanding FCC thin films. *Int. J. Plast.*, **22** (11), 2091–2117.

125 Anantheshwara, K. and Bobji, M.S. (2010) In situ transmission electron microscope study of single asperity sliding contacts. *Tribol. Int.*, **43** (5-6), 1099–1103.

126 Merkle, A.P., Erdemir, A., Eryilmaz, O.L., Johnson, J.A., and Marks, L.D. (2010) In situ TEM studies of tribo-induced bonding modifications in near-frictionless carbon films. *Carbon*, **48** (3), 587–591.

127 Lockwood, A.J., Wedekind, J., Gay, R.S., Bobji, M.S., Amavasai, B., Howarth, M. et al. (2010) Advanced transmission electron microscope triboprobe with automated closed-loop nanopositioning. *Meas. Sci. Technol.*, **21** (7)

128 Sumigawa, T., Murakami, T., Shishido, T., and Kitamura, T. (2010) Cu/Si interface fracture due to fatigue of copper film in nanometer scale. *Mater. Sci. Eng. A*, **527** (24-25), 6518–6523.

129 Wang, J.J., Lockwood, A.J., Peng, Y., Xu, X., Bobji, M.S., and Inkson, B.J. (2009) The formation of carbon nanostructures by in situ TEM mechanical nanoscale fatigue and fracture of carbon thin films. *Nanotechnology*, **20** (30)

130 Hirakata, H., Takahashi, Y., Van Truong, D., and Kitamura, T. (2007) Role of plasticity on interface-crack initiation from a free edge and propagation in a nano-component. *Int. J. Fract.*, **145**, 261–271.

131 Legros, M., Dehm, G., Balk, T.J., Arzt, E., Bostrom, O., Gergaud, P., Thomas, O., and Kaouache, B. (2003) Plasticity-related phenomena in metallic films on substrates, Proceedings, Materials Research Society Symposium, April 22–24, 2003, San Francisco, CA, USA, vol. 779, W4.2.1-12.

11
In-Situ Nanoindentation in the Transmission Electron Microscope
Andrew M. Minor

11.1
Introduction

11.1.1
The Evolution of *In-Situ* Mechanical Probing in a TEM

In-situ transmission electron microscopy (TEM) provides dynamic observations of the physical behavior of materials in response to external stimuli such as temperature, environment, stress, and applied fields. Since the basic mechanisms that determine a material's deformation behavior occur at nanometer length scales, mechanical testing methods for the electron microscope were among the first *in-situ* approaches to be developed [1, 2]. Starting during the late 1950s, *in-situ* straining stages were developed that led to the dynamic observation of dislocation motion in metals [3]. Subsequently, throughout the past 50 years much further development has been undertaken in the field of *in-situ* TEM mechanical testing [4], including the evolution of mechanical probing techniques such as *in-situ* nanoindentation.

Nanoindentation is perhaps the most widely used testing method for measuring the mechanical properties of materials with dimensions ranging from a few tens of nanometers to a few tens of microns [35]. When considering how to interrogate the mechanical behavior of materials at small length scales, several different methods can be utilized. For example, techniques such as wafer curvature [5, 6] bulge testing [7, 8] micro-beam bending [9], micro-tensile testing [10], micro-compression testing [11], and micro-electromechanical system (MEMS)-actuated tests [12] can each be used to measure the mechanical properties of thin films and small structures that have dimensions ranging from a few nanometers to a few microns. In contrast, nanoindentation addresses small length scales by using an equivalently small probe. A typical nanoindenter has a radius of curvature of 100 nm, which is approximately the same dimension as the thickness of an electron-transparent TEM sample. Thus, nanoindentation can be more broadly classified as a mechanical probing, where a mechanical test is necessarily localized to the dimensions of the probe and its displacement into a material.

In-situ Electron Microscopy: Applications in Physics, Chemistry and Materials Science, First Edition.
Edited by Gerhard Dehm, James M. Howe, and Josef Zweck.
© 2012 Wiley-VCH Verlag GmbH & Co. KGaA. Published 2012 by Wiley-VCH Verlag GmbH & Co. KGaA.

The successful implementation of a mechanical probing experiment inside a TEM dates back to Gane who, in 1970 [13], described the use of a mechanical contact probe inside an E.M.6 microscope. By building on the results of earlier investigations inside a scanning electron microscope [14], Gane used a piezo bimorph actuator to apply loads over a range from 2 μN to 10 mN. In this way, it proved possible to indent a gold surface with a tungsten tip, to blunt an aluminum tip, and even to compress submicron gold crystals. Although no images were recorded during these tests (only before and after), the results represented an extraordinary accomplishment in experimental mechanics.

Following the early successes of Gane, further progress in the development of contact probing holders for the transmission electron microscope was not made until the late 1980s and beyond. For the most part, these holders were used not directly for mechanical testing, but rather for *in-situ* scanning tunneling microscopy (STM) [15–22]. Subsequently, during the late 1990s, further developments led to the creation of holders and sample preparation schemes that allowed for controlled nanoindentations to be made within the transmission electron microscope [23, 24, 52]. Initially, these holders were only qualitative in nature, in that they could be used to perform controlled indentations using a piezoelectric motion, but the recording of quantitative force data proved to be elusive. Later, attempts were made [25, 64] to correlate load–displacement behavior with real-time images of the deformation response during *in-situ* nanoindentation in a transmission electron microscope. However, such attempts relied on the *ex post facto* determination of indenter displacement from sequential TEM images, as well as indirect correlations between the voltages applied to the piezoceramic actuator and the measurements of known bending moments. Unfortunately, this approach suffered from substantial uncertainties caused by nonlinearities in the piezoceramic response, which in turn resulted in inexact data with low temporal and load resolution. It was only in 2006 that a collaborative project between the National Center for Electron Microscopy (NCEM) at the Lawrence Berkeley National Laboratory and Hysitron, Inc. [35] led to a new *in-situ* TEM nanoindentation holder design that included a capacitive load sensor for quantitative force and displacement measurements during *in-situ* indentations [26] (with a force resolution of ∼0.2 μN and displacement resolution of ∼0.5 nm). Subsequently, by correlating the force–displacement response of the material with direct images of the microstructural response of a material in real time, it became possible to study the initial stages of nanoindentation-induced plasticity with high spatial and force resolution [27, 28]. Over the past few years, progression in the development of relevant instrumentation has indeed been rapid, such that today commercial *in-situ* TEM nanoindentation systems are available from Hysitron, Inc. and Nanofactory, among others.

11.1.2
Introduction to Nanoindentation

Indentation testing has, for many years, been used as a convenient method to characterize the hardness of a material, which is classically defined as a material's resistance to plastic deformation [29]. The advent of the microelectronics industry necessitated the development of characterization tools more appropriate to the size and scale of the thin films and small structures associated with microelectronic devices. By the 1980s, significant advances in nanoscale science and instrumentation had led to the first "instrumented" indentation techniques at submicron length scales. Pethica, Hutchings and Oliver [30] were the first to demonstrate a technique by which a continuous measurement of load and displacement could be provided during indentation, with resolutions in the sub-μN and the subnanometer regions. Since then, the technique of nanoscale hardness testing has become broadly referred to as "nanoindentation."

Conventional nanoindentation techniques have been developed over the past 25 years as a method to probe the mechanical properties of materials in the submicron size range [30, 31] that is a typical dimension of the thin films used in integrated circuits and MEMS. In a typical nanoindentation experiment, the sample surface is indented with a sharp diamond pyramid, which has a tip that can be characterized by a radius of curvature on the order of 100 nm. As a consequence, both elastic and plastic deformation occur in the sample, while the displacements imposed during nanoindentation are on the order of nanometers, and high-resolution load and displacement measurements are simultaneously recorded. In this way, it is possible to study atomistic-scale deformation processes such as dislocation nucleation events [32, 33], as well as pressure-induced phase transformations [34]. Although a multitude of behaviors have been investigated with conventional indentation techniques [35], the associated materials phenomena have more often been simulated with computational studies than directly observed using *in-situ* methods. During a typical nanoindentation experiment, the peak stresses occur below the surface, in volumes that are small enough to be often assumed as defect-free, perfect crystal [36]. Clearly, indentation into defect-free volumes cannot be accommodated by conventional plastic flow; rather, the nucleation of defects must occur first.

As an example, Figure 11.1 shows two of the most commonly measured discrete phenomena from a nanoindentation measurement. In Figure 11.1a, the load–displacement data from a 300 nm-thick Al film on Si and from a polished single crystal of Al are shown. The discrete displacement bursts that can be seen in the loading portions of the curves are typically observed during the nanoindentation of metals in the submicron region, and are commonly referred to as either "staircase-yielding" [37] or "pop-ins" [33, 38]. In Figure 11.1b, a conventional nanoindentation curve of single crystal <100> silicon is shown, demonstrating a different type of discrete behavior. In this case, while loading of the sample is smooth, upon unloading a displacement burst is seen in the unloading direction. This discrete behavior is typically associated with the nanoindentation of semiconductors, and is commonly referred to as a "pop-out" [39].

The pop-in phenomenon observed during the loading of a material at shallow depths signifies discrete yielding events, usually due to the nucleation of dislocations. Page and coworkers [40] have used TEM to investigate the post-indent microstructure around indentations where pop-ins had or had not occurred during the nanoindentation of sapphire. Post-indent TEM samples showed dislocation structures around

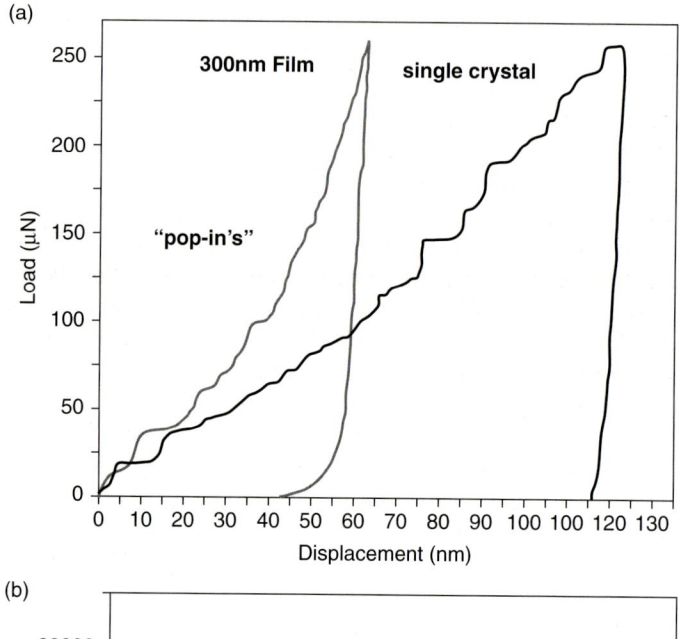

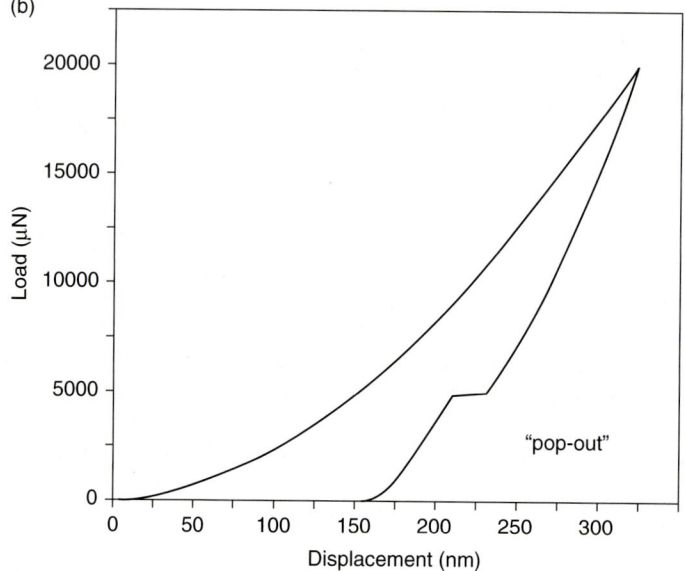

Figure 11.1 (a) Load versus displacement nanoindentation curve from a 300 nm-thick Al film (on Si) and single crystal Al <100>. Note the discrete displacement bursts during the loading portion of the curve; (b) Load versus displacement nanoindentation curve from a single crystal n-type <100> Si wafer. Note the discrete displacement burst on the unloading portion of the curve. See text for more detail.

indentations associated with discrete displacement bursts in the loading behavior, and no dislocations surrounding indents that exhibited superimposed loading and unloading (i.e., no pop-ins). Subsequently, Gerberich and coworkers [33] used atomic force microscopy (AFM) to show that no observable surface deformation resulted from indentations of Fe-3 wt% Si single crystals when there is no pop-in, but considerable surface deformation from indentations that displayed pop-in. Other studies have shown that the stress at which the initial displacement bursts occur are on the same order of the critical stress needed to nucleate dislocations in previously defect-free material [32, 41]. These measurements strongly support the idea that pop-ins are related to the nucleation of dislocations.

However, in some studies of pop-in behavior it has been argued whether or not displacement discontinuities are truly related to nucleation events. Other proposed rationales for such behavior have included oxide fracture or the activation of pre-existing dislocations [42]. Even though the peak stresses applied to a material during indentation occur below the surface [36], the in-plane stresses imposed at the surface are not zero. It has been argued that the presence of a native oxide layer can provide a defect source for the initiation of plasticity [42], or may act as a barrier to dislocations generated within the bulk. Typically, however, the presence of oxide layers is often ignored in analyses of load–displacement behavior measured during indentation.

The pop-out phenomenon observed during nanoindentation in semiconductors has been shown to be associated with phase transformations. Numerous nanoindentation studies of semiconductors such as Si [34, 39, 43, 44], Ge [45], and GaAs [45, 46] have shown pop-out behavior followed by indirect evidence of a phase transformation at the time of the pop-out phenomenon. The most common explanation for pop-outs is that the material undergoes a high-pressure phase transformation upon loading (thus densifying), followed by transformation back to a less-dense phase upon unloading. The transformation to a less-dense phase results in a large volume increase, which in turn causes a sudden displacement burst in the direction of unloading (a pop-out). For example, it is believed that silicon transforms to high-pressure metallic phases (e.g., β-Sn) during loading, as inferred by $in\text{-}situ$ Raman spectroscopy [34] and resistivity measurements [47]. Upon unloading, however, the high-pressure metallic phases are thought to transform to either a lower pressure metallic phase, or to undergo amorphization. Post-mortem TEM analysis has provided evidence of both bcc-R8 and amorphous silicon phases [48], where the product phase is determined by the unloading rate and the depth of indentation.

Conventional nanoindentation experiments have not been able to completely describe the discrete mechanisms of nanoscale plasticity, as direct observations are critical for establishing a one-to-one correlation with mechanisms. In a typical nanoindentation experiment, the contact areas are on the order of about 100 to 1000 nm^2 and, when combined with applied forces on the order of microNewtons to milliNewtons, this can result in stresses on the order of GigaPascals [32, 33, 49]. These stresses inside the small volume of perfect crystal during a nanoindentation experiment approach the ideal strength of a material. Thus, nanoindentation permits a direct probing of the mechanical behavior of a material at its point of elastic instability. Such behavior includes dislocation nucleation [32, 34] and pressure-induced phase transformations [34].

The ideal strength of a material is defined as the stress required to plastically deform an infinite, perfect crystal [50] although, in reality, crystals are neither infinite nor perfect. As a result, most materials deform at stress levels well below their ideal strengths, (e.g., for most metals $\tau_{yield} \approx (5\text{--}10\%)\tau_{ideal}$). This is because a typical solid will relieve an imposed stress through mechanisms of deformation that are activated at stresses much lower than its ideal strength.

While conventional nanoindentation tests are able to measure quantitatively the mechanical behavior of materials, the discrete deformation mechanisms that contribute to the measured behavior are rarely observed directly. Typically, the mode of deformation during a nanoindentation test is only studied *ex situ*, and *ex post facto*. Besides observing nanoindentation inside an electron microscope, *in-situ* observations of a material's deformation behavior can be obtained through indirect techniques such as *in-situ* Raman spectroscopy [34] and *in-situ* electrical resistivity measurements [51]. However, to truly observe the microstructural response of a material during indentation, the deformation must be imaged at a resolution on the same order as the size of the defects created. For this, a transmission electron microscope can provide subnanometer resolution, with an ability to image subsurface phenomena such as the creation of dislocations and the nucleation of phase transformations. The phenomenological interpretation of nanoindentation tests – and, indeed, the mechanical behavior of solids at their elastic limit – is a fundamental area of materials science that has benefitted from direct *in-situ* TEM experimental observations.

11.2
Experimental Methodology

A number of significant hurdles will be encountered when performing a nanoindentation test inside a transmission electron microscope. Among other constraints, it is necessary to:

- Place the laboratory inside the objective lens of a transmission electron microscope (which typically is 2–10 mm in height).
- Only work with electron-transparent samples.
- Maintain the sample in a vacuum environment.
- Irradiate the sample with high-energy electrons during the experiment.

In order to justify dealing with these obstacles, there must exist significant advantages to doing so. The primary motivation for running a nanoindentation experiment inside a transmission electron microscope is to achieve dynamic observations of the nanoindentation-induced deformation behavior. Often, the characterization of a sample "before" a nanoindentation test is either impractical or unobtainable, and an "after" image of an experiment is difficult to interpret without one. Thus, to run a nanoindentation test *in-situ* will allow for the acquisition of both "before", "after," and even "during" images of material deformation to provide a complete picture of a material's deformation response. In addition, there is a major advantage to being able to see what is occurring at small length scales. Often, an ability to address the relevant length scale can be difficult (contacting nanoparticles, small structures or even the surface cleanly), while imaging a test at the same length scale can be extremely advantageous for increasing the fidelity of data and helping with the interpretation of any artifacts.

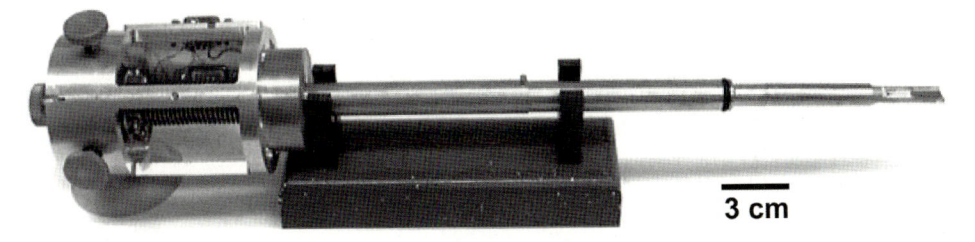

Figure 11.2 *In-situ* nanoindentation stage for the Jeol 200 CX transmission electron microscope. The coarse positioning screws can be seen on the base at left, the diamond tip and sample holder are on the far right end.

The *in-situ* nanoindentation experiments described in this chapter were made possible through the development of a series of sample stages for a transmission electron microscope. Starting with the indentation stage, as developed by Wall and Dahmen at the NCEM in Berkeley, California, in 1997 [52], a number of nanoindentation stages were designed and built for transmission electron microscopes at the NCEM, using similar designs. The initial experiments performed by Wall and Dahmen on Si were accomplished with a stage designed for use in a Kratos high-voltage microscope. The experiments described in this chapter were performed on subsequent stages designed for Jeol 200 kV and 300 kV transmission electron microscopes. The indentation stage used in the Jeol 200 kV microscope (see Figure 11.2) consists of a diamond indenter attached to a metal rod that is actuated by two mechanisms. For coarse positioning, the indenter can be moved in three dimensions by turning screws attached to a pivot at the end of the rod. For fine positioning, including the actual indentation, the indenter is moved in three dimensions with a piezoelectric ceramic crystal, which expands in response to an applied voltage. In order to be electrically conductive within the transmission electron microscope, the diamonds at the end of the rod must be doped with boron and attached with electrically-conductive epoxy. The diamond is mounted on the end of a rod that is, in turn, connected directly to the piezo-ceramic actuator, which controls its position in three dimensions. Further development using the same basic stage design was achieved through a collaborative research project with Hysitron, Inc., where a capacitive force sensor was incorporated behind the diamond indenter [53].

The *in-situ* nanoindentation experiments posed three significant constraints on the required geometry of the samples:

- The region of the sample to be imaged must be thin enough as to be electron-transparent (this constraint is common to all TEM investigations). The critical thickness is dependent on the material and the accelerating voltage of the microscope, but sample thicknesses of 200–300 nm were typically used.
- The electron-transparent part of the sample must be accessible to the diamond indenter in a direction normal to the electron beam (this constraint is unique to *in-situ* probing experiments).
- The sample must be mechanically stable such that indentation, and not bending, results from the indenter pressing on the thin region of the sample.

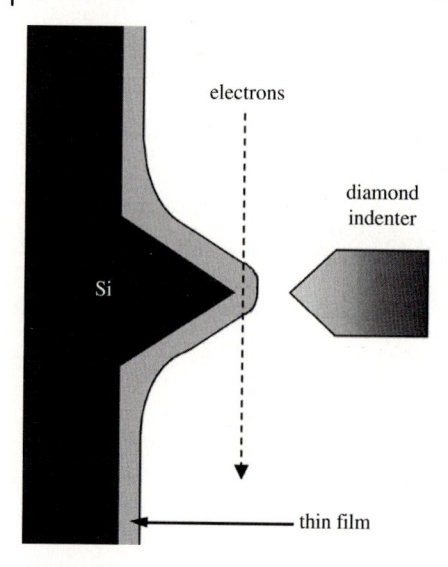

Figure 11.3 Schematic showing the geometric requirements for an *in-situ* nanoindentation sample. The sample must have its electron-transparent portion accessible to the diamond indenter, in a direction normal to the electron beam.

A schematic showing the experimental set-up in cross-section, and demonstrating the constraints on the sample design, is shown in Figure 11.3.

In order to fabricate samples that are electron-transparent, accessible to the diamond indenter, and also mechanically stable, it was necessary to design unique sample geometries for the *in-situ* nanomechanical probing experiments. In general, two different methods were used to meet these sample design constraints: (1) silicon wedge substrates were fabricated with bulk micromachining techniques; and (2) focused ion beam (FIB) -prepared samples were created from bulk materials. In the examples described in this chapter, the silicon wedge samples were either indented directly, or a thin film of Al was deposited on top of the wedge using physical vapor deposition (i.e., the evaporation of Al).

By using proven bulk micromachining and thin-film deposition techniques, the silicon wedge substrates could be produced in large volumes (about 800 per wafer) and with a repeatable quality. The final substrate design resulted in an electron-transparent area that was approximately 1.5 mm long and which was of a robust geometry for indentation. The microfabricated structures could then serve as a substrate for any material that could be deposited onto single crystal silicon in a thin-film form. The wedge structures were achieved using silicon-based lithographic techniques, including an anisotropic wet etch to give the wedges their basic geometry. Potassium hydroxide has been shown to be an extremely anisotropic etchant for Si [54, 55], such that the {100} and {110} planes are etched much faster than the {111} planes, which results in structures with walls defined by {111} planes. The final silicon wedge structure can then be used as a substrate upon which a thin film can be deposited for *in-situ* nanoindentation. The Si plateau on the tip of the wedge can be tailored to provide the correct thickness for the electron transparency of whichever material is deposited on top. In the case of Al, a plateau 150 nm wide with a 250 nm Al film deposited on top proved to be not only electron-transparent but also thick enough to be indented. A cross-section of the sharpest wedge to be fabricated, with a plateau of about 20 nm, is shown in Figure 11.4a, while an example of a wedge left with a thicker plateau of about 150 nm is shown in Figure 11.4b.

11.3
Example Studies

11.3.1
In-Situ TEM Nanoindentation of Silicon

The results obtained over 50 years of research into dislocation behavior in silicon support the conclusion that dislocations do not generally move during conventional mechanical testing at temperatures below 450 °C [56, 57]. Under extreme conditions, such as indentation loading, where localized stresses can approach the theoretical shear strength of the material, dislocation structures have been observed in silicon. Traditionally, these are thought to result from either block slip [58, 59] or phase transformations [25, 43, 48, 60] rather than from the nucleation and propagation mechanisms associated with conventional dislocation plasticity. However, the results of recent studies have shown evidence of room-temperature dislocation plasticity in silicon in the absence of phase transformations through the *post-mortem* TEM of shallow indentations [61].

While indentation-induced dislocation nucleation has been widely documented for metallic systems [62–64], the high Peierls barrier associated with covalently bonded materials tends to suppress dislocation activity. Previous studies have suggested that the indentation of silicon causes plastic deformation through a series of phase transformations[65]. Although, in some studies, dislocations have been observed [61, 66–68] or their presence inferred [69, 70], these are usually associated with the formation of new phases. The presence of dislocations is typically observed within a newly transformed metallic phase, or may be described as a consequence of the phase transformation. To date, however, a conclusive understanding of the dislocation formation has been lacking, due to an inability to directly observe the evolution of deformation.

Direct nanoscale observations of the mechanisms of deformation during the earliest stages of indentation in silicon were made possible through the technique of *in-situ* nanoindentation in the transmission electron microscope [71]. The *in-situ* experiments described here were performed on <100> n-type single crystal silicon samples that were fabricated lithographically (see Section 11.2). The wedge geometry allows the microfabricated silicon samples to provide for both electron-transparency and mechanical stability. Indentations were performed on two different wedge geometries, where the wedge was either terminated by a flat plateau about 150 nm

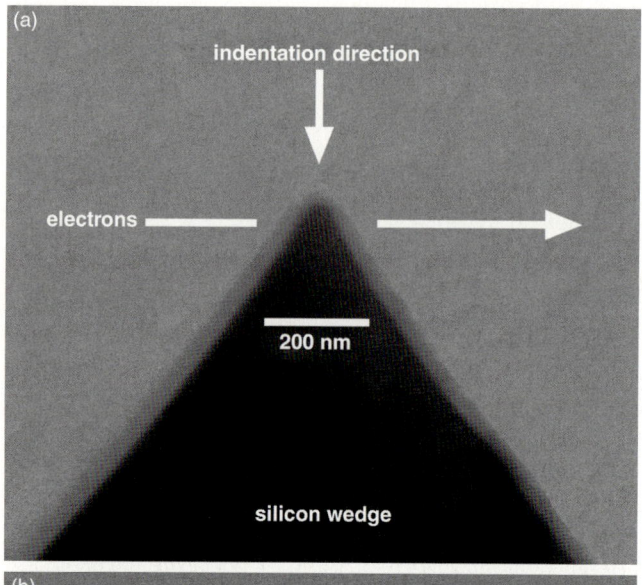

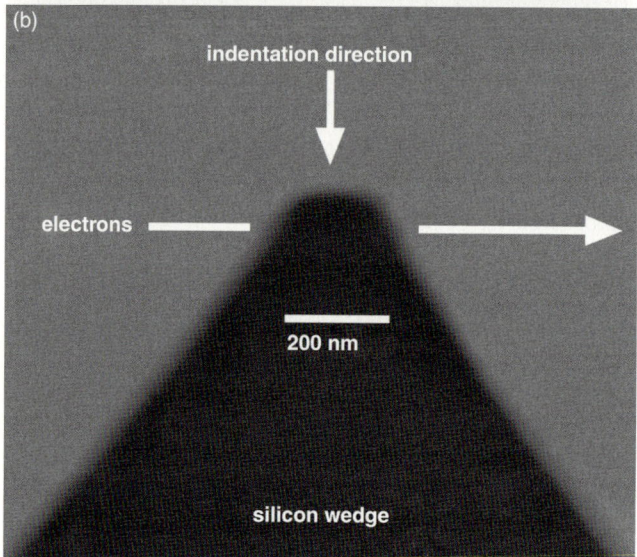

Figure 11.4 Scanning electron microscopy images of lithographically prepared silicon wedge samples in cross-section [71]. (a) The sharp geometry, with a plateau of ∼20 nm; (b) The blunt geometry, with a plateau of ∼150 nm. The cross-sections were prepared with a focused ion beam after a layer of Pt had been deposited for protection.

in width, or sharpened to a plateau width of approximately 20 nm (as shown in Figure 11.4).

In-situ nanoindentation experiments were performed on the silicon wedge samples to peak depths ranging between 50 and 200 nm. Indentations to depths greater than 200 nm were difficult to image *in-situ*, due to the inherent limitations in electron-transparency of the wedge geometry. During *in-situ* indentation the deformation is observed and recorded in real time, and diffraction patterns are taken directly after unloading. Subsequently, plastic deformation was found to proceed through dislocation nucleation and propagation in the diamond cubic lattice. A series of images taken during an indentation into the "blunt" geometry, which had a plateau of about 150 nm at the top of the wedge, is shown in Figure 11.5. The defect-free sample prior to indentation is shown in Figure 11.5a, while Figure 11.5b and c show the evolution of elastic strain contours as the indenter presses into the sample; no evidence of any plastic deformation is seen at this point. The elastic strain contours reveal the shape of the stress distribution in the sample (these are essentially the contours of principal stress). The nucleation and propagation of dislocations from the surface as deformation proceeds are shown clearly in Figure 11.5d and e. This particular indentation was taken to a peak depth of 54 nm, and resulted in the plastic zone shown in Figure 11.5f. The post-indentation selected area diffraction pattern of the indented region is shown in Figure 11.5g. Due to the high density of dislocations after indentation, a slight broadening of the diffraction spots was observed; such broadening was an expected consequence of a high density of dislocations created on multiple slip planes [1]. However, no additional diffraction spots or rings were present after indentation as compared to diffraction patterns taken prior to indentation. This indicated that no additional phases (either crystalline or amorphous) had formed. Figure 11.5h shows a dark-field condition using the $(0\bar{2}2)$ diffracted beam, and illustrates that at least one edge of the plateau is still continuous across the indented region.

In the case of deeper indentations, significant metal-like extrusions were also formed during indention. These large extrusions are shown in Figure 11.6a, which is a post-indentation TEM image of a 220 nm-deep indentation with the corresponding diffraction pattern. This metal-like deformation is clearly seen in Figure 11.6b, which is a plan view micrograph of the indent shown in Figure 11.6a, taken using field emission scanning electron microscopy (FESEM). The results of previous studies [66, 67, 72] have also described metal-like extrusions resulting from indentations into Si although, in all previous experimental cases, these extrusions were attributed to the flow of a transformed metallic phase (e.g., β-Sn or bcc-R8). As shown in Figure 11.6a, the diffraction pattern recorded after the indentation showed the extruded volume to be entirely single-crystal diamond cubic silicon in the same orientation as the remainder of the sample.

In the case of the "sharp" wedge geometry, the deformation behavior during indentation was identical to the above-described case, showing dislocation plasticity during loading. Upon unloading, however, fracture occurred in several of the indentations. Post-indentation TEM images of one such indentation into the sharp wedge geometry are shown in Figure 11.7. In Figure 11.7a, the fracture surface can be seen as smooth and noncrystallographic (within the resolution limit of the bright-field TEM image). During unloading two cracks were observed to nucleate at the surface and to propagate along a strain contour to the center, forming the semicircular fracture profile shown in Figure 11.7a. The dramatic contrast between the

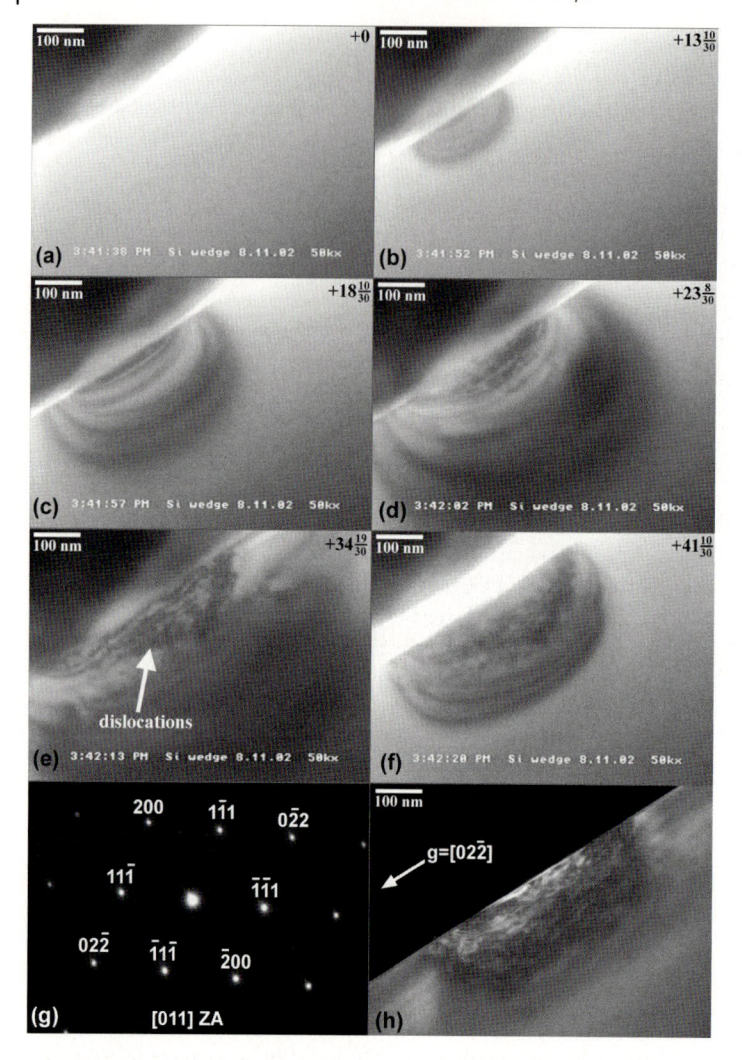

Figure 11.5 (a–f) A time series taken from a video of an *in-situ* nanoindentation into silicon <100> [71]. The diamond indenter is at the top left corner of each frame, and the silicon sample at the lower right. The time in seconds, from the start of the indentation, is shown at the top right corner of each frame. (a) Prior to indentation the silicon sample is defect-free; (b and c) The initial stage of indentation shows elastic strain contours resulting from the pressure applied by the indenter; (d and e) The dislocations can be seen to nucleate, propagate and interact as the indentation proceeds; (f) After a peak depth of 54 nm, the indenter is withdrawn, when the residual deformed region consists of dislocations and strain contours that are frozen in the sample; (g) A [011] zone axis electron diffraction pattern of the indented region directly after the *in-situ* indentation. With the exception of slight peak broadening, the diffraction pattern is identical to similar patterns taken prior to indentation, showing only single-crystal diamond cubic silicon with no additional phases. The tails seen on the diffraction spots pointing in the <200> direction are a geometric effect due to the wedge geometry, which exhibits a drastic change in thickness over a very short distance; (h) The same indentation in a $g = (02\bar{2})$ dark-field condition. Note the continuous surface across the indented region, indicating that the indentation left at least one side of the wedge intact.

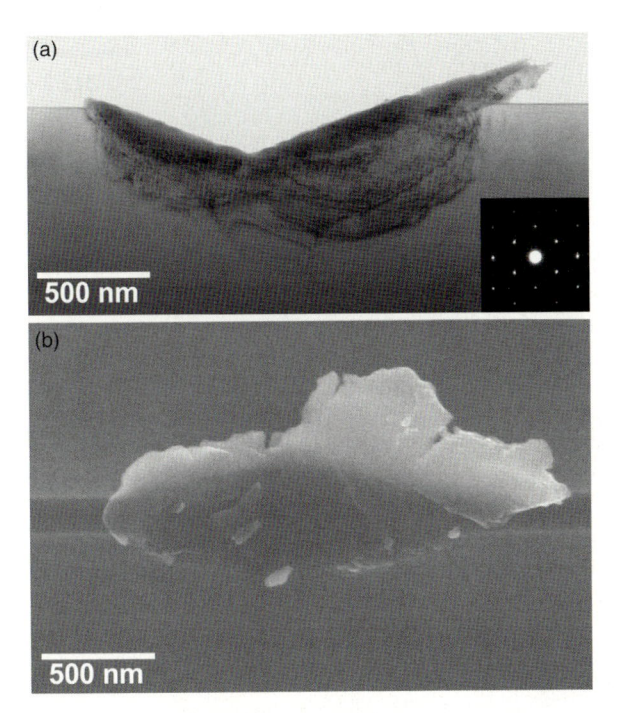

Figure 11.6 (a) Bright-field TEM image of an *in-situ* indentation into <100> silicon that was taken to a depth of 220 nm [71]. The image was taken in a kinematic condition in order to show the undeformed region surrounding the indented volume, which was heavily deformed through dislocation plasticity. The [110] zone axis electron diffraction pattern (inset) was taken after indentation, and shows only the presence of the diamond cubic phase of silicon. This diffraction pattern is identical to diffraction patterns taken before indentation, except for a slight broadening of the diffraction spots due to the heavily dislocated region of the indentation; (b) A plan-view scanning electron micrograph of the same indentation. Relatively large, metal-like extrusions can be seen surrounding the indentation. These extrusions can also be seen above the indented volume in panel (a) and, as shown by the inset diffraction pattern, are extrusions of single-crystal diamond cubic silicon.

crystallographic nature of the dislocations nucleated during indentation and the noncrystallographic nature of the fracture surface is shown in Figure 11.7b.

Previously, it has been shown that stresses near the ideal shear strength are achieved during indentation into defect-free *metal* systems, resulting in the nucleation of dislocations [33, 40]. However, the critical shear stress for dislocation nucleation in metallic materials is far less than that for silicon, owing to the high Peierls–Nabarro barrier of the latter (which is directly related to the strong covalent bonding in silicon). During conventional indentation experiments in silicon, it is generally believed that deformation is accommodated by phase transformation and/or fracture, rather than dislocation plasticity. The phase transformations that occur during indentation have also been observed during diamond anvil cell experiments, where high hydrostatic pressures are applied [73–76]. It follows that, during conventional indentation experiments in silicon, the critical pressure for phase

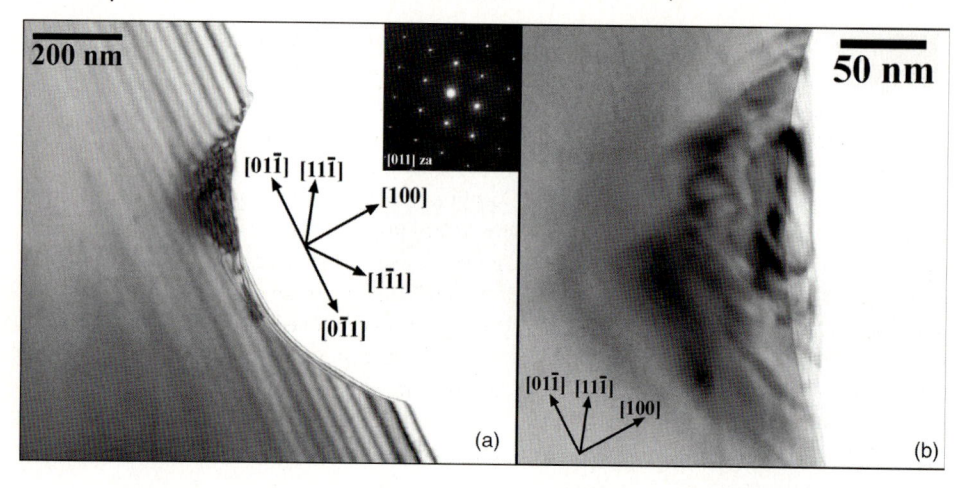

Figure 11.7 (a) Bright-field TEM image after an indentation of the sharp wedge geometry. The fracture surface is the semicircular border between the sample (on the left) and the vacuum (on the right). The image was taken in the [011] zone axis condition with the corresponding diffraction pattern inset. The stripes along the length of the sample are thickness contours resulting from the wedge-shaped geometry of the specimen; (b) A higher-magnification image of the region, showing residual dislocations adjacent to the fracture surface. The smooth fracture surface indicates a noncrystallographic fracture. The image was taken in the $g = [3\bar{1}1]$ condition [71].

transformation is reached before the shear stress reaches the critical value for dislocation nucleation. Presumably, this difference arises from the difference in sample geometry used in conventional instrumented nanoindentation tests and the *in-situ* nanoindentation tests described here.

Conventional indentation experiments are conducted on samples that have been well described as a flat half-space (e.g., a silicon wafer). In the case of the silicon wedge sample geometry, the sides of the wedge are of <111> orientation, making an angle of 70.5° to each other, and are separated by a flat plateau of <100> orientation. The plateau width, w, the wedge height, h, and the length of the wedge, L, have the relationship: $L \gg h \gg w$, such that the geometry is better described as a plane stress configuration than the plane strain configuration characterizing conventional indentation experiments. A plane stress configuration is defined by a lower hydrostatic component than a plane strain configuration. In the case of *in-situ* experiments, an increase in the ratio of maximum shear stress to hydrostatic pressure might allow the shear stress to reach the critical value for dislocation nucleation before the hydrostatic stress reaches the critical value for phase transformation. Additionally, the presence of the surfaces near the contact area (and thus the plastic zone) in the *in-situ* wedge samples would, presumably, make dislocation nucleation easier than in the half-space geometry. As the surfaces would provide a easy nucleation source, it is conceivable that the induced deformation can be completely accommodated by dislocation plasticity, such that the hydrostatic stresses required to nucleate phase transformations are never realized, although this effect requires further investigation. Most importantly, efforts to control the loading and unloading rate should be undertaken, since the loading rate has been shown to make a significant difference in the deformation pathway in silicon.

Fracture in silicon is widely observed in indentation experiments, and also in most mechanical tests. In the case of *in-situ* experiments, fracture events often appear to be a direct result of the presence of dislocations. In particular, a plastic zone characterized by indentation-induced dislocations can lead to a residual stress field upon unloading, resulting in fracture. As described by Lawn [58], a residual tensile field arises during unloading due to the accommodation of the plastically deformed volume and the surrounding elastic volume; this field is comprised of high-tensile stresses that drive fracture. Surprisingly, the path of crack propagation is noncrystallographic; indeed, rather than cleaving along the lowest energy planes, the crack will follow a strain contour associated with the residual stress field. This is energetically favorable if the stress along the strain contour is on the order of the theoretical cohesive strength of silicon.

11.3.2
In-Situ TEM Nanoindentation of Al Thin Films

The mechanical behavior of Al thin films has important relevance to the microelectronics industry, where Al metallization is used throughout a typical microchip. As the grain size of a metallic thin film typically scales with the thickness of that film, a variety of grain sizes can be studied by controlling the thickness of the film. The deformation behavior of submicron Al films is discussed in the following subsections.

In the case of the polycrystalline Al films described here, the films were deposited on silicon wedge substrates by evaporating 99.99% pure Al at 300 °C. Subsequently, a film thickness of approximately 250 nm gave the maximum thickness while maintaining electron transparency at 200 kV. It is preferable to use the maximum possible thickness that is still electron-transparent, as a thicker sample is less likely to bend during indentation. During *in-situ* nanoindentation of the Al films, the three-sided diamond pyramid indenter approaches the sample in a direction normal to the electron beam (Figure 11.8). The indentation is made in the cap of film on the flat top of the wedge, as shown in Figure 11.8.

The interpretation of conventional nanoindentation data is not always clear. For example, as most metals form native oxides, yielding under the nanoindenter may be governed by the fracture of an oxide film rather than the onset of plastic deformation in the material itself [42]. It is difficult to resolve these mechanisms *ex situ*, as the mechanisms associated with yielding may be only indirectly elucidated from quantitative load versus displacement behavior. Thus, one of the most significant advantages of performing *in-situ* nanoindentation within the transmission electron microscope is an ability to record the deformation mechanisms in real time, thereby avoiding the possible creation of artifacts from post-indentation sample preparation. However, as the Peierls barrier in Al is extremely low – and consequently the dislocation velocity is very fast – a video sampling rate of 30 frames per second

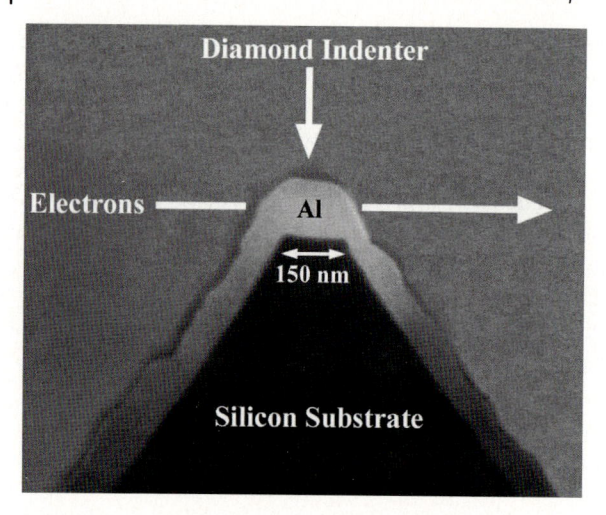

Figure 11.8 Cross-section of an *in-situ* nanoindentation sample. An Al film approximately 250 nm thick is deposited onto a wedge-shaped perturbation on a Si substrate [64].

would be too slow to capture the movements of individual dislocations. Hence, each video image captured during the *in-situ* experiments presented here is essentially a quasi-static image of the equilibrium configuration of defects. This point is illustrated in Figure 11.9, which shows a series of six images taken from a video during an *in-situ* nanoindentation experiment. In Figure 11.9a, the diamond is approaching an Al grain that is approximately 400 nm in diameter, while Figure 11.9b and c show images of the evolution of the induced stress contours during the initial stage of indentation, and correspond to purely elastic deformation in the absence of any pre-existing dislocations that could cause plasticity. The first indication of plastic deformation appears in Figure 11.9d, where the dislocations are nucleated and a set of prismatic loops is observed. Figure 11.9c and d represent consecutive frames of the video, and were recorded 1/30th of a second apart. As can be seen, the exact location of the nucleation event is not discernible, as the evolution of the dislocation configuration has already proceeded beyond the point at which that might be possible. The large increase in dislocation density that is achieved as deformation proceeds, and the dislocations tangle and multiply, are shown in Figure 11.9e and f.

The peak stresses (shear and hydrostatic) applied to a material during indentation occur below the surface [36]. Thus, elastic contact theory suggests that nucleation should occur from the bulk, below the indenter/sample interface. However, the in-plane stresses imposed at the surface are not zero. During a nanoindentation test into an initially defect-free solid volume, the initiation of defects necessary to accommodate the induced deformation can occur at a number of locations, including the surface, the interface of the native oxide layer with the bulk material, or the interior of the grain [42]. This competition for the location of the initial defect nucleation will depend on numerous material and indentation parameters, including the surface properties of the material and the resulting defects. Although the precise nucleation

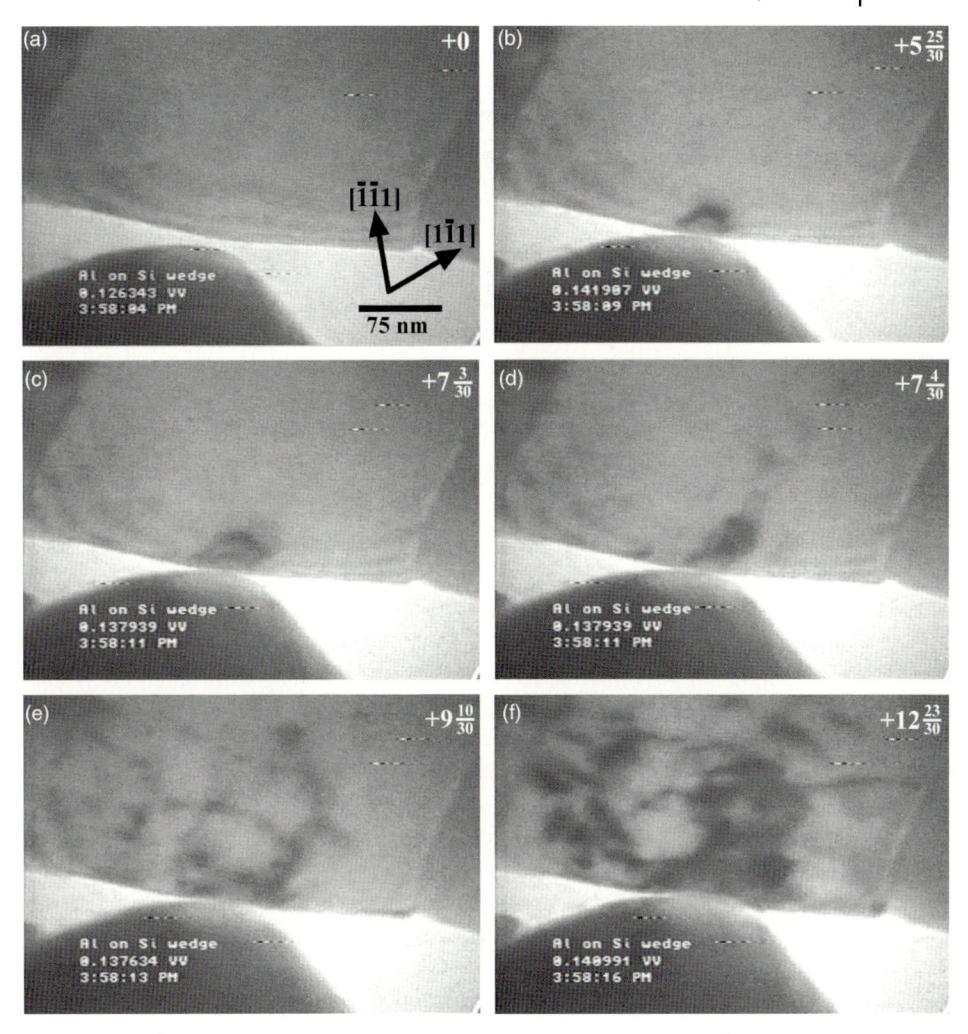

Figure 11.9 Time series of an Al grain showing the evolution of plastic deformation during an *in-situ* nanoindentation in the $[1\bar{1}1]$ direction [25]. The time elapsed from image (a) is given in seconds in the upper right corner of each frame; (b and c) Elastic deformation only; (d) The nucleation of dislocations is visible; (e and f) Characteristic resulting plastic deformation during deeper penetration and the pile-up of the dislocations at the grain boundaries and the substrate/film interface.

site may be difficult – if not impossible – to establish from such experiments, *in-situ* indentation experiments provide unique advantages over *ex-situ* TEM analyses of post-indent dislocation configurations. In addition to the difficulty in preparing TEM samples when indentations have been conducted, the strong image forces exerted on the dislocations by the free surface can lead to very different structures after the sample has been unloaded. Furthermore, being able to directly correlate the microstructural evolution in a material with the quantitative force responses can lead

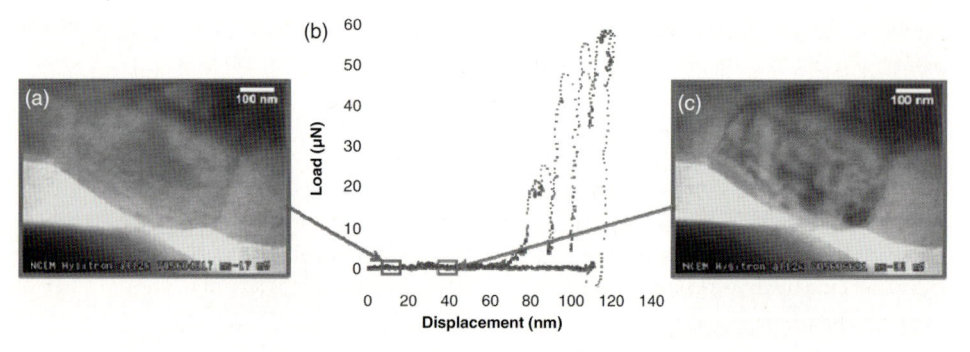

Figure 11.10 Quantitative *in-situ* TEM nanoindentation of an initially dislocation-free submicron Al grain using a Berkovich conductive diamond indenter [73]. (a) The diamond indenter approaches the defect-free Al grain from the bottom of the video frame; (b) Displacement-controlled force versus displacement curve; (c) The microstructure change resulting from the first dislocation burst, occurring before sustained contact that is denoted by the large increases in load that occur around 70 nm displacement in panel (b).

to valuable insight for the interpretation of *ex-situ* nanoindentation experiments. For instance, the results of a recent quantitative *in-situ* nanoindentation study of defect generation in Al thin films showed that the initiation of dislocation plasticity can occur before sustained contact loading, at force levels that are barely discernible during instrumented nanoindentation [73]. In fact, the stresses induced in the small load excursions were near the ideal strength of the material, enough to nucleate dislocations in a previously defect-free volume. In this study, the correlation of the *in-situ* video and the load versus displacement curve showed that the initiation of defects occurred prior to the first sustained load drop that would normally be interpreted as the yield point in a displacement-controlled experiment (Figure 11.10) [73]. The load drops normally associated with initial yielding in a displacement-controlled test actually occurred when the grain had a dislocation density of $10^{14}\,\mathrm{m}^{-2}$. Fundamentally, this demonstrated that yielding is a more complicated phenomenon than simply the nucleation of defects at near-ideal strength.

11.4
Conclusions

In this chapter, the evolution of *in-situ* nanoindentation, a description of the experimental methodology and examples from two different classes of materials – bulk single crystals and Al thin films – have been presented. Direct observations of each material system during indentation have provided a unique insight into the interpretation of *ex-situ* nanoindentation tests, as well as to the intrinsic mechanical behavior of nanoscale volumes of solids. The *in-situ* nanoindentation results presented in Section 3.1 have provided the first direct evidence of dislocation nucleation in single crystal silicon at room temperature. In contrast to the observation of phase transformations during conventional indentation experiments, the unique geometry employed for the *in-situ* experiments resulted in dislocation plasticity. The wedge geometry leads to a higher ratio of shear stress to hydrostatic pressure than that associated with conventional nanoindentation experiments, and may therefore lead to a change in the mode of plastic deformation. These observations have provided a crucial new insight into the failure of silicon-based microelectronic devices since, if silicon is subjected to high stresses, where the ratio of shear stress to hydrostatic pressure is large, then room-temperature nucleation and motion of dislocations is indeed possible. These observations suggest that room-temperature dislocation plasticity could be an important deformation mode even in ceramics, given sufficiently small volumes.

By using the *in-situ* indentation technique, details of indentations into Al thin films were presented in Section 3.2. These data included real-time observations of dislocation nucleation during the nanoindentation of an initially defect-free Al grain. Plastic deformation in the Al proceeded through the formation and propagation of prismatic loops punched into the material, and half-loops that emanated from the sample surface. In addition, quantitative tests showed that extensive yielding occurred before the onset of elastic loading, and that stresses close to the ideal strength of the material can be achieved, even in a sample with a high density of defects.

In addition to studying the characteristics of nanoindentation testing *in-situ*, nanomechanical probing inside the transmission electron microscope also represents a flexible experimental platform to study deformation in many different types of small volumes. Images from selected experimental techniques derived from *in-situ* nanoindentation, where nanoscale volumes can be deformed with the same holder using different shaped tips or sample geometries, are shown in Figure 11.11. For instance, with careful sample design, individual nanoparticles can be compressed with either a nanoindenter or a flat punch (Figure 11.11a) [77, 78]. In addition, by employing a FIB-machined tip, quantitative bending experiments can be conducted to accurately determine the elastic properties of a single nanowire (Figure 11.11b) [79]. Finally, FIB-structured nanopillars can be compressed in

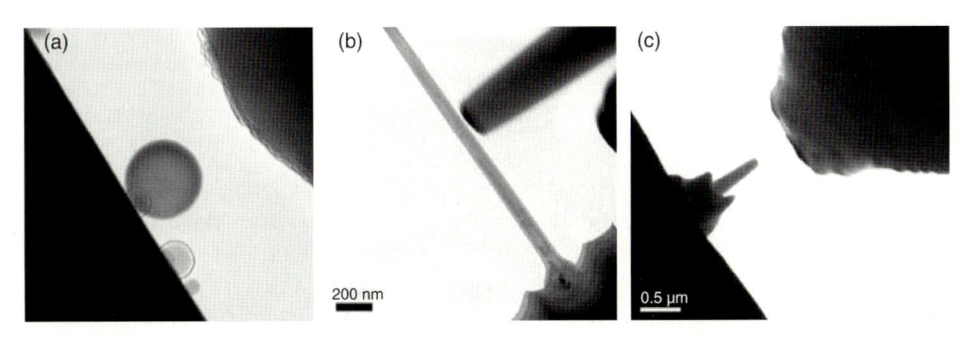

Figure 11.11 Variants of *in-situ* nanomechanical probing experiments. In all cases, the probe is approaching the sample from the upper right corner of the image. (a) Compression of a Si nanosphere; (b) Bending of a Si nanowire; (c) Nanocompression of the FIB-machined pillar with a flat punch.

the transmission electron microscope, leading to new insights into mechanical size effects and deformation in small volumes (Figure 11.11c) [80–83].

The experimental technique of *in-situ* nanoindentation in the transmission electron microscope has been shown to provide a unique capability for investigating the nanomechanical behavior of small solid volumes. Moreover, through quantitative *in-situ* testing, a direct correlation can be made between the microstructural evolution in a material and the load versus displacement signal generated by the indentation system. Such capability is essential in order to fully understand the mechanisms associated with indentation phenomena and the fundamental deformation behavior of materials.

Acknowledgments

These studies were performed at NCEM, which is supported by the Scientific User Facilities Division of the Office of Basic Energy Sciences, U.S. Department of Energy under Contract # DE-AC02-05CH11231. The author would also like to thank all of his collaborators who contributed to the content of this chapter, especially J.W. Morris, Jr, E.A. Stach, E.T. Lilleodden, Z.W. Shan, S.A. Syed Asif, and O.L. Warren.

References

1 Hirsch, P., Howie, A., Nicholson, R.B., Pashley, D.W., and Whelan, M.J. (1965) *Electron Microscopy of Thin Crystals*, Krieger Publishing Co., NY, NY.
2 Butler, E.P. (1979) *Rep. Prog. Phys.*, **42**, 833.
3 Wilsdorf, H.G.F. (1958) *Rev. Sci. Inst.*, **29**, 323.
4 Robertson, I., Ferreira, P., Dehm, G., Hull, R., and Stach, E.A. (2008) Visualizing the behavior of dislocations-seeing is believing. *MRS Bull.*, **33**, 122–131.
5 Flinn, P.A., Gardner, D.S., and Nix, W.D. (1987) Measurement and interpretation of stress in aluminum-based metallization as a function of thermal history. *IEEE Trans. Electron Devices*, **34**, 689.
6 Keller, R., Baker, S.P., and Artz, E. (1998) Quantitative analysis of strengthening mechanisms in thin Cu films: effects of film thickness, grain size, and passivation. *J. Mater. Res.*, **13**, 1307.
7 Javaraman, S., Edwards, R.L., and Hemker, K.J. (1998) Determination of the Mechanical Properties of Polysilicon thin films using Bulge Testing (eds R.C. Cammarata, M. Nastasi, E.P. Busso, and W.C. Oliver) *Thin-Films -Stresses and Mechanical Properties VII. Symposium, MRS Fall Meeting, Boston, MA, USA, 1–5 December 1997*. Materials Research Society, Warrendale, PA, USA, pp. 623–628.
8 Paviot, V.M, Vlassak, J.J., and Nix, W.D. (1995) Measuring the Mechanical Properties of thin Metal Films by Means of Bulge Testing of Micromachined Windows (eds S.P. Baker, C.A. Ross, P.H. Townsend, C.A. Volkert, and P. Borgesen) *Thin Films: Stresses and Mechanical Properties V. Symposium, MRS Fall Meeting, Boston, MA, USA, 28 November–2 December 1994*. Materials Research Society, Pittsburgh, PA, USA, pp. 579–584.
9 Weihs, T.P., Hong, S., Bravman, J.C., and Nix, W.D. (1988) Mechanical deflection of cantilever microbeams: A new technique for testing the mechanical properties of thin films. *J. Mater. Res.*, **3**, 931.

10 Hemker, K.J. and Last, H. (2001) Microsample tensile testing of LIGA nickel for MEMS applications. *Materials Science & Engineering A; Structural Materials: Properties, Microstructure and Processing*, Vol. A319–321, 12th International Conference on the Strength of Materials. ICSMA-12, Asilomar, CA, USA, 27 August–1 September 2000. Elsevier, pp. 882–886.

11 Uchic, M.D., Dimiduk, D.M., Florando, J.N., and Nix, W.D. (2004) *Science*, **305**, 986.

12 Muhlstein, C.L., Stach, E.A., and Ritchie, R.O. (2002) Mechanism of fatigue in micron-scale films of polycrystalline silicon for microelectromechanical systems. *Appl. Phys. Lett.*, **80**, 1532–1534.

13 Gane, N. (1970) *Proc. R. Soc. London, Ser. A*, **317**, 367.

14 Gane, N. and Bowden, F.P. (1968) *J. Appl. Phys.*, **39**, 1432.

15 Spence, J.C.H. (1988) *Ultramicroscopy*, **25**, 165.

16 Spence, J.C.H., Lo, W., and Kuwabara, M. (1990) *Ultramicroscopy*, **33**, 69.

17 Kuwabara, M., Lo, W., and Spence, J.C.H. (1989) *J. Vac. Sci. Technol. A*, **7**, 2745.

18 Lutwyche, M.I. and Wada, Y. (1995) *Appl. Phys. Lett.*, **66**, 2807.

19 Naitoh, Y., Takayanagi, K., and Tomitori, M. (1996) *Surf. Sci.*, **357**, 208.

20 Kizuka, T., Yamada, K., Deguchi, S., Naruse, M., and Tanaka, N. (1997) *J. Electron Microsc.*, **46**, 151.

21 Ohnishi, H., Kondo, Y., and Takayanagi, K. (1998) *Nature*, **395**, 780.

22 Svensson, K., Jompol, Y., Olin, H., and Olsson, E. (2003) *Rev. Sci. Instrum.*, **74**, 4945.

23 Stach, E.A., Freeman, T., Minor, A.M., Owen, D.K., Cumings, J., Wall, M.A., Chraska, T., Hull, R., Morris, J.W. Jr, Zettl, A., and Dahmen, U. (2001) Development of a nanoindenter for *in-situ* transmission electron microscopy. *Microsc. Microanal.*, **7** (6)

24 Bobji, M.S., Ramanujan, C.S., Pethica, J.B., and Inkson, B.J. (2006) *Meas. Sci. Technol.*, **17**, 1324.

25 Minor, A.M., Lilleodden, E.T., Stach, E.A., and Morris, J.W. (Oct 2002) *J. Electron. Mater.*, **31**, 958.

26 Minor, A.M., Syed Asif, S.A., Shan, Z.W., Stach, E.A., Cyrankowski, E., Wyrobek, T.J., and Warren, O.L. (2006) A new view of the onset of plasticity during the nanoindentation of aluminum. *Nat. Mater.*, **5**, 697–702.

27 Soer, W.A., DeHosson, J.T.H.M., Minor, A.M., Shan, Z., Syed Asif, S.A., and Warren, O.L. (2007) Indentation-induced plasticity in Al and Al-Mg thin films. *Appl. Phys. Lett.*, **90**, 181924–181933.

28 Sun, Y., Ye, J., Shan, Z., Minor, A.M., and Balk, T.J. (2007) Mechanical behavior of nanoporous gold thin films. *JOM*, **59** (9), 54.

29 Tabor, D. (1951) *Hardness of Metals*, Clarendon Press, Oxford, United Kingdom.

30 Pethica, J.B., Hutchings, R., and Oliver, W.C. (1983) Hardness measurement at penetration depths as small as 20nm. *Philos. Mag. A*, **48**, 593.

31 Oliver, W.C. and Pharr, G.M. (1992) An improved technique for determining hardness and elastic modulus using load and displacement sensing indentation experiments. *J. Mater. Res.*, **7**, 1564.

32 Gouldstone, A., Koh, H.-J., Zeng, K.-Y., Giannakopoulos, A.E., and Suresh, S. (2000) Discrete and continuous deformation during nanoindentation of thin films. *Acta Mater.*, **48**, 2277.

33 Gerberich, W.W., Nelson, J.C., Lilleodden, E.T., Anderson, P., and Wyrobek, J.T. (1996) Indentation induced dislocation nucleation: the initial yield point. *Acta Mater.*, **44**, 3585.

34 Domnich, V., Gogotsi, Y., and Dub, S. (2000) Effect of phase transformations on the shape of the unloading curve in the nanoindentation of silicon. *Appl. Phys. Lett.*, **76**, 2214–2216.

35 Gouldstone, A., Chollacoop, N., Dao, M., Li, J., Minor, A.M., and Shen, Y.L. (2007) Indentation across size scales and disciplines: Recent developments in experimentation and modeling. *Acta Mater.*, **55**, 4015–4039.

36 Johnson, K.L. (1987) *Contact Mechanics*, 2nd edn, Cambridge University Press, Cambridge.

37 Corcoran, S.G., Colton, R.J., Lilleodden, E.T., and Gerberich, W.W.

(1997) Nanoindentation Studies of Yield Point Phenomena on Gold Single Crystals (eds W.W. Gerberich, H. Gao, J.-E. Sundgren, and S.P. Baker) *Thin Films: Stresses and Mechanical Properties VI. Symposium, MRS Spring Meeting, San Francisco, CA, USA, 8–12 April 1996*. Materials Research Society, Pittsburgh, PA, USA, pp. 159–164.

38 Tangyunyong, P., Thomas, R.C., Houston, J.E., Michalske, T.A., Crooks, R.M., and Howard, A.J. (1993) Nanometer-scale mechanics of gold films. *Phys. Rev. Lett.*, **71**, 3319.

39 Domnich, V. and Gogotsi, Y. (2002) Phase Transformations in silicon under contact loading. *Rev. Adv. Mater. Sci.*, **3**, 1–36.

40 Page, T.F., Oliver, W.C., and McHargue, C.J. (1992) The deformation behavior of ceramic crystals subjected to very low load (nano)indentations. *J. Mater. Res.*, **7**, 450–473.

41 Pethica, J.B. and Oliver, W.C. (1988) Mechanical properties of nanometer volumes of material: use of the elastic response of small area indentations. Proceedings of the Materials Research Society Fall Meeting, Boston, MA, November 28–December 2, 1988, p. 13.

42 Gerberich, W.W., Kramer, D.E., Tymiak, N.I., Volinsky, A.A., Bahr, D.F., and Kriese, M.D. (1999) Nanoindentation-induced defect-interface interactions: phenomena, methods and limitations. *Acta Mater.*, **47**, 4115.

43 Pharr, G.M., Oliver, W.C., and Harding, D.S. (1991) *J. Mater. Res.*, **6**, 1129.

44 Weppelmann, E.R., Field, J.S., and Swain, M.V. (1993) *J. Mater. Res.*, **8**, 830.

45 Gogotsi, Y.G., Domnich, V., Dub, S.N., Kailer, A., and Nickel, K.G. (2000) Cyclic nanoindentation and Raman microspectroscopy study of phase transformations in semiconductors. *J. Mater. Res.*, **15**, 871–879.

46 Li, Z.C., Liu, L., Wu, X., He, L.L., and Xu, Y.B. (2002) Indentation induced amorphization in gallium arsenide. *Mater. Sci. Eng. A*, **337**, 21–24.

47 Bradby, J.E., Williams, J.S., and Swain, M.V. (2003) *In situ* electrical characterization of phase transformations in Si during indentation. *Phys. Rev. B.*, **67**, 085205.

48 Mann, A.B., van Heerden, D., Pethica, J.B., and Weihs, T.P. (2000) Size-dependent phase transformations during point loading of silicon. *J. Mater. Res.*, **15**, 1754–1758.

49 Krenn, C.R., Roundy, D., Cohen, M.L., Chrzan, D.C., and Morris, J.W. (2002) Connecting atomistic and experimental estimates of ideal strength. *Phys. Rev. B*, **65**, 134111/1-4.

50 Hirth, J.P. and Lothe, J. (1968) *Theory of Dislocations*, McGraw-Hill, New York.

51 Ruffell, S., Bradby, J.E., Williams, J.S., and Warren, O.L. (2007) An *in-situ* electrical measurement technique via a conducting diamond tip for nanoindentation in silicon. *J. Mater. Res.*, **22** (3), 578–586.

52 Wall, M. and Dahmen, U. (1997) *Microsc. Microanal.*, **3**, 593.

53 Warren, O.L., Shan, Z., Syed Asif, S.A., Stach, E.A., Morris, J.W. Jr, and Minor, A.M. (2007) *In situ* TEM nanoindentation. *Mater. Today*, **10** (4), 59–60.

54 Seidel, H., Csepregi, L., Heuberger, A., and Baumgartel, H. (1990) Anisotropic etching of crystalline silicon in alkaline solutions. *J. Electrochem. Soc.*, **137**, 3612–3632.

55 Fruhauf, J. and Hannemann, B. (2000) Wet etching of undercut sidewalls in {001}-silicon. *Sens. Actuators*, **79**, 55–63.

56 Alexander, H. and Haasen, P. (1968) *Solid State Phys.*, **22**, 27.

57 Sumino, K. (1994) *Handbook on Semiconductors*, Elsevier, Scientific Press, New York, pp. 73–181.

58 Lawn, B.R. (1993) *Fracture of Brittle Solids*, 2nd edn, Cambridge University Press, New York, pp. 249–306.

59 Hill, M.J. and Rowcliffe, D.J. (1974) *J. Mater. Sci.*, **9**, 1569.

60 Weppelmann, E.R., Field, J.S., and Swain, M.V. (1993) *J. Mater. Res.*, **8**, 830.

61 Saka, H., Shimatani, A., Suganuma, M., and Suprijadi (2002) *Philos. Mag.*, **82** (10), 1971–1981.

62 Page, T.F., Oliver, W.C., and McHargue, C.J. (1992) *J. Mater. Res.*, **7**, 450.

63. Gerberich, W.W., Nelson, J.C., Lilleodden, E.T., Anderson, P., and Wyrobek, P.J.T. (1996) *Acta Mater.*, **44**, 3585.
64. Minor, A.M., Morris, J.W. Jr, and Stach, E.A. (2001) *Appl. Phys. Let.*, **79**, 1625.
65. Clarke, D.R., Kroll, M.C., Kirchner, P.D., Cook, R.F., and Hockey, B.J. (1988) *Phys. Rev. Lett.*, **60**, 2156.
66. Bradby, J.E., Williams, J.S., Wong-Leung, J., Swain, M.V., and Munroe, P. (2000) *Appl. Phys. Lett.*, **77**, 3749.
67. Bradby, J.E., Williams, J.S., Wong-Leung, J., Swain, M.V., and Munroe, P. (2001) *J. Mater. Res.*, **16**, 1500.
68. Tachi, M., Suprijadi, S., Arai, S., and Saka, H. (2002) *Philos. Mag. Lett.*, **82**, 133.
69. Armstrong, R.W., Ruff, A.W., and Shin, H. (1996) *Mater. Sci. Eng. A*, **209**, 91.
70. Li, X., Diao, D., and Bhushan, B. (1997) *Acta Mater.*, **45**, 4453.
71. Minor, A.M. et al. (2005) Room temperature dislocation plasticity in silicon. *Philos. Mag.*, **85**, 323–330.
72. Perez, R., Payne, M.C., and Simpson, A.D. (1995) *Phys. Rev. Lett.*, **75**, 4748.
73. Wentorf, R.H. and Kasper, J.S. (1963) Two new forms of silicon. *Science*, **139**, 338–339.
74. Jamieson, J.C. (1963) Crystal structures at high pressures of metallic modifications of silicon and germanium. *Science*, **139**, 762–764.
75. Pilz, R.O., Maclean, J.R., Clark, S.J., Ackland, G.J., Hatton, P.D., and Crain, J. (1995) Structure and properties of silicon XII: A complex tetrahedrally bonded phase. *Phys. Rev. B*, **52**, 4072–4085.
76. Pfrommer, B.G., Cote, M., Louie, S.G., and Cohen, M.L. (1997) Ab initio study of silicon in the R8 phase. *Phys. Rev. B*, **56**, 6662–6668.
77. Deneen, J., Mook, W.M., Minor, A.M., Gerberich, W.W., and Carter, C.B. (2007) Fracturing a nanoparticle. *Philos. Mag.*, **87** (1), 29–37.
78. Shan, Z.W., Adesso, G., Cabot, A., Sherburne, M.P., Syed Asif, S.A., Warren, O.L., Chrzan, D.C., Minor, A.M., and Alivisatos, A.P. (2008) Ultrahigh stress and strain in hierarchically structured hollow nanoparticles. *Nat. Mater.*, **7**, 947–952.
79. Shan, Z.W., Cao, L.L., Gao, D., Minor, A.M., Asif, S.A.S., and Warren, O.L. (2007) Quantitatively exploring the mechanical behavior of Si nanowires inside a TEM. *Microsc. Microanal.*, **13** (2), 74–75.
80. Shan, Z.W., Mishra, R.K., Syed Asif, S.A., Warren, O.L., and Minor, A.M. (2007) Mechanical annealing in submicron-diameter Ni crystals. *Nat. Mater.*, **7**, 115–119.
81. Shan, Z.W., Li, J., Cheng, Y.Q., Minor, A.M., Syed Asif, S.A., Warren, O.L., and Ma, E. (2008) Plastic flow and failure resistance of metallic glass Insight from *in-situ* compression of nanopillars. *Phys. Rev. B*, **77**, 155419–155426.
82. Ye, J., Mishra, R.K., and Minor, A.M. (2008) Relating nanoscale plasticity to bulk ductility in aluminum alloys. *Scr. Mater.*, **59** (9), 951–954.
83. Shriram, V., Yang, J.-M., Ye, J., and Minor, A.M. (2008) Determining the stress required for deformation twinning in nanocrystalline and ultrafine-grained copper. *JOM*, **60** (9), 66.

Part IV
Physical Properties

12
Current-Induced Transport: Electromigration
Ralph Spolenak

12.1
Principles

Electromigration is a process of directional material flow induced by an electric current. Consequently, as the electric current densities in microelectronic devices are typically very high, electromigration constitutes a major concern with regards to the reliability of such microelectronics systems. On this basis, electromigration has for almost 40 years undergone extensive investigations, as it can cause deleterious effects on materials such as aluminum and copper, and their alloys, and more recently, also on the lead-free solder materials that are used for packaging.

The basic mechanism that underlies electromigration is an interaction of electrons with any defects in a material (the "wind force") on one hand, and the Coulomb interaction between the externally applied electric field and ionized atoms in the material on the other hand. In the case of microelectronic materials, the "wind force" is dominant since, as the electromigration force F_{EM} is weak, it merely biases the diffusion processes in a material and becomes observable only at elevated temperatures ($T_{test} > 100\,°C$) and high current densities j ($j > 10^5\,A\,cm^{-2}$). The electromigration force is defined as follows:

$$F_{EM} = Z^* e \varrho j \qquad (12.1)$$

where Z^* is the effective charge that includes "wind" and Coulomb forces, e is the elementary charge, and ϱ is the resistivity of the conductor line. The effective charge is a material property which can vary for the different alloying components within the same alloy. Thus, the presence of an alloying element may also cause a significant reduction in the effective charge of the host material. During the early 1970s, this was the prime motivation for introducing Cu as an alloying element into aluminum interconnects [1, 2], when it was shown that the presence of Cu would strongly reduce the effective charge for aluminum, and thus result in greatly increased lifetimes of the interconnects. The median time to failure (MTF) of interconnects was determined by testing parallel line arrays of identical interconnects, until all had failed; this was typically carried out at elevated temperatures and current densities.

In-situ Electron Microscopy: Applications in Physics, Chemistry and Materials Science, First Edition.
Edited by Gerhard Dehm, James M. Howe, and Josef Zweck.
© 2012 Wiley-VCH Verlag GmbH & Co. KGaA. Published 2012 by Wiley-VCH Verlag GmbH & Co. KGaA.

This empirical approach led to the formulation of Black's law [3]:

$$MTF = Aj^{-n}e^{-\frac{E_A}{kT}} \tag{12.2}$$

where E_A is the activation energy for electromigration, k is Boltzman's constant, and T is the absolute temperature. Although such an approach led to a rapid optimization of alloying and processing conditions, it provided very little insight into the fundamental mechanisms involved.

On examining these fundamental mechanisms, it is advantageous to consider first the electromigration-induced drift rates:

$$v = \frac{D}{kT}F_{EM} = \frac{D}{kT}eZ^*\varrho j \tag{12.3}$$

where D is the effective diffusivity of atoms exposed to the electromigration force. On analyzing Equation 12.3 further, it can be seen that the rate at which electromigration occurs is equally dependent on two parameters: (i) how rapidly the atoms can move, as described by the *diffusivity* (this term depends exponentially on temperature via one or several activation energies); and (ii) the strength of the driving force.

The diffusivity depends heavily on the diffusion path. According to the geometry and structure of a wire, and also the material from which the wire is constructed, these diffusion paths may be either the volume of the material, the grain-boundaries, dislocation cores, interfaces, and surfaces. For example, in wide aluminum lines the dominant diffusion path would be the grain-boundaries, whereas in copper the dominant diffusion path would usually be the top interface between the copper line and a glassy dielectric. All of these diffusion paths are very similar to those identified for diffusional creep.

It is important to note that an electromigration-induced material transport in itself does not lead to damage, regardless of whether the rates are high or low. Damage usually accumulates at sites of flux divergence, which can range from differences in surface diffusivities from grain to grain, over a change in diffusion path (i.e., from grain-boundary diffusion to interfacial diffusion), to diffusion barriers as found in so-called Blech [4] segments and dual damascene structures. Blech segments are finite conductor lines that are deposited on a continuous conductor of a significantly lower conductivity. In this case, most of the current will flow through the high-conductivity segments, causing edge displacement on the cathode side and hillock formation on the anode side.

The driving force can, fundamentally, be altered only by a change in the material system (e.g., from Cu to Al) or by the addition of alloying elements (e.g., Cu in Al or Sn in Cu). However, in encapsulated interconnects additional driving forces may arise during the electromigration process. If a material is transported in a system of constant volume, then the local stress states will also be changed. Typically, material is transported from the cathode to the anode side, causing tensile and compressive stress states, respectively. The resulting gradient in stress subsequently counteracts the electromigration force, and will eventually cause the electromigration process to be halted, provided that stresses are not released by the formation of hillocks or voids. Consequently, shorter lines will exhibit higher stress gradients for given maximal stresses, and are less prone to electromigration damage [4–6]; this occurrence is commonly known as the Blech [4] effect. In some cases, an additional driving force may also originate from thermal gradients, and either assist or counteract the electromigration process.

Consequently, voids can nucleate only at locations of high tensile stress, which must in turn be a local flux divergence. Voids have also been found to migrate during the early stages of electromigration, so that the interaction of motion with the microstructure present also becomes important (an overview of this situation is available in Ref. [7]).

In-situ experiments were then conducted which focused on an understanding of the fundamentals of the electromigration process, and which usually included local probes. The following aspects were investigated as a function of time and temperature:

- Edge drift rates, using transmission electron microscopy (TEM), scanning electron microscopy (SEM) and optical microscopy.
- Compositional changes, using energy-dispersive X-ray (EDX) spectroscopy in TEM and SEM, and X-radiography.
- Changes in surface morphology, using scanning probe techniques.
- Microstructural changes, using TEM, electron backscatter diffraction (EBSD) in the scanning electron microscope, X-radiography, and focused ion beam (FIB).
- Void formation, growth and propagation kinetics, using SEM and TEM.
- Stress evolution, using convergent electron beam diffraction (CBED) in the transmission electron microscope, various X-ray techniques, and Raman spectroscopy.

Each of these aspects is described in detail in the following subsections.

12.2
Transmission Electron Microscopy

Whilst initial interrupted *ex-situ* experiments were carried out to investigate the location of void nucleation and the kinetics of void growth, propagation, and shape change, it soon became clear that only *in-situ* studies could provide the necessary insight (Table 12.1). Thus, the following investigations were carried out using TEM:

- *Imaging:* void nucleation and propagation, edge drift rates.
- *Selected area diffraction:* correlation of voids and the orientation of boundary grains.
- *CBED:* local analysis of mechanical stresses

12.2.1
Imaging

Surprisingly, the first *in-situ* electromigration studies were carried out in the transmission electron microscope during the early 1970s, and have been conducted

Table 12.1 Comparison of the different *in-situ* methods.

Properties	Electron microscopy								X-radiography			Special techniques		
	SEM			TEM										
	IM	EDX	EBSD	IM	EDX	SAED	CBED	IM	Topog.	Diff.	Raman	SPM	FIB
Resolution	2 nm	0.5 mm	20 nm	1 Å	100 nm	50 nm	50 nm	50 nm	0.5 μm	0.5 μm	300 nm	2 nm	5 nm
Time scale	s	min	s	s	min	Min	min	min	min	hrs	min	min	s
Cost	+	+	+	++	++	++	++	+++	+++	+++	+	+	++
Detection limit	–	0.1 wt%	–	–	0.1 wt%	–	–	–	–	–	–	–	–
Strain resolution	–	–	10^{-3}	–	–	10^{-3}	10^{-4}	–	–	10^{-4}	10^{-4}	–	–
Availability	+++	++	++	++	++	++	+	+	+	+	++	+++	++
Effects													
Edge displacement	+	–	–	+	–	–	–	+	–	–	–	+	+
Void observation	+	–	–	+	–	–	–	++	–	–	–	+	+
Elemental distribution	(+)	+	–	(+)	+	–	–	–	–	–	–	–	–
Surface topography	(+)	–	–	–	–	–	–	–	(+)	–	–	+	(+)
Stress measurement	–	–	(+)	–	–	(+)	+	–	(+)	+	(+)	–	–
Orientation	(+)	–	+	+	–	+	+	–	–	+	–	–	+

IM, Imaging; SAED, Selected area electron diffraction; Diff., Diffraction; SPM, Surface probe methods; Topog., Topography. Other abbreviations are as indicated in the text. +++, very high; ++, high; +, moderate; (+), potentially possible.

with ever-increasing experimental sophistication to the present day. In the early studies, as performed by Horowitz and Blech [1] in 1972, evolution of the void volume was monitored as a function of time during electromigration loading. However, instead of an Al–Cu alloy, an Al/Cu/Al trilayer was investigated, which was roughly equivalent to a concentration of Al–6 wt% Cu. The opening voids gave an excellent contrast in the TEM images, the time evolution of which could be quantified. Subsequently, transport rates and activation energies were determined. The activation energy was effectively doubled compared to pure aluminum, and the transport rate was reduced by a factor of between 10 and 100. Another notable study was carried out by Klein [8] in 1973, whereby *in-situ* TEM studies to investigate the electromigration behavior of Au by observing the growth rates of hillocks were substantiated by the observation of hillocks in *ex-situ* SEM observations. A Blech-type structure [4] was applied in the transmission electron microscope to determine the evolution of voids on the cathode side. In another early study conducted by Lobotka and Vávra [9], attempts were made to elucidate the microstructural changes in Al–Cu alloys as a function of the applied current characteristics, differentiating between alternating current (AC), direct current (DC), noise current, and their combination. Subsequently, it was shown that nondirectional currents (AC, noise) would lead to Joule heating of the thin film, which resulted in a grain boundary decoration by Al_2Cu precipitates, whereas an additional DC loading resulted in Cu transport. This constituted further progress in *in-situ* observations, where not only the strong contrast between voids and interconnect material was used, but also diffraction contrast to investigate the grain structure and elemental contrast to pinpoint Cu-rich zones.

Another important innovation in *in-situ* TEM imaging emerged with a strong reduction in interconnect width. When a width of less than 1 µm was reached, at least Al became electron-transparent for high-voltage transmission electron microscopes (~2 MeV). Okabayashi and coworkers [10, 11] performed *in-situ* experiments in a side-view Blech structure on 700 nm-wide lines; the time evolution of such a void within a Blech-type structure is shown in Figure 12.1. In contrast to the behavior of the cathode edge, voids within a Blech segment may even heal during the electromigration process. This is due to a local heating effect, when the current passes through the TiN shunt layer at the bottom below the void when, if the void is sufficiently small, any prior damage can be reversed. The void was shown to close within minutes at 350 °C and at a current density (j) of $6.5\,MA\,cm^{-2}$, and disappeared entirely at a later stage (see Figure 12.1).

While Blech structures and wide segments have provided information concerning drift rates and void growth kinetics that was essential for wide lines (the line width being significantly larger than the grain size), the shape of voids became critical for narrow lines, in which a single grain usually spans the width of a line (bamboo or near-bamboo structures). Again, the void shape evolution could only be observed using *in-situ* methods, an example being the study conducted by Riege *et al.* [12]. In this case, it was found that voids would nucleate long before a macroscopic (open circuit) failure of the device, such that void growth, motion and shape – but not void nucleation – would be changed and have a critical influence on electromigration

(a) t=53m56s

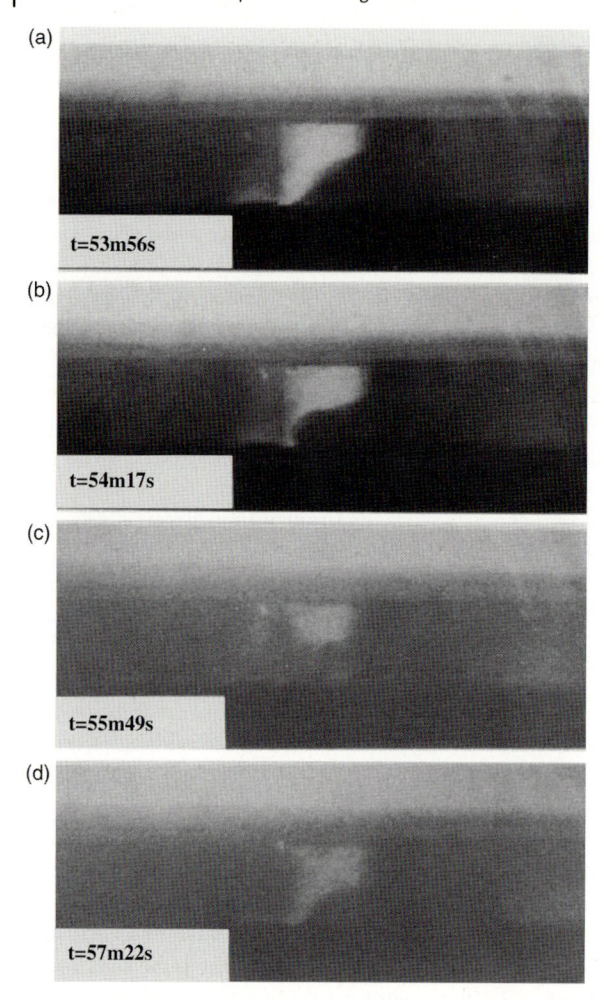

(b) t=54m17s

(c) t=55m49s

(d) t=57m22s

Figure 12.1 Video-taped SVTEM images showing mass transport from the cathode-side Al segment through gap 3 to the anode-side segment. The electron flow is from the right-hand to left-hand side, and the interconnect height is 500 nm. Figure reproduced with permission from Ref. [10]; © 1996, AIP (for explanation of acronym see [22]).

failure. Specifically, the interaction between voids and grain boundaries was found to have a critical influence on the void shape. Notably, if oriented along the current direction, a void with the same volume but with a high aspect ratio would be "harmless"; however, if oriented across the current direction, the void would cause an open circuit.

Further systematic studies on conventional test structures that were prepared for TEM [13] showed there to be a correlation between the Cu concentration in Al–Cu alloys and damage formation. Voids would be formed preferentially in regions of Cu depletion, whereas hillocks were observed in regions with a high density of Al_2Cu

precipitates. In addition, it was shown that grain boundaries should, ideally, intersect the line edges at normal angles in order to minimize the interaction between grain boundaries and voids.

When the material system was changed from aluminum to copper during the late 1990s, many of the *in-situ* studies that had been conducted by that time could no longer be carried out, due mainly to the reduced electron transparency for copper (an effect of a higher Z number) without thinning of the specimens. Thus, the details of novel *in-situ* studies have become available only recently. Chen and coworkers [13–15] further pushed the development of the transmission electron microscope to its limits by observing electromigration in high-resolution mode, and found the copper diffusion to be anisotropic and most pronounced on {111} planes in <110> directions, by observing the drift of atomic steps at grain boundaries.

During the past decade, electromigration has been recognized not only as a damage-generating mechanism in microelectronics, but also as a means of fabricating nanometer-sized gap structures. This has recently been demonstrated by Heersche and coworkers [16] (see Figure 12.2), whereby three different samples were exposed to a DC current; this eventually led to failure and thus to gap formation of nanometer scale. Typically, the edges of the gap were found usually to be asymmetric, except in the case of the second sample, when the current was switched off prior to failure. This phenomenon may have implications in the emerging field of *plasmonics*, where light confined by surface plasmons may be used to transport information rather than electrons.

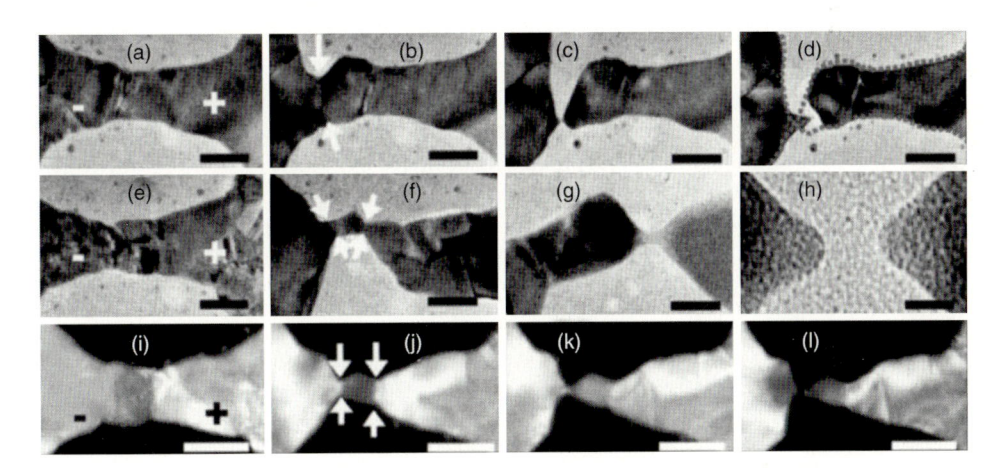

Figure 12.2 Active breaking of three devices (three rows, respectively). (a, e, and i) The wires before electromigration; (b, f, and j) The initial stage of the electromigration, in which grains fuse and voids begin to form along grain boundaries; (c and k) The last frame before breaking occurs; (d and l) The corresponding subsequent frame, (sampling frequency of 20 Hz). The contour of panel (c) is indicated in panel (d) by a dotted line. After frame (g), the active breaking was stopped and the device broken without a current being applied (h); (i–l) Frames taken in dark-field mode. Scale bars: (a–e) and (i–l) = 50 nm; (f) = 25 nm; (g) = 10 nm; (h) = 5 nm. The arrows indicate grain boundaries. Figure reproduced with permission from Ref. [16]; © 2007, AIP.

12.2.2
Diffraction

In addition to imaging, where voids, grains, and precipitates were visualized, electron diffraction has also been employed in *in-situ* studies, although only in very few cases. The reason for this is that an accurate diffraction orientation usually requires a double-tilt transmission electron microscope holder, which is incompatible with the feed-through necessary for an *in-situ* experiment. In one of the studies conducted by Chen et al. [14], forbidden reflections of 1/3{422} character were observed in a Cu(111) grain and attributed to a significant thinning of this specific grain by surface diffusion, which is an important diffusion pathway for copper electromigration.

12.2.3
Convergent Beam Electron Diffraction (CBED): Measurements of Elastic Strain

One further extension of these methods is that of CBED, in which the sample is illuminated with a highly focused but convergent electron beam, resulting in so-called HOLZ (High Order Laue Zone) patterns on an area detector. Given that the inelastically diffracted signal is suppressed by an energy filter, these patterns are highly sensitive to variations in lattice parameter, yielding a strain resolution of 2×10^{-4}. An aluminum conductor line is shown in Figure 12.3 (in side view), where the measured spots are marked by circles. In this case, a stress gradient as predicted by Blech [4] and Korhonen [17] could be verified, and found to be superimposed on a strong variation in stress state from grain to grain (cf. Section 12.4.4).

Each of the measurement points in Figure 12.3 results in a pattern such as that shown in Figure 12.4a, where the line pattern is subsequently transformed into spot pattern through a Hough transformation (Figure 12.4b). A spot pattern can be fitted with a much higher accuracy than a line pattern. Due to the high complexity of this method, in terms of instrumentation as well as data analysis, only a few *in-situ* measurements have been carried out to date. Nonetheless, the procedure is unique in terms of its local resolution of stress analysis in conjunction with highly accurate microstructure characterization.

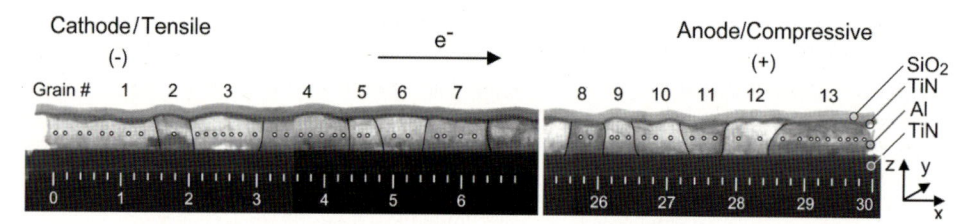

Figure 12.3 Composite TEM image of the anode and cathode ends of the 30 μm-long Al segment chosen for CBED analysis. The measurement positions are indicated as circles within the grains. Figure reproduced with permission from Ref. [49]; © 2005, MRS (Materials Research Society).

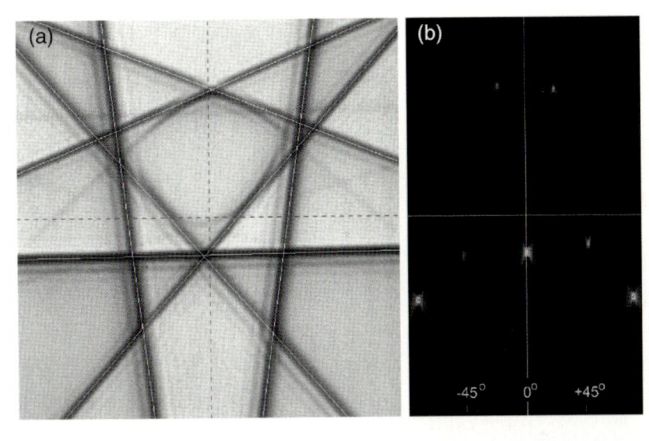

Figure 12.4 (a) Central portion of a [1 3 3] CBED pattern acquired in the center of the line; (b) The corresponding Hough transform. Because of the unidirectional strain, the symmetry of the pattern is broken with respect to the vertical line. Figure reproduced with permission from Ref. [62]; © 2000, Elsevier.

12.3
Secondary Electron Microscopy

12.3.1
Imaging

Today, the most popular *in-situ* microscopy technique is that of SEM. Such popularity is due mainly to the ease of sample preparation, the (usually) large sample chamber that allows for even complete probe stations to be placed into the microscope, and the useful additional characterization tools such as EDX analysis and EBSD. In terms of imaging, SEM has been used for:

- edge drift monitoring;
- void motion, growth and shape change;
- hillock evolution; and
- precipitate observation.

In contrast to TEM observations, grain structures may not usually be observed with SEM, while precipitates are only poorly visible by backscatter electron imaging (BEI) due to a strongly enhanced atomic number contrast.

An example of how drift velocities in Cu–Sn alloys were monitored is shown in Figure 12.5, where a Blech-type structure was imaged *in-situ* and the edge location quantified. Edge drift was found to occur within a similar incubation time as for the Al–Cu system; this corresponded to the time required for a critical distance to be depleted from the alloying element Sn, which was transported selectively by electromigration before any copper electromigration occurred. Measurements taken at different temperatures led to the determination of activation energies for the electromigration process, which were found to increase with Sn concentration.

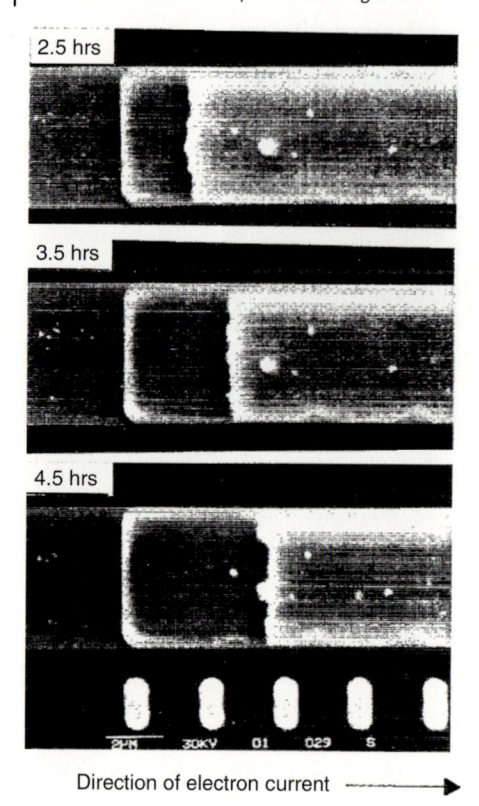

Figure 12.5 Mass depletion of 0.1 at different times (hours) of electromigration testing with current density of $-2.1 \times 10^6\,A\,cm^{-2}$ at 400 °C. Figure reproduced with permission from Ref. [23]; © 1995, AIP.

In pure imaging mode, the observation of voids was the second most intensively studied aspect of *in-situ* experiments, a point reflected by the extremely high number of reports made in this area. As this aspect is most important for bamboo or near-bamboo lines, the first reports had originated during the early 1990s, at which time the line dimensions became sufficiently small to allow for such microstructures. Although the details of narrow lines were first reported in 1992 [18], subsequent *ex-situ* (e.g., Ref. [19]) studies pinpointed the importance of the shape change of voids in providing an understanding of electromigration failure mechanisms. This, in turn, led to an entire series of studies [12, 20–27] in which void evolution during electromigration was investigated. Simulation studies could then be used to pinpoint which void shape (respectively void–microstructure arrangement) was the most susceptible to damage [28, 29].

One major issue for SEM investigations is the surface sensitivity of the interaction, specifically when detecting secondary electrons for imaging (SEI mode). Although, initially, only unpassivated structures could be investigated, it was soon realized that the passivation layers used in microelectronics have a major influence on the

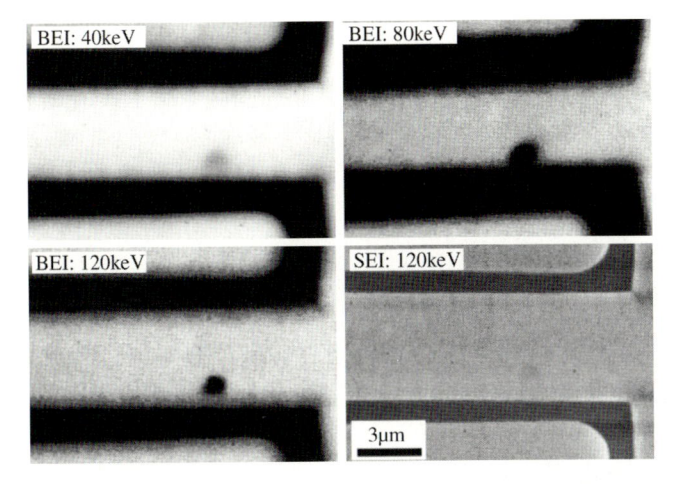

Figure 12.6 As the accelerating voltage increases, the ability to image voids under passivation increases with backscattered electron imaging (BEI). In secondary electron imaging (SEI) mode, the buried voids are not visible. Figure reproduced with permission from Ref. [63]; © 2000, AIP.

electromigration behavior. In aluminum alloys, the effect is mostly in terms of mechanical constraint, whereas in copper alloys the quality of the top interface critically influences the electromigration rates. Doan *et al.* solved this issue by converting a transmission electron microscope to a scanning electron microscope, by equipping it with both secondary electron and backscatter electron detectors. The high accelerating voltage thus available, of up to 120 kV, allowed penetration of the passivation layer. An example of an aluminum interconnect passivated by a 200 nm SiO_2 layer is shown in Figure 12.6, where a void that was barely visible at 40 keV in BEI mode became darker at 80 keV and exhibited a strong contrast at 120 keV. An image at the latter voltage, on the other hand, obtained by secondary electrons rendered the void invisible. These studies led to the successful detection of void formation, growth, motion and shape change processes under much more realistic conditions.

12.3.2
Elemental Analysis

It was soon recognized that the correct alloying elements could provide a significant slowing of electromigration damage in interconnect materials [2, 23]. In order to understand these alloying effects, it was necessary to monitor the distribution of the alloying elements during the electromigration process. This, again, was most easily accomplished with SEM utilizing EDX analysis, such that the composition could be monitored with a local resolution of less than 1 μm. For most alloying elements, it was found that the alloying elements had to be depleted over at least a critical distance corresponding to the Blech length before electromigration could occur in the matrix itself (e.g., Ref. [30]). In addition, care had to be taken that alloying elements with limited solubilities were selected, in order to minimize the increase in resistivity. Comparisons for alloying elements in aluminum and copper are available in Refs [30–32]. Local elemental analysis has also been carried out using X-radiography, and is described below [33, 34].

12.3.3
Electron Backscatter Diffraction (EBSD)

During recent years, the addition of EBSD detectors to the scanning electron microscope has become increasingly common. This allows for the qualitative characterization of defect densities, and the quantitative determination of orientation. In conjunction with *in-situ* tests, EBSD also allows for any correlation between orientation and damage evolution to be assessed. A correlation between the fastest diffusion paths (large-angle grain boundaries) as well as blocking grains with damage sites has been observed [35, 36]. Surprisingly, however, the orientation of hillocks was found not to correlate with the (111) texture of the surrounding grains.

12.4
X-Radiography Studies

Traditionally, X-radiography investigations exhibit a poor local resolution, and have mostly been used for phase analysis, texture measurements, and residual stress analysis. However, with the advent of third-generation synchrotron sources, X-ray focusing became viable due to their great brilliance and a primary spot size of several hundreds of microns. Thus, techniques such microscopy in two and three dimensions (tomography), in addition to spectral analysis combined with imaging and local stress measurements by diffraction, have become possible.

12.4.1
Microscopy and Tomography

In 2001, Schneider *et al.* [37] first demonstrated the feasibility of *in-situ* experiments on Cu interconnects that could be observed in plane-view through several microns of Si and passivation layers. The lateral resolution of approximately 70 nm could be implemented by using Fresnel zone plates with exposure times of 30 s. Future developments will most likely arrive at resolutions of about 30 nm, with exposure times on the order of seconds. The ability to combine such measurements with sample rotation led to provision of tomographic images, as demonstrated again by Schneider *et al.* [38] and by Levine *et al.* [39]. As the process of rotation and realignment is time-consuming, tomography is currently not feasible for *in-situ* experiments, however. Thin samples may also be analyzed using TEM, as shown recently by Ercius *et al.* [40], who utilized high-aperture annular dark-field detectors to avoid a strong diffraction contrast in crystalline samples. In this case, the TEM resolution in three dimensions was found to be better than 30 nm [40], but was limited by the specimen thickness. In X-ray tomography, the three-dimensional resolution was shown to be 140 nm [39], and limited by the resolution of the Fresnel zone plate applied. Notably, tomography allows for the post-mortem analysis of the exact void shape and the location of failure.

12.4.2
Spectroscopy

Microscopic X-radiographic techniques have also been coupled to spectroscopic procedures. For example, Solak *et al.* [33] used a scanning photoemission spectro-microscope and the Advanced Light Source in Berkeley with a lateral resolution of 100 nm and an energy resolution of 300 meV. An Al–Cu interconnect, before and after electromigration damage, is shown in Figure 12.7, where the images acquired at different energies exhibited different contrast mechanisms. For example, the Al_2Cu precipitates appeared white at energies above 49 eV, but dark below this value (Figure 12.7a).

In Figure 12.7b, however, a change in contrast was apparent, even in the electromigration damage (voids), at an energy of 42 eV. An interesting observation was also the decoration of grain boundaries with Cu atoms (as indicated by dashed circles), which is a consequence of the electromigration process. Spectroscopy thus allows for an unambiguous identification of the defect type.

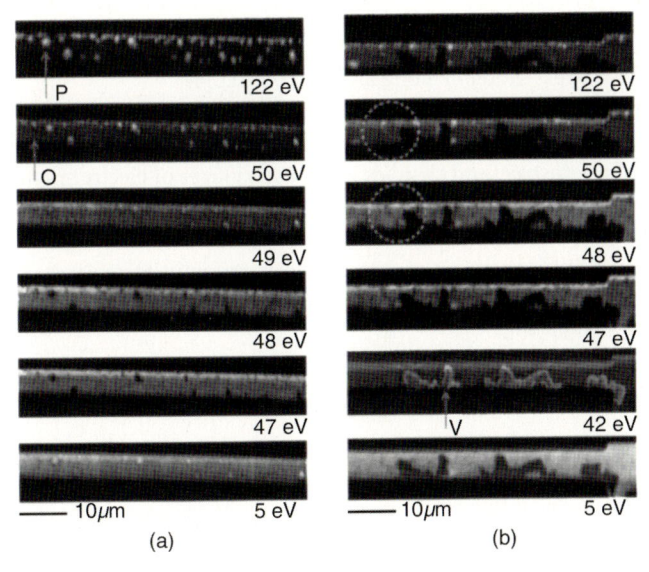

Figure 12.7 Micrographs of a 5 μm-wide Al(Cu) line. (a) Before and (b) after electromigration testing at the indicated photoelectron energies by scanning photoemission spectro-microscopy. Figure reproduced with permission from Ref. [33]; © 1999, AIP.

12.4.3
Topography

The measurement of X-ray topography – namely, a local monitoring of the intensity of (004) Si intensities – indicates (quantitatively) the stresses created by X-rays, and can be carried out very quickly. The change in X-ray intensity is proportional to the stress level. In the first *ex-situ* experiments, performed during the 1970s [41], a gradient in stress was reported in the Blech structures. The first transient experiments were carried out by Wang *et al.* [42], using a scanning beam, and showed the stresses to be correlated also to the local copper concentration, while a stress gradient was found to develop in the copper-depleted regions. Solak *et al.* improved on this method by using an imaging Fresnel zone plate-based technique with a resolution of better than 500 nm that allowed the visualization of a stress gradient over a length of 100 µm. A similar approach, of using the substrate as a probe for stresses, was also applied by Ma *et al.* [43], utilizing Raman spectroscopy. Here, the Raman peak shift in the Si adjacent to an Al interconnect was monitored with a resolution of better than 1 µm during an electromigration experiment. The stress sensitivity of this probe was comparable to that of the X-ray probe. Again, this required modeling of the stress state to induce the stress state present in the interconnect.

12.4.4
Microdiffraction

The direct observation of a local stress state within the interconnect itself is only possible by the diffraction of focused X-rays. Although this can only be achieved at synchrotron sources, the following insights have been acquired:

- The observation of a stress gradient along an interconnect [44].
- The correlation of stress state by microbeam diffraction and Cu concentration by micro-X-ray fluorescence [34].
- The observation of microstructural changes in the early stages of electromigration – that is, before any damage or resistance change could be observed, in Al [45, 46] and Cu [47] interconnects.
- The observation of a large variation of stress state from grain to grain (e.g., Ref. [48]) which is comparable to observations in the transmission electron microscope, using CBED [49].
- The study of electromigration in Sn, where a reorientation of high-resistance grains was observed by an electromigration-induced grain-growth process [50].

It is important to note that the first two observations were made with a white beam with a size of 10 µm in reflection. A single Laue peak was analyzed with an energy-dispersive detector to determine the lattice spacing of a highly fiber-textured (111) interconnect, with the average of several grains being measured.

For the latter three observations, a Laue beam was utilized with a beam size of 0.5 to 1 µm. In this case, the diffracted signal from single grains was analyzed utilizing an area detector, and small deviations from the perfect cubic symmetries were used [51] to determine five deviatoric components of the strain tensor. In addition, directional peak broadening can be correlated to the formation of geometrically necessary dislocations or small-angle grain boundaries, which serve as sinks and sources for atoms during the electromigration process [30].

12.5
Specialized Techniques

12.5.1
Focused Ion Beams

A focused ion beam (FIB) can be used for both imaging and sample modification. For example, the TEM test structure shown in Figure 12.3 was cut using a FIB, so as to remove any material that might obstruct the electron beam, and also to thin the sample to the desired thickness. In that process, Ga is usually incorporated into the test structure, though this may have an adverse effect on the results (e.g., cf. Ref. [52]). A case in which the need arose to investigate damage evolution in a buried structure is shown in Figure 12.8, where even high-voltage SEM (as shown in Figure 12.6) could not penetrate to sufficient depth. In this case, the eventual failures are shown at the end of the electromigration test although, interestingly, voids were also found close to the diffusion barrier, but not always exactly at its location. In addition to the Ga contamination, changes in stress state due to the creation of a free surface had also to be taken into account.

As FIB images exhibit a high grain contrast due to the ion-channeling effect, grain structures can very easily be imaged. This was demonstrated during the early 1990s, when void formation was correlated with the microstructure, namely a triple point by using *in-situ* FIB observation [53]. The opposite effect – the absence of any orientation contrast – is shown in Figure 12.9, where a fast diffusion section had been introduced into a single crystal Al line by nano-indentation. As this segment was longer than the Blech length, damage could be observed in the form of a slit-like void. The absence of any contrast as a function of sample tilt relative to the ion beam indicated that the fast diffusion segment was due to local increases in dislocation density and, consequently, to pipe diffusion rather than to the formation of grain boundaries.

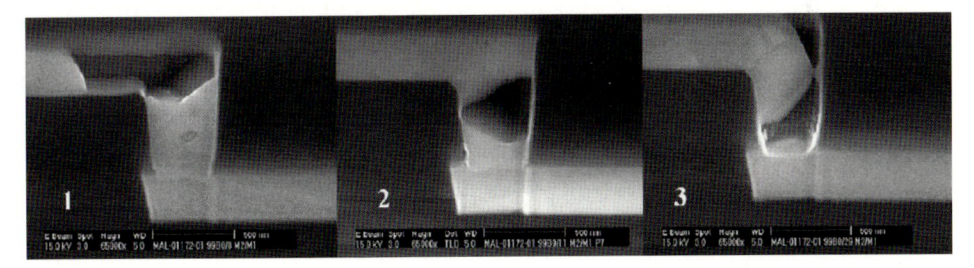

Figure 12.8 Typical failure features in a dual damascene structure. The stack was opened using a focused ion beam (FIB) before testing to allow for *in-situ* observations with SEM. Figure reproduced with permission from Ref. [64]; © 2002, Elsevier.

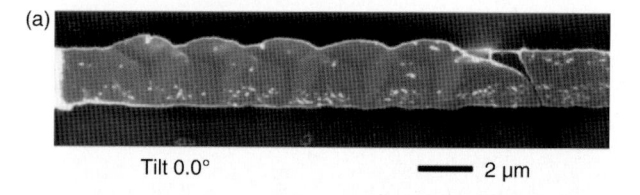

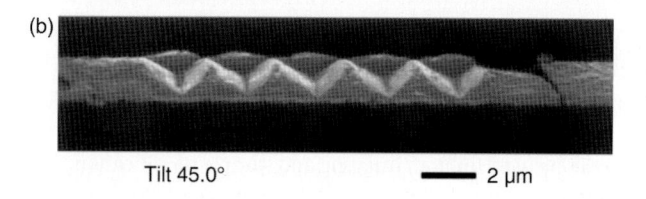

Figure 12.9 Focused ion beam (FIB) images from a row of five indentations in a single-crystal conductor line after electromigration testing. The fatal void can be seen to the right of the indentations. No orientation contrast is seen at tilt angles of (a) 0° or (b) 45°. Note that, in order to obtain these images, the native Al_2O_3 layer was removed by sputtering using the FIB. In Figure 12.9a the bright spots are the remaining oxide, which has not yet been sputtered away. Figure reproduced with permission from Ref. [65]; © 2000, Elsevier.

Further studies on single-crystal lines have been reported elsewhere [54, 55].

12.5.2
Reflective High-Energy Electron Diffraction (RHEED)

In RHEED, two electron beams rather than a single beam (as in a standard scanning electron microscope) were used. The first beam was used to image the sample and its surface topography by secondary electrons, while the second beam struck the sample at angles of between 2° and 3°, with the diffracted beam being detected on a fluorescent screen. Subsequently, the diffraction spots originating from a single grain were used for imaging, such that the shape of single grains could by determined simply by scanning the beam. During the 1990s Masu et al. [56] showed that, in copper, voids were formed at the grain boundaries and that the in-plane <110> direction should be aligned with the electron flow in order to minimize electromigration-induced damage.

12.5.3
Scanning Probe Methods

As one of the diffusion paths in electromigration may also be the sample surface, it is also interesting to investigate any surface topography changes that might occur during the electromigration process. This can best be achieved by using scanning probes and materials that are not easily oxidized; thus, the most prominent studies were carried out on silver and gold. The results of a study conducted by Levine et al. [57] (also during the early 1990s) are shown in Figure 12.10. In this case, small nodules were seen to form during electromigration, but to disappear at later stages

(a)

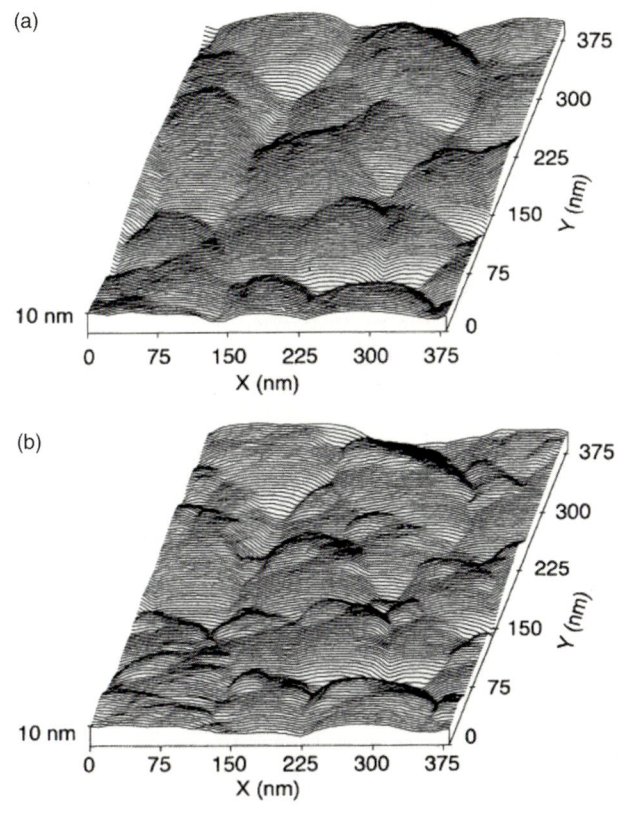

Figure 12.10 Linescans of topographic STM scans (a) before and (b) after the formation of nodules. Figure reproduced with permission from Ref. [57]; © 1993, APS (American Physical Society).

(Figure 12.10b). The authors attributed such changes to a surface electromigration of the contaminants in the silver alloy, such as carbon, while grain-boundary grooving and grain growth during the electromigration process were also observed. Notably, none of these observations could have been made using conventional SEM.

A more direct approach was taken by de Pablo *et al.* [58], who investigated the electromigration behavior of Au at high current densities ranging between 1 and 10 MA cm^{-2}. The authors observed a thinning on the cathode side and an increased roughness, while the electromigration was found to be more prominent in the middle of a 5 μm-wide segment than at its edges.

12.6
Comparison of *In-Situ* Methods

Each of the methods described above has made a significant contribution to electromigration research, and in the following subsection they are compared and their differences highlighted. This may also serve as a summary for the phenomena observed to date, and should help to formulate an idea of the future of the field, and which *in-situ* methods may become most suited to these investigations. In summarizing the techniques, certain points related to the electromigration become clear:

i) **How rapidly does electromigration occur?** This issue has been addressed by observing edge drift rates in Blech structures, as well as quantifying void growth rates. Both, SEM and TEM observations have been very important in this regard.

ii) **Which diffusion path is dominant?** Here, a correlation between damage and microstructure is instrumental (SEM + EBSD, TEM + diffraction, Microdiffraction), and in some cases even the local stress state [59] may reveal information about the diffusion path. Indirectly, it can be deduced by determining activation energies, which is possible by applying the techniques listed in (i).

iii) **How are alloying elements redistributed by electromigration?** Alloying elements, if localized, can be found using imaging techniques (SEM, TEM), although for quantitative analysis an energy-dispersive analysis of the X-rays emitted by the excited atoms is required. Such excitation can be achieved by electrons, as well as X-rays. The local resolution depends not only on probe size but also sample thickness.

iv) **What are the sites of highest flux divergencies?** Flux divergencies can be identified by the occurrence of damage, as well as the local evolution of stresses. Consequently, imaging techniques and micro Laue diffraction have been the methods of choice.

v) **How do stresses affect the electromigration process?** Stresses will pinpoint the locations of flux divergence, and also regulate the electromigration flux. In some instances, they may lead to "immortal" structures [5, 6]. Ideal probes are TEM + CBED and micro-Laue diffraction; for global stress gradients, broader X-rays, X-ray topography and Raman spectroscopy have been used.

It must be borne in mind that, to date, over 4000 reports have been made on electromigration, with a current rate of about 200 per year, and research effort in this area having been intensified over the past five years. With regards to *in-situ* studies, however, only 200 reports have been identified to date, of which about 40 have centered on electron microscopy studies; the impact of these studies significantly outweighs their numbers, however. For future studies, either new probes must be developed (e.g., focused positrons to measure vacancy concentrations or local mass spectroscopy to investigate self-diffusion) or the local resolution must be pushed to ever-lower limits.

Recent developments in the microelectronics industry have favored TEM techniques, notably because as the line width has fallen below 100 nm, samples have been rendered electron-transparent without further preparation, while the dielectric layer consists of a porous material that exhibits a very high compliance. Thus, the effect of stresses due to rigid encapsulation are reduced to a point where sample preparation has only a minor effect on changes in stress states.

While some of the above-described techniques have been developed simply as a proof of concept, many breakthroughs in electromigration studies continue to rely on large numbers of investigations, allowing these phenomena to be explored in systematic fashion. Indeed, such studies will surely continue to have a significant impact on the understanding of electromigration, with current "hot" topics being: (i) the influence of the microstructure of the seed layer on electromigration processes in copper interconnects [60]; (ii) the utilization of electromigration in nanotechnology (e.g., [16]); and (iii) electromigration in soldered joints [61].

Clearly, the application of *in-situ* electron microscopy techniques has had – and will continue to have – a profound impact on research into electromigration.

References

1 Horowitz, S.J. and Blech, I.A. (1972) *Mater. Sci. Eng.*, **10**, 169.
2 Blech, I.A. (1977) *J. Appl. Phys.*, **48**, 473–477.
3 Black, J.R. (1969) *IEEE Trans. Electron Devices*, **ED16**, 338.
4 Blech, I.A. and Meieran, E.S. (1969) *J. Appl. Phys.*, **40**, 485.
5 Gan, C.L., Thompson, C.V., Pey, K.L., Choi, W.K., Tay, H.L., Yu, B., and Radhakrishnan, M.K. (2001) *Appl. Phys. Lett.*, **79**, 4592–4594.
6 Gan, C.L., Thompson, C.V., Pey, K.L., and Choi, W.K. (2003) *J. Appl. Phys.*, **94**, 1222–1228.
7 Arzt, E., Kraft, O., Spolenak, R., and Joo, Y.C. (1996) *Z. Metallkd.*, **87**, 934–942.
8 Klein, B.J. (1973) *J. Phys. F. Met. Phys.*, **3**, 691.
9 Lobotka, P. and Vavra, I. (1981) *Phys. Status Solidi A*, **63**, 655–661.
10 Okabayashi, H., Kitamura, H., Komatsu, M., and Mori, H. (1996) *Appl. Phys. Lett.*, **68**, 1066–1068.
11 Okabayashi, H., Komatsu, M., and Mori, H. (1996) *Jpn. J. Appl. Phys. 1*, **35**, 1102–1106.
12 Riege, S.P., Prybyla, J.A., and Hunt, A.W. (1996) *Appl. Phys. Lett.*, **69**, 2367–2369.
13 Shih, W.C. and Greer, A.L. (1997) *Thin Solid Films*, **292**, 103–117.
14 Chen, K.C., Liao, C.N., Wu, W.W., and Chen, L.J. (2007) *Appl. Phys. Lett.*, **90**.
15 Liao, C.N., Chen, K.C., Wu, W.W., and Chen, L.J. (2005) *Appl. Phys. Lett.*, **87**.
16 Heersche, H.B., Lientschnig, G., and O'Neill, K. (2007) *Appl. Phys. Lett.*, **91**.
17 Korhonen, M.A., Borgesen, P., Tu, K.N., and Li, C.Y. (1993) *J. Appl. Phys.*, **73**, 3790–3799.
18 Besser, P.R., Madden, M.C., and Flinn, P.A. (1992) *J. Appl. Phys.*, **72**, 3792–3797.
19 Arzt, E., Kraft, O., Nix, W.D., and Sanchez, J.E. (1994) *J. Appl. Phys.*, **76**, 1563–1571.
20 Rosenberg, R. and Berenbau, L. (1968) *Appl. Phys. Lett.*, **12**, 201.
21 Flinn, P.A., Madden, M.C., and Marieb, T.N. (1994) *MRS Bull.*, **19**, 51–55.
22 Jo, B.H. and Vook, R.W. (1995) *Thin Solid Films*, **262**, 129–134.
23 Lee, K.L., Hu, C.K., and Tu, K.N. (1995) *J. Appl. Phys.*, **78**, 4428–4437.
24 Kraft, O. and Arzt, E. (1997) *Acta Mater.*, **45**, 1599–1611.
25 Kraft, O., Sanchez, J.E., Bauer, M., and Arzt, E. (1997) *J. Mater. Res.*, **12**, 2027–2037.
26 Lau, J.T., Prybyla, J.A., and Theiss, S.K. (2000) *Appl. Phys. Lett.*, **76**, 164–166.
27 Bohm, J., Volkert, C.A., Monig, R., Balk, T.J., and Arzt, E. (2002) *J. Electron. Mater.*, **31**, 45–49.
28 Kraft, O. and Arzt, E. (1995) *Appl. Phys. Lett.*, **66**, 2063–2065.
29 Kraft, O., Mockl, U.E., and Arzt, E. (1995) *Qual. Reliab. Eng. Int.*, **11**, 279–283.
30 Spolenak, R., Kraft, O., and Arzt, E. (1998) *Microelectron. Reliab.*, **38**, 1015–1020.
31 Spolenak, R. (1999) *Alloying Effects in Electromigration*, University of Stuttgart, Stuttgart, p. 152.

32 Barmak, K., Cabral, C., Rodbell, K.P., and Harper, J.M.E. (2006) *J. Vac. Sci. Technol. B*, **24**, 2485–2498.

33 Solak, H.H., Lorusso, G.F., Singh-Gasson, S., and Cerrina, F. (1999) *Appl. Phys. Lett.*, **74**, 22–24.

34 Kao, H.K., Cargill, G.S., Giuliani, F., and Hu, C.K. (2003) *J. Appl. Phys.*, **93**, 2516–2527.

35 Wetzig, K., Buerke, A., Wendrock, H., and von Glasow, A. (1999) *Materialprufung*, **41**, 418–421.

36 Buerke, A., Wendrock, H., and Wetzig, K. (2000) *Cryst. Res. Technol.*, **35**, 721–730.

37 Schneider, G., Hambach, D., Niemann, B., Kaulich, B., Susini, J., Hoffmann, N., and Hasse, W. (2001) *Appl. Phys. Lett.*, **78**, 1936–1938.

38 Schneider, G., Meyer, M.A., Denbeaux, G., Anderson, E., Bates, B., Pearson, A., Knochel, C., Hambach, D., Stach, E.A., and Zschech, E. (2002) *J. Vac. Sci. Technol. B*, **20**, 3089–3094.

39 Levine, Z.H., Kalukin, A.R., Kuhn, M., Frigo, S.P., McNulty, I., Retsch, C.C., Wang, Y.X., Arp, U., Lucatorto, T.B., Ravel, B.D., and Tarrio, C. (2000) *J. Appl. Phys.*, **87**, 4483–4488.

40 Ercius, P., Weyland, M., Muller, D.A., and Gignac, L.M. (2006) *Appl. Phys. Lett.*, **88**.

41 Blech, I.A. and Herring, C. (1976) *Appl. Phys. Lett.*, **29**, 131–133.

42 Wang, P.C., Noyan, I.C., Kaldor, S.K., Jordan-Sweet, J.L., Liniger, E.G., and Hu, C.K. (2001) *Appl. Phys. Lett.*, **78**, 2712–2714.

43 Ma, Q., Chiras, S., Clarke, D.R., and Suo, Z. (1995) *J. Appl. Phys.*, **78**, 1614–1622.

44 Wang, P.C., Cargill, G.S., Noyan, I.C., and Hu, C.K. (1998) *Appl. Phys. Lett.*, **72**, 1296–1298.

45 Valek, B.C., Bravman, J.C., Tamura, N., MacDowell, A.A., Celestre, R.S., Padmore, H.A., Spolenak, R., Brown, W.L., Batterman, B.W., and Patel, J.R. (2002) *Appl. Phys. Lett.*, **81**, 4168–4170.

46 Valek, B.C., Tamura, N., Spolenak, R., Caldwell, W.A., MacDowell, A.A., Celestre, R.S., Padmore, H.A., Braman, J.C., Batterman, B.W., Nix, W.D., and Patel, J.R. (2003) *J. Appl. Phys.*, **94**, 3757–3761.

47 Budiman, A.S., Nix, W.D., Tamura, N., Valek, B.C., Gadre, K., Maiz, J., Spolenak, R., and Patel, J.R. (2006) *Appl. Phys. Lett.*, **88**.

48 Spolenak, R., Brown, W.L., Tamura, N., MacDowell, A.A., Celestre, R.S., Padmore, H.A., Valek, B., Bravman, J.C., Marieb, T., Fujimoto, H., Batterman, B.W., and Patel, J.R. (2003) *Phys. Rev. Lett.*, **90**.

49 Nucci, J., Kramer, S., Arzt, E., and Volkert, C.A. (2005) *J. Mater. Res.*, **20**, 1851–1859.

50 Wu, A.T., Tu, K.N., Lloyd, J.R., Tamura, N., Valek, B.C., and Kao, C.R. (2004) *Appl. Phys. Lett.*, **85**, 2490–2492.

51 Tamura, N., MacDowell, A.A., Spolenak, R., Valek, B.C., Bravman, J.C., Brown, W.L., Celestre, R.S., Padmore, H.A., Batterman, B.W., and Patel, J.R. (2003) *J. Synchrotron Radiat.*, **10**, 137–143.

52 Kiener, D., Motz, C., Rester, M., Jenko, M., and Dehm, G. (2007) *Mater. Sci. Eng. A – Struct.*, **459**, 262–272.

53 Kumikawa, M.I. and Komoda, H. (1992) *Jpn. J. Appl. Phys. 2*, **31**, L1147–L1149.

54 Shingubara, S., Nakasaki, Y., and Kaneko, H. (1991) *Appl. Phys. Lett.*, **58**, 42–44.

55 Joo, Y.C. and Thompson, C.V. (1997) *J. Appl. Phys.*, **81**, 6062–6072.

56 Masu, K., Hiura, Y., Tsubouchi, K., Ohmi, T., and Mikoshiba, N. (1991) *Jpn. J. Appl. Phys. 1*, **30**, 3642–3645.

57 Levine, L.E., Reiss, G., and Smith, D.A. (1993) *Phys. Rev. B*, **48**, 858–863.

58 de Pablo, P.J., Asenjo, A., Colchero, J., Serena, P.A., Gomez-Herrero, J., and Baro, A.M. (2000) *Surf. Interface Anal.*, **30**, 278–282.

59 Spolenak, R., Tamura, N., and Patel, J. (2006) X-ray microdiffraction as a probe to reveal flux divergences in interconnects. American Institute of Physics (AIP) Conference Proceedings, Dresden, Germany, vol. 817, pp. 288–295.

60 Hu, C.K. (2007) MRS Fall Meeting 2007, Boston, Symposium M - Materials and

Hyperintegration Challenges in Next Generation Interconnect Technology, C. K. Hu, Electromigration Reliability in Nanoscale Cu Interconnects.

61 Nah, J.W., Ren, F., Tu, K.N., Venk, S., and Camara, G. (2006) *J. Appl. Phys.*, **99**.

62 Kramer, S., Mayer, J., Witt, C., Weickenmeier, A., and Ruhle, M. (2000) *Ultramicroscopy*, **81**, 245–262.

63 Doan, J.C., Lee, S., Lee, S.H., Meier, N.E., Bravman, J.C., Flinn, P.A., Marieb, T.N., and Madden, M.C. (2000) *Rev. Sci. Instrum.*, **71**, 2848–2854.

64 Meyer, M.A., Herrmann, M., Langer, E., and Zschech, E. (2002) *Microelectron. Eng.*, **64**, 375–382.

65 Baker, S.P., Joo, Y.C., Knauss, M.P., and Arzt, E. (2000) *Acta Mater.*, **48**, 2199–2208.

13
Cathodoluminescence in Scanning and Transmission Electron Microscopies
Yutaka Ohno and Seiji Takeda

13.1
Introduction

Cathodoluminescence (CL) is a phenomenon of light emission that is induced by electron irradiation. Typically, cathodoluminescent light is emitted from a region in which the electrons are irradiated, while the optical parameters – such as the photon energy, intensity, and polarization – will vary depending on the electronic structure in that region. Consequently, if CL spectroscopy is performed in either a scanning or a transmission electron microscope, it is possible to examine the electronic and atomic structures, simultaneously, in small regions observed *in-situ* by employing either scanning electron microscopy (SEM) or transmission electron microscopy (TEM), in both of which techniques electrons are irradiated.

CL spectroscopy in SEM and TEM enables an assessment to be made in detail of the atomistic structure in small regions, with an extremely higher spectral resolution (typically higher than a few meV) compared to other spectroscopic techniques conducted in the scanning or transmission electron microscope, including electron energy-loss spectroscopy (EELS), X-ray emission spectroscopy (XES), and energy-dispersive X-ray spectroscopy (EDS). For example, the structural and compositional variation – as well as the concentration and distribution of defects – in inhomogeneous and heterogeneous materials can be determined using either spatial- or depth-resolved CL techniques. In favorable cases, it is possible to detect compositional fluctuations with a resolution as low as $10^{14}\,\mathrm{cm}^{-3}$. Likewise, electronic properties such as localized energy levels and carrier capture cross-sections, both of which are connected with carrier lifetimes and diffusion lengths, can be analyzed by using spectral- or time-resolved CL techniques. This is especially beneficial when studying energy levels in a low-energy range, between the band edges of semiconducting materials (of the order of 10^0 eV at most), which dominate the electronic properties of the device products made from these materials. It is also interesting to note that the atomistic structure of defects and nanostructures inside a material – which is greatly affected by the surrounding material – can be examined directly by using CL in the scanning or transmission electron microscope. Indeed, as the miniaturization and

In-situ Electron Microscopy: Applications in Physics, Chemistry and Materials Science, First Edition.
Edited by Gerhard Dehm, James M. Howe, and Josef Zweck.
© 2012 Wiley-VCH Verlag GmbH & Co. KGaA. Published 2012 by Wiley-VCH Verlag GmbH & Co. KGaA.

integration of such electronic devices has continued, CL in these electron microscopes has become established as an indispensable micro-characterization technique.

Based on the pioneering SEM-CL studies conducted in 1969 [1], and on TEM-CL studies conducted in 1978 [2], more than 5000 reports have been made of CL being applied to the scanning or transmission electron microscope. In this chapter, a brief summary is provided of the principles that underlie the generation and interpretation of CL signals, and recent CL assessments of various electronic structures are reviewed. Attention is also focused on studies with semiconducting materials, which represent one of the most important issues in recent CL-SEM and -TEM investigations. A detailed description of the principles, and reviews of CL studies conducted before 1990, are reviewed elsewhere [3].

13.2
Principles of Cathodoluminsecence

13.2.1
The Generation and Recombination of Electron-Hole Pairs

Electrons, when irradiated into a material, are able to undergo both elastic and inelastic scattering. The irradiated material is excited via inelastic electron scattering, with such excitation resulting in the formation of X-rays, Auger electrons, secondary electrons, electron-hole pairs, and so forth. In order to analyze the cathodoluminsecent light, which is emitted via a recombination of electron-hole pairs, it is first necessary to understand how the electron-hole pairs are generated and recombined.

When a beam of electrons is irradiated into a material, each electron changes its direction via an elastic scattering, and reduces its kinetic energy via an inelastic scattering. As a result of these scattering processes, the original trajectories of the electrons are randomized such that the electron beam is able to penetrate the material. The extent of such penetration depends on the atomic number of atoms Z in the material; for thick solid materials, the shape varies from that of a pear for smaller atomic numbers, through a sphere for $15 < Z < 40$, to a hemisphere for larger atomic numbers. The lateral maximum radius of an electron penetration range is phenomenologically expressed as [4]

$$R_{\text{thick}} = 0.5\left(0.0276 A/\varrho Z^{0.889}\right) E_b^{1.67} \cdot (\mu m), \tag{13.1}$$

where E_b is the electron energy (in keV), A is the atomic weight (in g mol^{-1}), and ϱ is the material density (in g cm^{-3}). For a thin solid material, through which most incident electrons can transmit, the shape of the electron penetration range is conical and the lateral maximum radius, at the electron exit surface, is written as [5]

$$R_{\text{thin}} = 3.12(Z/E_b)(\varrho/A)^{0.5} t^{1.5} \cdot (\mu m), \tag{13.2}$$

where t is the thickness of the material (in μm).

Electron-hole pairs are generated inside an electron penetration range (a so-called generation volume). The generated electrons and holes are able to diffuse into a material, with the distribution of the electron-hole pairs being dominated by the diffusion of minority carriers. The stationary density of minority carriers at **r** $\Delta n(\mathbf{r})$ obeys the differential equation of continuity,

$$D\nabla^2[\Delta n(\mathbf{r})] - \Delta n(\mathbf{r})/\tau(\mathbf{r}) + g(\mathbf{r}) = 0, \tag{13.3}$$

where D is the diffusion constant of minority carriers, τ is the mean recombination lifetime, and g is the generation rate of electron-hole pairs.

The second term in Equation 13.3 is the recombination rate of electron-hole pairs at **r**. $1/\tau(\mathbf{r}) = 1/\tau_{rad}(\mathbf{r}) + 1/\tau_{non}(\mathbf{r})$, in which τ_{rad} and τ_{non} are, respectively, the mean lifetime for radiative recombination and that for nonradiative recombination. The number of photons emitted per unit time at **r** is $\Delta n(\mathbf{r})/\tau_{rad}(\mathbf{r})$.

13.2.2
Characteristic of CL Spectroscopy

Cathodoluminescent light is emitted via various radiative electronic transitions, one of which involves a recombination between an electron in the conduction band and a hole in the valence band (a band-to-band transition), which is typical in direct-gap semiconductors at high temperatures. At a temperature T, at which kT (k is Boltzmann's constant) is smaller than the excitonic binding energy, an excitonic level for free excitons is formed just below the conduction band, such that decay of the excitons (the free exciton transition between the excitonic levels and the valence band) results in a CL emission. When a defect (including a dopant) exists, it may induce donor and acceptor levels. In this case, the generated electrons and holes become trapped at the levels, such that cathodoluminescent light may be emitted via a donor-to-valence band transition, a conduction band-to-acceptor transition, and a donor-to-acceptor transition. These are the typical transitions in indirect-gap semiconductors at high temperatures. Similar to the case of free excitons, the excitonic level bound to a donor or an acceptor is formed at low temperatures, and cathodoluminescent light may be emitted via the decay of the excitons (bound exciton transitions). Likewise, when an impurity with incomplete inner shells (such as a rare-earth ion or a transition metal) exists, its excitation and radiative deexcitation (an inner shell transition) results in a cathodoluminescent emission. Reviews of these detailed recombination processes are available in Ref. [6] for inner shell transitions, and in Ref. [7] for the other transitions.

The photon energy of cathodoluminescent light emitted via an electronic transition is expressed as $\Delta E + \delta$, where ΔE corresponds to the energy difference between the energy levels concerning the transition, and δ is a positive value for a donor-to-acceptor transition [7] (it is zero for the other transitions). The energy levels in a material will vary, depending on the structure and composition [8], the morphology [9], and the structure of any existing defects [10]; hence, the parameters can be assessed by examining the photon energies of the cathodoluminescent lights emitted from the material.

The intensity of a cathodoluminescent light emitted from a generation volume is proportional to $\int_0^\infty \Delta n(\mathbf{r})/\tau_{\text{rad}}(\mathbf{r})d\mathbf{r}$. The CL intensity measurements can be applied to assess the density and distribution of both radiative [11] and nonradiative [12, 13] point defects. In addition, time-resolved CL intensity measurements allow the carrier dynamics to be assessed, such as excited carrier paths towards lower energy states, as well as carrier lifetimes [14, 15].

Certain extended defects, such as twin boundaries [16], dislocations [17, 18], platelets [19] and strains induced by a uniaxial stress [20], as well as low-dimensional nanostructures such as nanowires [21, 22], form anisotropic electronic states, and electronic transitions via the states result in the emission of polarized cathodoluminescent lights. CL polarization analyses may also be helpful when assessing, quantitatively, the atomistic structure of such defects and nanostructures.

13.2.3
CL Imaging and Contrast Analysis

A CL image – that is, a CL intensity map – is obtained by detecting the intensity of a cathodoluminescent light as the electron beam scans a raster over a specimen. An intensity map for a cathodoluminescent light with a specific energy is referred to as a "monochromatic" CL image, while a map showing the intensity of the overall cathodoluminescent lights is termed a "panchromatic" CL image.

When a microstructure exists in which the radiative or nonradiative recombination lifetime differs from that of the surrounding homogeneous material, such as a defect, then a contrast arises in the CL images. This contrast will vary depending on the structure, size and location of the microstructure, as well as on the experimental conditions such as E_b, I_b, and T. For example, the distribution of defects in degraded devices can be determined by the imaging technique [23–25].

Theories of CL contrast due to localized defects are provided (e.g., Ref. [26] as a simplified theory in which the point source of electron-hole pairs is assumed, and Refs [27, 28] as more generalized theories). In particular, the dislocations have been extensively examined, as they may cause electronic recombination due to the dislocation cores, as well as to Cottrell atmospheres of point defects around the cores, where long-range strain fields are introduced. In addition to the distribution of dislocations, the minority carrier lifetime and diffusion length around a dislocation are assessed with the theories.

13.2.4
Spatial Resolution of CL Imaging and Spectroscopy

The spatial resolution of CL can be determined by two factors: (i) the size of generation volume; and (ii) the diffusion length of minority carriers, which are typically the order of 100 nm and 100–1000 nm (depending on T), respectively. The density of the minority carriers decreases rapidly with an increasing distance from the generation point of carriers. Previously, theoretical calculations with correct carrier lifetimes and diffusion constants have suggested that the range in which minority carriers exist is twice as large as the generation volume, at most. Therefore, the lateral spatial resolution can be approximated to the size of generation volume; $2R_{thick}$ for thick solid materials (for SEM-CL), and $2R_{thin}$ for thin solid materials (for TEM-CL).

Various attempts have been made to improve the lateral spatial resolution. For example, CL imaging – when operated at a low acceleration voltage – is potentially a viable method, as R_{thick} decreases with decreasing E_b (Equation 13.1) and the carrier diffusion effect is generally suppressed due to a short lifetime of nonradiative recombination at the electron entrance surface; this method has been attempted in SEM [29]. Another possible method would be to employ CL imaging and spectroscopy by near-field detection [30, 31], which enables the detection of cathodoluminescent light that is emitted from a selected region, close to the electron entrance surface, in the generation volume. By applying these methods, a spatial resolution of less than a few hundred nanometers – comparable to the resolution of TEM-CL – is achievable for SEM-CL.

The atomistic structure of a region at an arbitrary depth, from a subjacency of the electron entrance surface to more than a few thousand nanometers, can be assessed by using CL spectroscopy with various E_b, as the size of generation volume will depend on E_b (Equation 13.1). Both, structural and compositional variations at interfaces inside the materials can be examined using this depth-resolved technique [32].

13.2.5
CL Detection Systems

Currently, various types of CL detection system have been designed [3], with some being available commercially. Typically, the cathodoluminescent light is collected with an ellipsoidal mirror inserted into the electron microscope; the cathodoluminescent light reflected by the mirror is then guided into a monochromator, either directly or via an optical fiber, and then collected into a photodetector, such as a charge-coupled device and a photomultiplier tube. The CL detection systems are designed so as not to interfere with other *in-situ* electron microscopy systems, such as specimen holders used at high and low temperatures, EELS, XES, and EDS systems. Some systems, to be used for novel *in-situ* measurements, such as four-probe electrical measurement in SEM [33], environmental TEM [34], and near-field optical measurements in TEM [35, 36], can be installed simultaneously with the CL systems (Figure 13.1).

13.3
Applications of CL in Scanning and Transmission Electron Microscopies

The application of CL in SEM and TEM has been used to assess the atomistic structure of various types of electronic material, including covalent, ionic, and molecular crystals. Some CL-based assessments of semiconducting materials in electronic devices, using this approach, are reviewed in the following subsections.

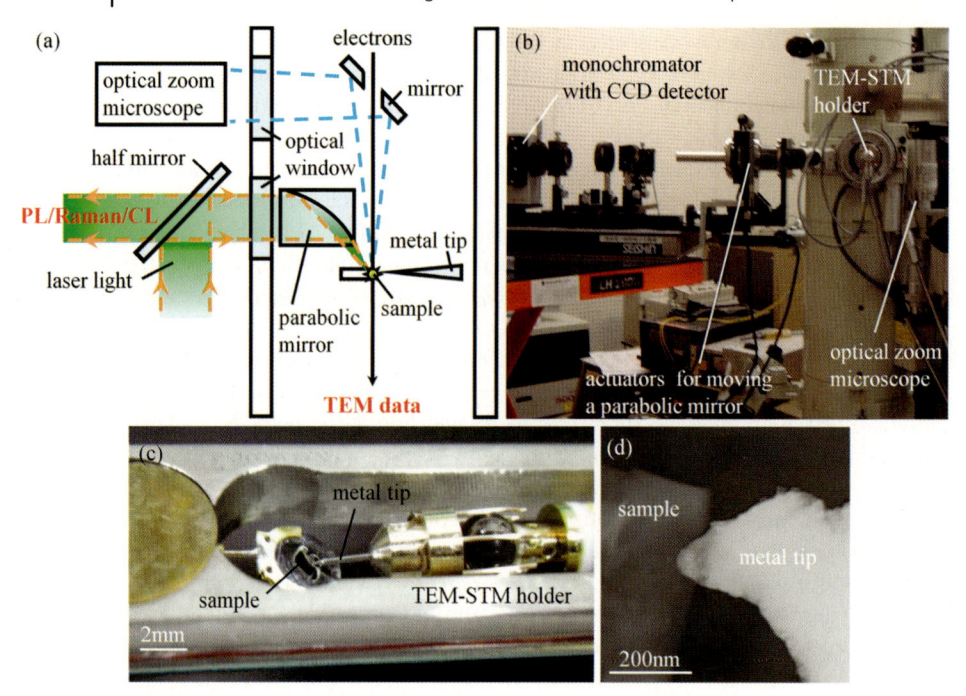

Figure 13.1 (a) Schematic and (b) external views of an apparatus for *in-situ* near-field optical measurements in the transmission electron microscope [35]. (c, d) Optical and TEM images, respectively, of a specimen and a metal tip in the microscope holder.

13.3.1
Assessments of Group III–V Compounds

13.3.1.1 Nitrides

Since the successful creation of the room-temperature (RT) blue light-emitting diode, nitride-based nanostructures have undergone intensive examination, with CL in both SEM and TEM having contributed considerably in advancing device fabrication and performance.

Due to a lack of any suitable growth substrate, there exist mismatches of lattice constant and thermal expansion coefficients between a GaN epilayer and its substrate, such that the mismatches result in the generation of a number of dislocations (between 10^6 and 10^{10} cm^{-2}), although the epilayer is known to have a high efficiency of light emission. It is well known that dislocations act as nonradiative recombination centers in other Group III–V [37, 38] and Group II–VI [17, 39] semiconductor devices, and that the device lifetime is affected by dislocations even of the order of 10^4 cm^{-2}. Consequently, the optical properties of dislocations in GaN have been a subject of controversy. The structure and optical properties of individual dislocations, whether grown-in [40, 41] and indentation-induced [42, 43], may be examined using CL in combination with TEM, and it has been shown that the dislocations can serve as recombination active centers. Dislocations (except for

60° dislocations on basal planes) act as nonradiative recombination centers [42], and the recombination activity is high for edge dislocations but low for screw dislocations [40]. By analyzing the CL contrast of individual dislocations, however, it is possible to estimate the diffusion length of minority carriers [41, 44]. Although the diffusion length depends on the dopant concentration and T, it is smaller in comparison with the other semiconductors, at about 80–160 nm for undoped GaN and about 60 nm for Si-doped GaN (with a concentration of 3×10^{18} cm^{-3}), even at low temperatures. The interpretation here is that, as the diffusion length is smaller than the nearest-neighbor distances of dislocations, GaN can tolerate a high density of dislocations.

Most nitride-based devices consist of multilayers of various compositions, such as AlInGaN grown on a sapphire substrate. A heterogeneous distribution of In atoms in InGaN [45, 46] and Al atoms in AlGaN [47], due to kinetic phase segregation, has been observed using SEM-CL, and it has been proposed that a random localization of excitons at such compositional fluctuations would prohibit nonradiative recombination at dislocations. On the other hand, the results of time-resolved CL studies have suggested that valleys are formed in an InGaN layer, and that the energetic screening of dislocations in the valleys may provide a high emission efficiency [15]. Any defects, except for dislocations close to GaN/sapphire [8, 48] and InGaN/GaN [49] interfaces, as well as in GaN [50] and InGaN [51] epilayers, may be examined using depth-resolved CL. Epitaxial laterally overgrown GaN are often used as a substrate for device fabrication, as the density of the threading dislocations is reduced from 10^{10} to 10^6 cm^{-2}. For the overgrown GaN on oxides [10] and on tungsten [52], the growth mechanism and the microstructure may be examined in detail by using SEM-CL imaging.

Some attempts have been made to fabricate self-organized nanostructures by selective epitaxial growth [53, 54], and this has been shown capable of enabling a three-dimensional control of nanostructures as nanowires and quantum dots (QDs) in ordinary Group III–V compounds (see Section 13.3.1.2).

13.3.1.2 III–V Compounds Except Nitrides

Following the successful fabrication of GaAs-based quantum wells, the creation of more low-dimensional structures – that is, quantum wires and QDs – has been attempted, mainly because such structures have the potential to enhance device performance. Moreover, their electron confinement permits the production of quantum electronic states which, in turn, provide an important system for fundamental physics.

Recently, much effort has been focused on growing nanostructures by spontaneous selective epitaxial growth, rather than by artificial methods such as lithography, as the self-organized nanostructures are free from nonradiative recombination at etched interfaces. One concept of self-organization is to employ differences in the surface migration of Group III atoms, or reactive species, on different growth planes. The atomistic structures of nanowires self-organized in grooves with {111}A sidewalls [55] or on mesas with {001} side-walls [56], may be assessed using SEM-CL imaging. Minority carrier paths and lifetimes in a wire may also be determined using time-resolved CL [14].

During a heteroepitaxial growth in the Stranski–Krastanov mode, strained clusters are formed spontaneously on a substrate, with CL imaging revealing that such clusters have a discrete density of states in zero-dimensional quantum structures (e.g., InAlAs dots on AlGaAs [57] and InAs dots on GaAs [58]). Such self-organized QDs have been investigated on the basis of their potential application in laser diodes and optical memory, while SEM-CL has been used to assess the inhomogeneity of the QDs in terms of their height and lateral size [59]. The carrier dynamics in a QD, as well as a wetting layer surrounding the dot, may also be revealed using temperature-dependent TEM-CL [60].

Recently, CL has been used to demonstrate other types of self-organized nanostructure that exhibit CL emissions. For example, when the twinning of AlGaAs was examined using TEM-CL, multiple twin boundaries – when arranged quasi-periodically – were seen to act as a superlattice and to emit an intense monochromatic light polarized parallel to the boundaries [16]. The details have also been discussed of self-organized nanowire allays [61] and quantum discs [19] in GaInP (Figure 13.2).

13.3.2
Group II–VI Compounds and Related Materials

With a wide band gap, a large exciton binding energy, and a low-power threshold for optical pumping, Group II–VI compounds and related materials would be expected to be used for short-wavelength optoelectronic devices and solar cells. It is well known that dislocations can easily be introduced into a Group II–VI-based device during its fabrication and operation, in comparison with in GaN [62], due partially to electronic excitation effects such as a recombination-enhanced dislocation motion [63], and these will have a severe adverse effect on the device's performance. Consequently, many recent CL studies have focused on defect assessment and the fabrication of nanostructures in which dislocations are difficult to introduce.

13.3.2.1 Oxides
ZnO, with a wide direct gap of 3.37 eV and large exciton binding energy of 60 meV at room temperature, has emerged as a promising analog to GaN, with excitonic emissions having been demonstrated at elevated temperatures of up to 550 K [64]. Similar to GaN, dislocations on basal planes act as nonradiative recombination centers with a large recombination activity when they are introduced at room temperature [65, 66] and during crystal growth [67]. In contrast to GaN, the recombination activity is decreased drastically when dislocations are introduced at elevated temperatures above 923 K, due to the interaction of point defects with dislocations via the migration of point defects [68]. This characteristic of ZnO may be an advantage over GaN, since all emissions in GaN are suppressed with the introduction of dislocations, even at elevated temperatures [69]. The defect level of screw dislocations and of 60° dislocations, respectively, was estimated at 0.9 and 0.3 eV in depth, using TEM-CL under light illumination [70]; moreover, it was proposed that a defect level of 60° dislocations in compound semiconductors would

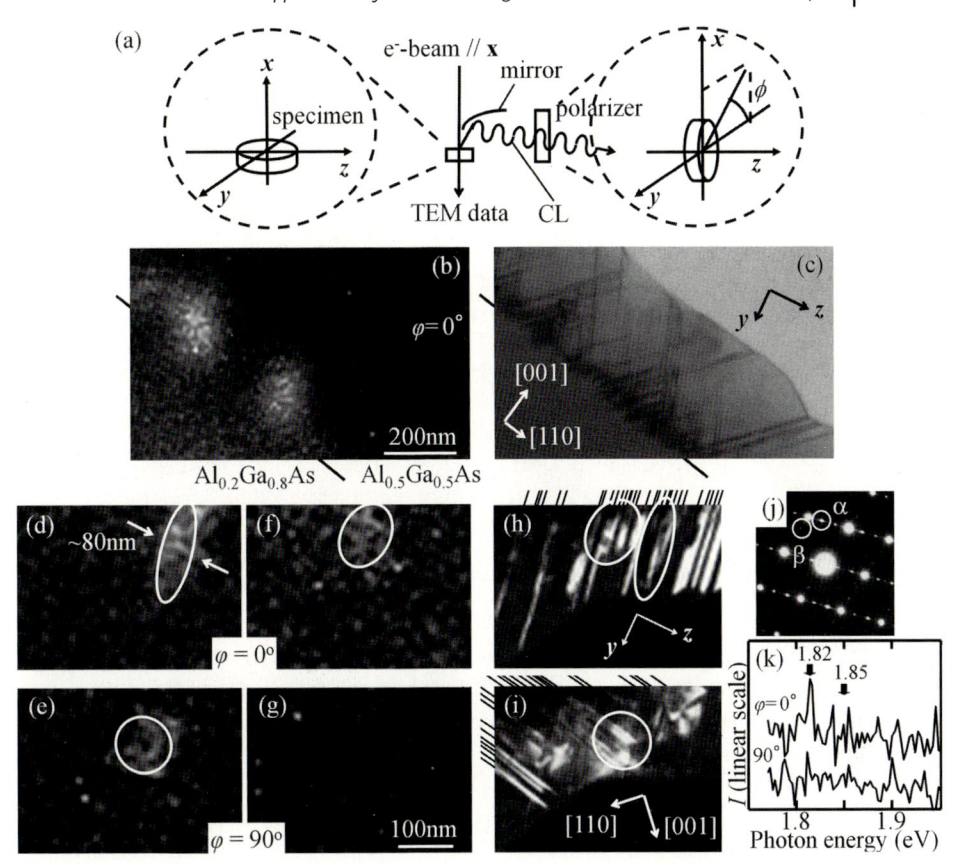

Figure 13.2 (a) The experimental set-up for quantitative analyses of polarized cathodoluminescent light. When such light is polarized in the x–y plane, the observed intensity is maximum or minimum, respectively, for $\phi = 0°$ or $90°$; (b) A panchromatic CL intensity map and (c) the corresponding bright-field TEM image; (d–g) Monochromatic CL intensity maps and (h–i) the corresponding dark-field TEM images, in which the locations of twin boundaries are indicated with the solid lines. The photon energy of the cathodoluminescent light is: (d, e) 1.82 eV or (f, g) 1.85 eV. $\phi =$ (d, f) $0°$ or (e, g) $90°$. Panels (h) and (i) were, respectively, recorded with a twin spot of α or β in (j); (k) Polarized CL spectra for $\phi = 0°$ and $90°$, obtained from the square area of panel (h) [16].

be governed by the band gap energy, irrespective of the crystal structure (wurtzite or zinc blende) and the group (II–VI or III–V) [35] (Figure 13.3).

ZnO nanowires and nanowalls may be fabricated, and the atomistic structures of individual nanowires determined, by using SEM-CL imaging. Typically, there will be no strain in a nanowire (ca. 40 nm diameter) [71], whereas an inhomogeneous strain and impurity distributions would be observed in a microwire (ca. 5000 nm diameter) [72]. The optical properties have also been examined of individual nanowires of Group III [73, 74] and Group IV [75–77] oxides.

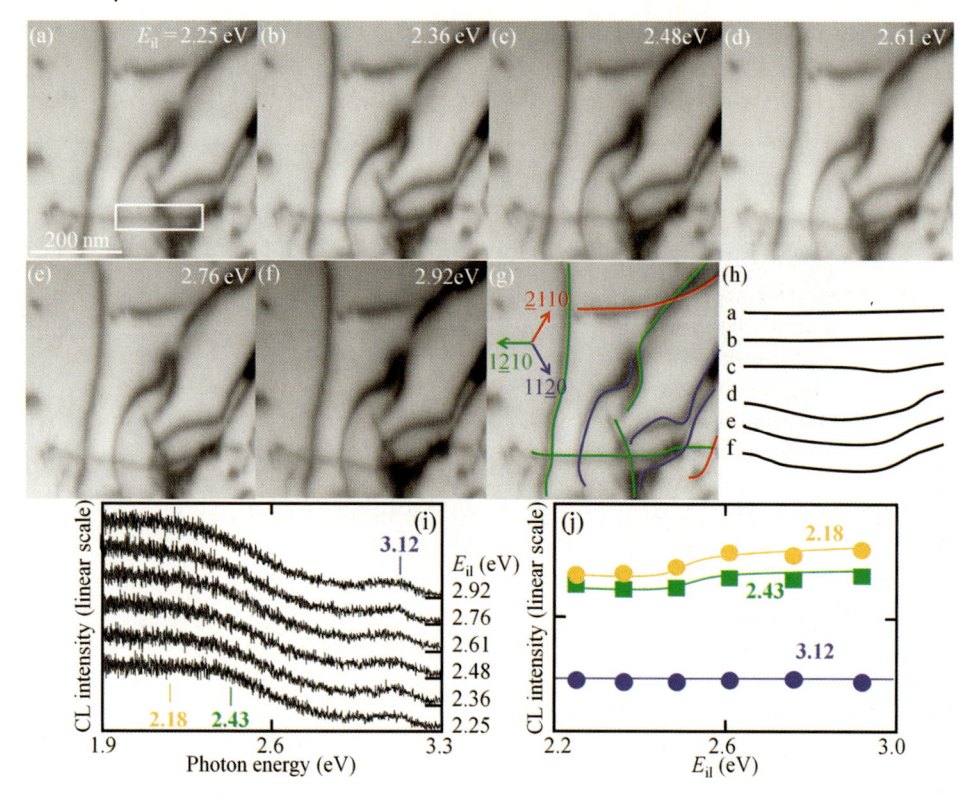

Figure 13.3 TEM images of dislocations in a specimen deformed at 1023 K after the illumination of a monochromatic light with photon energy of (a) 2.25 eV, (b) 2.36 eV, (c) 2.48 eV, (d) 2.61 eV, (e) 2.76 eV, or (f) 2.92 eV (g = 11–20); (g) A schematic view of the dislocations with the Burgers vector of [1–210] (green curves), [11–20] (blue curves), and [–2110] (red curves) in panel (a). (h) Variation of the line shape of the edge dislocation in the squared area in panel (a) during light illumination; (i) CL spectra of a specimen deformed at 923 K, obtained at temperature of 120 K, after the illumination of a monochromatic light. The photon energy is 2.25, 2.36, 2.48, 2.61, 2.76, or 2.92 eV from the bottom to the top; (j) The CL intensity at 3.12, 2.43, or 2.18 eV versus the photon energy. Reproduced with permission from Ref. [70]; © 2009, Wiley-VCH Verlag GmbH & Co. KGaA.

13.3.2.2 Group II–VI Compounds, Except Oxides

The distribution of dislocations in degraded ZnSe light-emitting devices can be assessed by using SEM-CL imaging [23, 25], and the atomistic structure of individual dislocations [18, 78] examined. The dynamic nature of a dislocation in an electronic excitation condition – that is, under the simultaneous illumination of light and electrons – may be examined *in-situ* over a wide temperature range (35 to 600 K) [79], using TEM-CL under light illumination [80].

Although selenide-based solar cells have a high efficiency of light emission, they still include many defects such as dislocations, stacking faults, and grain boundaries. The optical properties of defects have been assessed in both CdTe/CdS [81–83] and CuInGaSe$_2$ [84, 85].

13.3.3
Group IV and Related Materials

Among Group IV compounds, Si is the most important semiconductor although, because Si is an indirect-gap semiconductor, very few CL assessments have been made with this material. Emission lines related to dislocations [86] have been examined using SEM-CL and found to be independent of metal impurities [87]; rather, they arose due to glide dislocations, Lomer–Cottrell dislocations, and jogs [88]. The recombination activity of dislocations interacted with metal impurities has also been investigated [89], while the atomistic structure of a Si/SiO$_2$ interface [90] and of a SiO$_2$ gate of a dynamic random-access memory [91] have been assessed using SEM-CL. Si nanostructures such as porous-Si [92, 93] and Si nanowires [21, 94] have also been investigated, based on their potential as light-emitting materials.

Both, SiC and diamond have been considered attractive semiconducting materials for devices operated at high power and at high temperatures. In this respect, dislocations in 4H-SiC, related to the degradation of 4H-SiC devices, have been examined [95, 96], while discussions have also been undertaken into the optical properties of defects in diamond, such as dislocations [17, 97], stacking faults [97], as well as the distribution of dopant atoms [11] and strains [98].

13.4
Concluding Remarks

The application of CL in combination with SEM and TEM represents a powerful – and unique – approach to examining, *in-situ*, the atomistic structure of small regions inside materials, with high spectral resolution. Moreover, this *in-situ* spectroscopy technique, when conducted within the electron microscope, may lead to the discovery and assessment of a variety of novel nanomaterials.

References

1 Williams, P.M. and Yoffe, A.D. (1969) Monochromatic cathodoluminescence image in scanning electron microscope. *Nature*, **221** (5184), 952.
2 Petroff, P.M., Lang, D.V., Logan, R.A., and Strudel, J.L. (1978) Scanning transmission electron microscopy techniques for simultaneous electronic analysis and observation defects in semiconductors. *Scanning Electron Microsc.*, **1**, 325–332.
3 Yacobi, B.G. and Holt, D.B. (eds.) (1990) *Cathodoluminescence Microscopy of Inorganic Solids*, Plenum Press, New York, pp. 89–120.

4. Kanaya, K. and Okayama, S. (1972) Penetration and energy-loss theory of electrons in solid targets. *J. Phys. D: Appl. Phys.*, **5** (1), 43.
5. Goldstein, J.I. (1979) Principles of thin film X-ray microanalysis, in *Introduction to Analytical Electron Microscopy* (eds J.J. Hren, J.I., Goldstein and D.C. Joy), Plenum Press, New York, pp. 83–120.
6. Henderson, B. and Imbusch, G.F. (eds) (1989) *Optical Spectroscopy of Inorganic Solids*, Clarendon Press, Oxford, pp. 315–386.
7. Yu, P.Y. and Cardona, M. (eds) (2001) *Fundamentals of Semiconductors*, Springer, New York, pp. 345–426.
8. Sun, X.L., Goss, S.H., Brillson, L.J., Look, D.C., and Molnar, R.J. (2002) Depth-dependent investigation of defects and impurity doping in GaN/sapphire using scanning electron microscopy and cathodoluminescence spectroscopy. *J. Appl. Phys.*, **91** (10), 6729–6738.
9. Herman, M.A., Bimberg, D., and Christen, J. (1991) Heterointerfaces in quantum wells and epitaxial growth processes: Evaluation by luminescence techniques. *J. Appl. Phys.*, **70** (2), R1–R52.
10. Bertram, F., Riemann, T., Christen, J., Kaschner, A., Hoffmann, A., Thomsen, C., Hiramatsu, K., Shibata, T., and Sawaki, N. (1999) Strain relaxation and strong impurity incorporation in epitaxial laterally overgrown GaN: Direct imaging of different growth domains by cathodoluminescence microscopy and micro-Raman spectroscopy. *Appl. Phys. Lett.*, **74** (3), 359–361.
11. Graham, R.J., Shaapur, F., Kato, Y., and Stoner, B.R. (1994) Imaging of boron dopant in highly oriented diamond films by cathodoluminescence in a transmission electron microscope. *Appl. Phys. Lett.*, **65** (3), 292–294.
12. Ohno, Y. and Takeda, S. (1996) Study of electron-irradiation-induced defects in GaP by *in-situ* optical spectroscopy in a transmission electron microscope. *J. Electron Microsc.*, **45** (1), 73–78.
13. Ohno, Y., Kawai, Y., and Takeda, S. (1999) Vacancy-migration-mediated disordering in CuPt-ordered (Ga, In)P studied by *in-situ* optical spectroscopy in a transmission electron microscope. *Phys. Rev. B.*, **59** (4), 2694–2699.
14. Merano, M., Sonderegger, S., Crottini, A., Collin, S., Renucci, P., Pelucchi, E., Malko, A., Baier, M.H., Kapon, E., Deveaud, B., and Ganiere, J.D. (2005) Probing carrier dynamics in nanostructures by picosecond cathodoluminescence. *Nature.*, **438** (7067), 479–482.
15. Sonderegger, S., Feltin, E., Merano, M., Crottini, A., Carlin, J.F., Sachot, R., Deveaud, B., Grandjean, N., and Ganiere, J.D. (2006) High spatial resolution picosecond cathodoluminescence of InGaN quantum wells. *Appl. Phys. Lett.*, **89** (23), 232109-1–232109-3.
16. Ohno, Y., Yamamoto, N., Shoda, S., and Takeda, S. (2007) Intense monochromatic light emission from multiple nanoscale twin boundaries in indirect-gap AlGaAs epilayers. *Jpn. J. Appl. Phys.*, **46** (35), L830–L832.
17. Yamamoto, N., Spence, J.C.H., and Fathy, D. (1984) Cathodluminescence and polarization studies from individual dislocations in diamond. *Philos. Mag. B.*, **49** (6), 609–629.
18. Mitsui, T. and Yamamoto, N. (1997) Distribution of polarized-cathodoluminescence around the structural defects in ZnSe/GaAs(001) studied by transmission electron microscopy. *J. Appl. Phys.*, **81** (11), 7492–7496.
19. Ohno, Y. (2005) Polarized light emission from antiphase boundaries acting as slanting quantum wells in GaP/InP short-period superlattices. *Phys. Rev. B.*, **72** (12), 121307 (R)/1-4.
20. Tang, Y., Rich, D.H., Lingunis, E.H., and Haegel, N.M. (1994) Polarized-cathodoluminescence study of stress for GaAs grown selectively on patterned Si(100). *J. Appl. Phys.*, **76** (5), 3032–3040.
21. Ozaki, N., Ohno, Y., Kikkawa, J., and Takeda, S. (2005) Growth of silicon nanowires on H-terminated Si {111} surface templates studied by transmission electron microscopy. *J. Electron Microsc.*, **54** (S1), i25–i29.

22 Yamamoto, N., Bhunia, S., and Watanabe, Y. (2006) Polarized cathodoluminescence study of InP nanowires by transmission electron microscopy. *Appl. Phys. Lett.*, **88** (15), 153106-1–153106-3.

23 Saijo, H., Ning, G., Yabuuchi, Y., Takahashi, Y., Isshiki, T., Shiojiri, M., and Ogawa, K. (1994) Analytical color fluorescence electron-microscopy observation of laser-diodes. *J. Electron Microsc.*, **43** (2), 77–83.

24 Cheng, Y.M., Herrick, R.W., Petroff, P.M., Hibbsbrenner, M.K., and Morgan, R.A. (1995) Degradation studies of proton-implanted vertical cavity surface emitting Lasers. *Appl. Phys. Lett.*, **67** (12), 1648–1650.

25 Bonard, J.M., Ganiere, J.D., Vanzetti, L., Paggel, J.J., Sorba, L., and Franciosi, A. (1998) Combined transmission electron microscopy and cathodoluminescence studies of degradation in electron-beam-pumped $Zn_{1-x}Cd_xSe/ZnSe$ blue-green lasers. *J. Appl. Phys.*, **84** (8), 1263–1273.

26 Jakubowicz, A. (1986) Theory of cathodoluminescence contrast from localized defects in semiconductors. *J. Appl. Phys.*, **59** (6), 2205–2209.

27 Schreiber, J. and Hildebrandt, S. (1994) Basic dislocation contrasts in SEM-CL/EBIC on III–V semiconductors. *Mater. Sci. Eng. B.*, **24** (1–3), 115–120.

28 Holt, D.B. and Napchan, E. (1994) Quantitation of SEM EBIC and CL signals using Monte Carlo electron-trajectory simulations. *Scanning*, **16** (2), 78–86.

29 Myhajlenko, S. (1991) Low-voltage scanning electron microscopy cathodoluminescence observations of gallium-arsenide. *Scanning Microsc.*, **5** (3), 603–610.

30 Cramer, R.M., Sergeev, O.V., Heiderhoff, R., and Balk, L.J. (1999) Spectrally resolved cathodoluminescence analyses in the optical near-field. *J. Microsc.*, **194**, 412–414.

31 Pastre, D., Bubendorff, J.L., and Troyon, M. (2000) Resolution in scanning near-field cathodoluminescence microscopy. *J. Vac. Sci. Technol. B*, **18** (3), 1138–1143.

32 Brillson, L.J., and Viturro, R.E. (1988) Low-energy cathodoluminescence spectroscopy of semiconductor interfaces. *Scanning Microsc.*, **2** (2), 789–799.

33 Lin, X., He, X.B., Lu, J.L., Shi, D.X., and Gao, H.J. (2005) Four-probe scanning tunneling microscope with atomic resolution for electrical and electro-optical property measurements of nanosystems. *Chin. Phys.*, **14** (8), 1536–1543.

34 Yoshida, H., Shimizu, T., Uchiyama, T., Kohno, H., Homma, Y., and Takeda, S. (2009) Atomic-scale analysis on the role of molybdenum in iron-catalyzed carbon nanotube growth. *Nano Lett.*, **9** (11), 3810–3815.

35 Ohno, Y., Yonenaga, I., and Takeda, S. (2011) In-situ analysis of optoelectronic properties of semiconductor nanostructures and defects in transmission electron microscopes, in *Optoelectronic Devices and Properties* (ed. O. Sergiyenko), InTech, pp. 241–262.

36 Kizuka, T. and Oyama, M. (2011) Individual cathodoluminescence and photoluminescence spectroscopy of zinc oxide nanoparticles in combination with in situ transmission electron microscopy. *J. Nanosci. Nanotechnol.*, **11** (4), 3278–3283.

37 Petroff, P.M., Logan, R.A., and Savage, A. (1980) Nonradiative recombination at dislocations in III–V compound semiconductors. *Phys. Rev. Lett.*, **44** (4), 287–291.

38 Wang, J.N., Steeds, J.W., and Woolf, D.A. (1992) The study of misfit dislocations in $In_xGa_{1-x}As/GaAs$ strained quantum-well structures. *Philos. Mag. A.*, **65** (4), 829–839.

39 Mitsui, T., Yamamoto, N., Tadokoro, T., and Ohta, S. (1996) Cathodoluminescence image of defects and luminescence centers in ZnS/GaAs(100). *J. Appl. Phys.*, **80** (12), 6972–6979.

40 Yamamoto, N., Itoh, H., Grillo, V., Chichibu, S.F., Keller, S., Spence, J.S., DenBaars, S.P., Mishra, U.K., and Nakamura, S., and Salviati, G. (2003) Cathodoluminescence characterization of dislocations in gallium nitride using a transmission electron microscope. *J. Appl. Phys.*, **94** (7), 4315–4319.

41 Nakaji, D., Grillo, V., Yamamoto, N., and Mukai, T. (2005) Contrast analysis of dislocation images in TEM–cathodoluminescence technique. *J. Electron Microsc.*, **54** (3), 223–230.
42 Albrecht, M., Strunk, H.P., Weyher, J.L., Grzegory, I., Porowski, S., and Wosinski, T. (2002) Carrier recombination at single dislocations in GaN measured by cathodoluminescence in a transmission electron microscope. *J. Appl. Phys.*, **92** (4), 2000–2005.
43 Lei, H., Leipner, H.S., Schreiber, J., Weyher, J.L., Wosinki, T., and Grzegory, I. (2002) Raman and cathodoluminescence study of dislocations in GaN. *J. Appl. Phys.*, **92** (11), 6666–6670.
44 Pauc, N., Phillips, M.R., Aimez, V., and Drouin, D. (2006) Carrier recombination near threading dislocations in GaN epilayers by low voltage cathodoluminescence. *Appl. Phys. Lett.*, **89** (16), 161905/ 1-3.
45 Chichibu, S., Wada, K., and Nakamura, S. (1997) Spatially resolved cathodoluminescence spectra of InGaN quantum wells. *Appl. Phys. Lett.*, **71** (16), 2346–2348.
46 Khatsevich, S., Rich, D.H., Zhang, X., Zhou, W., and Dapkus, P.D. (2004) Temperature dependence of excitonic recombination in lateral epitaxially overgrown InGaN/GaN quantum wells studied with cathodoluminescence. *J. Appl. Phys.*, **95** (4), 1832–1842.
47 Riemann, T., Christen, J., Kaschner, A., Laades, A., Hoffmann, A., Thomsen, C., Iwaya, M., and Kamiyama, S. (2002) Direct observation of Ga-rich microdomains in crack-free AlGaN grown on patterned GaN/sapphire substrates. *Appl. Phys. Lett.*, **80** (17), 3093–3095.
48 Goss, S.H., Sun, X.L., Young, A.P., Brillson, L.J., Look, D.C., and Molnar, R.J. (2001) Microcathodoluminescence of impurity doping at gallium nitride/sapphire interfaces. *Appl. Phys. Lett.*, **78** (23), 3630–3632.
49 Pereira, S., Correia, M.R., Pereira, E., O'Donnell, K.P., Trager-Cowan, C., Sweeney, F., and Alves, E. (2001) Compositional pulling effects in $In_xGa_{1-x}N$/GaN layers: A combined depth-resolved cathodoluminescence and Rutherford backscattering/channeling study. *Phys. Rev. B*, **64** (20), 205311-1–205311-5.
50 Salviati, G., Albrecht, M., Zanotti-Fregonara, C., Armani, N., Mayer, M., Shreter, Y., Guzzi, M., Melnik, Y.V., Vassilevski, K., Dmitriev, V.A., and Strunk, H.P. (1999) Cathodoluminescence and transmission electron microscopy study of the influence of crystal defects on optical transitions in GaN. *Phys. Status Solidi A*, **171** (1), 325–339.
51 Wu, X.H., Elsass, C.R., Abare, A., Mack, M., Keller, S., Petroff, P.M., DenBaars, S.P., Speck, J.S., and Rosner, S.J. (1998) Structural origin of V-defects and correlation with localized excitonic centers in InGaN/GaN multiple quantum wells. *Appl. Phys. Lett.*, **72** (6), 692–694.
52 Kaschner, A., Hoffmann, A., Thomsen, C., Bertram, F., Riemann, T., Christen, J., Hiramatsu, K., Sose, H., and Sawaki, N. (2000) Micro-Raman and cathodoluminescence studies of epitaxial laterally overgrown GaN with tungsten masks: A method to map the free-carrier concentration of thick GaN samples. *Appl. Phys. Lett.*, **76** (23), 3418–3420.
53 Liu, Q.K.K., Hoffmann, A., Siegle, H., Kaschner, A., Thomsen, C., Christen, J., and Bertram, F. (1999) Stress analysis of selective epitaxial growth of GaN. *Appl. Phys. Lett.*, **74** (21), 3122–3124.
54 Petersson, A., Gustafsson, A., Samuelson, L., Tanaka, S., and Aoyagi, Y. (1999) Cathodoluminescence spectroscopy and imaging of individual GaN dots. *Appl. Phys. Lett.*, **74** (23), 3513–3515.
55 Gustafsson, A., Samuelson, L., Malm, J.O., Vermeire, G., and Demeester, P. (1994) Cathodoluminescence of single quantum wires and vertical quantum wells grown on a submicron grating. *Appl. Phys. Lett.*, **64** (6), 695–697.
56 Rajkumar, K.C., Madhukar, A., Rammohan, K., Rich, D.H., Chen, P., and Chen, L. (1993) Optically active three-dimensionally confined structures

realized via molecular beam epitaxial growth on nonplanar GaAs (111)B. *Appl. Phys. Lett.*, **63** (21), 2905–2907.

57 Leon, R., Petroff, P.M., Leonard, D., and Fafard, S. (1995) Spatially-resolved visible luminescence of self-assembled semiconductor quantum dots. *Science*, **267** (5206), 1966–1968.

58 Grundmann, M., Christen, J., Ledentsov, N.M., Bohrer, J., Bimberg, D., Ruvimov, S.S., Werner, P., Richter, U., Gosele, U., Heydenreich, J., Ustinov, V.M., Egorov, A.Y., Zhukov, A.E., Kopev, P.S., and Alferov, Zh.I. (1995) Ultranarrow luminescence lines from single quantum dots. *Phys. Rev. Lett.*, **74** (20), 4043–4046.

59 Rich, D.H., Zhang, C., Mukhametzhanov, I., and Madhukar, A. (2000) Cathodoluminescence wavelength imaging of μm-scale energy variations in InAs/GaAs self-assembled quantum dots. *Appl. Phys. Lett.*, **76** (24), 3597–3599.

60 Akiba, K., Yamamoto, N., Grillo, V., Genseki, A., and Watanabe, Y. (2004) Anomalous temperature and excitation power dependence of cathodoluminescence from InAs quantum dots. *Phys. Rev. B.*, **70** (16), 165322/ 1-9.

61 Rich, D.H., Tang, Y., and Rin, H.T. (1997) Linearly polarized and time-resolved cathodoluminescence study of strain-induced laterally ordered $(InP)_2/(GaP)_2$ quantum wires. *J. Appl. Phys.*, **81** (10), 6837–6852.

62 Yonenaga, I., Koizumi, H., Ohno, Y., and Taishi, T. (2008) High-temperature strength and dislocation mobility in the wide band-gap ZnO: Comparison with various semiconductors. *J. Appl. Phys.*, **103** (9), 093502/ 1-4.

63 Maeda, K. and Takeuchi, S. (eds) (1996) *Enhancement of Dislocation Mobility in Semiconducting Crystals by Electronic Excitation*, North-Holland, Amsterdam.

64 Bagnall, D.M., Chen, Y.F., Zhu, Z., Yao, T., Shen, M.Y., and Goto, T. (1998) High temperature excitonic stimulated emission from ZnO epitaxial layers. *Appl. Phys. Lett.*, **73** (8), 122077-1–122077-3.

65 Bradby, J.E., Kucheyev, S.O., Williams, J.S., Jagadish, C., Swain, M.V., Munroe, P., and Phillips, M.R. (2002) Contact-induced defect propagation in ZnO. *Appl. Phys. Lett.*, **80** (24), 4537–4539.

66 Takkouk, Z., Brihi, N., Guergouri, K., and Marfaing, Y. (2005) Cathodoluminescence study of plastically deformed bulk ZnO single crystal. *Phys. B.*, **366** (1-4), 185–191.

67 Sieber, B., Addad, A., Szunerits, S., and Boukherroub, R. (2010) Stacking faults-induced quenching of the UV luminescence in ZnO. *J. Phys. Chem. Lett.*, **1** (20), 3033–3038.

68 Ohno, Y., Koizumi, H., Taishi, T., Yonenaga, I., Fujii, K., Goto, H., and Yao, T. (2008) Optical properties of dislocations in wurtzite ZnO single-crystals introduced at elevated temperatures. *J. Appl. Phys.*, **104** (7), 073515-1–073515-6.

69 Yonenaga, I., Ohno, Y., Taishi, T., Tokumoto, Y., Makino, H., Yao, T., Kamimura, Y., and Edagawa, K. (2011) Optical properties of fresh dislocations in GaN. *J. Cryst. Growth*, **318** (1), 415–417.

70 Ohno, Y., Taishi, T., and Yonenaga, I. (2009) In-situ analysis of optoelectronic properties of dislocations in ZnO by TEM observations. *Phys. Status Solidi A*, **206** (8), 1904–1911.

71 Fan, H.J., Scholz, R., Zacharias, M., Gosele, U., Bertram, F., Forster, D., and Christen, J. (2005) Local luminescence of ZnO nanowire-covered surface: A cathodoluminescence microscopy study. *Appl. Phys. Lett.*, **86** (2), 023113-1–023113-3.

72 Nobis, T., Kaidashev, E.A., Rahm, A., Lorenz, M., Lenzner, J., and Grundmann, M. (2004) Spatially inhomogeneous impurity distribution in ZnO micropillars. *Nano Lett.*, **4** (5), 797–800.

73 Magdas, D.A., Cremades, A., and Piqueras, J. (2006) Growth and luminescence of elongated In_2O_3 micro- and nanostructures in thermally treated InN. *Appl. Phys. Lett.*, **88** (11), 113107-1–113107-3.

74 Gao, H.J., Duscher, G., Kim, M., Pennycook, S.J., Kumar, D., Cho, K.G., and Singh, R.K. (2000) Cathodoluminescent properties at nanometer resolution through Z-contrast scanning transmission electron

microscopy. *Appl. Phys. Lett.*, **77** (4), 594–596.

75 Calestani, D., Lazzarini, L., Salviati, G., and Zha, M. (2005) Morphological, structural and optical study of quasi-1D SnO_2 nanowires and nanobelts. *Cryst. Res. Technol.*, **40** (10–11), 937–941.

76 Maestre, D., Cremades, A., and Piqueras, J. (2005) Growth and luminescence properties of micro- and nanotubes in sintered tin oxide. *J. Appl. Phys.*, **97** (4), 044316-1–044316-4.

77 Fernandez, I., Cremades, A., and Piqueras, J. (2005) Cathodoluminescence study of defects in deformed (110) and (100) surfaces of TiO_2 single crystals. *Semicond. Sci. Technol.*, **20** (2), 239–243.

78 Shreiber, J., Hilpert, U., Horing, L., Worschechm, L., Konig, B., Ossau, W., Waag, A., and Landwehr, G. (2000) Luminescence studies on plastic stress relaxation in ZnSe/GaAs(001). *Phys. Status Solidi B*, **222** (1), 169–177.

79 Ohno, Y. (2005) Photoinduced stress in a ZnSe/GaAs epilayer containing 90° α partial dislocations. *Appl. Phys. Lett.*, **87** (18), 181909-1–181909-3.

80 Ohno, Y., and Takeda, S. (1995) A new apparatus for in-situ photoluminescence spectroscopy in a transmission electron microscope. *Rev. Sci. Instrum.*, **66** (10), 4866–4869.

81 Sochinskii, N.V., Serrano, M.D., Dieguez, E., Agullorueda, F., Pal, U., Piqueras, J., and Fernandez, P. (1995) Effect of thermal annealing on Te precipitates in CdTe wafers studied by Raman scattering and cathodoluminescence. *J. Appl. Phys.*, **77** (6), 2806–2808.

82 Toda, A., Nakano, K., and Ishibashi, A. (1998) Cathodoluminescence study of degradation in ZnSe-based semiconductor laser diodes. *Appl. Phys. Lett.*, **73** (11), 1523–1525.

83 Jahn, U., Okamoto, T., Yamada, A., and Konagaki, M. (2001) Doping and intermixing in CdS/CdTe solar cells fabricated under different conditions. *J. Appl. Phys.*, **90** (5), 2553–2558.

84 Hetzer, M.J., Strzhemechny, Y.M., Gao, M., Contreras, M.A., Zunger, A., and Brillson, L.J. (1995) Direct observation of copper depletion and potential changes at copper indium gallium diselenide grain boundaries. *Appl. Phys. Lett.*, **86** (16), 162105-1–162105-3.

85 Romero, M.J., Ramanathan, K., Contreras, M.A., Al-Jassin, M.M., Noufi, R., and Sheldon, P. (2003) Cathodoluminescence of Cu(In, Ga)Se_2 thin films used in high-efficiency solar cells. *Appl. Phys. Lett.*, **83** (23), 4770–4772.

86 Higgs, V., Lightowlers, E.C., and Tajbakhsh, S. (1992) Cathodoluminescence imaging and spectroscopy of dislocations in Si and $Si_{1-x}Ge_x$ alloys. *Appl. Phys. Lett.*, **61** (9), 1087–1089.

87 Sekiguchi, T., Kveder, V.V., and Sumino, K. (1994) Hydrogen effect on the optical activity of dislocations in silicon introduced at room temperature. *J. Appl. Phys.*, **76** (12), 7882–7888.

88 Sekiguchi, T., and Sumino, K. (1996) Cathodoluminescence study on dislocation in silicon. *J. Appl. Phys.*, **79** (6), 3253–3260.

89 Lee, W., Chen, J., Chen, B., Chang, J., and Sekiguchi, T. (2011) Cathodoluminescence study of dislocation-related luminescence from small-angle grain boundaries in multicrystalline silicon. *Appl. Phys. Lett.*, **94** (11), 112103-1–112103-3.

90 Schafer, J., Young, A.P., Brillson, L.J., Niimi, H., and Lucovsky, G. (1998) Depth-dependent spectroscopic defect characterization of the interface between plasma-deposited SiO_2 and silicon. *Appl. Phys. Lett.*, **73** (6), 791–793.

91 Yoshikawa, M., Matsuda, K., Yamaguchi, Y., Matsunobe, T., Nagasawa, Y., Fujino, H., and Yamane, T. (2002) Characterization of silicon dioxide film by high spatial resolution cathodoluminescence spectroscopy. *J. Appl. Phys.*, **92** (12), 7153–7156.

92 Cullis, A.G., Canham, L.T., Williams, G.M., Smith, P.W., and Dosser, O.D. (1994) Correlation of the structural and optical properties of luminescent, highly oxidized porous silicon. *J. Appl. Phys.*, **75** (1), 493–501.

93 Itoh, M., Yamamoto, N., Takemoto, K., and Nittono, O. (1996) Cathodoluminescence

imaging of n-type porous silicon. *Jpn. J. Appl. Phys.*, **35** (8), 4182–4186.

94 Dovrat, M., Arad, N., Zhang, X.H., Lee, S.T., and Sa'ar, A. (2007) Optical properties of silicon nanowires from cathodoluminescence imaging and time-resolved photoluminescence spectroscopy. *Phys. Rev. B*, **75** (20), 205343-1–205343-5.

95 Chen, B., Sekiguchi, T., Ohyanagi, T., Matsuhata, H., Kinoshita, A., and Okumura, H. (2009) Electron-beam-induced current and cathodoluminescence study of dislocation arrays in 4*H*-SiC homoepitaxial layers. *J. Appl. Phys.*, **106** (7), 074502-1–074502-4.

96 Miao, R.X., Zhang, Y.M., Tang, X.Y., and Zhang, Y.M. (2011) Investigation of luminescence properties of basal plane dislocations in 4H-SiC. *Acta Phys. Sin.*, **60** (3), 037808.

97 Graham, R.J. and Ravi, K.V. (1992) Cathodoluminescence investigation of impurities and defects in single crystal diamond grown by the combustion-flame method. *Appl. Phys. Lett.*, **60** (11), 1310–1312.

98 Burton, N.C., Steeds, J.W., Meaden, G.M., and Shreter, Y.G. (1995) Strain and microstructure variation in grains of CVD diamond film. *Diamond Rel. Mater.*, **4** (10), 1222–1234.

14
In-situ TEM with Electrical Bias on Ferroelectric Oxides
Xiaoli Tan

14.1
Introduction

Many materials respond to an external electric field/current by changing their microstructures. Transmission electron microscopy (TEM) allows the microstructure of inorganic materials to be examined with spatial and chemical resolutions on the atomic level; consequently, attempts have been made to apply electrical signals to TEM specimens during imaging, in the hope of revealing the dynamic processes that occur at the nanometer scale. Although such electrical *in-situ* TEM technique was first demonstrated during the early 1960s, when Blech and Meieran passed an electrical direct current through an Al thin film to observe the electromigration process [1], only a handful of reports based on this technique have been made to date. Whilst most of these have involved electromigration in interconnecting lines [1–3] and domain switching in ferroelectrics [4–8], the application of the technique has been expanded more recently to studies of other phenomena, such as the breakdown of the grain boundary conducting barrier [9] and the mechanical fatigue of metallic thin films [10]. In this chapter, attention is focused on the electric field-induced phenomena in perovskite ferroelectric ceramics. A summary is also provided of the author's previous contributions to the field [11–19].

Ferroelectric ceramics have found widespread application in capacitors, transducers, actuators, energy storage devices, and random access memories, due to their unique dielectric, piezoelectric, and ferroelectric properties. When these ferroelectric materials are subjected to strong external electric fields, the primary response of their microstructure is a polarization switching through domain wall motion [20]. When the applied field is sufficiently high, but lower than the dielectric breakdown strength, then phase transitions can be triggered [21–26]. Among the electric field-induced phase transitions, the antiferroelectric-to-ferroelectric transition provides the physics-based foundation for the application of antiferroelectrics in energy-storage and large displacement control devices [21, 22], whereas the relaxor-to-ferroelectric transition couples the polar order with the underlying chemical order [25, 26]. In addition, when driven hard with bipolar electric fields for extended cycles,

In-situ Electron Microscopy: Applications in Physics, Chemistry and Materials Science, First Edition.
Edited by Gerhard Dehm, James M. Howe, and Josef Zweck.
© 2012 Wiley-VCH Verlag GmbH & Co. KGaA. Published 2012 by Wiley-VCH Verlag GmbH & Co. KGaA.

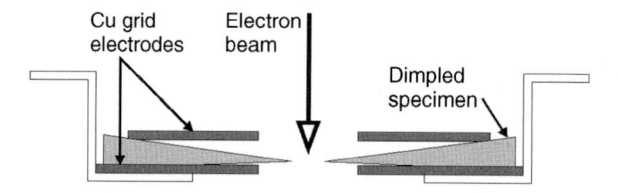

Figure 14.1 The configuration used by previous research groups to study electric field-induced domain switching in ferroelectric ceramics.

ferroelectric materials may lose their physical integrity due to grain boundary cracking [27–29]. Studies of the unique electric field *in-situ* TEM technique have, for the first time, been applied to directly visualize the dynamic process of the polarization switching of nanometer-sized domains [11] and the cavitation of grain boundaries in ferroelectric ceramics [14], the fracture of domain walls in ferroelectric single crystals [13], the evolution of incommensurate modulations in antiferroelectric ceramics [15, 16], and the growth of polar nanoregions in relaxor ferroelectric ceramics [18, 19].

Previous investigations of ferroelectrics with electric field *in-situ* TEM techniques by other research groups have been limited to the polarization switching of ferroelectric domains [4–8]. The electric field was applied to the specimen by two parallel TEM copper grids through a configuration, as shown in Figure 14.1. In this case, the nominal electric field is parallel to the electron beam and, therefore, negligible disturbance to the TEM imaging can be achieved. One major drawback of this configuration, however, is that the actual electric field in the thin area of the specimen is far less than the nominal field. Furthermore, the magnitude of the actual field is difficult to estimate due to the presence of vacuum gaps between the copper grid electrodes and the specimen.

Thus, the electrode configuration was modified by depositing gold films directly onto the specimen surface, to better quantify the actual field (Figure 14.2). In this configuration, the presence of the central hole disturbed the electric field within the specimen. According to previous analytical solutions for the electric field disturbed by a perpendicular insulating flaw, the direction of the actual field at the flaw tip would remain unchanged, but the magnitude would be intensified [30, 31], with the intensification ratio depending on the geometry of the flaw and the dielectric permittivity of the specimen. When the flaw is a circular hole penetrating through the thickness, and the dielectric permittivity of the ferroelectric material is much greater than that of the material filled in the flaw, then the intensification ratio would be 2. The specimen geometry used in the present study is shown in Figure 14.2a, where the central perforation can be considered as the insulating flaw and, in most cases, the perforation would take a circular shape. After having loaded the specimen into the transmission electron microscope chamber, the flaw is then filled with a vacuum, with a relative permittivity of 1. The ferroelectric ceramic materials studied here have a relative dielectric permittivity greater than 300 at room temperature [11–19]. To a rough approximation, the intensification ratio can be taken as 2 at the tip of the central perforation if the effect of the thickness variation due to the

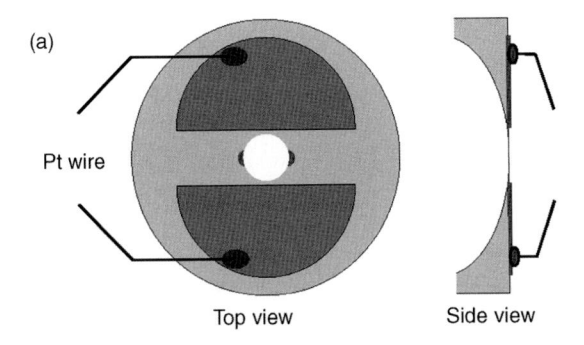

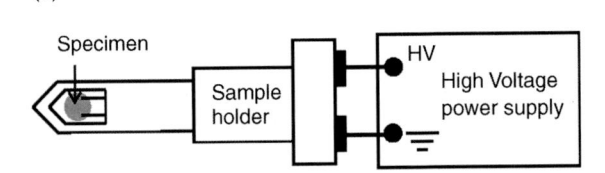

Figure 14.2 The electric field *in-situ* TEM experimental set-up used in these studies. (a) The specimen geometry shows the two half-circle-shaped gold electrodes and the central perforation. The two small dark areas at the edge of the perforation exaggerate and highlight the two tips of the circular flaw. In this study, all *in-situ* TEM observations were made at electron transparent areas at these two sites; (b) The test system comprised of a special transmission electron microscope holder and a high-voltage power supply.

mechanical dimpling is ignored. In Figure 14.2a, the two tips of the central perforation are exaggerated and highlighted by the two dark areas. Circularly perforated TEM specimens without cracks were selected for all *in-situ* studies, with observations being carried out at these two sites. In this way, the direction of the actual electric field could be determined by identifying the direction of the nominal field, while the magnitude of the actual field could be estimated by doubling the nominal electric field. All of the field levels stated for the *in-situ* TEM experiment in this chapter refer to this actual field.

14.2
Experimental Details

The TEM specimens were prepared using conventional procedures, including polishing, dimpling, and ion milling. Half-circle-shaped gold electrodes were coated on the flat surfaces of the 3 mm disks, as shown in Figure 14.2a, and a special mask was then used to form the central gap. The electroded specimens were first examined using optical microscopy, and crack-free specimens were loaded to the special transmission electron microscope holder with two electrical feedthroughs. The specimens were then fixed to the bottom of the sample cup, using insulating varnish, and connections made from the electrical leads on the holder tip to the gold electrodes on the specimen, through thin platinum wires. A high-voltage source was connected to the outside terminals of the sample holder, as shown in Figure 14.2b. Both, static and cyclic electric fields with amplitudes of up to $80\,kV\,cm^{-1}$ were applied to the specimens during the *in-situ* studies; a frequency of 30 Hz was used for the applied cyclic electric fields. As the applied field was perpendicular to the optic axis of the transmission electron microscope, this caused an electron beam deflection which was compensated for by adjusting the beam deflection on the microscope. The TEM experiments were carried out using a single-tilt specimen holder on a Phillips CM-12 microscope operating at 120 kV for the studies described in Sections 14.3–14.5, with a double-tilt holder on a Phillips CM-30 microscope operating at 300 kV for the studies in Section 14.6, and a single-tilt holder on a JEOL 2010 microscope operating at 200 kV for those in Section 14.7.

14.3
Domain Polarization Switching

The electric field *in-situ* TEM technique with the modified electrode configuration was first tested with a commercial lead zirconate titanate (PZT) ceramic (PZT EC65; EDO Corporation, Salt Lake City, UT, USA) [11]. The ceramic has an average grain size of 8 μm, and a mixed nanometer-sized and micrometer-sized domain structure. It contained 4% porosity, with most pores less than 30 μm in size. At room temperature, the ceramic has a tetragonal crystal structure and a ferroelectric coercive field of $8\,kV\,cm^{-1}$.

The regular lamellar 90° ferroelectric domains in this tetragonal ceramic were disrupted by proprietary chemical dopants. These disrupted polar domains appeared as nanometer-sized dark patches in the bright-field TEM images [7]. As the PZT EC65 ceramic shows normal ferroelectric behavior and displays a tetragonal perovskite symmetry, these disrupted domains are believed to be 90° ferroelectric domains. The diffraction contrast mechanism in TEM is hence originated from the difference in polar axis across the domain wall, and from the strain field associated with the domain wall. The evolution of the morphology of the nanometer-sized domains under external electric field for the very first $1\frac{1}{4}$ cycles was recorded, and is shown in Figure 14.3. A diagram of the field strength E, with time, t, is inserted in the figure for clarity. The static fields were increased stepwise and images were taken at field levels of 0, +40, +80, 0, −40, −80, and +80 kV cm^{-1}, respectively. The bright-field images were recorded from the interior of a single grain with its $<1\bar{1}1>$ direction parallel to the electron beam under multi-beam conditions. The white arrow in Figure 14.3a indicates the in-plane <110> crystallographic direction, while the white arrows in Figure 14.3b, c, e and f denote the directions of the applied electric fields.

Initially, at the thermally depoled state, the nanometer-sized domains showed some extent texture (Figure 14.3a). When the electric field was increased to $+40\,kV\,cm^{-1}$, however, these short-range features broke up and were reorganized (Figure 14.3b).

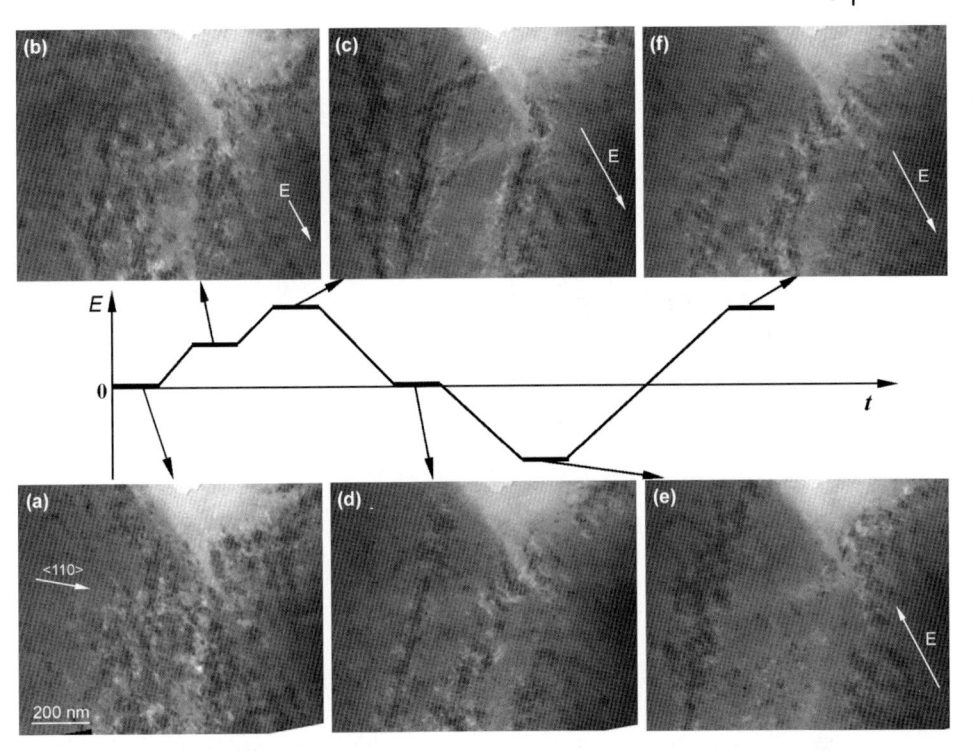

Figure 14.3 Bright-field TEM images of the morphology of nanometer-sized ferroelectric domains in the PZT EC65 ceramic at different field levels: **B**//<1$\bar{1}$1>; multibeam condition. (a) The initial morphology; (b) $+40\,\text{kV}\,\text{cm}^{-1}$; (c) $+80\,\text{kV}\,\text{cm}^{-1}$; (d) $0\,\text{kV}\,\text{cm}$; (e) $-80\,\text{kV}\,\text{cm}^{-1}$; (f) $+80\,\text{kV}\,\text{cm}^{-1}$. The in-plane <110> direction is shown in panel (a). The white arrows in panels (b), (c), (e), and (f) indicate the field direction.

At $+80\,\text{kV}\,\text{cm}^{-1}$, these irregular-shaped domains were forced to align along the {110} plane; a nearly contrast-free zone appeared between these aligned domains, as shown in Figure 14.3c. The aligned domains remained during the second quarter of the electrical cycle when the field was decreased to zero (see Figure 14.3d); however, upon reversing the electric field to $-80\,\text{kV}\,\text{cm}^{-1}$, these aligned domains were disrupted (Figure 14.3e). Returning the external field to $+80\,\text{kV}\,\text{cm}^{-1}$ led to a partial recovery of the domain structure at the previous application of the $+80\,\text{kV}\,\text{cm}^{-1}$ level (Figure 14.3f).

For the tetragonal PZT-EC65 ceramic examined in these studies, proprietary dopants were added to disrupt the regular micrometer-sized domains in order to optimize the piezoelectric properties. After thermal annealing, the ceramic lost piezoelectricity but remained ferroelectric. The electric field-induced domain switching involved both 180 and 90° polarization reorientation. The nearly contrast-free zone between the aligned dark patches in Figure 14.3c suggested that a coalescence of the nanometer-sized domains occurred under external fields.

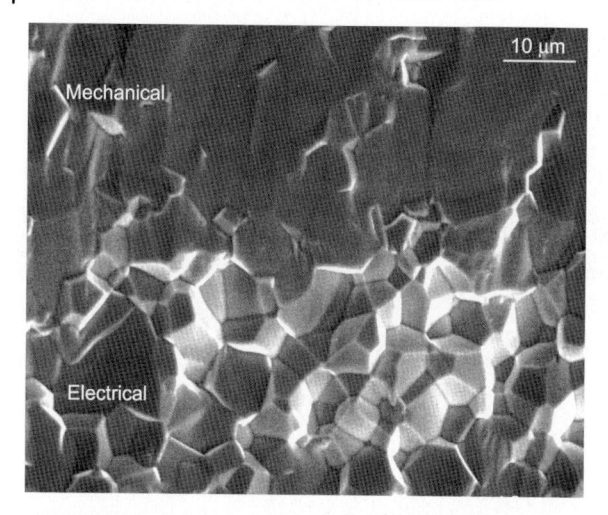

Figure 14.4 Scanning electron microscopy image of the fracture surface of the PZT EC 65 piezoelectric ceramic. The ceramic fractured in an intergranular manner under electrical loadings, but in a transgranular manner under mechanical loadings.

14.4
Grain Boundary Cavitation

Extensive repeated domain polarization switching may cause the failure of ferroelectric ceramics, one form of which is grain boundary cracking in polycrystalline materials that have been subjected to an extended cycling of bipolar electric fields [27–29]. It is interesting to note that the grain boundaries in these ceramics are not mechanically weak, and that transgranular fracture features are typically observed under mechanical loadings [27, 28]. The features revealed following scanning electron microscopy (SEM) of a fracture surface generated by electrical and mechanical loadings in sequence in the PZT EC65 ceramic are shown in Figure 14.4. In this case, an abrupt change in the fracture mode became evident when the crack growth driving force was switched from cyclic electric fields to mechanical bending. The electric field *in-situ* TEM technique was applied to investigate the field-induced grain boundary fracture in this ferroelectric ceramic [14].

Conventional TEM examinations revealed that a thin amorphous layer could exist along the grain boundaries, while pockets of amorphous phase accumulations were identified at the triple junctions of grain boundaries. Subsequent energy-dispersive spectrometry (EDS) analyses of the amorphous phase demonstrated a composition, the major components of which included Pb, Si, and Al. The abundance of Pb was believed to have resulted from an excess of PbO being added to compensate for vaporization losses, while Si and Al were among the impurities segregated to the grain boundaries. As noted in previous *in-situ* TEM studies, the amorphous phase was considered to have caused the electric field-induced grain boundary cracking [14].

14.4 Grain Boundary Cavitation | 327

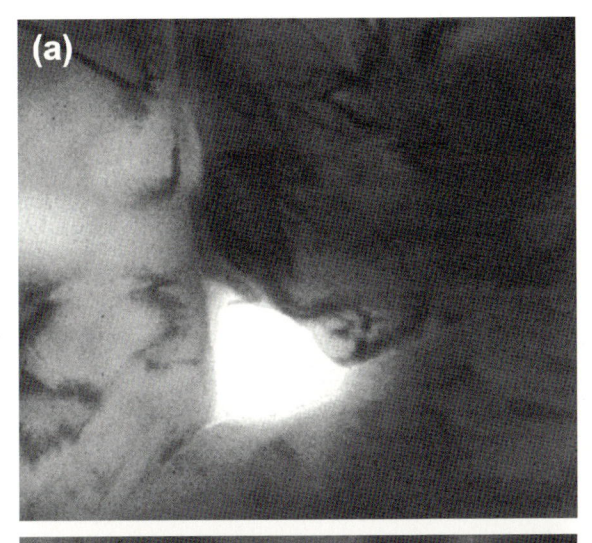

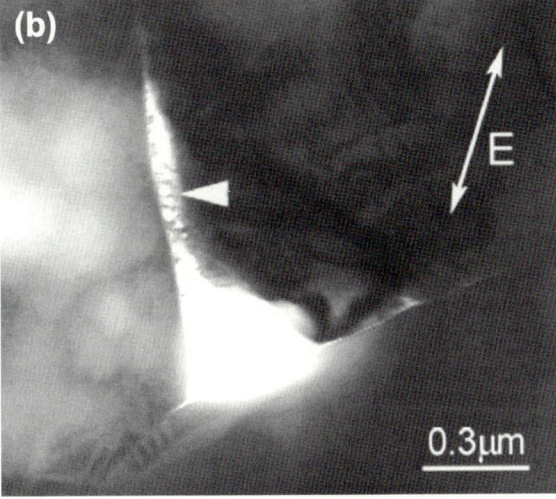

Figure 14.5 Bright-field TEM images revealed that the electric field-induced fracture initiated from a pore at the triple junction of grain boundaries. (a) The initial morphology of the pore; (b) After 50 000 cycles of electric field of $\pm 48\,\text{kV cm}^{-1}$. The bright arrow indicates the field direction. The microcrack was initiated through a cavitation process.

The development of microcracks from a pore in the PZT EC65 ceramic after repeated electrical cycling at a frequency of 30 Hz is shown in Figure 14.5. The pore, which was located at the triple junction of grain boundaries and found about 5 μm from the central perforation, was considered to be a sintering defect rather than due to the transmission electron microscope sample preparation. The initial morphology of the pore before application of the electric field is shown in Figure 14.5a, as an irregular bright triangle with the corners pointing along three grain boundaries.

However, when cyclic electric fields were applied along the direction of the arrow shown in Figure 14.5b, the three grain boundaries became inclined to the electric field. The morphology of the pore after cycling for 50 000 cycles at a field-amplitude of ±48 kV cm^{-1} is shown in Figure 14.5b. A comparison of Figure 14.5a and b indicates that cracks developed from the corners of the pore and propagated along the grain boundaries. The field-induced cracking occurred along all three grain boundaries, irrespective of their orientations, although one grain boundary experienced the most crack growth. The strong tendency of grain boundary cracking is consistent with the intergranular nature of the electric field-induced crack growth observed in the same material of bulk form [28].

A close examination of the cracks revealed that cavities were formed along the grain boundary, as indicated by the bright triangle in Figure 14.5b. The cavities had diameters of 30–40 nm, and remained isolated in part of the grain boundary but coalesced into a microcrack at another part of the boundary. The field-induced crack growth process appeared to consist of cavitation, microcrack formation, and coalescence, and this was verified by a series of images taken from a growing crack (see Figure 14.6). Initially, a grain boundary crack was induced from the edge of the perforation in the TEM specimen, and seen to extend to point "A" after 32 000 cycles at ±36 kV cm^{-1}. Ahead of the crack tip, local damage was found at point "B", the triple junction of the grain boundaries (Figure 14.6a) although, as the electric cycling continued, the crack tip advanced to point "C" (Figure 14.6b). During the same period, the damage at point "B" was extended into a microcrack and the portion of the grain boundary between the crack tip ("C") and the microcrack ("B") was thinned, indicating that damage had occurred over that portion of the grain boundary. The form of the damage is shown in Figure 14.7, magnified from the area marked by the square window in Figure 14.6b. The thinned section shown in Figure 14.7 revealed, again, an almost completely cracked area with cavities. Upon further electrical cycling, the portion of the grain boundary from point "C" to "B" was broken and the crack tip moved ahead to beyond point "B" (Figure 14.6c). Such a cycle of cavity nucleation, growth and coalescence was repeated as electrical cycling continued, and resulted in stable crack growth along the grain boundary phase over many grains.

It is widely believed that the electric field-induced fracture in ferroelectric ceramics is caused by mismatch stresses resulting from incompatible piezoelectric strain [27–29]. Under this assumption, the grain boundary cavitation process found in this *in-situ* TEM study was unusual because it occurred at room temperature, where the grain boundary phase would be expected to behave in a highly brittle manner under mechanical stresses. In fact, the cavitation of the glassy phase suggested that a local melting of the grain boundary phase may have taken place during electrical cycling. The field-induced grain boundary fracture process was interpreted as follows. The amorphous phase was expected to have a much lower dielectric permittivity than the ferroelectric grain interior. Hence, the applied electric field was highly intensified at grain boundaries due to the field-splitting effect, and this led to local dielectric breakdown that, in turn, generated transient heat. Due to its low melting point, the amorphous phase melted for a transient

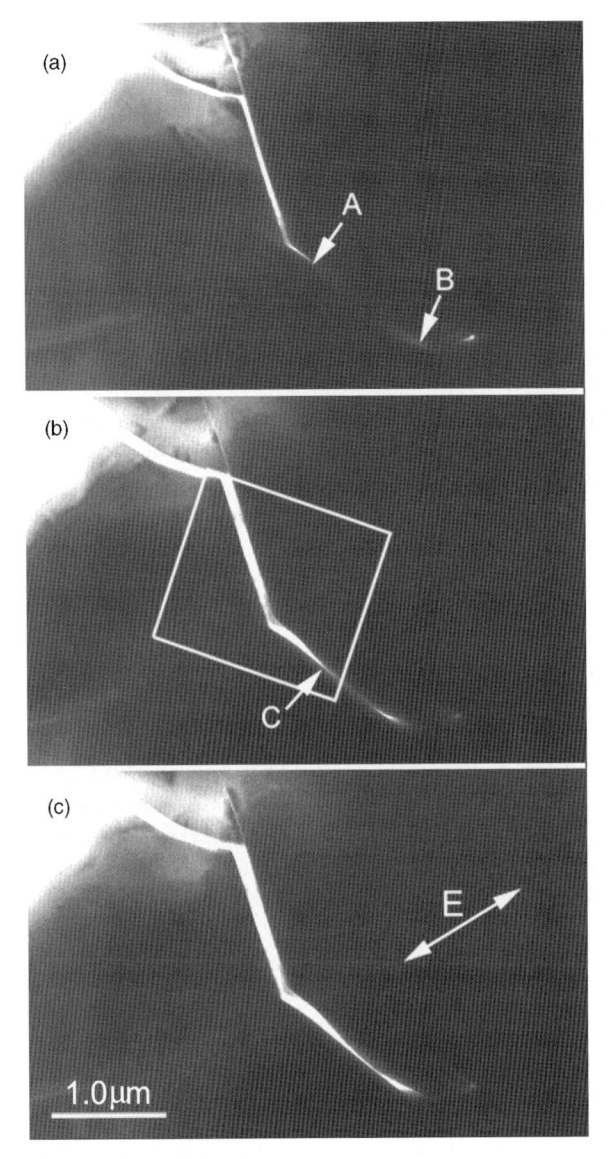

Figure 14.6 Bright-field TEM images reveal the electric field-induced crack growth behavior. (a) After 32 000 cycles at $\pm 36\,\text{kV cm}^{-1}$; (b) After 22 000 cycles at $\pm 56\,\text{kV cm}^{-1}$; (c) After 11 000 cycles at $\pm 60\,\text{kV cm}^{-1}$. The direction of the electric field is indicated by the bright arrow in panel (c).

period. It is suggested that the grain boundary cavities developed under incompatible piezoelectric stress during the transient heating. On the next pulse of the dielectric breakdown, additional cavities were added and the remaining ligaments were further weakened, leading to a coalescence of the cavities and advance of the grain boundary crack.

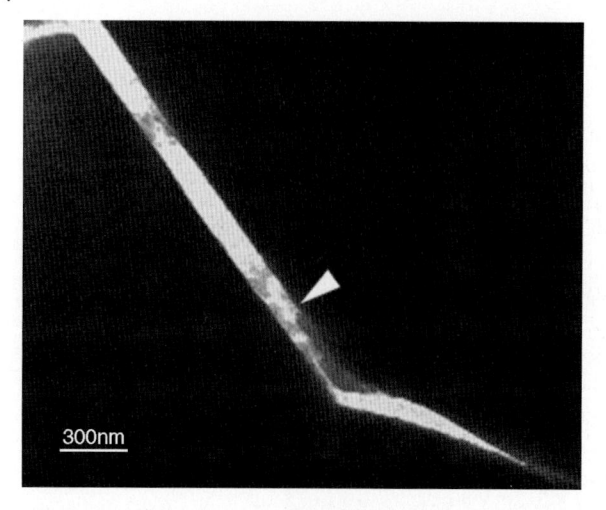

Figure 14.7 Bright-field TEM image of the boxed area in Figure 14.6b. The bright triangle indicates the cavities in the glass ligament behind the crack tip.

The details of a semi-quantitative analysis of the observed field-induced fracture were provided elsewhere [14], where the local melting event was supported by the estimation of several key parameters. In addition, the scenario of grain boundary phase melting pictured by the electric field *in-situ* TEM study was in good agreement with the findings of previous studies conducted in bulk dielectric ceramics. Beauchamp [32] noted a correlation of dielectric breakdown strength with grain size in MgO, and also observed – using optical microscopy – that dielectric breakdown would always be initiated at the grain boundaries, most often at triple junctions of grain boundaries. When Ling and Chang [33] performed *in-situ* optical microscopy study on the failure mechanism of $BaTiO_3$ multilayer capacitors, they observed local melting and crack formation in the $BaTiO_3$ ceramic. Supportive evidence that was more directly related to the current *in-situ* TEM study was provided by Kanai and coworkers [34], who evaluated the electric field-induced failure in the relaxor ferroelectric ceramics of $(Pb_{0.875}Ba_{0.125})$ $[(Mg_{1/3}Nb_{2/3})_{0.5}(Zn_{1/3}Nb_{2/3})_{0.3}Ti_{0.2}]O_3$ in the bulk form. In the Pb-rich ceramics, continuous amorphous layers were observed along the grain boundaries and pockets of the amorphous phase at triple junctions of grain boundaries with TEM, and the electric field-induced intergranular fracture at accelerated conditions ($35\,kV\,cm^{-1}$, $85\,°C$, 95% relative humidity) with SEM. Most importantly, the exposed grain boundaries on the fracture surface were found to be full of cavities with diameters of about 100 nm. Considering that dielectric breakdown had occurred during these tests, and that the PbO-containing amorphous phase had a low melting point and low glass transition temperature, these cavities may well have been the result of local melting.

14.5
Domain Wall Fracture

The use of piezoelectric single crystals avoids such grain boundary fracture and, in addition, much higher piezoelectric properties are to be expected. Unfortunately, the ability to grow PZT single crystals has proven markedly intractable, with recent studies of $Pb(Zn_{1/3}Nb_{2/3})O_3$–$PbTiO_3$ and $Pb(Mg_{1/3}Nb_{2/3})O_3$–$PbTiO_3$ having yielded large single crystals with compositions close to the rhombohedral/tetragonal morphotropic phase boundary in both systems. Yet, these crystals exhibited ultrahigh piezoelectric properties [23].

The *in-situ* TEM technique has been applied to a $0.65Pb(Mg_{1/3}Nb_{2/3})O_3$–$0.35PbTiO_3$ single crystal with tetragonal crystal structure to examine the response to external fields [13]. The TEM specimen was prepared from a {010} thin slice, and the electrodes were so deposited that the nominal field was parallel to the <001> direction. As shown in Figure 14.8a, the crystal specimen in this area contained lamellar domains with widths of 0.1 to 0.5 μm. Convergent beam electron diffraction analysis indicated that these were 90° domains with in-plane polarization vectors, and the domain walls were shown to be parallel to the {101} plane.

Within the area shown in Figure 14.8a, a crack was also generated during ion milling; this pre-existing crack was about 3 μm long and inclined to the domain walls (the dark arrow in Figure 14.8a denotes the crack tip). First, static electric fields were applied to the specimen while the specimen was imaged in the transmission electron microscope. The direction of the applied electric field was parallel to the <001> direction (as indicated in Figure 14.8b), and the field was increased gradually until crack growth was detected. When the crack growth was observed, the electric field was turned off immediately. Photographic images recorded from two successive applications of the static electric field are shown in Figure 14.8b and c; here, the arrows indicate the locations of the crack tip as it moved ahead. On the first application of the electric field, the crack growth was detected when the actual applied field reached $20\,kV\,cm^{-1}$. The resultant crack configuration is shown in Figure 14.8b, where the original crack had grown forward for about a couple of domain spacings, and then abruptly altered its direction to propagate along a direction close to the domain wall. This observation demonstrated that the static electric field, and not only cyclic fields, is capable of inducing fracture in piezoelectric crystals.

The crack growth induced by the static electric field was both stable and repeatable. When the electric field was shut down, the crack growth stopped, but when the static field was resumed the crack grew again when the field reached $20\,kV\,cm^{-1}$. The crack growth was confined to the {101} plane (as shown in Figure 14.8c). The repeatability of the critical electric field required to induce crack growth in the two consecutive electric loadings indicated that the field-induced fracture process was not strongly dependent on the domain reconfiguration in the crystal caused by the prior application and subsequent removal of the electric field. The area around the crack

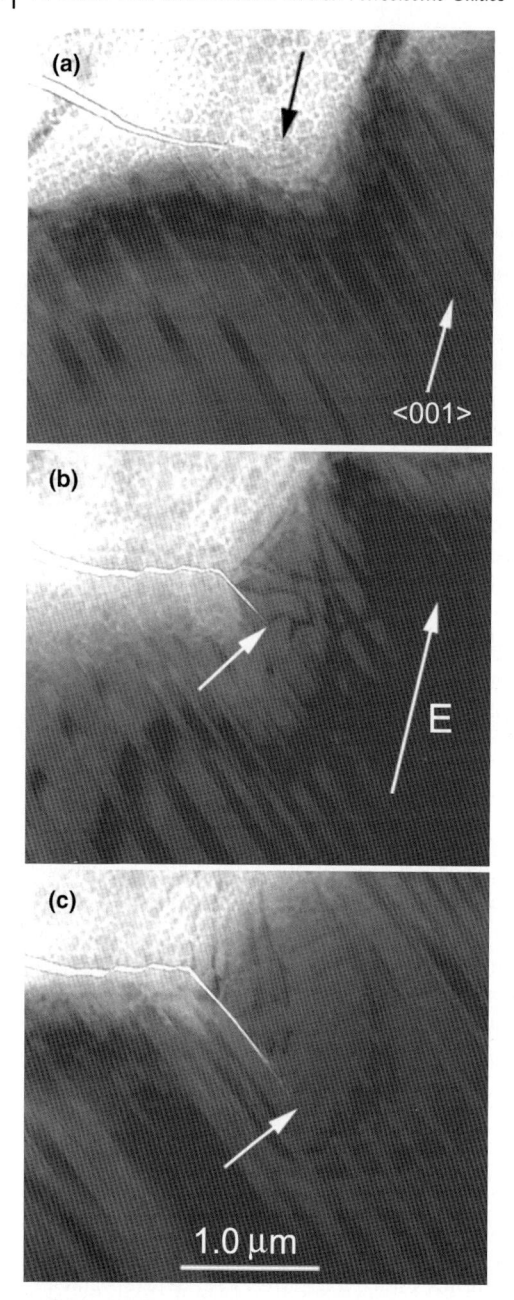

Figure 14.8 Bright-field TEM images showing the electric field-induced domain wall fracture in a $0.65Pb(Mg_{1/3}Nb_{2/3})O_3$–$0.35PbTiO_3$ single crystal with tetragonal structure: $\mathbf{B}//<010>$; multibeam condition. (a) The pre-existing crack and the regular 90° domain stripes. The in-plane <001> direction is indicated by the bright arrow. The domain wall is parallel to the {101} plane; (b) Crack growth after the first application of a static field of $20\,kV\,cm^{-1}$. The field was applied along the <001> direction; (c) Crack growth after a second application of the same field.

tip was examined at higher magnifications, whereby it was confirmed that the crack had propagated exactly along the domain wall.

Electric field-induced crack growth was also performed in a bulk $0.65\text{Pb}(\text{Mg}_{1/3}\text{Nb}_{2/3})\text{O}_3$–$0.35\text{PbTiO}_3$ single crystal in order to determine whether the results obtained from the *in-situ* TEM study were representative. In this case, the crystal was cut to dimensions of $6.0 \times 3.0 \times 0.8$ mm, with surfaces parallel to the pseudo-cubic {010} planes. Two major surfaces of the crystal (6.0×3.0 mm) were polished with 1 μm particle size diamond paste. In order to minimize the residual stresses introduced by grinding and polishing, the crystal was then annealed at 300 °C for 3 h. Au films were evaporated onto the two side surfaces (6.0×0.8 mm) as electrodes, in order to provide fields along the <001> direction. The crystal was then loaded mechanically by pressing a Vickers diamond indentor against the {010} surface at 0.5 N for 5 s. The indentor was so oriented that two cracks emitting from the indentation corners would be perpendicular to the field direction, while the other two were parallel to the field direction, as schematically shown in Figure 14.9a. The Au film electrodes were then connected to a high-voltage power supply, and cyclic sinusoidal electric fields of 30 Hz were applied to the crystal along the <001> direction. Optical microscopy with cross-polarized light revealed an electric field-induced growth of two cracks that were perpendicular to the field. After the initial self-similar extension along the {001} plane, the crack deflected by about 45° to follow the {101} domain wall. The tip of the growing crack (seen optical microscopy) after cycling at a nominal field of $\pm 4.0\,\text{kV}\,\text{cm}^{-1}$ for 190 000 cycles is shown in Figure 14.9b. Therefore, the electric field-induced crack growth behavior in the bulk crystal is the same as that in the thin crystal, observed using TEM.

Although the grain boundary is eliminated from single crystals, the domain wall, separating volumes with uniform polarizations, has been revealed to be a preferred pathway for electric field-induced fracture. This is due to the incompatible strain at the domain wall. Consider two adjacent 90° domains A and B with the polarization vector parallel to the [1] direction. The applied static electric field is parallel to the polarization of the domain A, but perpendicular to the polarization of the adjacent domain B. The piezoelectric strain developed in the domain A without constraints can be expressed as:

$$\begin{pmatrix} 0 & 0 & d_{31} \\ 0 & 0 & d_{31} \\ 0 & 0 & d_{33} \\ 0 & d_{15} & 0 \\ d_{15} & 0 & 0 \\ 0 & 0 & 0 \end{pmatrix} \times \begin{pmatrix} 0 \\ 0 \\ E \end{pmatrix} = \begin{pmatrix} d_{31} E \\ d_{31} E \\ d_{33} E \\ 0 \\ 0 \\ 0 \end{pmatrix} \qquad (14.1)$$

334 | *14 In-situ TEM with Electrical Bias on Ferroelectric Oxides*

(a)

<001>
<100>
Au film

(b)

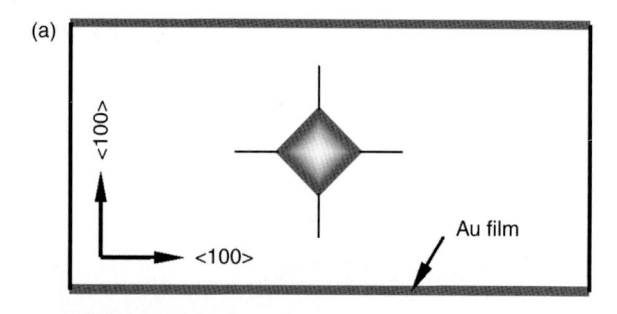

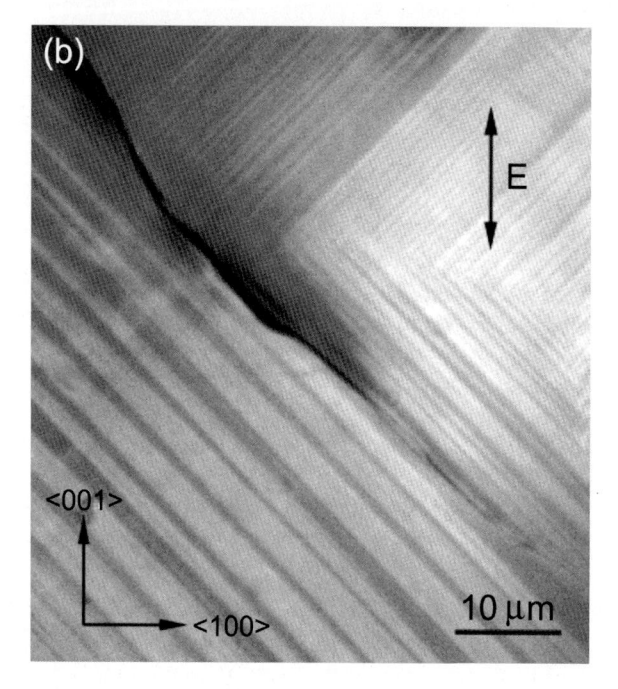

E

<001>
<100>

10 μm

Figure 14.9 Confirmation of the electric field-induced domain wall fracture in a bulk 0.65Pb$(Mg_{1/3}Nb_{2/3})O_3$–0.35PbTiO$_3$ single crystal. (a) Specimen geometry of the bulk crystal. The crack was introduced by a Vickers indentation; (b) Optical microscopy image showing the tip of the growing crack after 190 000 cycles at $\pm 4.0\,\text{kV}\,\text{cm}^{-1}$. The field was applied along the <001> direction at a frequency of 30 Hz.

where d_{ij} is the piezoelectric coefficients and E is the applied electric field. The piezoelectric strain generated in the domain B without constraints is expressed as

$$\begin{pmatrix} 0 & 0 & d_{31} \\ 0 & 0 & d_{31} \\ 0 & 0 & d_{33} \\ 0 & d_{15} & 0 \\ d_{15} & 0 & 0 \\ 0 & 0 & 0 \end{pmatrix} \times \begin{pmatrix} E \\ 0 \\ 0 \end{pmatrix} = \begin{pmatrix} 0 \\ 0 \\ 0 \\ 0 \\ d_{15}E \\ 0 \end{pmatrix} \qquad (14.2)$$

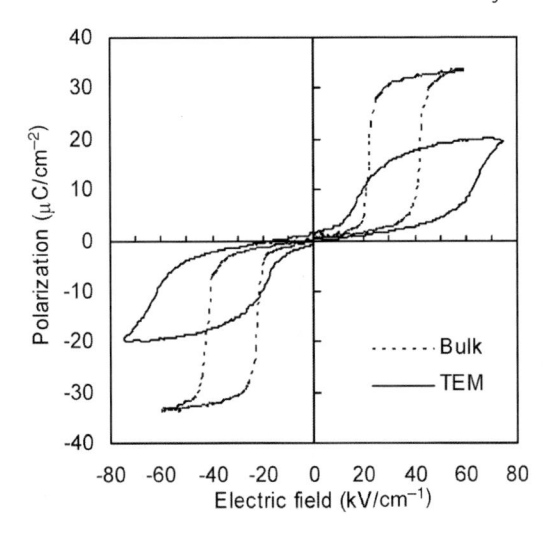

Figure 14.10 Electric field-induced antiferroelectric ↔ ferroelectric phase switching characterized by polarization measurement at 4 Hz in the PZST 45/6/2 polycrystalline ceramic. The hysteresis loop with the dashed line was obtained from a standard sample with 12 mm diameter and 300 μm thickness. The solid line loop line was measured from a dimpled TEM specimen electroded in the configuration shown in Figure 14.2a.

Obviously, the piezoelectric strains at the domain wall are not compatible to each other, and elastic energy will be built up in the close vicinity of the wall. In addition, the 90° domain walls are planar defects and have interfacial energies. Fracture along the domain walls would require a lower energy than along other crystallographic planes. Consequently, the domain wall is a preferred pathway for electric field-induced crack growth.

14.6
Antiferroelectric-to-Ferroelectric Phase Transition

Recently, antiferroelectric ceramics have been attracting increasing attention due to their potential applications in energy-storage devices [21, 22, 35]. The physics mechanism for the energy storage/release is the antiferroelectric ↔ ferroelectric phase switching, controlled by external electric fields. These ceramics are based on the prototypical antiferroelectric $PbZrO_3$ with various dopants such as La, Nb, Sn, and Ti [21, 22, 35–40]. The dopants serve multiple functions, including: adjusting the critical field level for the phase switching; enhancing the energy storage volume density; and facilitating ceramic processing. Consequently, the antiferroelectric polycrystalline ceramics of $Pb_{0.99}Nb_{0.02}[(Zr_{1-x}Sn_x)_{1-y}Ti_y]_{0.98}O_3$ (abbreviated as PZST $100x/100y/2$) were examined with the electric field *in-situ* TEM technique.

In the modified antiferroelectric ceramics, the free energies of the antiferroelectric and the ferroelectric states were so close that only a moderate electric field (<100 kV cm^{-1}) could trigger the phase transition. The field-induced transition in PZST 45/6/2, as revealed by polarization measurements and displaying the characteristic double hysteresis loops at room temperature, is shown in Figure 14.10. The dashed curve was obtained from a standard bulk sample with a diameter of 12 mm and a thickness of 300 µm, with an electric field applied along the thickness direction. The solid curve represents the result from a dimpled TEM specimen with an electrode configuration shown in Figure 14.2a. For the standard bulk sample, the measurement showed a critical field of ~40 kV cm^{-1} for the antiferroelectric-to-ferroelectric phase switching, and ~20 kV cm^{-1} for the backward switching.

As a result of the small difference in the free energy, there is a strong competition between the antiferroelectric ordering and the ferroelectric ordering in the material, and this leads to the formation of incommensurate modulations [36–40]. The modulation appears as regular fringes along {110} planes in the dark-field image, and as satellite spots in the selected area electron diffraction pattern. The modulation in the PZST 45/6/2 ceramic with a wavelength about 2.1 nm is shown in Figure 14.11. The satellites can be expressed as $ha^* + kb^* + lc^* \pm \frac{1}{n}\{a^* + b^*\}$, with $n = 7.29$ in Figure 14.11b.

An *in-situ* TEM study was carried out on specimens of PZST 45/6/2 and PZST 42/4.5/2 polycrystalline ceramics [15, 16]. In the following, results from the PZST 42/4.5/2 ceramic specimen are presented. In this specimen, one of the grains was found to have its [001] axis parallel to the electron beam without secondary tilting; this grain was about 5 µm in size. The [001] zone-axis diffraction pattern was recorded at different field levels to monitor the evolution of the incommensurate modulation. The change of satellite spots in the [001] zone-axis diffraction pattern under static electric field is depicted in Figure 14.12. Initially, this grain displayed two sets of satellite reflections; as evident in Figure 14.12a, the satellite spots were strong in intensity and slightly elongated in shape. The static electric fields were then applied and increased stepwise, with the applied field direction determined as being along the gray line in Figure 14.12b. At an actual electric field of 60 kV cm^{-1}, the horizontal set of satellite spots almost completely disappeared. However, for the vertical set many satellite spots also disappeared, while the remaining spots exhibited severe streaking (Figure 14.12b).

The bright rectangular boxes in Figure 14.12a and b enclose two fundamental reflections of 120 and 210. The appearance of the satellite spots of these two fundamental reflections at a series of field levels of 0, 8, 20, 30, 40, and 60 kV cm^{-1} is shown in Figure 14.13. It is evident that satellite reflections in both sets began to grow weaker at an electric field of 8 kV cm^{-1}, which was far below the macroscopic critical field for the antiferroelectric-to-ferroelectric phase transition. The satellite spots in the horizontal set had almost completely disappeared at 40 kV cm^{-1}, and became slightly diffuse before their final disappearance. In contrast, the satellite reflections in the vertical set began to show severe streaking at 20 kV cm^{-1}, and their presence persisted up to 60 kV cm^{-1}. This demonstrates an orientation dependence of their response to the external field.

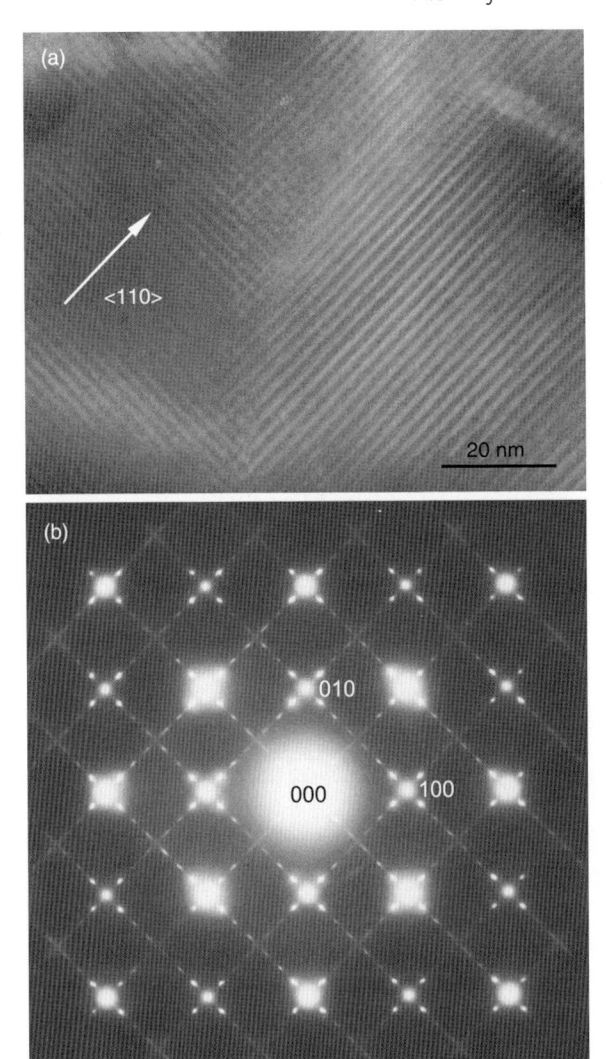

Figure 14.11 The incommensurate modulation observed in the PZST 45/6/2 ceramic by TEM. (a) Dark-field image formed with the 110 fundamental spot and its four satellite spots: **B**//<001>; multibeam condition. The regular fringes with wavelength about 2.1 nm are parallel to the {110} planes; (b) The selected area electron diffraction pattern with a zone axis of <001>. The diffraction pattern is indexed on the basis of a pseudocubic structure.

The value of n in $ha^* + kb^* + lc^* \pm \frac{1}{n}\{a^* + b^*\}$ was measured from electron diffraction patterns at different field levels, and is plotted in Figure 14.14. At each field level, at least 10 pairs of satellites were measured; for diffused satellite spots, the measurement was performed at the point with the highest intensity. A significant difference in n occurred across a 90° domain wall within a single grain. At zero field,

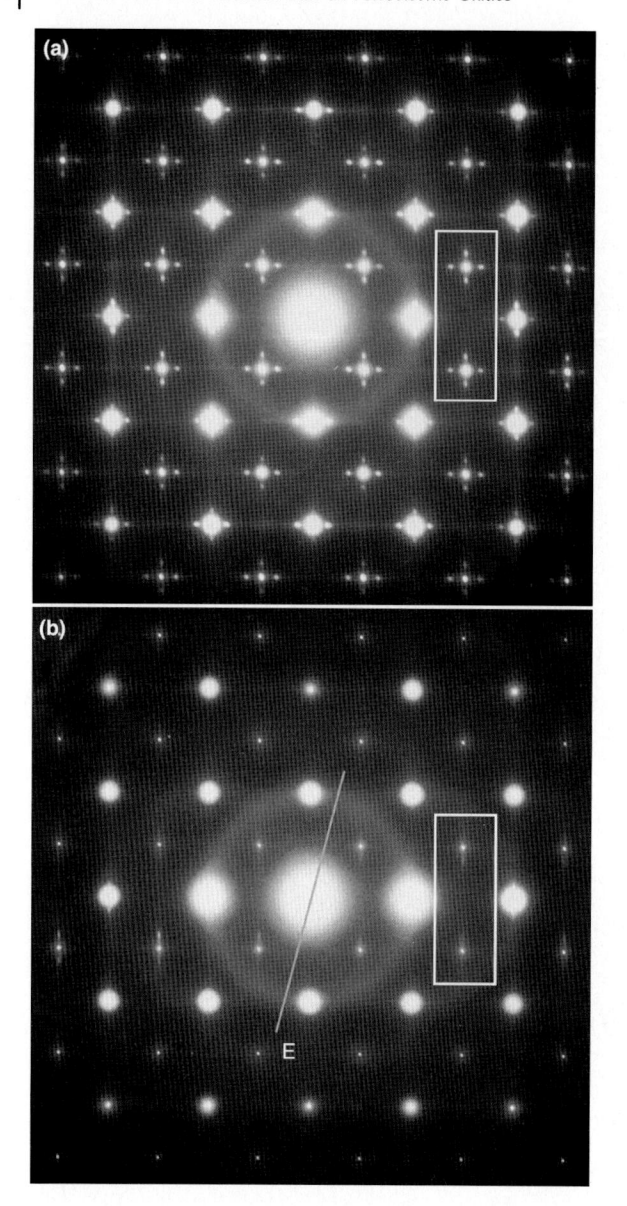

Figure 14.12 Change of the satellite reflections in the <001> zone axis selected area diffraction pattern under external electric field. (a) E = 0; (b) E = 60 kV cm^{-1}. The gray line indicates the direction of the applied field. The two rectangular boxes enclose the 120 and 210 fundamental reflections.

$n = 6.78$ for the horizontal set of satellite spots, corresponding to a wavelength of 1.97 nm, and $n = 7.67$ for the vertical, corresponding to a wavelength of 2.23 nm. It should be noted here that the grain was about 5 μm in size and the material had been annealed at 1300 °C for 6 h after hot pressing. Consequently, residual stresses rather

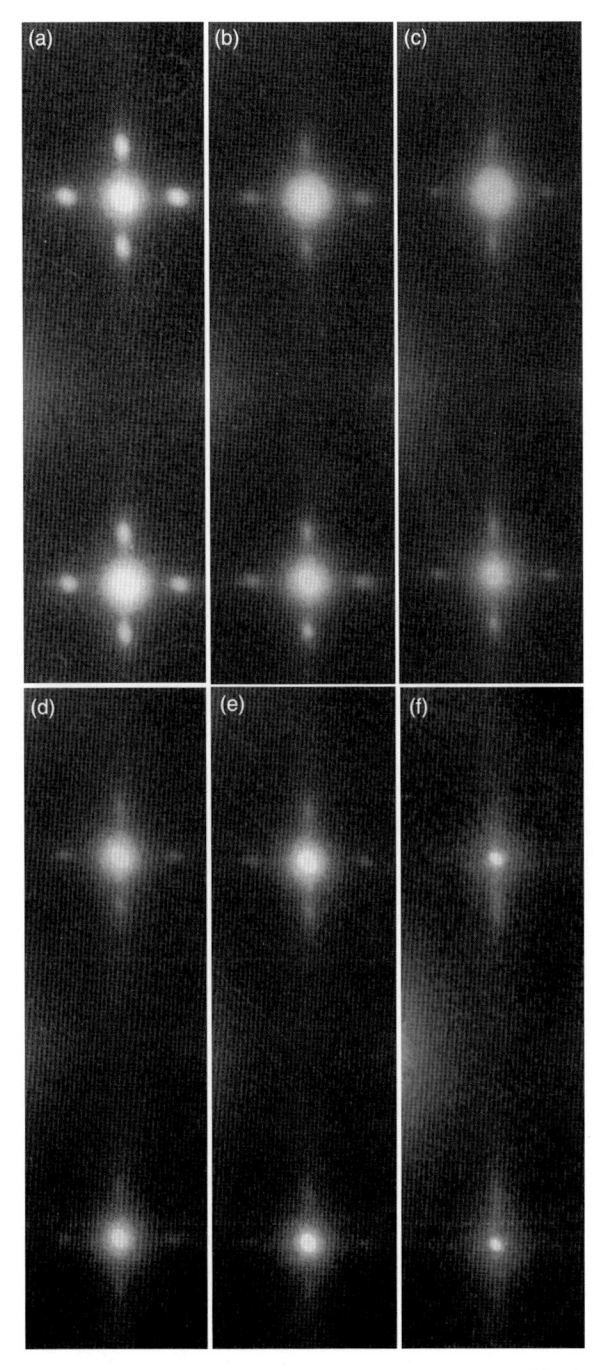

Figure 14.13 Evolution of the satellite reflections surrounding the 120 and 210 fundamental reflections. (a) E = 0; (b) E = 8 kV cm^{-1}; (c) E = 20 kV cm^{-1}; (d) E = 30 kV cm^{-1}; (e) E = 40 kV cm^{-1}; (f) E = 60 kV cm^{-1}.

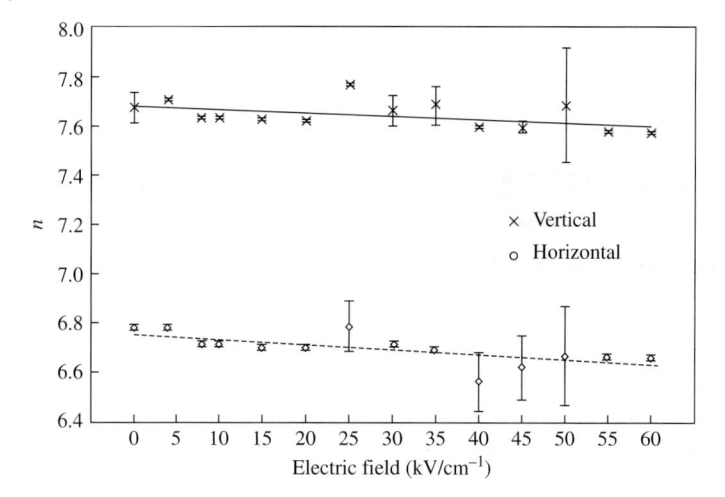

Figure 14.14 Change of n value with applied electric fields for the vertical set and horizontal set of satellite reflections shown in Figures 14.12 and 14.13.

than chemical heterogeneities would most likely be responsible for the significant difference in wavelength.

Another evident feature in Figure 14.14 was the negligible change in the incommensurate modulation wavelength under strong applied electric fields. A linear fit to the mean value of n gave a slope of -0.0061 Å per kV cm^{-1} for the horizontal set, and -0.0035 Å per kV cm^{-1} for the vertical set. The higher rate of change for the horizontal set was in accordance with an evolution of the satellite spots of that set under electric fields. In Figure 14.13, these spots were seen to disappear at a lower field level, with the wavelength in real space decreasing slightly with increasing electric fields for both sets. At 40 kV cm^{-1}, this was recorded as 1.95 nm (a 1.2% decrease from the initial wavelength) for the horizontal set, and as 2.22 nm (a 0.7% decrease from the initial wavelength) for the vertical set.

The presence of incommensurate modulations in antiferroelectric ceramics is believed to be a result of the competition between the ferroelectric and the antiferroelectric ordering [39]. A continuous increase in the modulation wavelength with increasing temperature was observed previously in ceramics with compositions similar to the PZST 45/6/2 and PZST 42/4.5/2 examined [38]. This increase was as high as 40% (from 2.1 nm to ca. 2.94 nm) when the temperature was increased from room temperature to 150 °C [38]. However, observations on the electric field-induced transition of the incommensurate modulation failed to reproduce the huge change in wavelength. Instead, the change in wavelength under electric fields was minimal (<2%), even at fields close to the critical field for the antiferroelectric-to-ferroelectric transition. Consequently, the results of the electric field *in-situ* TEM study suggested the occurrence of a different mechanism for the electric field-induced phase transition compared to the thermally induced study. Taken together, the results from these TEM studies suggested that the incommensurate modulation in modified perovskite antiferroelectric oxides existed in the form of a transverse Pb-cation displacement wave. In fact, the observed incommensurate modulation is a mixture of commensurate modulations, where modulations in the antiferroelectric 90° domains consisted of 180° domain slabs with thicknesses of three or four layers of the pseudo-tetragonal {110} plane [16].

14.7
Relaxor-to-Ferroelectric Phase Transition

The dielectric property of $Pb(Mg_{1/3}Nb_{2/3})O_3$, the prototype relaxor ferroelectric compound, displays a high relative permittivity, a broad dielectric peak, and a strong frequency dispersion [25, 26, 41, 42]. The origin of the relaxor ferroelectric behavior can be traced back to the structure of the compound, where nanometer-scale (<5 nm) 1:1 cation ordering exists on the B-site of the ABO_3 perovskite structure [43, 44]. In addition to the nanoscale chemical ordering, this compound also contains nanoscale polar ordering [45]. Despite extensive theoretical and experimental studies having been conducted over several decades, however, the relationship between these two types of nanoscale ordering remains unclear [25, 46].

The polar nanoregions in $Pb(Mg_{1/3}Nb_{2/3})O_3$ can grow into micrometer-sized ferroelectric domains when driven by external electric fields at temperatures below −60 °C, which corresponds to a field-induced first-order phase transition [47, 48]. Measurements of the field-induced polarization and the X-ray diffraction peak shifts under bias at low temperatures have suggested that the phase transition takes place abruptly after an incubation period [48]. It is also inferred that the polar nanoregions become coarser during the incubation period; however, the morphological evolution of the polar nanodomains during the phase transition has yet to be directly imaged at the nanometer scale.

For many years, TEM has been used to image both polar and chemical nanodomains in relaxor ferroelectric compounds [43, 44]. Clearly, a transmission electron microscope equipped with a specimen holder with electrical bias and cooling capabilities would be ideal for studying the electric field-induced relaxor to normal ferroelectric phase transition. The acquisition of such a unique specimen holder with liquid nitrogen cooling allowed the *in-situ* TEM technique to be extended to polycrystalline specimens of a Sc-doped $Pb(Mg_{1/3}Nb_{2/3})O_3$ ceramic [18, 19]. In this case, the Sc was introduced by forming a solid solution of $0.92Pb(Mg_{1/3}Nb_{2/3})O_3$–$0.08Pb(Sc_{1/2}Nb_{1/2})O_3$ (hereafter abbreviated as PSMN8). Sc-doping has been shown to greatly enhance the cation ordering, and to slightly enhance the polar ordering in $Pb(Mg_{1/3}Nb_{2/3})O_3$ ceramic [49, 50]. Since, at low doping levels large cation-ordered domains can be obtained, while the relaxor behavior is retained, this facilitates a direct observation of the interaction between the chemical domains and the growing polar domains.

The *in-situ* TEM study was carried out by field-cooling (at $2\,°C\,min^{-1}$) a PSMN8 TEM specimen under a static electric field of $10\,kV\,cm^{-1}$, from room temperature to predetermined temperature points [18, 19]. When one grain with its <110> direction close to the electron beam was focused on at room temperature (as shown in

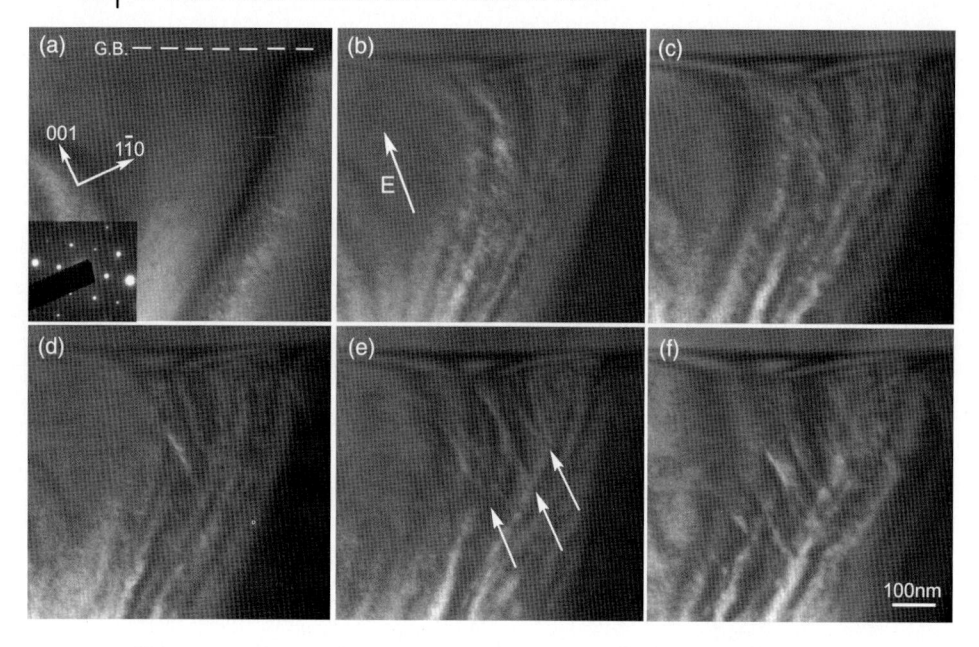

Figure 14.15 The morphological evolution of the polar nanodomains during field cooling under 10 kV cm^{-1}, as revealed by the *in-situ* TEM technique. The applied field direction is shown by the arrow in panel (b). (a) The initial polar nanoregions at room temperature. The inset shows the <110> zone-axis selected area electron diffraction pattern; (b) −50 °C; (c) −55 °C; (d) −70 °C; (e) −90 °C; (f) −90 °C after 30 min.

Figure 14.15), a very faint contrast of polar nanoregions was noted, in association with bending contours (Figure 14.15a). The inset in Figure 14.15a shows the <110> zone-axis electron diffraction pattern, where the $1/2$(111) superlattice diffraction spots are clearly seen, while the in-plane directions <001> and <1$\bar{1}$0> are indicated by bright arrows. The grain boundary is also denoted by the bright dashed line at the top of the image.

A fixed static electric field of 10 kV cm^{-1}, with the direction indicated by the bright arrow in Figure 14.15b, was then applied. The direction of the applied field was very close to the <001> direction, but no detectable morphological changes of the polar nanoregions were observed at room temperature under the static field. However, when the specimen was cooled to −50 °C, clear changes were noted in the area close to the grain boundary (see Figure 14.15b), where clustering of the polar nanoregions had occurred. The coalescence of the polar nanodomains continued during further cooling to −55 °C (Figure 14.15c). Up to this temperature, the morphology of the nanodomains had retained an irregular shape, without well-defined domain walls, but when the temperature reached −70 °C there was a dramatic change in the domain morphology (Figure 14.15d), with large ferroelectric domains (>200 nm) with flat domain walls close to the {1$\bar{1}$0} plane appearing in close vicinity to the grain boundary. Further cooling to −90 °C led to the growth of existing large domains and the appearance of new large domains (Figure 14.15e). When the temperature was held constant at −90 °C for 30 min, further growth of the large domains was noted (Figure 14.15f).

The results shown in Figure 14.15 indicate that the field-induced relaxor-to-normal-ferroelectric phase transition involves two stages: (i) a gradual coalescence of polar nanodomains; and (ii) an abrupt formation of long-range ferroelectric domains. This process is very similar to that in <111>-oriented $Pb(Mg_{1/3}Nb_{2/3})O_3$ single crystals, as depicted by X-ray diffraction and polarization measurements [48]. It should be noted that the current observations were made with an individual grain in a polycrystalline ceramic and that, in contrast, polarization measurements from polycrystalline ceramic samples are unable to demonstrate such a two-stage process. Clearly, it is at this point where TEM can make unique contributions.

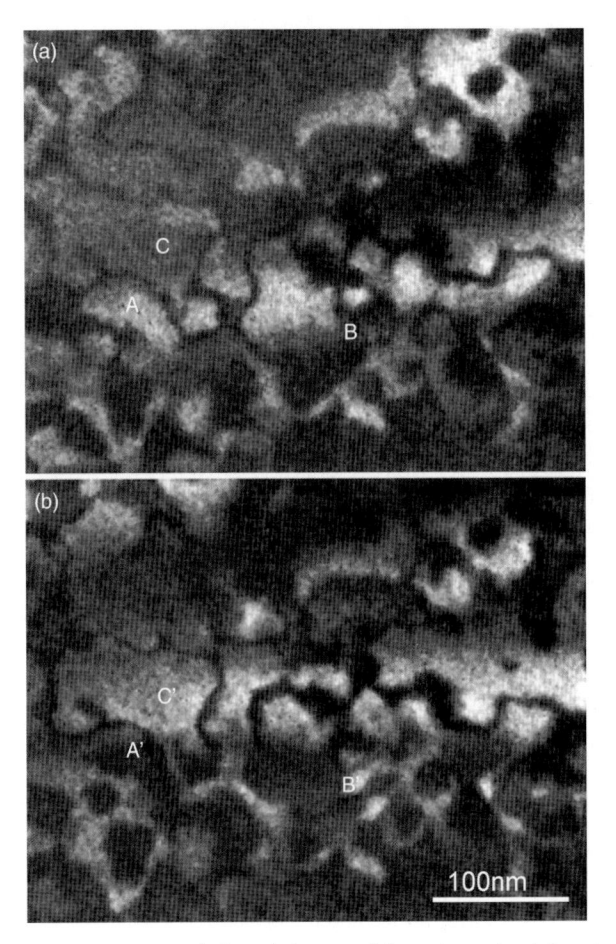

Figure 14.16 Morphological changes of the cation-ordered domains during field-cooling. (a) 10 kV cm^{-1} at room temperature; (b) 10 kV cm^{-1} at −65 °C. See text for details.

During the field-cooling process shown in Figure 14.15, cation-ordered domains in the same grain were also recorded under each field level, through dark-field imaging with the $^1/_2\{111\}$ superlattice diffraction. The original morphology of these chemical domains at room temperature, and that at $-65\,°C$ under $10\,kV\,cm^{-1}$, are shown in Figure 14.16. Here, it can be seen that the cation-ordered domains in PSMN8 are on the order of 100 nm, and much larger than that in pure $Pb(Mg_{1/3}Nb_{2/3})O_3$. The comparison between Figure 14.16a and b reveals that both the size and the morphology of the cation-ordered domains remain unchanged during the field-induced relaxor-to-normal ferroelectric phase transition process. However, a closer examination indicates that there are obvious changes in the contrast of some cation-ordered domains. For example, domain A in Figure 14.16a has a bright contrast, and this changes to a dark domain A′ in Figure 14.16b; likewise, the dark domain B in Figure 14.16a turns into a bright domain B′ in Figure 14.16b. Many other cation-ordered domains with unchanged contrast have been identified (e.g., C in Figure 14.16a versus C′ in Figure 14.16b).

The persistence of the morphology of cation-ordered domains during the electric field-induced relaxor-to-ferroelectric phase transition is anticipated, because the diffusion activities of B-site cations are extremely limited at these low temperatures. The change in contrast of some chemical domains was not initially expected, however, and can be explained as follows. In PMN-based complex perovskite oxides, the B-site cations are packed periodically on {111} planes in the ordered domains. Therefore, the cation-ordered domains have four variants: $^1/_2(111)$; $^1/_2(\bar{1}11)$; $^1/_2(1\bar{1}1)$; and $^1/_2(\bar{1}1\bar{1})$. During electron diffraction, dynamical conditions may be considered which permit the double diffraction route [40], for example

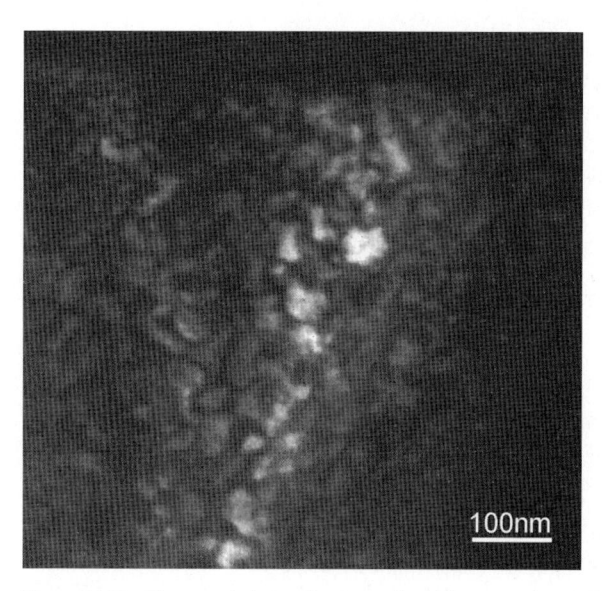

Figure 14.17 The morphology of cation-ordered domains during field-cooling.

$$\left\{\frac{\bar{1}\bar{1}1}{2\,2\,2}\right\} + \{110\} = \left\{\frac{1\,1\,1}{2\,2\,2}\right\} \tag{14.3}$$

That is to say, the $1/2(111)$ superlattice diffraction spot used for dark-field imaging may have contributions from cation-ordered domain variants of both $1/2(111)$ and $1/2(\bar{1}\bar{1}1)$. Upon the application of external electric fields, the double diffraction route may be modified and, as a consequence, the $1/2(\bar{1}\bar{1}1)$ domain variant will turn into a dark contrast, such as domain A in Figure 14.16a. At the same time, other domain variants, $1/2(\bar{1}11)$ and/or $1/2(\bar{1}1\bar{1})$, may become favored for double diffraction to the $1/2(111)$ superlattice spot. This will lead to a brightening in the contrast of these domains, such as domain B in Figure 14.16a.

As noted above, the interactions between cation-ordered chemical domains and growing polar domains would be of great interest. In the PSMN8 ceramic under examination here, large cation-ordered domains facilitate the direct observation of such interactions under electric fields. A dark-field micrograph of the cation-ordered domains in exactly the same area within the same grain as in Figure 14.15 is shown in Figure 14.17. A close comparison of Figure 14.15d and f with Figure 14.17 indicates that, neither the initiation of large polar domains, nor the advancement of the walls of the polar domains, demonstrates any clear correlation with the underlying cation-ordered domains. In other words, the chemical domains appear not to have any strong interactions with the large ferroelectric domains in the 4 atom% Sc-doped $Pb(Mg_{1/3}Nb_{2/3})O_3$ ceramic.

Acknowledgments

The author thanks J.K. Shang, Z. Xu, H. He, and W. Qu for their significant contributions to the studies described here. Financial support from the National Science Foundation of the USA through the CAREER grant DMR-0346819 is acknowledged. The *in-situ* TEM experiments were conducted at the Center for Microanalysis of Materials, University of Illinois at Urbana-Champaign, and at the Materials & Engineering Physics Program, Ames Laboratory, which is supported by the Department of Energy of U.S., Office of Basic Energy Sciences, under Contract No. DE-AC02-07CH11358.

References

1 Blech, I.A. and Meieran, E.S. (1967) *Appl. Phys. Lett.*, **11**, 263.
2 Berenbaum, L. (1971) *J. Appl. Phys.*, **42**, 880.
3 Mori, H., Okabayashi, H., and Komatsu, M. (1997) *Thin Solid Films*, **300**, 25.
4 Yamamoto, N., Yagi, K., and Honjo, G. (1980) *Phys. Status Solidi A*, **62**, 657.
5 Snoeck, E., Normand, L., Thorel, A., and Roucau, C. (1994) *Phase Trans.*, **46**, 77.
6 Krishnan, A., Bisher, M.E., and Treacy, M.M.J. (1999) Ferroelectric thin films VII. Proceedings, MRS Symposium,

November 30–December 3, 1998, Boston, USA, vol. 541, p. 475.
7. Randall, C.A., Barber, D.J., and Whatmore, R.W. (1987) *J. Microsc.*, **145**, 275.
8. Qi, X.Y., Liu, H.H., and Duan, X.F. (2006) *Appl. Phys. Lett.*, **89**, 092908.
9. Johnson, K.D. and Dravid, V.P. (1999) *Appl. Phys. Lett.*, **74**, 621.
10. Tan, X., Du, T., and Shang, J.K. (2002) *Appl. Phys. Lett.*, **80**, 3946.
11. Tan, X. and Shang, J.K. (2001) *Mater. Sci. Eng.*, **A314**, 157.
12. Xu, Z., Tan, X., Han, P., and Shang, J.K. (2000) *Appl. Phys. Lett.*, **76**, 3732.
13. Tan, X., Xu, Z., Shang, J.K., and Han, P. (2000) *Appl. Phys. Lett.*, **77**, 1529.
14. Tan, X. and Shang, J.K. (2002) *Philos. Mag. A*, **82**, 1463.
15. He, H. and Tan, X. (2004) *Appl. Phys. Lett.*, **85**, 3187.
16. He, H. and Tan, X. (2005) *Phys. Rev. B*, **72**, 024102.
17. Tan, X., He, H., and Shang, J.K. (2005) *J. Mater. Res.*, **20**, 1641.
18. Qu, W., Zhao, X., and Tan, X. (2006) *Appl. Phys. Lett.*, **89**, 022904.
19. Qu, W., Zhao, X., and Tan, X. (2007) *J. Appl. Phys.*, **102**, 084101.
20. Lines, M.E. and Glass, A.M. (1977) *Principles and Applications of Ferroelectrics and Related Materials*, Clarendon Press.
21. Pan, W., Zhang, Q., Bhalla, A., and Cross, L.E. (1989) *J. Am. Ceram. Soc.*, **72**, 571.
22. Yang, P. and Payne, D.A. (1992) *J. Appl. Phys.*, **71**, 1361.
23. Park, S.E. and Shrout, T.R. (1997) *J. Appl. Phys.*, **82**, 1804.
24. Wada, S., Suzuki, S., Noma, T., Suzuki, T., Osada, M., Kakihana, M., Park, S.E., Cross, L.E., and Shrout, T.R. (1999) *Jpn. J. Appl. Phys.*, **38**, 5505.
25. Ye, Z.G. (1998) *Key Eng. Mater.*, **155-156**, 81.
26. Zhao, X., Qu, W., Tan, X., Bokov, A., and Ye, Z.-G. (2007) *Phys. Rev. B*, **75**, 104106.
27. Cao, H. and Evans, A.G. (1994) *J. Am. Ceram. Soc.*, **77**, 1783.
28. Shang, J.K. and Tan, X. (2001) *Mater. Sci. Eng.*, **A301**, 131.
29. White, G.S., Raynes, A.S., Vaudin, M.D., and Freiman, S.W. (1994) *J. Am. Ceram. Soc.*, **77**, 2603.
30. McMeeking, R.M. (1989) *J. Appl. Math. Phys.*, **40**, 615.
31. Suo, Z. (1991) *Smart Structures and Materials*, vol. 24/ AMD vol. 123, G.K. Haritos and A.V. Srinivasan (eds), American Society of Mechanical Engineers, New York, p. 1.
32. Beauchamp, E.K. (1971) *J. Am. Ceram. Soc.*, **54**, 484.
33. Ling, H.C. and Chang, D.D. (1989) *J. Mater. Sci.*, **24**, 4128.
34. Kanai, H., Furukawa, O., Nakamura, S., and Yamashita, Y. (1993) *J. Am. Ceram. Soc.*, **76**, 459.
35. Campbell, C.K., van Wyk, J.D., and Chen, R. (2002) *IEEE Trans. Compon. Packag. Technol.*, **25**, 211.
36. Chang, Y., Lian, J., and Wang, Y. (1985) *Appl. Phys. A*, **36**, 221.
37. Speck, J.S., De Graef, M., Wilkinson, A.P., Cheetham, A.K., and Clarke, D.R. (1993) *J. Appl. Phys.*, **73**, 7261.
38. Xu, Z., Viehland, D., and Payne, D.A. (1995) *J. Mater. Res.*, **10**, 453.
39. Viehland, D., Dai, X.H., Li, J.F., and Xu, Z. (1998) *J. Appl. Phys.*, **84**, 458.
40. Knudsen, J., Woodward, D.I., and Reaney, I. (2003) *J. Mater. Res.*, **18**, 262.
41. Smolensky, G.A. (1970) *J. Phys. Soc. Jpn*, **28**, 26.
42. Cross, L.E. (1987) *Ferroelectrics*, **76**, 241.
43. Chen, J., Chan, H.M., and Harmer, M.P. (1989) *J. Am. Ceram. Soc.*, **72**, 593.
44. Hilton, A.D., Barber, D.J., Randall, C.A., and Shrout, T.R. (1990) *J. Mater. Sci.*, **25**, 3461.
45. Burns, G. and Dacol, F.H. (1983) *Solid State Commun.*, **48**, 853.
46. Burton, B.P., Cockayne, E., and Waghmare, U.V. (2005) *Phys. Rev. B*, **72**, 064113.
47. Ye, Z.G. and Schmid, H. (1993) *Ferroelectrics*, **145**, 83.
48. Dkhil, B. and Kiat, J.M. (2001) *J. Appl. Phys.*, **90**, 4676.
49. Farber, L. and Davies, P.K. (2003) *J. Am. Ceram. Soc.*, **86**, 1861.
50. Zhao, X., Qu, W., He, H., Vittayakorn, N., and Tan, X. (2006) *J. Am. Ceram. Soc.*, **89**, 202.

15
Lorentz Microscopy
Josef Zweck

15.1
Introduction

Following its invention in 1931 by Ruska and Knoll [1, 2], the transmission electron microscope became available on a commercial basis within a comparatively short time. As with any other new investigational technique, research groups worldwide rapidly began to exploit the benefits of the new instrument, and to seek alternative uses beyond those for which the microscope had initially been designed. The primary purpose of the microscope was as a magnifying instrument with previously unheard of powers of magnification and resolution, but which was also capable of generating electron beam diffraction patterns. However, as electrons are known to be deflected by magnetic fields due to the Lorentz force – a property that is applied when using magnetic fields to form "lenses" to focus electron beams – an additional, perhaps more obvious, possibility might be to use the microscope to image magnetic structures within magnetic specimens.

The first attempts to achieve this goal were made in 1959 and 1960 by Hale and Boersch [3–5], when the magnetic fields present within a specimen were used to deflect an electron beam which penetrated it. This deflection could, in turn, easily be converted to a contrast not of the specimen itself, but rather to image a plane either above (underfocus, $-\Delta f$) or below (overfocus, $+\Delta f$) the specimen. At this point, the deflected beams would partially overlap with any undeflected or differently deflected portions of the electron beam, so as to increase the beam intensity on a local basis but to leave other areas with a reduced intensity. This concept is shown schematically in Figure 15.1.

This rather simple technique of defocused imaging was complemented by the Foucault technique [4, 6–10], and later extended by the introduction of more sophisticated and advanced techniques that included electron beam holography [11–16] and differential phase contrast (DPC) imaging [17, 18]. During recent years, the transport-of-intensity equation (TIE) [19–21] and the Fresnel technique have considerably enhanced the value of the technique, whereas energy loss magnetic dichroism (EMCD) [22] has still to demonstrate an ability to create images of micromagnetic structures.

In-situ Electron Microscopy: Applications in Physics, Chemistry and Materials Science, First Edition.
Edited by Gerhard Dehm, James M. Howe, and Josef Zweck.
© 2012 Wiley-VCH Verlag GmbH & Co. KGaA. Published 2012 by Wiley-VCH Verlag GmbH & Co. KGaA.

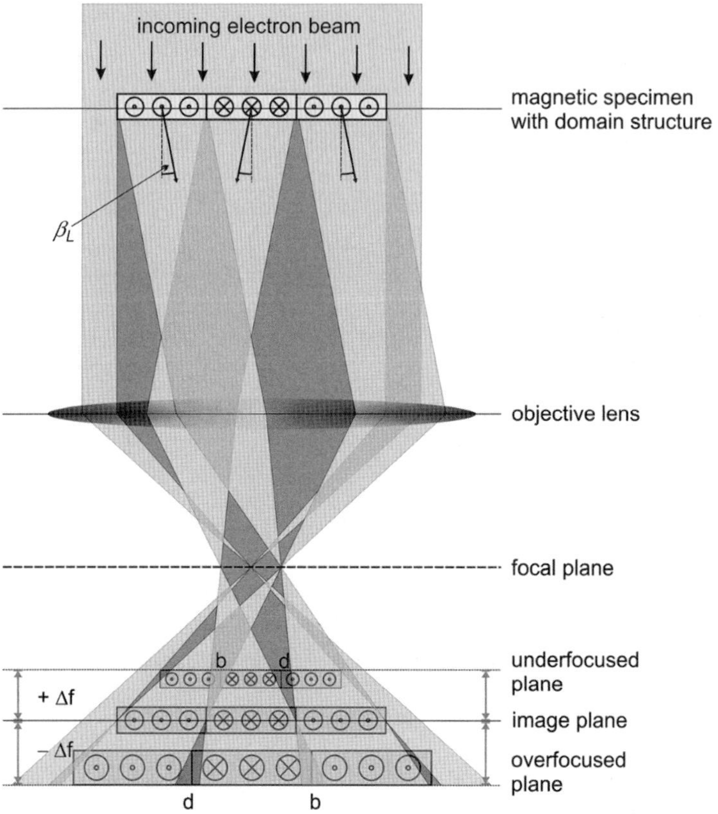

Figure 15.1 Schematic of the Fresnel imaging mode for magnetic imaging. The incoming electron beam is deflected in the specimen by the Lorentz angle, β_L. In the regular (Gaussian) image plane no contrast is apparent. However, for under- and overfocused images indicated by planes at $+-\Delta f$ a contrast appears where domain walls exist in the specimen. Bright areas with an increased electron intensity are denoted by "b," while dark areas (reduced intensity) are marked with "d."

When imaging with transmission electron microscopy (TEM), it is clear that to use a greatly excited (magnetic) objective lens would contradict the requirements needed to visualize undisturbed micromagnetic structures. Rather, an objective lens field (usually in the 1–2 Tesla range) would modify or even erase all of the information that is being sought. The most common ways of overcoming this problem are to use the microscope either in "low-magnification" mode, or with a Lorentz lens. In low-magnification mode, the objective lens is switched off while the diffraction lens acts as a remote, long-focal length lens that, due to its distance from the specimen, creates only a negligible magnetic field at the specimen's location. However, in this case both the magnification and resolution will be very limited.

The use of a Lorentz lens (see Figure 15.2) will circumvent this situation. The Lorentz lens is a specially designed, long-focal length lens that allows magnifications of approximately ×50 000 times to be achieved in Gaussian focus, at a

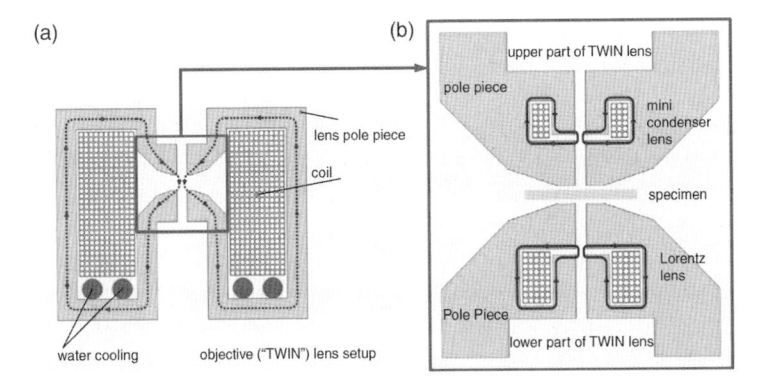

Figure 15.2 Details of the "TWIN" lens set-up. (a) The complete objective lens; (b) The specimen area magnified. The upper pole-piece of the objective is embedded in the minicondenser lens, while the long focal length Lorentz lens is incorporated into the lower pole-piece. In Lorentz mode operation, the regular objective is turned off, and the Lorentz lens is used instead. The magnetic fields which serve as a lens are shown in blue and are rather far from the specimen, leaving the specimen area virtually field-free.

resolution of approximately 2 nm. Moreover, even when the objective lens is turned off a residual field can be found at the specimen's location, which stems from other lenses and correctors further away. One rather simple method of reducing this residual field involves the use of a minicondenser lens, for which an excitation is possible with either a "regular" or an "inverted" current (at least in the case of a Philips/FEI microscope). In this case, a magnetically soft specimen is placed in the microscope's specimen holder, and the specimen is continuously tilted back and forth. In this way the minicondenser excitation ensures that no magnetic features (e.g., domain walls or magnetic ripple structures) will be changed, and that any residual fields are counteracted by the minicondenser. If these conditions can be satisfied, then the residual field will be less than the coercivity field H_c of the specimen being examined. When doing this with a permalloy ($Ni_{81}Fe_{19}$) specimen, the measured residual field was 2 kA m^{-1} (25 Oe) at 300 kV, and corresponded well with H_c for this type of specimen. Even better results may be achieved when, instead of a soft magnetic specimen, a specialized specimen holder with a magnetic field sensor at the specimen's location is used. Whilst the magnetic field may be very low at such a location, those using a Lorentz or diffraction lens should be made aware of the high stray fields that the specimen might encounter during its transfer into the microscope column. Typically, it is advisable to turn all such "lenses" off during any transfers so as to ensure that, during insertion into the microscope, the specimen is not affected in terms of its magnetic state.

With care having been taken to ensure that the specimen is unaltered by the microscope unintentionally, consideration can be given as to how the micromagnetic state of the specimen can be modified in a controlled manner, in readiness for the measurements.

15.2
The *In-Situ* Creation of Magnetic Fields

The main challenge when conducting *in-situ* magnetizing experiments in the transmission electron microscope is to be able to apply a magnetic field to the specimen in precisely the desired direction and at the correct amplitude, while at the same time causing an absolutely minimal disturbance to the optical quality of the magnetic lenses (ideally, no disturbance at all). As the interaction between the specimen's induction with the electron beam is determined by the in-plane component $\vec{B}_\perp$, which is perpendicular to the electron beam's path causing the Lorentz force $\vec{F}_L$

$$\vec{F}_L = e \cdot \vec{v} \times \vec{B}_\perp, \tag{15.1}$$

where e is the electron charge and $\vec{v}$ is the electron velocity, the primary interest involves changing the in-plane component of the specimen's induction (methods of achieving this are discussed in more detail below). Due to the very nature of TEM, all specimens capable of being investigated must be thin (generally <100 nm thickness, depending on the accelerating voltage used and the imaging detail required). This means that, in general, the shape anisotropy implied by the specimen's geometry will force the local magnetic moments to lie within the plane of the specimen. Any deviation from a planar alignment will cause a demagnetizing field which, in turn, will counteract the cause that tries to align the local moments out of plane. As is well-known to people working in the field of magnetism, demagnetizing fields are generally difficult (often impossible) to calculate. However, in the case of an ellipsoid a solution can be found more easily [23], with the most common specimens (quasi-infinite thin films, particles with $\frac{thickness}{diameter} \ll 1$, etc.) being successfully approximated by these ellipsoids. In general, these specimens will have their magnetic moments – and thus also their magnetic domains – lying in-plane. This may lead to a common situation that in-plane external fields should be applied in order to modify and manipulate the specimen's magnetization. Yet, in fact, very few situations occur where this is not true – that is, for specimens with a very strong perpendicular magnetic anisotropy [24], for those with a high aspect ratio $\frac{thickness}{diameter} \gg 1$, or for those with specific magnetic features such as Bloch walls, Bloch lines, and magnetic vortices [23, 25–27] which have local moments pointing partly or entirely out-of-plane. For these features, the use of a perpendicular magnetic field may be advantageous for specific investigations.

If there is a need to change the z (out-of-plane) component of the specimen's induction, it is possible simply to utilize a regular objective lens that creates a magnetic field perpendicular to the specimen's plane, and thus parallel to the incoming electron beam. If this lens is only slightly excited, then a weak magnetic field perpendicular to the specimen's plane can be achieved. It may be recalled that, in order for a magnetic field to act as a lens, there must be an inhomogeneity along z, as approximated by Glaser's "Glockenfeld" (bell-shaped field) [28, 29]:

$$B_z(z) = \frac{B_0}{1+(z/a)^2} \tag{15.2}$$

where B_0 is the maximum field at the lens center, and $2a$ is the full-width at half-maximum (FWHM) of the field distribution along the z-axis. Although this seems to imply that the z position of the specimen within the magnetic pole-piece would be critical, for a standard objective lens the homogeneous magnetic field created by the pole-pieces (which do not act as a lens) is much larger than the inhomogeneous part. Hence, for practical purposes the inhomogeneity may be neglected when using the objective lens to magnetize the specimen.

When creating in-plane magnetic fields, a variety of magnetizing stages have been employed, some of which are described in the following subsections.

15.2.1
Combining the Objective Lens Field with Specimen Tilt

This is the simplest and most widely used technique to create in-plane magnetic fields at the specimen's location. The standard objective lens is excited to a small degree to generate a magnetic field $\vec{B}_z = (0, 0, B_z)$. The specimen is then tilted and the corresponding component B_{ip} of the field, which lies now in the plane of the specimen, can be used to manipulate the micromagnetic features present. Clearly, when the tilting angle α is known, B_{ip} is given by a simple trigonometric relationship (see Figure 15.3):

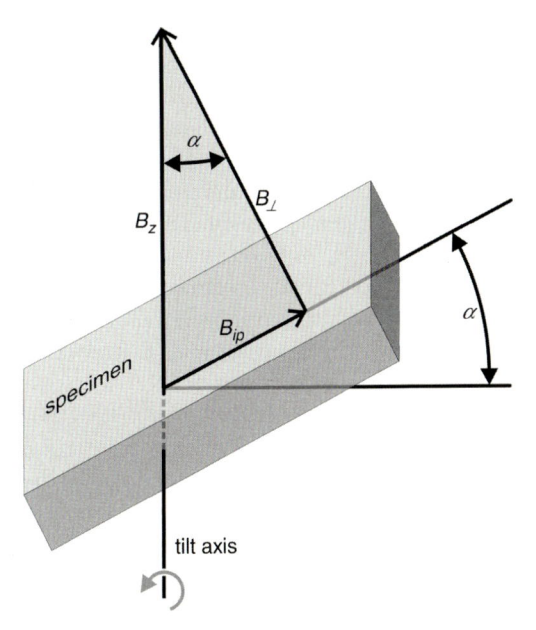

Figure 15.3 The creation of an in-plane field component B_{ip} in a tilted specimen from an existing vertical component B_z.

$$B_{ip} = B_z \cdot \sin(\alpha) \tag{15.3}$$

However, it must always be borne in mind that there is a remaining vertical component B_z which may, in certain cases, cause unintended side effects. This may be the case when the magnetic structure already contains perpendicular components that may now be either enhanced or suppressed, depending on the direction of B_z. Notably, when the specimen itself is not entirely flat, a too-high perpendicular component may lead to distortions in the domain walls and an unexpected behavior of magnetic features, such as vortices [30]. Hence, it is advisable to keep the vertical field as low as possible in order to avoid artifacts. Special care must also be taken if a double tilt holder is used, to vary not only the strength but also the direction of the in-plane components. As it is often the case that the motorized drives of a double tilt holder do not operate at the same speed for each tilt axis, special care must also be taken when, for example, a computer-assisted rotation of an in-plane field vector must be carried out at a constant field strength. If the drives were to operate at different speeds, the specimen may experience larger fields than desired between the initial and final positions, and this may lead to erroneous results in the presence of a hysteretic behavior.

15.2.2
Magnetizing Stages Using Coils and Pole-Pieces

One perhaps rather obvious method of creating a magnetic field at the specimen's location is to use a soft magnetic pole-piece, around which a coil is wound to create a stray field at the pole-piece's end points. This has been achieved with one pair of pole-pieces [31], where the direction of the applied magnetic field is limited to the axis defined by the pole-pieces (see Figure 15.4).

In a more complex set-up, two pairs of pole-pieces were used, arranged at right angles, so as to allow in-plane fields to be applied in any direction, as required by the correct excitation of the respective coils (see Figure 15.5). However, in the case of the design shown in Figure 15.5, due to the length/diameter ratio the demagnetizing fields were found to be so strong that fields of only $\pm 5 \, \text{kA m}^{-1}$ could be achieved at the specimen's location. Consequently, another design was developed which yielded a higher field, the main idea being to prevent the existence of demagnetizing fields which weaken the achievable field strength due to "magnetic charges" that appear at interfaces between the pole-piece material and its surroundings.

Since, in a transmission electron microscope holder it is generally not possible simply to increase the length of the pole-pieces, due to limitations in available space, an alternative approach would be to use closed-circuit pole-pieces that would allow the field to emerge only where it is desirable. A schematic representation of such a set of pole-pieces is shown in Figure 15.6, with a technical drawing in Figure 15.7.

Due to the closed-circuit design, the achievable field strengths range from approximately $\pm 6 \, \text{kA m}^{-1}$ (along the holder's axis) to $\pm 18 \, \text{kA m}^{-1}$ (perpendicular to the holder's axis). This is achieved by including approximately 1600 windings each for the two field directions, at currents of 60 mA and 125 mA, respectively. The field

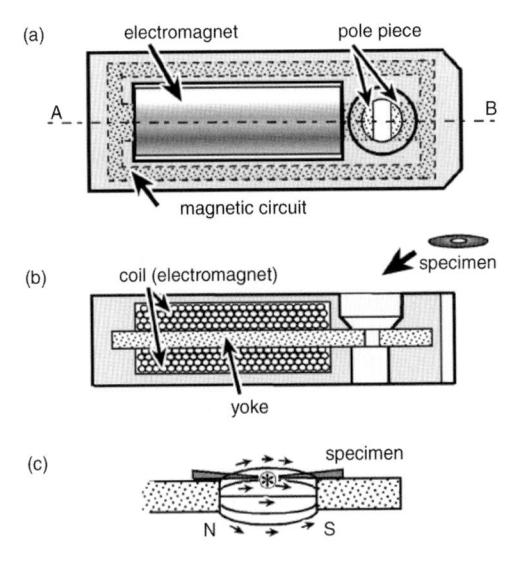

Figure 15.4 Magnetizing stage with one direction of magnetization. (a) Top view of the coil, yoke, and specimen position. The magnetic field is applied along the line A–B; (b) Side view of the magnetizing holder; (c) Schematic of magnetic stray fields emerging from the pole-piece and interacting with the specimen. This figure is taken from [31].

strength was measured versus the coil current, and shown to be linear over a sufficiently large range (see Figure 15.8).

The main disadvantage of this (or any) holder design which uses pole-pieces is that, in order to create a sufficiently large field in the gap region, the pole piece must have a

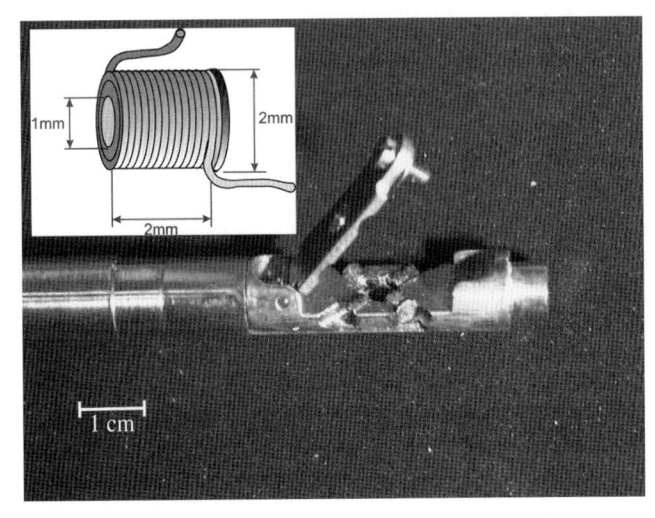

Figure 15.5 View of a specimen holder with two pairs of pole-pieces at right angles for full two-dimensional in-plane magnetic field capability. The inset shows the dimensions of the coil and yoke used.

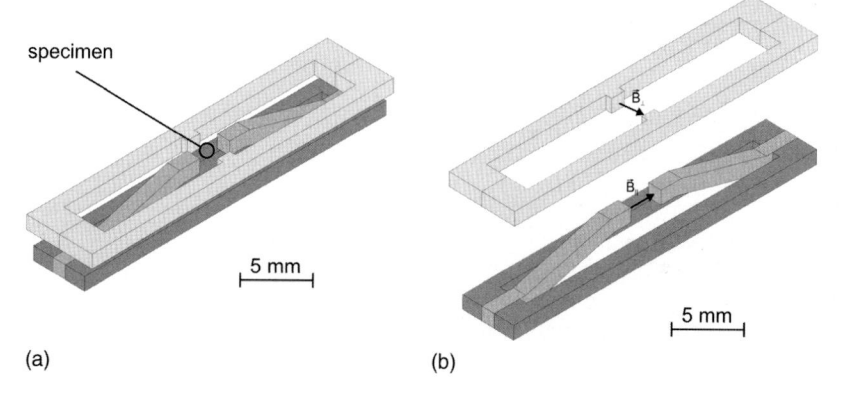

Figure 15.6 Construction drawing of a set of pole pieces with full two-dimensional field capability. (a) The assembled pole-piece system; (b) Exploded view of the upper and lower pole pieces. Note the guiding pole-piece which transfers the field generated in the lower system into the upper level. The individual fields generated by the two sets are indicated by arrows.

certain diameter that is capable of carrying the magnetic flux required. For the design shown in Figure 15.6, a gap width of 1 mm was chosen, in combination with a pole-piece of dimensions of 1×1 mm. This means, that over an extended length t along the z-direction, the electron beam will experience a magnetic field perpendicular to its trajectory, causing a deflection of the beam by the Lorentz angle:

$$\beta_L = \frac{eB_\perp t}{mv} = \frac{eB_\perp t}{[2m_0 E(1 + E/2E_0)]^{1/2}} \tag{15.4}$$

where e is the charge of the electron, t is the thickness of field region, $B_\perp$ is a constant component of induction perpendicular to the beam trajectory over thickness t, mv is

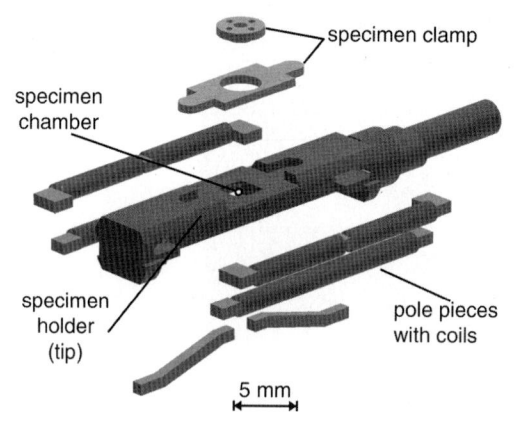

Figure 15.7 Technical drawing of the two-dimensional field holder's tip area with closed magnetic circuits (cf. Figure 15.6).

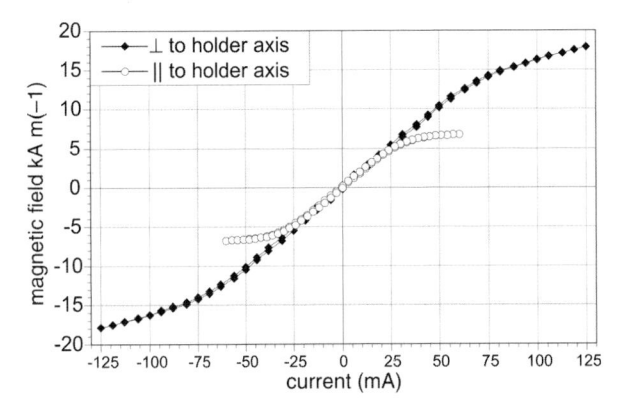

Figure 15.8 Magnetic field values achievable by the two-dimensional holder positioned perpendicular to (filled symbols) and parallel to (open symbols) the holder's axis.

the modulus of the electron's relativistic momentum, m_0 is the electron rest mass, E_0 is the rest energy of an electron, and E is the total energy of an electron. For the dimensions given above, it is clear that the field between the pole-pieces is neither homogeneous nor confined to the area between the pole-pieces within the gap. This situation is shown in Figure 15.9, where the extension of the in-plane field component along the z-axis is shown.

Clearly, a much more realistic approach would be to integrate along z over the field distribution to yield β_L

$$\beta_L = \int_{z_{min}}^{z_{max}} \frac{eB_\perp(t)dt}{mv} = \int_{z_{min}}^{z_{max}} \frac{eB_\perp(t)dt}{[2m_0 E(1+E/2E_0)]^{1/2}} \tag{15.5}$$

with a suitable choice for z_{min}, z_{max}.

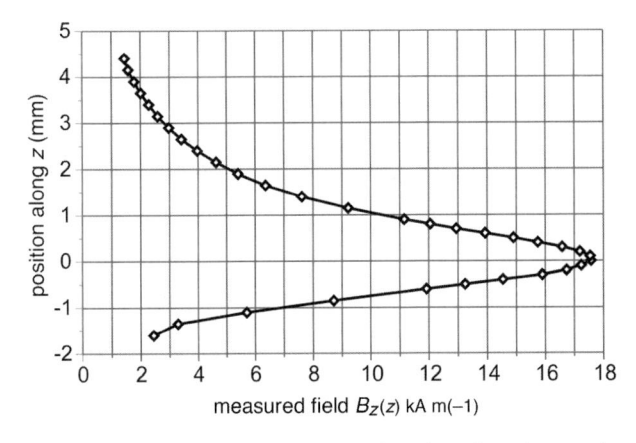

Figure 15.9 Distribution of in-plane field B_z along the z-direction for a Philips CM30 (300 kV) transmission electron microscope with a TWIN lens configuration. The center of the pole-pieces is at position 0.

It is, in fact, the considerable extension of the in-plane field that causes deflections of the beam that are sufficiently large to deflect the beam well away from the optic axis to be of further use. Hence, when the design shown in Figure 15.6 is used, the situation is limited not by the achievable field but rather by the quantity $\int_{z_{min}}^{z_{max}} B_\perp(t)dt$. For a given field strength $B_\perp(t)$, the thickness of the pole-piece is the limiting parameter, and to avoid this limitation other designs are needed.

15.2.3
Magnetizing Stages Without Coils

15.2.3.1 Oersted Fields

The creation of an Oersted magnetic field is still achieved by passing an Amperian current through a wire, after which the magnetic field at the specimen's location can be calculated using the law of Biot–Savart.

As noted above, one serious disadvantage of the coil–pole-piece design is the fact that the electrons will be deflected by large angles, due to the large gap region, although to overcome this problem an alternative magnetizing holder was designed [32]. In this case (cf. Figure 15.10), two gold wires (each of 100 μm diameter) carry identical currents such that each wire creates an Oersted field that adds up above and below the two wires to a purely horizontal field in, for example, the $\pm B_x$ direction. As the upper and lower horizontal fields are only 200 μm apart in a vertical direction, and the whole set-up is completely symmetric, the deflection of an electron caused by the upper field is on the one hand not very strong, but on the other hand is counteracted by the second, oppositely directed, field. Hence, the electron will be deflected twice by the same Lorentz angle β_L, but with opposite sign, which simply causes only a very slight displacement. Clearly, the holder design does not allow for large direct currents, and so offers only limited continuous field strengths in the range of about $20\,\mathrm{Oe\,A^{-1}} = 1.6\,\mathrm{kA\,m^{-1}}$. For short current pulses, the heating was

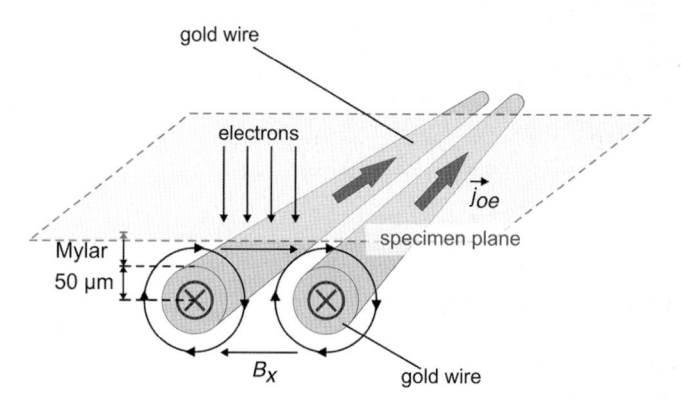

Figure 15.10 The magnetizing holder [32]. Two gold wires carry the field-generating current j_{Oe}, which creates a horizontal magnetic field B_x in the specimen's plane and a negative field $-B_x$ in a corresponding plane below the wires. The Mylar film is used as a spacer to keep the specimen in the optimum position.

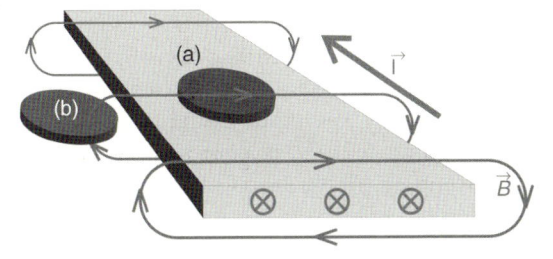

Figure 15.11 The generation of *in-situ* magnetic fields using thin-film conductors. The peripheral magnetic field can be used to apply an in-plane field for magnetic particles placed either (a) on the film or (b) vertical fields when placed beside the film on the supporting Si_3N_4 membrane.

found to be tolerable, and field strengths of up to about 300 Oe = 23.8 kA m^{-1} were achieved. Consequently, the holder would be suitable especially when studying switching processes, that can be achieved by the rapid application of high magnetic fields.

A rather similar approach involves the use of a thin conducting film on a Si_3N_4 membrane, where it must be ensured that the combination of film plus membrane remains sufficiently electron-transparent. Depending on the direction of the magnetic field required, magnetic particles may be placed directly onto or beside the conductor (see Figures 15.11 and 15.12). If a current is passed through the conductor, its Oersted field will penetrate the particles and influence their micromagnetic balance accordingly. However, as the horizontal fields cancel out exactly and, due to the thickness of the conducting film being only about 50 nm, the effect of the beam shift is generally not observable. If the conducting film is shaped in such a way as to act as a strip line, it can be used to apply high-frequency fields, depending also on the

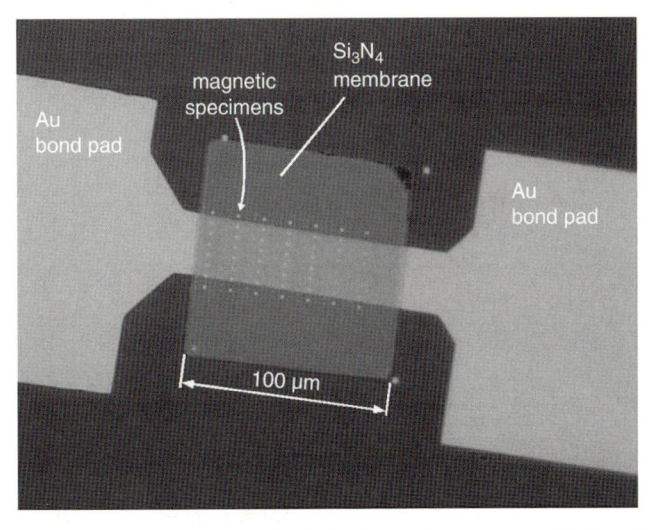

Figure 15.12 Light micrograph of a thin Au conductor across a thin film (30 nm) Si_3N_4 membrane. Magnetic particles can be seen on and beside the conductor strip.

feedthrough of the signal into the microscope. To date, frequencies of up to 300 MHz have been successfully demonstrated [38]. Although, clearly, the achievable magnetic fields are small due to the limited currents that can be used in a thin-film conductor, this situation is at least partially compensated by the fact that the specimen is in the immediate vicinity of the field-creating device, so that it experiences the maximum obtainable field achievable with this set-up.

Finally, it should be noted that the passing of high current densities through the thin-film conductor will lead rapidly to damage of the conductor, due to electron migration.

15.2.3.2 Spin Torque Applications

The recent interest in "spintronics" [33–37] has led to the recognition of spin torque effects that may serve to influence micromagnetic structures. In "spintronics," the main goal is either to move a magnetic domain wall between two predefined positions, or to invert the magnetization of a specific volume of magnetic material. In both cases, this would clearly allow the construction of "magnetic bits" that have the potential to form high-density memory devices which will consume less energy and operate at higher frequencies compared to the currently used dynamic random-access memory (DRAM) cells. The underlying principle of switching micromagnetic volumina can, of course, also be applied to affect magnetic specimens *in-situ*, while being investigated in an electron microscope. To demonstrate this principle, a magnetic material with a magnetic domain wall separating two oppositely magnetized areas should be considered, as shown in Figure 15.13. A spin-polarized electron [34] (i.e., an electron from the majority carrier band) which travels across the magnetic domain wall will experience a torque $\vec{M}$ to realign it with the new prevailing direction for majority band electrons [34]. Due to the conservation of total momentum, however, the change in momentum $\frac{d\vec{L}}{dt} = \vec{M}$ must be compensated by an additional momentum change of equal size (but of opposite sign) within the system. This compensating momentum change is generally absorbed by the surrounding spins, and causes only a slight disturbance that is generally barely noticeable.

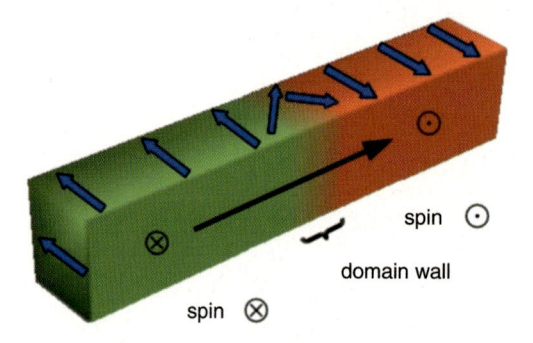

Figure 15.13 Magnetic specimen with two opposite magnetic domains, separated by a 180° domain wall. A spin-polarized electron [34] traveling across the domain wall experiences a torque and, in turn, exerts a torque on the spins in the domain wall.

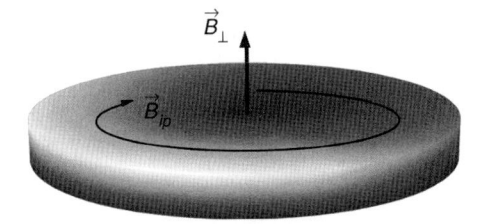

Figure 15.14 Schematic of a magnetic disk-shaped specimen with flux closure (vortex) configuration and in-plane induction B_{ip}. The induction $B_\perp$ at the disk's center, the vortex core, is oriented perpendicular (arbitrarily up or down) to the specimen plane.

However, if the current densities involved are high (in the range of 10^6 to 10^8 A cm^{-2}), the torque may be large enough actually to reorient the local spins, causing domain wall movement or, due to the existing effective local magnetic field, to excite an oscillatory motion of spins. The resulting motion of the spins can be calculated from the generalized Thiele equation [39]. For the purpose of the magnetic *in-situ* manipulation of micromagnetic structures and their simultaneous observation, this means that the use of a spin-polarized current, when passed through a magnetic specimen rather than through a nearby conductor, will cause changes that can be used to determine certain *in-situ* magnetic properties.

An example of this can be provided by considering a thin, cylinder-shaped magnetic disk with a typical diameter of 1 μm and a thickness of several tens of nanometers. These disks are well-known to form a magnetic flux-closure pattern with a so-called "vortex" at its very center [25–27, 40]; a schematic of such a specimen is shown in Figure 15.14. These magnetic flux closure patterns with a centered vortex produce distinct contrast patterns when imaged with the Fresnel imaging technique.

The observable contrasts are explained in detail in Figure 15.15. If a parallel electron beam is transmitted through a magnetic disk with a flux closure pattern and a vortex at its center, the contrast observable in the defocused image plane below the specimen will depend on the chirality of the flux closure. This determines if the electrons will be deflected by the Lorentz force either towards the center of the disk (the flux is oriented counterclockwise, ccw) or away from the disk center (the flux is oriented clockwise, cw). Correspondingly, a bright/dark spot is observed at the center, and a dark/bright fringe at the specimen's perimeter for ccw/cw orientation, respectively. Thus, the position of the vortex center is easily visible from the bright or dark spot. It should be noted that if the defocus is changed from overfocus to underfocus, then the contrasts will invert. It should also be noted that the vortex core is oriented parallel to the electron beam, and therefore does not contribute to the contrast; rather, the contrast simply represents the chirality of the flux orientation.

Now, if a current is injected into the magnetic disk, the current will normally be polarized within a few nanometers, due to the existence of minority and majority spin bands in ferromagnetic materials. The two bands will be energetically shifted by $\Delta E = 2\mu_B |\vec{B}|$, and therefore the density of states at the Fermi energy level will be different for the majority and minority carriers, leading to a more or less pronounced spin polarization:

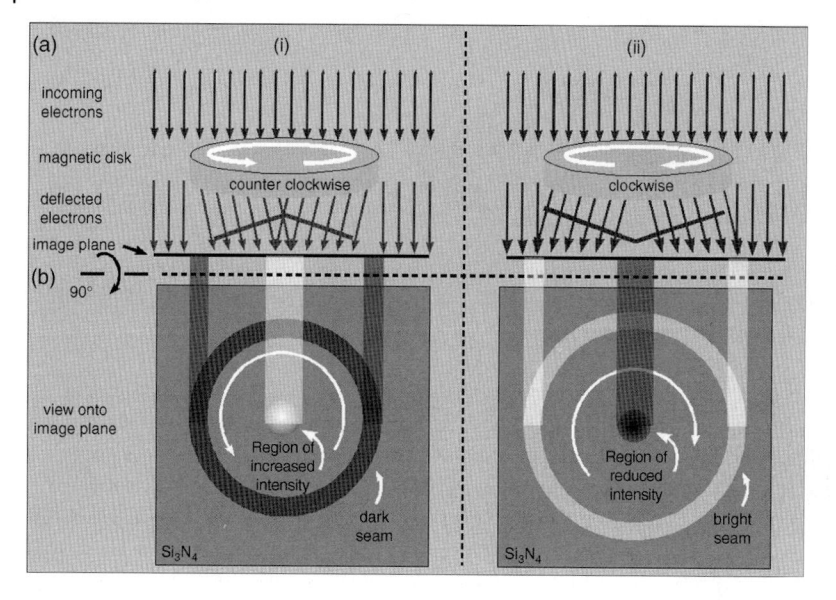

Figure 15.15 Contrast formation for magnetic disks with counterclockwise (row i) and clockwise (row ii) flux closure pattern. (a,i) A parallel electron beam is deflected towards the disk center by the Lorentz force and creates a bright center area. The perimeter of the specimen becomes dark due to the deflection of rim electrons; (b,i) A sketch of the visual appearance of the disks, imaged in underfocus condition; (a,ii) For clockwise flux closure, electrons are deflected outwards, creating a bright rim and a dark center area; (b,ii) A sketch of the visual appearance of the disks, imaged in underfocus condition. Note that the contrast is inversed when passing from underfocus to overfocus.

$$P = \frac{n_{up} - n_{down}}{n_{up} + n_{down}} \tag{15.6}$$

that depends on the material's density of states at the Fermi level. Here, n_{up}, n_{down} represent the numbers of carriers with their spins being in the "up" or "down" state, respectively. This spin-polarized current will proceed towards the vortex and interact with it, causing the above-mentioned torque. A local tilting of the vortex's induction in the effective overall field of the magnetic specimen will then lead to a precessional motion of the vortex core, with the precession frequency being determined by the magnetic properties of the specimen and by its geometric shape. Consequently, if an alternating current is used, with a frequency close to the resonance condition, it would be possible to excite the vortex to a stationary precessional motion. For known geometric parameters, it is then possible to determine the magnetic properties of the specimen. Due to the complex behavior of the vortex motion, no further explanation nor detailed description will be provided here, as this is beyond the scope of the chapter.

To summarize, it may be stated that the spin-polarized current interacts with the vortex core, exciting it to oscillations in its eigenfrequency. If an alternating current with a correct frequency is applied, however, a resonant behavior of the vortex core is expected, leading to a stationary circular motion of the vortex core. The first

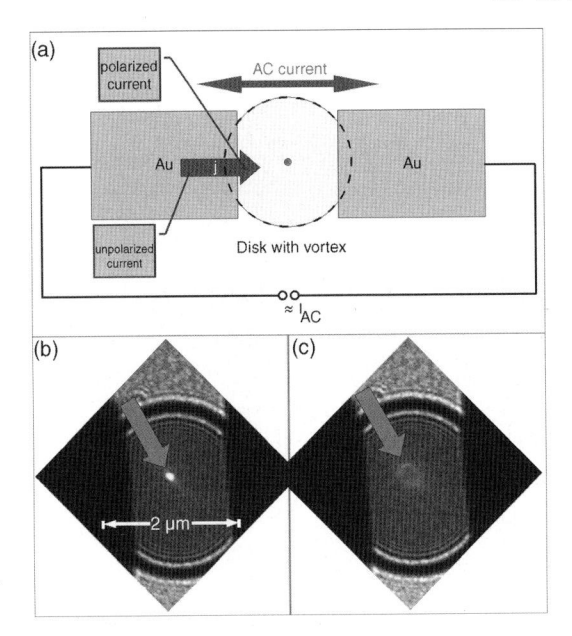

Figure 15.16 (a) Schematic of the experimental set-up; (b) Magnetic disk between two Au leads in a non-resonant condition. The stationary vortex position is marked by an arrow; (c) Same as in panel (b), but with the alternating current tuned to resonance (ca. 83 MHz). The previously sharp vortex image is spread into a circular ring pattern.

experimental results obtained for a $Ni_{81}Fe_{19}$ specimen between two gold leads are shown in Figure 15.16b and c. In this case, the experimental set-up is shown in Figure 15.16a, while Figure 15.16b shows the frequency off-resonance of the alternating current, where a vortex can be clearly seen in Fresnel imaging mode (indicated by an arrow). For a resonant condition, the vortex begins to precess, and while the precession frequency is in the range of several hundreds of MHz, the time required to achieve a complete circle is very small compared to the average exposure time of approximately 1 s. Hence, the precessional motion of the vortex is recorded as a bright circular pattern (see Figure 15.16c).

15.2.3.3 Self-Driven Devices

In certain cases, it may be desirable to determine the magnetic fields created by devices that can be excited electrically, such as read/write heads for storage technology. In these cases, the devices must be mounted in the microscope's specimen holder and connected to an electric feedthrough; this enables an excitation by the application of an electric current. Since, in this case, interest is not targeted at the reaction of a magnetic specimen to an external field, but rather to the shape and extension of the field itself, it is possible to employ nonstandard imaging techniques. Hence, by using a local probe technique it is possible to sense the deflection of the electron beam by the fields created by the device, and thus to measure indirectly the spatial distribution of the generated fields. This situation is shown in Figure 15.17,

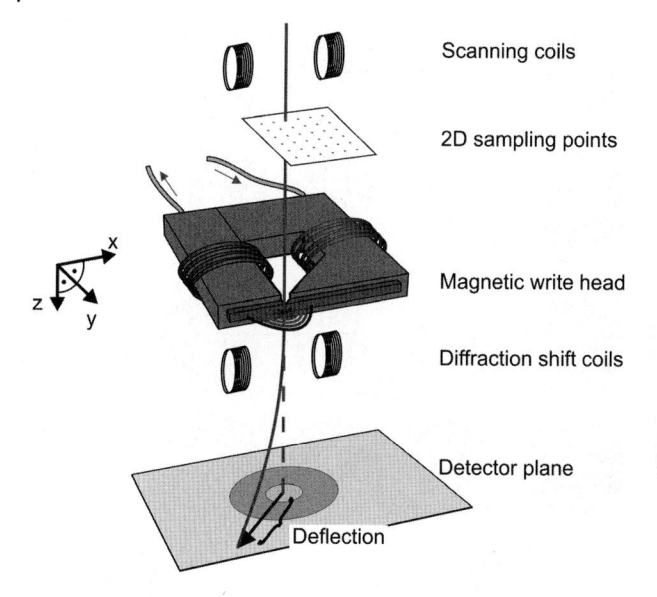

Figure 15.17 An electron beam is moved by scanning coils on a two-dimensional sampling grid across the area of magnetic stray field emerging from the magnetic write head. The deflection of the beam due to Lorentz force is monitored in the detection plane, either by a position-sensitive device or by the current required for the descan coils in x- and y-directions to re-center the beam onto the detector.

where a magnetic write head is mounted in the specimen holder and excited using the electric leads. The emerging magnetic stray field from the gap region is monitored by moving an electron beam step-by-step through the stray field area, using the scanning transmission electron microscope coils. The emerging stray fields deflect the electron beam due to the Lorentz force acting on the beam electrons, and this leads to a shift of the beam in the detection plane. In this case, the deflection can be measured either by using a position-sensitive device or simply by re-centering the beam onto the bright-field detector of the microscope, using the diffraction shift coils. The currents required for the diffraction shift coils to re-center the beam in two dimensions provide a direct measure of the local direction and strength of the stray fields in the specimen area.

15.3
Examples

15.3.1
Demagnetization and Magnetization of Ring Structures

An example of a magnetizing experiment is shown in Figure 15.18, for microstructured magnetic rings made of permalloy ($Ni_{81}Fe_{19}$), produced on an electron-

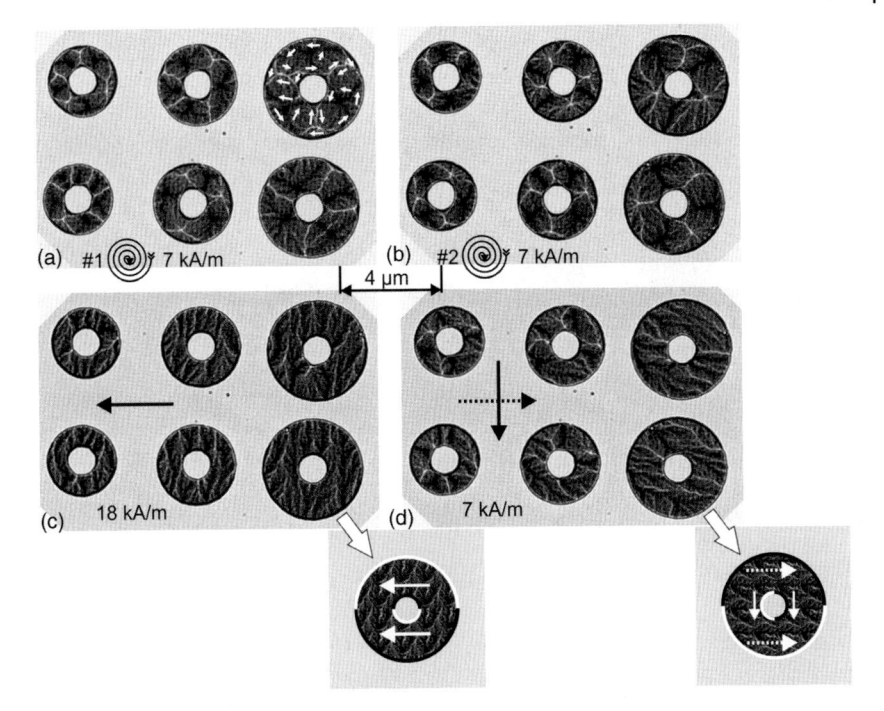

Figure 15.18 (a, b) Remanent states of magnetic rings after a spiralic demagnetization, showing similar but different remanent states. In (a) the local magnetic configuration is indicated by white arrows. The starting field was 7 kA m^{-1}, and while the field was rotated, its amplitude was continuously decreased; (c) With a magnetic field of 18 kA m^{-1} applied in the direction indicated by an arrow. The magnetic contrasts visible within the rings are due to imperfect magnetization, known as magnetization ripple [23]; (d) After initial magnetization with 18 kA m^{-1} along the dotted direction, the external field was reduced to 7 kA m^{-1} and rotated by 90°. Two schematic detail images show, as examples, the dominant internal induction's direction as indicated by arrows and derived from the inner and outer ring's seam colors (see text for details).

transparent Si_3N_4 membrane, using electron beam lithography. The two upper images (Figure 15.18a and b) show the remanent state of the rings after a spiralic demagnetization has been performed. To achieve this, an initial field of 7 kA m^{-1} was applied and, while continuously rotating the field's direction, this was gradually reduced to zero. Hence, the momentary magnetic field vector's tip moved on a spiral, effectively demagnetizing the specimen as much as possible. As a consequence, vortex-like structures could be found within the rings, with an alternating sense of rotation of subsequent vortices; this is indicated in one ring in Figure 15.18a by white arrows. As can be seen, the remanent states differ in their individual structure after subsequent demagnetization cycles. Figure 15.18c shows the situation following the application of a magnetic field of 18 kA m^{-1} along the direction indicated by the arrow. The rings appear to be predominantly magnetized along the given direction, as can be seen from the brightness of the inner and outer ring diameters. These edge brightnesses correspond well to the induction indicated by arrows in the schematic drawing below (Figure 15.18c). The feather-like structure within the ring is due to the so-called "magnetic ripple," and indicates that the material is not completely saturated. In Figure 15.18d, the same magnetic field as in Figure 15.18c was applied initially, but in the opposite direction (dotted arrow). Subsequently, the field was reduced to 7 kA m^{-1} and rotated though 90° (solid arrow). From the edge seam's brightnesses it can be seen that, while the outer seam brightness indicates an overall induction along the direction given by the dotted white arrows in the schematic drawing below Figure 15.18d, that of the seams of the inner diameter indicate an induction rotated by 90° (solid white arrows). This indicates that the rotation of the specimen's induction occurs initially closer to the inner ring area, whereas the outer ring area remains predominantly unrotated.

15.3.2
Determination of Wall Velocities

When using a magnetizing set-up according to Ref. [32], it is possible to apply high-frequency alternating current fields to specimens, and this allows, for example, the determination of magnetic domain wall velocities. As the switching time $\tau = \frac{1}{\nu}$ for frequencies $\nu > 10$ Hz is considerably smaller than the average exposure time for an image, this will lead to a double exposure of the domain wall's position with positive and negative magnetic fields applied. At this point, it is more convenient to use rectangular field amplitudes rather than sinusoidal amplitudes, to provide a better determination of the final positions of the walls. From the separation of walls for various exciting frequencies, the domain wall velocities can then easily be derived; for the example given in Figure 15.19 this was approximately 45 m s^{-1}. For two different excitation frequencies of 50 Hz and 8 MHz (Figure 15.19a and b, respectively) the distance traveled by the domain walls between the switching time τ was determined in the boxed areas, and the velocity then calculated.

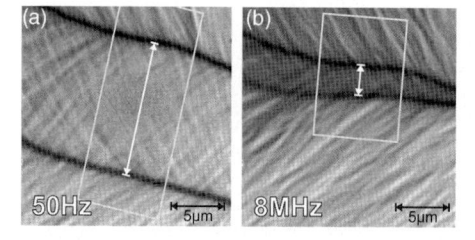

Figure 15.19 Determination of domain wall separations for two different excitation frequencies for (a) 50 Hz and (b) 8 MHz within the boxed areas. The resulting velocity was approximately 45 m s^{-1}.

15.3.3
Determination of Stray Fields

An example of the determination of stray fields created by externally driven devices – namely, those close to the write surface of a magnetic write head – is shown in Figure 15.20 for two different configurations. In Figure 15.20b, the stray field in the gap region of the head appears rather broad, mainly because for the small regions close to the gap, the ferrite material is incapable of maintaining the high induction required for writing purposes. Consequently, there is a leakage of magnetic flux also in the vicinity of the gap area. In an attempt to confine the magnetic field more closely to the gap area (which would lead to a higher writing density), the ferrite is coated by functional layers that possess a higher susceptibility, but at the cost of a worse high-frequency performance. As can be seen in Figure 15.20a, the flux is in fact confined to a smaller area close to the original gap. However, at the interface regions between the

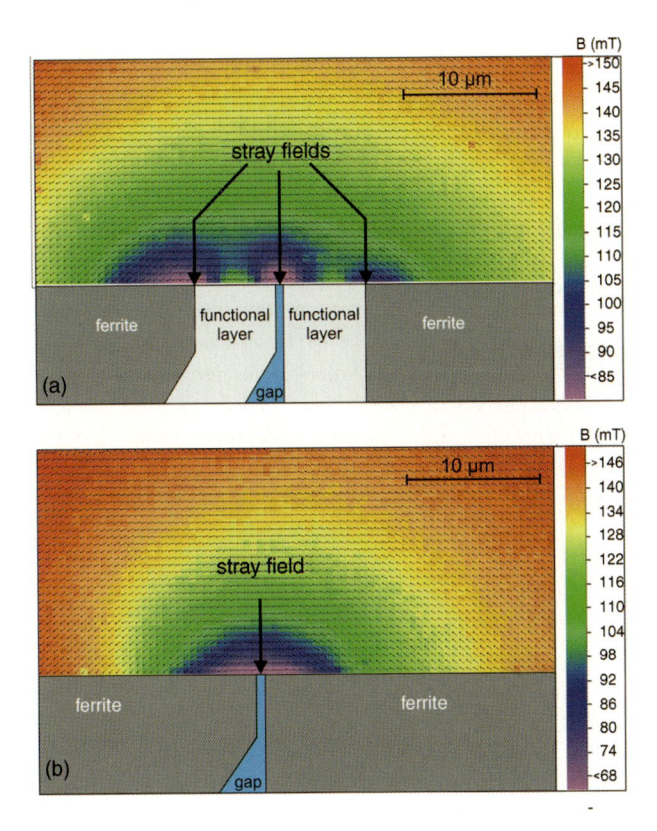

Figure 15.20 Magnetic stray field maps for two different types of write head. (a) A so-called metal-in-gap head, which uses special functional layers to create more localized stray fields; (b) The conventional design. Although this works in principle (as can be seen from the color coding of the field strengths), additional stray fields that degrade the overall function emerge at the interfaces between the ferrite and functional layers.

ferrite and the functional layers, the match of magnetic properties is insufficient to prevent additional flux leakage there, such that there is no significant improvement.

15.4
Problems

Although some problems directly related to the creation of the magnetic fields have been addressed above, it should be noted here that further problems require consideration. However, not all such problems have yet been considered in sufficient detail.

The most obvious problem is a heating of the electric leads that carry the current. Such heating will generally cause drift problems that are not severe for low magnifications and when (relatively remote) coils are used to create the fields. However, heating becomes increasingly problematic when the leads are thin and close to the specimen, as in the case of a strip line geometry (see Section 15.2.3.1, where a thin lead has a specimen atop, or in Section 15.2.3.2, spin torque effects). The problem becomes even greater if the leads are placed directly onto a thin, electron-transparent membrane of Si_3N_4, which is a poor conductor of heat. Consequently, the generation of even moderate amounts of Joule's heat may have major adverse effects that must be overcome.

A further problem, that is directly connected to the set-up described here, is the onset of electromigration. Typically, the aim is to pass as much current through the lead as possible, without causing heating problems, in order to achieve as high a field as possible. However, such high current densities may easily achieve values at which severe electromigration will begin (see Figure 15.21).

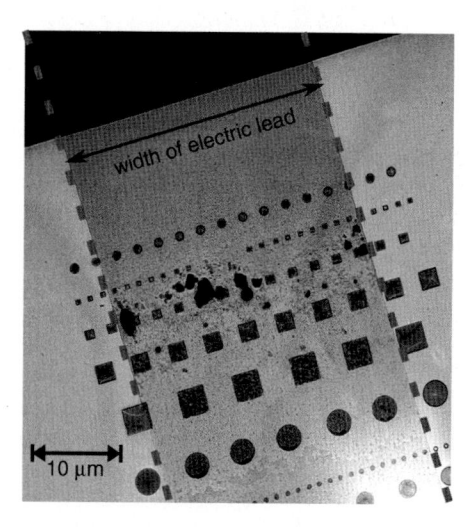

Figure 15.21 Effect of electromigration on magnetic leads and specimens. The boundary of the current-carrying lead on the transparent membrane is indicated by two dashed lines. As can be clearly seen, the lead and the magnetic specimen both suffered from electromigration effects; whilst this caused a spotty texture within the lead area, the magnetic specimens outside the leads were left unharmed.

Finally, the problem of electron beam–specimen interaction remains. This difficult and as-yet unresolved problem relates to determining the temperature increase within a patterned specimen when supported only by a thin membrane and illuminated by an electron beam.

15.5
Conclusions

The possibility of creating controllable *in-situ* magnetic fields within the electron microscope column, while simultaneously employing electron optics in normal fashion, has been demonstrated. Although, when possible, the use of a so-called Lorentz lens is favorable, it is not strictly required for many purposes. Various techniques for the creation of magnetic fields have been described and demonstrated with examples of static fields, magnetic field pulses, and continuous high-frequency excitations. Clearly, TEM in Lorentz imaging mode may be applied to high-resolution *in-situ* investigations and is – in combination with a wealth of supporting analytical tools – ideally suited to a host of research projects involving magnetic materials and/ or novel spintronic devices.

Acknowledgments

The author thanks his coworkers and former students, C. Dietrich, J. Gründmayer, M. Müller, S. Otto, P. Sellmann, and T. Uhlig for their contributions. He would also like to thank various national and international funding institutions which have, in different ways, contributed to the studies described here by their financial support, including the Deutsche Forschungsgemeinschaft (DFG), the German Bundesministerium für Bildung und Forschung (BMBF), the European Commission, and the University of Regensburg. Thanks are also due to FEI company for sharing certain confidential manufacturing details that enabled construction of the microscope holders.

References

1 Ruska, E. and Knoll, M. (1931) Die Magnetische Sammelspule für Schnelle Elektronenstrahlen (The magnetic focusing coil for fast electron beams). *Z. Techn. Physik*, **12**, 389–400.
2 Knoll, M. and Ruska, E. (1932) Das Elektronenmikroskop (The electron microscope). *Z. Phys.*, **78**, 318–339.
3 Hale, M.E., Fuller, H.W., and Rubinstein, H. (1959) Magnetic domain observations by electron microscopy. *J. Appl. Phys.*, **30**, 789–790.
4 Boersch, H. and Raith, H. (1959) Elektronenmikroskopische Abbildung Weißscher Bezirke in dünnen Ferromagnetischen Schichten (Electron microscopic imaging of Weiß domains in thin ferromagnetic films). *Naturwissenschaften*, **46**, 574.

5 Boersch, H., Raith, H., and Wohlleben, D. (1960) Elektronenoptische Untersuchungen Weißscher Bezirke in dünnen Eisenschichten (Electron optical investigations of Weiß domains in thin iron films). *Z. Phys.*, **159**, 388–396.

6 Fuller, H.W. and Hale, M.E. (1960) Determination of magnetization distribution in thin films using electron microscopy. *J. Appl. Phys.*, **31**, 238–248.

7 Chapman, J.N. (1984) The investigation of magnetic domain structures in thin foils by electron microscopy. *J. Phys. D: Appl. Phys.*, **17**, 623–647.

8 Salling, C., Schultz, S., and McFadyen, I. (1991) Measuring the coercivity of individual submicron ferromagnetic particles by Lorentz microscopy. *IEEE Trans. Magn.*, **27**, 5184–5189.

9 Chapman, J.N., Johnston, A.B., Heyderman, L.J., McVitie, S., Nicholson, W.A.P., and Bormans, B. (1994) Coherent magnetic imaging by TEM. *IEEE Trans. Magn.*, **30** (6 pt 1), 4479–4484.

10 Chapman, J.N., Johnston, A.B., and Heyderman, L.J. (1994) Coherent Foucault imaging – a method for imaging magnetic domain structures in thin films. *J. Appl. Phys.*, **76** (9), 5349–5355.

11 Lau, B. and Pozzi, G. (1978) Off-axis electron micro-holography of magnetic domain walls. *Optik*, **51** (3), 287–296.

12 Tonomura, A., Matsuda, T., Endo, J., Arii, T., and Mihama, K. (1980) Direct observation of fine-structure of magnetic domain walls by electron holography. *Phys. Rev. Lett.*, **44** (21), 1430–1433.

13 Tonomura, A. (1983) Observation of magnetic domain structure in thin ferromagnetic films by electron holography. *J. Magn. Magn. Mater.*, **31–34** (Pt 2), 963–969.

14 Dunin-Borkowski, R.E., McCartney, M.R., and Smith, D.J. (1998) Towards quantitative electron holography of magnetic thin films using in situ magnetization reversal. *Ultramicroscopy*, **74**, 61–73.

15 Lichte, H. (2002) Electron interference: mystery and reality. *Philos. Trans. Roy. Soc. A*, **360** (1794), 897–920.

16 Lehmann, M. and Lichte, H. (2002) Tutorial on off-axis electron holography. *Microsc. Microanal.*, **8**, 447–466.

17 Dekkers, N.H. and de Lang, H. (1974) Differential phase contrast in a STEM. *Optik*, **41** (4), 452–456.

18 Chapman, J.N., McFadyen, I.R., and McVitie, S. (1990) Modified differential phase contrast Lorentz microscopy for improved imaging of magnetic structures. *IEEE Trans. Magn.*, **26** (5), 1506–1511.

19 Bajt, S., Barty, A., Nugent, K.A., McCartney, M., Wall, M., and Paganin, D. (2000) *Ultramicroscopy*, **83** (1–2), 67–73.

20 Allen, L.J. and Oxley, M.P. (2001) Phase retrieval from series of images obtained by defocus variation. *Opt. Commun.*, **199**, 65–75.

21 Volkov, V.V. and Zhu, Y. (2004) Lorentz phase microscopy of magnetic materials. *Ultramicroscopy*, **98**, 271–281.

22 Schattschneider, P., Rubino, S., Hébert, C., Rusz, J., Kuneš, J., Novák, P., Carlino, E., Fabrizioli, M., Panaccione, G., and Rossi, G. (2006) Detection of magnetic circular dichroism using a transmission electron microscope. *Nature*, **441**, 486–488.

23 Hubert, A. and Schäfer, R. (1998) *Magnetic Domains*, Springer.

24 Köhler, M., Schweinböck, T., Schmidt, T., Zweck, J., Fischer, P., Schütz, G., Eimüller, T., Guttmann, P., and Schmahl, G. (2000) Imaging of sub-100-nm magnetic domains in atomically stacked Fe(001)/Au(001) multilayers. *J. Appl. Phys.*, **87** (9 Pt 1-3), 6481–6483.

25 Raabe, J., Pulwey, R., Sattler, R., Schweinbock, T., Zweck, J., and Weiss, D. (2000) Magnetization pattern of ferromagnetic nanodisks. *J. Appl. Phys.*, **88** (7), 4437–4439.

26 Schneider, M., Hoffmann, H., and Zweck, J. (2000) Lorentz microscopy of circular ferromagnetic permalloy nanodisks. *Appl. Phys. Lett.*, **77**, 2909–2911.

27 Shinjo, T., Okuno, T., Hassdorf, R. et al. (2000) Magnetic vortex core observation in circular dots of Permalloy. *Science*, **289**, 930–932.

28 Glaser, W. (1941) Strenge Berechnung Magnetischer Linsen mit

unsymmetrischer Feldform nach H=H0/$[1 + (z/a)_2]$ (Strict calculation of magnetic lenses with non-symmetric field shape like H=H0/$[1 + (z/a)_2]$). *Z. Phys.*, **117**, 285–315.

29 Reimer, L. (1997) *Transmission Electron Microscopy*, 4th edn, Springer, Berlin.

30 Dietrich, Ch., Hertel, R., Huber, M., Weiss, D., Schäfer, R., and Zweck, J. (2008) Influence of perpendicular magnetic fields on the domain structure of permalloy microstructures grown on thin membranes. *Phys. Rev. B*, **77**, 174427–174428.

31 Inoue, M., Tomita, T., Naruse, M., Akase, Z., Murakami, Y., and Shindo, D. (2005) Development of a magnetizing stage for in situ observations with electron holography and Lorentz microscopy. *J. Electron. Microsc.*, **54** (6), 509–513.

32 Yi, G., Nicholson, W.A.P., Lim, C.K., Chapman, J.N., McVitie, S., and Wilkinson, C.D.W. (2004) A new design of specimen stage for in situ magnetising experiments in the transmission electron microscope. *Ultramicroscopy*, **99**, 65–72.

33 Vanhaverbeke, A., Bischof, A., and Allenspach, R. (2008) Control of domain wall polarity by current pulses. *Phys. Rev. Lett.*, **101**, 107202–107202-4.

34 Parkin, S.S.P., Hayashi, M., and Thomas, L. (2008) Magnetic domain-wall racetrack memory. *Science*, **320** (5873), 190–194.

35 Serrano-Guisan, S., Rott, K., Reiss, G., Langer, J., Ocker, B., and Schumacher, H.W. (2008) Biased quasiballistic spin torque magnetization reversal. *Phys. Rev. Lett.*, **101**, 087201–087201-4.

36 Zhang, S. and Li, Z. (2004) Roles of nonequilibrium conduction electrons on the magnetization dynamics of ferromagnets. *Phys. Rev. Lett.*, **93**, 127204–127204-4.

37 Vanhaverbeke, A. and Viret, M. (2007) Simple model of current-induced spin torque in domain walls. *Phys. Rev. B*, **75**, 024411–024411-5.

38 Gründmayer, J. and Zweck, J. (2006) Imaging of fast magnetization processes within patterned magnetic materials in a Lorentz transmission electron microscope. Proceedings of the 16th International Microscopy Congress, Sapporo, Japan, vol. 3: Materials Science, p. 1541.

39 Guslienko, K.Yu. (2006) Low-frequency vortex dynamic susceptibility and relaxation in mesoscopic ferromagnetic dots. *Appl. Phy. Lett.*, **89**, 022510.

40 Cohen, M.S. (1965) Lorentz microscopy of small ferromagnetic particles. *J. Appl. Phys.*, **36** (5), 1602–1611.

Index

a

aberration-correcting electron optics 51
aberration correctors 94, 131, 142
aberration-induced emittance growth 79
adatoms
– density of surface sinks for 100
– effective charge of silicon 113
– electromigration 115
– electromigration of 115
– migration length of 116
– step bunching and the electromigration 111
adsorbed atoms, kinetics 99
aerosol Au particle reaction with disilane 182
agglomeration 126, 130, 183
Al(Cu) line, micrographs of 293
Al_2O_3 particles, secondary electron SEM images 32
Al sample, stress–strain measurements 246
Al thin films, mechanical behavior of 269
aluminum conductor line 288
ambient environment reaction
– with various components 154
ambient gas, interactions 154
amorphous C 193
amorphous material, crystallization 151
analytical TEM 66, 67
annular dark field (ADF) imaging 91
annular dark-field STEM (ADF-STEM) 64
– thickness and temperature dependence 65
anti-bunch formation 111
antiferroelectric ceramics 322, 335, 336, 340
antiferroelectric-to-ferroelectric transition 321, 336, 340
arbitrary waveform generator (AWG) cathode laser system 80, 81
Arrhenius dependence 109
Arrhenius law 130

astigmatism 11, 12, 50, 51. *See also* spherical aberration
– influence, schematic drawing 12
atomic defects 125
– agglomeration 126, 130, 183
– in solids 126
atomic force microscopy (AFM) 100, 230, 258
– tip 231
atomic number 23
– contrast 21
Au crystal, plastic deformation 137
Au films
– deformations 244
– dislocation nucleation 244
– electrodes 333
Auger electrons (AEs) 3, 20, 25, 44, 66, 304
Auger electron spectroscopy (AES) 205
Au nanoparticles
– ETEM image sequence 153, 154

b

backscattered electron imaging (BEI) 291
back-scattered electrons (BSEs) 20–22, 27, 32
– characteristics 20
– detection efficiency 15
– efficiency 21, 22, 24
– emission 13
– escape depth 22
– scintillator detectors 16
– signal from 14
backscattered-limited resolution 27
band-to-band transition 305
$BaTiO_3$ ceramic, crack formation 330
Bauschinger effect 246
BCF theory 116, 117
beam effects 174
beam electrons, threshold energy 129
beam paths 79

Berkovich conductive diamond indenter   272
Bethe formula   18
Black's law   282
Blech segments   282
Blech-type structure   285
– drift rates   298
body-centered cubic (BCC)   83
Boltzman's constant   18, 282, 305
bottom-up design process   211
bound exciton transitions   305
Bragg angle   33
Bragg contrast   109
Bragg-diffracted beam   48
Bragg diffraction contrast   102
Bragg's law   33, 34
Bremsstrahlung radiation   26
bright-field (BF) detector   56
bright-field (BF) electron diffraction   73
bright-field STEM (BF-STEM)   64
bright-field TEM images
– cavities, in glass ligament   330
– of Cu(100) surfaces   202
– electric field-induced crack growth behavior   329
– electric field-induced domain wall fracture   332
– electric field-induced fracture   327
– morphology of nanometer-sized ferroelectric domains   325
– sharp wedge geometry   268
BSEs. *See* back-scattered electrons (BSEs)
Burgers circuit   54
Burgers vector   54, 55, 235, 243
Burton–Cabrera–Frank (BCF) theory   109

c

calibration
– of growth environment   173
– precise
– – high voltage of microscope   63
– – pulse arrival times   74
carbon nanotubes   126
– formation   155
– irradiation   135
carbon replica grating, bright-field images   77
catalytic processes   162
cathode ray tube computer monitor-based system   13
cathodoluminescence (CL)   36, 303, 305
– contrast, theories   306
– monochromatic image   306
– panchromatic image   306
– principles of   304
– – CL spectroscopy, characteristic   305, 306

– – detection systems   307
– – electron-hole pairs, generation/recombination   304, 305
– – imaging, and contrast analysis   306
– – spatial resolution   306, 307
– room-temperature (RT) blue light-emitting diode   308
– in SEM and TEM   303
– – applications   307–313
cavities, in glass ligament   330
CBED. *See* convergent electron beam diffraction (CBED)
[1 3 3] CBED pattern   289
central laser initiation point   87
ceria–zirconia mixed oxides   151
cerium–zirconium oxides   162
channeling contrast   31
characterization technique   156
charge-coupled device (CCD) cameras   31, 77, 82, 227
chemical reactions
– observation   145–165
– types   146
chemical reaction, types   150–154
chemical vapor deposition (CVD)   172
Child–Langmuir effect   92
chromatic aberrations   11, 94. *See also* Spherical aberration
CL. *See* cathodoluminescence (CL)
CNT
– formation   160, 161, 164
– growth, low-magnification images   155
– nucleation   160
coefficient of emissivity   106
complex wave function   46, 54, 58
composite TEM image, of anode and cathode ends of Al segment   272
compositional contrast   21
concentric multi-shell fullerene clusters   134
confocal laser microscopy   211
consecutive reflection electron microscopy   110
contour plot of copper islands after deposition   186
contrast mechanisms   28–31
– channeling contrast   31
– composition contrast   31
– topographic contrast   28–30
contrast transfer function   49
conventional
– detector   29
– imaging techniques (*See* conventional TEM)
– indentation techniques   257
– nanoindentation techniques   257, 260

– optical microscopy   3
– scanning electron microscopes   7
– structuring techniques   133
conventional TEM   39, 44, 90, 94, 172
– of defects in crystals   54, 55
– straining stage   229, 242
convergence angle   49, 62, 63, 65, 91, 93, 94
convergent electron beam diffraction (CBED)   61–63, 267, 272, 278, 283, 284, 288, 289, 294, 298
– characterization of amorphous structures by diffraction   63
– large-angle convergent beam electron diffraction   63
– scattering geometry   62
copper electromigration, diffusion pathway   288
copper grid electrodes   322
corrosive gase   164
Coulomb field   26
Coulomb forces   281
coulombic interactions   90
Coulomb interaction   281
crack growth   326, 328, 331–333, 335
crystal growth
– electrochemical   183
– experiments   172–175
– from liquid phase   183, 187
– reactions   184
– from vapor phase   183
crystallization processes   90, 94
crystal surface, evolution   99
C_s-corrector   40, 51
Cu/α-Al$_2$O$_3$ interface (IF)   239
– dislocation network   238
Cu–Au alloy   202
– oxidation   158
Cu–Au(100) oxidation   199
Cu reflections   202
Cu–Sn alloys   289
– drift velocities   289

d

dark-field (DF) electron diffraction   73
dark-field (DF) TEM images   197
– Cu$_2$O dark-field images   197
de Broglie wavelength   40
defective carbon nanotube, reconstruction   134
deflection system   13
deformation
– behavior   265
– mechanisms   230
dehydroxylation   152, 153

deliquescence   153
dendritic transition   202
denuded zones   116
detection quantum efficiency (DQE)   157
device fabrication process   153
diamond crystal, nucleation   135
dielectric breakdown   330
differential cryogenic pumping device   105
differential Howie–Whelan equations   55
differential phase contrast (DPC) imaging   347
differentially pumped system, drawback   149
differential phase contrast (DPC) technique   56
differential pumping device
– design   105
– schematic drawing   105
differential pumping system, schematic flow-chart   150
differential scanning calorimetry (DSC)   88
diffraction pattern
– by adjusting microscope.s projector lens system   59
– Al film possess   246
– energy-filtered   60
– selected area electron diffraction patterns   83
– by using EBSD detector   31
diffractograms   51, 52
diffusional processes   244
diffusion-induced grain boundary   244
diffusion processes   281
diffusivity   282
digermane (Ge$_2$H$_6$)   176
3-D islands   114
dislocation mechanisms, controlling plasticity   240
dislocation nucleation   244
dislocations, TEM images   231
dispersive X-ray spectroscopy   34–36
displacement threshold   135
2-D negative islands, nucleation   118
doped ceria (CeO$_2$)   162
double-walled carbon nanotubes   137
3-D oxide islands, shape dynamics   205
3-D shadowing effect   29
DTEM. See dynamic transmission electron microscopy (DTEM)
dual damascene structure, typical failure features   295
dynamical image contrast formation   73, 75
dynamical scattering theory   46
dynamic transmission electron microscopy (DTEM)   71, 72, 74, 76

374 | *Index*

- aberration correction   93, 94
- acquiring high time resolution movies   81, 82
- applications   82
- arbitrary waveform generation laser system   80
- crystallization under far-from-equilibrium conditions   88–90
- current performance   74, 75
- diffusionless first-order phase transformations   82–85
- electron sources and optics   75–80
- experimental applications   82–88
- global space charge   90, 91
- next-generation   91–94
- novel electron sources   91, 92
- observing transient phenomena in reactive multilayer foils   85–88
- pulse compression   93
- relativistic beams   92, 93
- single-shot work   72–82
- space charge effects in single-shot   90, 91
- stochastic blurring   91
- time resolution   82

e

edge dislocation core model   49
EELS detector   162
efflorescence phenomena   153
elastically scattered electrons   53
elastic contact theory   270
elastic deformation   270
elastic scattering   16
- angles   17
- cross-section   45
- direction via   304
- of electrons   42, 44, 66
- inversely proportional to   18
- multiple   17
electric field-induced
- antiferroelectric ↔ ferroelectric phase switching   335
- crack growth   328, 329, 333
- domain switching   325
- domain wall fracture, confirmation of   334
- phase transitions   321
-- relaxor-to-ferroelectric   344
electric field *in-situ* TEM experimental set-up   323
electrochemical deposition
- of Cu onto polycrystalline Au electrode   185
- and polymer growth, from liquid precursors   173
electrochemical liquid cells   184

electrode configuration   322
electromagnetic lenses   5, 9–13
electromigration   281. *See also* transmission electron microscopy (TEM)
- damage   291
- diffusion paths   296
- *ex-situ* experiments   283
- failure mechanisms   290
- focused ion beam (FIB)   295, 296
- induced drift rates   282
- induced material transport   282
- *in-situ* experiments   283
- *in-situ* methods   298
-- comparison   284, 297–299
- parameters   282
- phenomenon   111
- process   297, 298
- scanning probe methods   296, 297
- testing, mass depletion   290
electron backscatter diffraction (EBSD)   31–34, 212, 217, 283, 292, 298
- scans for determining local crystal orientations   212, 214
electron beam   6, 134
- schematic drawing   4
electron beam effects   157
electron beam-induced current (EBIC)   36
electron beam-induced decomposition (EBID)   160
electron beam-induced voltage (EBIV)   36
electron beam spot   138
electron current density   12
electron detectors   13–16
- Everhart–Thornley detector   13, 14
- in-lens/through-the-lens detectors   16
- scintillator detector   15, 16
- solid-state detector   16
electron diffraction   158, 191, 344
- intensity calculations   115
electron-diffraction patterns   161
electron diffraction techniques   59–61
- fundamentals   59–61
electron–electron interactions   72
electron–electron scattering   127
electron energy-loss spectroscopy (EELS)   65, 66, 127, 191, 303
electron guns   6–9
electron-hole pairs   304, 305
electron hologram   45
electronic drift compensation   131
electron interaction constant   57
electron irradiation   131–140, 139
electron–matter interaction   16–28, 26
electron microscopy

– in electron energy loss spectroscopy
  (EELS)   65, 66, 127, 191, 303
– electron irradiation   131, 132
– ion irradiation   132
– sample heating, benefits   106
– setup in   131, 132
electron–optical system   4
electron penetration range   305
electron probe   9, 157
electrons   125, 126
– beam rays, schematic representation   103
– energy distribution   21
– experiments   132–141
– inelastic electron scattering   128
– scattering   128, 129
– setup in electron microscope   131, 132
– sources, properties   9
– source, type   6, 7
– structure factor   45
– trajectories, Monte Carlo simulation   19
electron–specimen interaction constant   47
electrons scattering processes   44
electron transmission microscopy,
  advanced   39
electron-transparent film/polyimide area   242
electron-transparent windows
– principle   146, 147
– schematic representation   147
electrostatic deflector array   82
electrostatic potential barrier, schematic
  drawing   7
energy-dispersive X-ray spectroscopy (EDS)   9,
  34, 65, 66, 127, 145, 193, 283, 303, 326
energy-filtered diffraction pattern   60
energy-filtered TEM (EFTEM) images   66
energy-filtered transmission electron
  microscopy (EFTEM) imaging   163
energy loss magnetic dichroism (EMCD)   347
environmental cells   192, 193
environmental scanning electron
  microscopes   6
environmental transmission electron
  microscopes (ETEM)s   149, 156, 159
epitaxial nucleation, of oxide islands   203
epitaxial silicon carbide   107
escape depth   20
Everhart–Thornley (ET) detector   13, 14, 28,
  29
Ewald sphere   60
– construction diagram   46
– curvature   47
excitation error   47, 54, 60
ex-situ measurements, on test samples   174
ex-situ nanoindentation experiments   272

f

fabricating devices   90
face-centered cubic (fcc) metal films   233
fast Fourier transforms (FFTs)   161
fcc single crystals   241
Fermi level   25
ferroelectric ceramics   321, 322
– electric field-induced domain switching
  in   322
– electric field-induced fracture   328
– field-induced grain boundary   326
ferroelectric domains   343
ferroelectric material, dielectric permittivity
  of   322
ferroelectric oxides
– *in-situ* TEM technique   321, 322
– – antiferroelectric-to-ferroelectric phase
    transition   335–341
– – domain polarization switching   324–326
– – domain wall fracture   331, 335
– – experiment   323, 324
– – grain boundary cavitation   326, 330
– – relaxor-to-ferroelectric phase
    transition   341–345
ferroelectric single crystals   146
fiber-based electrooptical modulator   80
field-cooling   341
– cation-ordered domains, morphological
  changes   343
– cation-ordered domains, morphology   344
field emission gun (FEG)   8, 10, 50, 126
field emission scanning electron microscopy
  (FESEM)   264
field-induced crack growth process   328
field-induced relaxor-to normal-ferroelectric
  phase transition   343
film deposition   231
fine-grained microstructures   89
fluctuation electron microscopy (FEM)   63
fluorescence process   26
flux divergencies   298
focused ion beam (FIB)   229, 295
– cutting   241
– grain contrast   295
– milling   214
– prepared samples   262
– scanning electron microscope   233
– single-crystal conductor line   296
– structuring   133
foreshortening effect   104
Foucault imaging mode   56
Fourier coefficients   44
Fourier components   45
Frank–Read dislocation   237

Frenkel pair 129
frequency-tripled laser pulses 74
Fresnel contrast 57
Fresnel imaging mode 56
Fresnel imaging, schematic of 348
Fresnel zone plate-based technique 292, 294
full-width at half-maximum (FWHM) electron pulse 73, 351
fusion reactors 125

g
GaAs-based quantum wells, fabrication 309
Ga^+ ions 176
GaN epilayer 308, 309
gas-handling system 194, 195
gas-injection system 148
gas injector, schematic diagram 148, 149
gas–solid interactions 148
Gatan GIF Quantum series, of imaging energy filters 157
Gatan TV system 107
Gaussian distribution 12
global space charge (GSC) 90
gold electrodes 324
grain boundaries 282, 333, 342
– cavities 329
– crack 329
– with Cu atoms 293
– diffusion 282
– grooving 297
– motion 244
– plasticity by motion 244, 245
– triple junction 327
graphite–diamond interface, electron irradiation 140
graphitic carbon 134
grid/support materials reaction
– with sample/with each other 154, 155
growth rate 174–176, 178, 179

h
hardening 242
hardness 222
hexagonal close packing (HCP) 83
high-angle annular dark-field (HAADF) detector 64
high-current pulsed electron probes 72
high-energy electron 25, 26
higher order Laue zone (HOLZ) lines 62
highest spatial frequency 53
high oxygen flux 117
high-pressure impulse loading 92
high-quantum efficiency photocathode 78
high-resolution focal series 58

high-resolution images 30, 156, 158–161, 229
high resolution SEM 8
high-resolution TEM (HRTEM) 42, 48–53, 101
– experimental conditions 43
– image-formation process 48
– images 45, 47, 49, 51, 53, 54
– simplified ray diagram of image 50
high spatial resolution 100
high-temperature microscopy 131
high-voltage accelerator design 92
high-voltage electron microscope 93
holography 57
HOLZ reflections 62
Hough transformation 288, 289
Howie–Whelan differential equations 48
HRTEM. See high-resolution TEM (HRTEM)
hydroxylation 152, 153

i
image formation process 13
InAs dots 310
incident wave vector 61
incoherent aberrations 53
indentation-induced dislocation nucleation 263
inelastic scattering 16, 17, 61, 66, 127, 128
– cross-section 18
– effects 20
– electron beam by the oxygen gas 204
– of ions 129
– multiple 18
– processes 18
infrared spectroscopy 145
in-lens detector system 27, 29, 30
– advantage 23
– schematic drawing 15
– SEM image 30
in-lens/through-the-lens detectors 16
inline electron holography 57–59
inline holograms 58
– advantages 58
in-situ bending device, with vertically aligned loading axes 215
in-situ 3-D imaging techniques 205
in-situ electron irradiation 141, 142
in-situ imaging 163
in situ ion irradiation experiments 132
in-situ loading samples in tension, compression, and bending
– of macroscopic samples
– – applications of 216, 217
– – dynamic loading 216
– – static loading 214, 215

– of micron-sized samples 217, 218
– – applications of *in-situ* testing of small-scale samples 220–222
– – *in-situ* microindentation and nanoindentation 222, 223
– – static loading 218–220
in-situ nanoindentation 255, 263
– experimental methodology 260–263
– *in-situ* mechanical probing 255, 256
– nanoindentation 256–260
– – Al thin films, *in-situ* 269–272
– – silicon, *in-situ* 263–269
– sample 262
– – cross-section of 270
in-situ nanomechanical probing experiments 262
in-situ oxidation
– effect of electron beam irradiation 195
– experiments using window environmental cells 193
– importance of 192
– study of surface oxidation 192
in-situ Raman spectroscopy 259, 260
in-situ TEM 88
– growth experiments 173
in-situ TEM straining experiments 227
– instrumented stages 230–233
– mechanical straining 229, 230
– MEMS/NEMS devices 230–233
– size-dependent dislocation plasticity 239
– – grain size heterogeneities, influences 245–247
– – plasticity by grain boundaries motion 244, 245
– – plasticity in geometrically confined single crystal fcc metals 241–243
– – transitions in dislocation plasticity 243, 244
– thermally strained metallic films, dislocation mechanisms
– – basic concepts 233–235
– – nucleation and multiplication in thin films 236–239
– – polycrystalline Cu films, diffusion-induced dislocation plasticity 239
– – in single crystalline films 235, 236
– thermal straining 228, 229
in-situ UHV-REM technique 114, 115
in-situ ultrahigh-vacuum (UHV) environmental transmission electron microscopy (TEM) 191
in-situ visualization, of oxidation processes 203
instrumentation, and basic electron optics 40–42

interaction energy 114
interaction volume 18
– schematic drawing 20
inverse pole figure (IPF) maps, of polycrystalline aluminum sample 217
ion-channeling effect 295
ion–electron scattering 129
ion irradiation 125, 126, 132, 140, 141
– experiments 127
– physics 126–129
– radiation defects in solids 129, 130
ion irradiation facility, setup 133
ions scattering 129
irradiation principles 126
island nucleation 116
isothermal annealing 113
isotropic atomic scattering factors 44

j

JEOL 200 CX transmission electron microscope, *in-situ* nanoindentation stage 261
JEOL 2000FX microscope platform 72, 73
Johnson–Mehl–Avrami–Kolomogrov (JMAK) semi-empirical formulae 88

k

kinematical approximation 60
kinematical scattering theory 46
kinetic theory of gases 193
Kossel cones 33
Kramers relation 35
Kratos high-voltage microscope 261

l

large-angle convergent beam electron diffraction (LACBED) technique 42, 63
large-bore lens 79
latex sphere 59
lattice distortion 54
lattice imaging 136
Laue condition 61
Lawrence Livermore National Laboratory (LLNL) 72, 157
– dynamic transmission electron microscope 73, 75
lead zirconate titanate (PZT) ceramic 324
– bright-field TEM images 325
– EC65 ceramic 324, 325
– – microcracks development 327
– – nanometer-sized ferroelectric domains, morphology of 325
– – scanning electron microscopy image 326
lens system 11

light elements analysis 34
liquid cell biological imaging 94
liquid crystal display (LCD) computer monitor-based system 13
liquid-phase growth processes 183
– electrochemical nucleation 184
– growth in TEM system 184–187
– observing liquid samples using TEM 183, 184
Lomer–Cottrell dislocations 313
Lorentz force 10, 55, 56
Lorentz lenses 55
Lorentz microscopy 55–57, 56, 347-367
– coils/pole-pieces, magnetizing 352–356
– domain wall separations 364
– dynamic randomaccess memory (DRAM) cells 358
– electromigration, effect of 366
– electron beam 350, 362
– Fresnel imaging, schematic of 348
– in-plane field component, creation 351
– *in-situ* magnetic fields, generation of 357
– *in-situ* magnetizing experiments 350, 351
– magnetic disk 361
– – contrast formation 360
– – schematic of 359, 361
– magnetic field values 355
– magnetic rings, remanent states 363
– magnetic specimen 358
– magnetic stray field maps 365
– magnetizing holder 356
– – technical drawing of 354
– – view of 353
– magnetizing stages without coils
– – oersted magnetic field 356–358
– – self-driven devices 361, 362
– – spin torque applications 358–361
– objective lens field with specimen tilt, combining 351, 352
– problems solving 366-367
– ring structures, demagnetization/magnetization of 362–364
– stray fields, determination of 365, 366
– thin Au conductor, light micrograph of 357
– TWIN lens set-up 349
– TWINlens configuration 355
– use of 348
– wall velocities, determination of 364
low-energy electron diffraction (LEED) 100
– applications 101
low-vapor-pressure liquids 6, 173
low-voltage scanning electron microscopy (SEM) 101

m

magnesium oxide (MgO) 152
mean free path (MFP) 17
mechanical annealing 242
median time to failure (MTF) 281
MEMS/NEMS-based tensile testing
– of nanocrystalline Al and Au 245
MEMS/NEMS devices 231
– drawbacks 232
– limitations 233
– TEM straining stage 231, 232
metal-induced crystallization 172
metal matrix compound (MMC) 217
micro-bending beam 218
micro-compression 241
microelectromechanical system (MEMS) technology 146, 174, 227
– actuated tests 255
– applications 88
microindentation 222, 223
micro-Laue diffraction 298
microscopic x-radiographic techniques 293
micro-tensile testing 241
micro-x-ray fluorescence 294
miniaturization 217
minority carriers, diffusion constant 305
Mo laser mirror 73
molecular beam epitaxy (MBE) 172
– methods 99
molecular dynamics (MD) 63
monochromatic light, TEM images of dislocations 312
Monte Carlo simulation 19
– electron trajectory simulations 18
Mott formula 44
Mott scattering 17
movie mode technology 81
multi-walled carbon nanotube 139
– electron irradiation 136

n

nanoampere electron beam 76
nanocrystalline Ni films, *ex-situ* TEM studies 145
nanocrystalline Ti film, experimental isothermal phase diagram 84
nanocrystallites, deformation 135
nano-electromechanical systems (NEMS) 227
nanoelectronics technology 99
nanoindentation 222, 223, 255
– *ex-situ* experiments 272
– *in-situ* nanoindenter 219
– load *vs.* displacement curve 258
nanoreactor, schematic cross-section 148

nanoscale synthesis processes, robust scaling 145
nanostructre, growth 153, 154
nanotechnology 180
nanowire formation 178
nanowires, synthesis temperature 158
National Television System Committee (NTSC) 156
natural oxide films 107
Nb-doped lead zirconium titanate (PZT) 151
negative C_s imaging (NCSI) conditions 51
neodymium-doped yttrium aluminum garnet (YAG) lasers 74
Ni micro-compression pillars, stress–strain curves 241
Ni pillars, *in-situ* TEM compression 242
NiTi pulsed laser-induced crystallization process 88
nitridation 152
nonacarbonyldiiron [$Fe_2(CO)_9$]
– electron beam-induced decomposition (EBID) 160
noncrystallographic fracture 268
nonvanishing excitation error 46
nucleate phase transformations 268
nucleation 153, 154
nucleation barrier energy 84
nucleation kinetics
– of Ge islands on Si(001) 176
– in nanostructures 180–183
nucleation processes 115

o

off-axis and inline electron holography 57–59
off-axis electron holography 57–59
off-axis hologram 58
optical microscopy 26, 333
– disadvantages of 211
– image 334
(111)-oriented Al films
– cross-sectional thermal straining experiments of 235
(100)-oriented Au films, transition 243
Ostwald ripening 139
oxidation phenomena 196
– of $Nb_{12}O_{29}$ 204
– nucleation and initial oxide growth 197, 198
– pathways 203–205
– surface reconstruction 196
oxidation reactions 150, 151
oxidation/reduction cycles 151
oxide nanostructures
– growth mechanisms for 153

oxide nuclei, orientations of 202, 203
oxygen pressure 202, 203

p

partial spatial coherence 40
Pati–Cohen model 84
$Pb(Mg_{1/3}Nb_{2/3})O_3$
– dielectric property of 341
– polar nanoregions 341
– polycrystalline specimens 341
PbO-containing amorphous phase 330
Peierls–Nabarro barrier 267
phase distortion function 49, 50
phase shift 50, 51, 57–59, 109
phase transformations 82, 141, 151, 160, 182, 260, 263, 267, 268, 272
phosphorescent screen 31
photocathode source 91
photomultiplier systems 13–15
photon energy 305
photons emission 25, 26
picosecond-nanometer resolution single-shot imaging 93
piezo-ceramic actuator 261
piezo-driven *in-situ* fatigue testing device 216
piezoelectric coefficients 334
piezoelectric single crystals 146
– uses 331
piezoelectric strain 333, 334
pinning phenomenon 110
plain excitation error 48
Planck's constant 25
plasmonics 287
plastic deformation 241, 245, 263
– Al grain, time series of 271
point analysis 35
polarized cathodoluminescent light, quantitative analyses 311
polar nanodomains, morphological evolution 342
polycrystalline Al films 269
– stress values 236
polycrystalline ceramic, polarization measurements 343
polycrystalline Cu films 239
– dislocations emission 240
polycrystalline films 235, 236, 237
– flow stresses 237
polycrystalline reactant materials 85
polycrystalline tantalum, back-scattered electron image 33
polycrystalline thin films, thermal stress measurements of 236
polyimide, single-crystal Al film 242

polymerization 151, 152
pressure-induced phase transformations 257
primary electrons (PEs) 20
probe aperture-dependent semi-convergence angle 12
projector lens system 42
proportionality constant 45
protective oxide films 202
PSMN8 cation-ordered domains 344
PSMN8 ceramic 345
pulsed electron diffraction data 86
pulsed laser-induced crystallization process 88
pump laser 89
PZST 45/6/2 ceramic
– field-induced transition 336
– incommensurate modulation 337

q
qualitatively imaging 227
quantitative *in-situ* TEM nanoindentation 272
quantum dots (QDs) 309
– growth kinetics 176, 177
quasi-coherent approximation 49, 53

r
radiofrequency (RF)-based photoguns 92
Raman peak shift 294
Raman spectroscopy 145, 157, 294
rapid material processes 81
rapid solid-state chemical reactions in reactive multilayer foils (RMLFs) 85
reaction front morphology, snap-shot images 87
reaction rates (kinetics) 164
reciprocal lattice vectors 45, 46, 54, 61, 62
recording media 156, 157
reduced density function (RDF) 63
reduction (redox) reactions 150, 151
reflection coefficient 118
reflection electron microscopy 99–107
– consequent set 117
– epitaxial growth 115, 116
– extreme sensitivity 104
– high sensitivity 103
– images 113
– images, features 103
– monatomic steps 109–111
– silicon substrate preparation 107–109
– step bunching 111–114
– stepped silicon images 112
– surface patterns formation 99–102
– surface reconstructions 114, 115

– thermal oxygen etching 116–118
reflective high-energy electron diffraction (RHEED) 100–102, 107, 108, 113, 116, 118, 296. *See also* electromigration
– disappearance 116
– oscillations 118
residual gas analyzer (RGA) detector 195
reverse Monte Carlo (RMC) simulations 63
reversible switching, using O_2 181
Rose criterion analysis 76, 77
rules of momentum conservation 127, 128
Rutherford scattering 16–18

s
sample normal vector 22
sample temperature 174
satellite reflections
– changes 338
– electric fields for 340
– evolution 339
satellites electronic, components in 125
scanning electron microscopy (SEM) 3, 145, 211, 326
– auger electrons (AEs) 25
– backscattered electrons (BSEs) 20–22, 27
– components 4–16
– – schematic drawing 5
– contrast mechanisms 28–31
– deflection system 13
– dispersive X-ray spectroscopy 34–36
– electromagnetic lenses 9–13
– electron backscattered diffraction (EBSD) 31–34
– electron detectors 13–16
– electron guns 6–9
– electron–matter interaction 16–28
– emission of photons 25, 26
– emission of X-rays 25, 26
– images of polycrystalline aluminum sample 213
– for *in-situ* testing 212
– interaction volume, and resolution 26–28
– for microstructural characterization 3
– other signals 36
– preparation of specimen 212
– secondary electrons (SEs) 22–25, 27
– technical requirements 212–214
– *vs.* optical microscopy 212
– X-rays 27, 28
scanning probe microscopy (SPM) 145
scanning transmission electron microscopy (STEM) 39–41, 64, 65, 131, 155
– imaging, advantages 65
– imaging modes

– – annular dark-field STEM (ADF-STEM) 64
– – bright-field STEM (BF-STEM) 64
– and Z-contrast 63–66
scanning tunnel microscopy (STM) 100, 187
scattering processes 44, 126, 304
scattering vector, function 83
Schottky effect 6, 7
Schottky emitters 9, 10, 41, 53
Schwoebel effect 111
scintillator detector 15, 16
secondary electron imaging (SEI) mode 290, 291
secondary electron microscopy
– electron backscatter diffraction (EBSD) 31, 34, 217, 283, 292, 298
– elemental analysis 291, 292
– imaging 289–291
secondary electrons (SEs) 22–25, 27
– emission 13
secondary electron yield 24
selected area (SA) aperture 42
selected-area electron diffraction (SAED) 73, 77, 152, 159, 202, 203, 284
– patterns from oxidized surfaces 202
selected area electron diffraction patterns (SAEDPs) 83
self-cleaning process 8
self-organization effects 125
self-organization processes 130
semiconductors 173
semi-quantitative analysis 34, 35, 330
shadowing effects 28
shape transition
– bright-field image of a Cu(110) film oxidized at 201
– Cu(200) dark-field images 200
– during oxide growth in alloy oxidation 199–202
shot noise 76
signal-to-noise ratios (SNRs) 27, 40, 71, 78, 79, 83
silicide formation 172
silicon
– images 108
– *in-situ* indentation 267
– *in-situ* nanoindentation 266
– nanostructures 313
– plateau 263
– surface morphology 114
– technology 231
– thermal etching, 2-D mechanism 118
– wedge samples
– – *in-situ* nanoindentation experiments 264

– – scanning electron microscopy images of 264
Si nanowires 153
Si nanowires, nucleation and growth of 176
single pump-probe snapshot 81
single-shot approach 71, 72
single-shot bright-field series, change in grain morphologies 85
single-shot DTEM *vs.* conventional continuous-wave (CW) TEM 78
single-shot electron diffraction 86
– data 83
single-tilt TEM straining stage, optical image of 230
single-walled nanotubes 136
– electron irradiation 138
SiN thin films 193
size-dependent dislocation plasticity 239–247
– Cu film, dislocations emission 240
solid energy diagram, schematic drawing 17
solid-phase chemical reactions 145
solids
– defects formation 129, 130
– defects migration 130
– energetic particles in, scattering 126, 127
– radiation defects in 129, 130
solid-state detector 16
solid-state reactions 164
spatial coherence 49, 53, 58, 78, 79, 93
spectroscopic techniques 150, 162
specula-reflected electron beam
– temporal dependences of intensity 118
spherical aberration 11, 50, 51, 55, 94
– coefficient 50
split-off beam 74
sputtering effects 139
stainless steel sample, single-shot pulsed image 76
standard pumping system 105
STEM. *See* scanning transmission electron microscopy (STEM)
STEM-EELS
– advantages 67
– maps 66, 67
step bunching phenomenon 111–114
step shape meandering 110
stochastic blurring 91
strain relaxation 175
Stranski–Krastanov mode 310
stress-driven grain boundary motion
– in nanocrystalline Al 245
stress–strain curves 241
stroboscopic approach, refined to subpicosecond time resolution 71

structural diagnostic methods 100
– requirements 100
structural modification 158–161
surface and environmental conditions 193, 194
surface chemistry 173
surface defects, on surface oxidation 198, 199
surface morphology instability phenomenon 113
surface phase transitions 111
surface-sensitive techniques 126

t

Ta disk cathodes 78
temperature-resolved high-resolution imaging 158, 159
temporal coherence 40
temporal resolution 164
tensile testing 244
terraces 109
thermal annealing 105
thermal conductivity 127
thermal cycles 165
thermal diffuse scattering (TDS) 64
– reduction 64
thermal dislocation network 237
thermal etching, 2-D mechanism 118
thermal field emitter (TFE) 8
thermally strained metallic films, dislocation mechanisms
– concepts 233–235
– nucleation, and multiplication in thin films 236–239
– polycrystalline Cu films, diffusion-induced dislocation plasticity 239
– in single crystalline films 235, 236
thermal straining experiments 229
thermionic cathode 13
thermionic electron guns 7, 8
thermionic source 6
thermogravimetric analysis (TGA) 197
thin-film deposition techniques 262
thin polyimide layer, causing fracture 242
threading dislocation deposition 233, 234
three-lens system, demagnification 10
time-resolved diffraction 85
time-resolved experiments, in dynamic transmission electron microscope 74
time-resolved high-resolution images 161
time-temperature-transformation (TTT) curve 84
topographic contrast contributions 31
topographic STM scans, linescans of 297
transfer cross-coefficient (TCC) 53
transmission electron microscopy (TEM) 3, 39, 45, 50, 59, 125, 127, 145, 172, 227, 321. *See also* scanning transmission electron microscopy (STEM)
– ambient environment reaction with various components 154
– analytical 66, 67
– application 101
– available information under reaction conditions 157–164
– basics 39
– chamber 322
– chemical changes 161–163
– chemical reactions observation 145–165
– chemical reaction types suitable for 150–154
– conventional 39, 44, 90, 94, 172
– – of defects in crystals 54, 55
– – straining stage 229, 242
– convergent beam electron diffraction (CBED) 61, 63, 283, 288, 289
– diffraction 288
– electrochemical nucleation and growth in 184–187
– electron diffraction 158
– experimental setup 154–157
– grid/support materials reaction with the sample/with each other 154, 155
– high-resolution imaging 158–161
– hydroxylation and dehydroxylation 152, 153
– imaging 283–287
– independent verification of results and electron beam effects 157
– *in-situ* deformation studies 227
– instrumentation 146–150
– limitations and future developments 164
– nitridation 152
– nucleation and growth of nanostructre 153, 154
– observing liquid samples using 183, 184
– oxidation and reduction (redox) reactions 150, 151
– phase transformations 151
– polymerization 151, 152
– principles 41
– reaction rates (kinetics) 164
– recording media 156, 157
– resolution 184
– selecting appropriate characterization technique 156
– spatial resolution 71
– structural modification 158–161
– temperature and pressure considerations 155, 156

transport-of-intensity equation (TIE)   347
tungsten filament, schematic drawings   8
turbomolecular pump (TMP)   149
two-dimensional rocking curves   63

u

ultra-high-vacuum (UHV)   173, 174
– conditions   24, 99, 100
– level   6
– scanning electron microscopes   25
– UHV-REM experiments   119
– UHV-REM system   104
– UHV-REM technique   108, 109
ultrahigh-vacuum reflection electron microscopy (UHVREM)   102, 104
ultra-high vacuum (UHV) TEM systems   149
universal loading device
– placed in scanning electron microscope chamber   213
– for tension, compression/fatigue tests on small samples   220

v

vacuum system   5, 102. See also ultra-high-vacuum (UHV)
vapor–liquid–solid growth of nanowires   177–180
vapor-liquid-solid (VLS) mechanism   160
vapor-phase growth processes   175, 176
vapor–solid–sold (VSS) mechanism   160
Vickers diamond indentor   333
video-taped SVTEM images   286
visible light emission   26

w

wavelength-dispersive spectrometer (WDS)   34
– disadvantage   34
weak-beam dark-field (WBDF) TEM   42, 43
weak phase object approximation   47

wedge-shaped cross-sectional sample   228
Wehnelt cylinder   6, 7
wind force   281

x

Xe crystals   140, 141
– nucleates   141
x-radiography studies   292
– microdiffraction   294, 295
– microscopy, and tomography   292, 293
– spectroscopy   293
– topography   294
x-ray diffraction   145
x-ray emission spectroscopy (XES)   303, 307
x-ray energy-dispersive spectroscopy (EDS)   161
x-ray photoelectron spectroscopy (XPS)   205, 206
x-ray pump-probe techniques   71
x-rays   27, 28
– diffraction   343
– dot images   35
– emission   25, 26
– energies   35
– energy regions   36
– mean free path   27
– source   27
– spectrum   26

y

Young's modulus   222

z

Z-contrast imaging   64
zero-order Laue zone (ZOLZ)   61–63
Ziegler–Natta catalyst   152
ZnO nanowires   311
ZnSe light-emitting devices   312